terra australis 34

Terra Australis reports the results of archaeological and related research within the south and east of Asia, though mainly Australia, New Guinea and island Melanesia — lands that remained terra australis incognita to generations of prehistorians. Its subject is the settlement of the diverse environments in this isolated quarter of the globe by peoples who have maintained their discrete and traditional ways of life into the recent recorded or remembered past and at times into the observable present.

List of volumes in Terra Australis

terra australis 34

PEOPLED LANDSCAPES

Archaeological and Biogeographic Approaches to Landscapes

EDITED BY SIMON G. HABERLE & BRUNO DAVID

Australian
National
University

E PRESS

ANU
E PRESS

© 2012 ANU E Press

Published by ANU E Press
The Australian National University
Canberra ACT 0200 Australia
Email: anuepress@anu.edu.au
Web: http://epress.anu.edu.au

National Library of Australia Cataloguing-in-Publication entry

Title: Peopled landscapes : archaeological and biogeographic approaches to landscapes /
              edited by Simon G. Haberle & Bruno David.

ISBN: 9781921862717 (pbk.) 9781921862724 (ebook)

Notes: Includes bibliographical references.

Subjects: Human ecology--Australasia.
              Landscape assessment--Australasia.
              Landscape changes--Australasia.
              Nature--Effect of human beings on--Australasia.

Contributors:
              Haberle, Simon G.
              David, Bruno, 1962-

Dewey Number: 333.72099

Series Editor: Sue O'Connor

Copy editing: Kirsten Lawson

Typesetting and design: Toby Wood

Cover image: Looking southwest from the entrance of Bridgewater South Cave across the southern sections of Bridgewater Lakes towards Cape Duquesne, 10 February 2008 (Photo: Ian J. McNiven). Archaeological excavations at the cave by Harry Lourandos (1983) in the 1970s demonstrated Aboriginal occupation back to c.13,000 years ago. Palynological coring of the lakes by Lesley Head (1988) in the 1980s demonstrated over 7500 years of vegetation change, some of which reflected low-level Aboriginal landscape burning.

Head, L. 1988. Holocene vegetation, fire and environmental history of the Discovery Bay region, south-western Victoria. *Australian Journal of Ecology* 13(1):21-49.

Lourandos, H. 1983. Intensification: A late Pleistocene - Holocene archaeological sequence from southwestern Victoria. *Archaeology in Oceania* 18:81-94.

Back cover map: Hollandia Nova. Thevenot 1663 by courtesy of the National Library of Australia.
Reprinted with permission of the National Library of Australia.

Terra Australis Editorial Board: Sue O'Connor, Jack Golson, Simon Haberle, Sally Brockwell, Geoffrey Clark

Contents

Introduction

1

Peopled landscapes: The impact of Peter Kershaw on Australian Quaternary science

Bruno David
School of Geography and Environmental Science, Monash University, Clayton, Victoria
bruno.david@monash.edu

Simon G. Haberle
The Australian National University, Canberra, ACT

Donald Walker
The Australian National University, Canberra, ACT

"*I don't think the human mind can comprehend the past and the future. They are both just illusions that can manipulate you into thinking there's some kind of change.*"

Bob Dylan (Ft. Lauderdale Sun-Sentinel Interview, 28 September 1995).

Introduction

The way we view the Australian landscape at the start of the 21st century is notably different to how we viewed it in the late 20th century, and Peter Kershaw has had a most significant role in this. One of the key elements to Peter's intellectual contribution lies in the discovery that the Australian landscape is more changeable and dynamic than was previously imagined, and in particular more deeply influenced by human history than we could have then known. The notion that people arriving in an uninhabited landscape over 40,000 years ago so fundamentally changed fire regimes, and that Aboriginal people have continued to shape the environment through ongoing landscape firing practices ever since – and as a consequence affected the whole ecology of an island continent – is one that shook the scientific community and forced a rethink of the way we view the long-term history and present environmental state of Australia. The impacts of this on the broader Australian and international community have contributed to a critical rethinking of political paradigms and conservationist policies and their articulation with Indigenous perspectives on landscape, especially as these relate to fire management (e.g. Hale and Lamb 1997); indeed, in the mid-1990s the Australian Conservation Foundation's general approach to landscape management radically shifted to take better account of Indigenous concerns, together with Indigenous voices a result of the accumulated wisdom of previous years of palaeoecological research that by then demanded consideration of the role of people in landscape management. These were precisely the kinds of issues spurred by Peter's findings in north Queensland and elsewhere, research results that had shown to be critical to understanding the Australian environment today as in the past. Thus while throughout his career Peter himself has been more directly concerned with gathering evidence towards establishing the facts of Australia's landscape history, those results have influentially fed back into community perspectives, political dialogue and policy-making. Peter's work continues to engage both the scientific community and the public in an ongoing debate over the role of people in shaping the environment.

The investigation that contributed most to the shift in our understanding of the role of people in Australian landscapes was based around a pollen record from Lynch's Crater (Figure 1) in northeast Australia (Kershaw 1974). Peter began working on this long environmental record for his PhD thesis (Kershaw 1973) in the Department of Biogeography and Geomorphology at the Australian National University (ANU) in the early 1970s, at a time when palaeoecological research in Australia was in its infancy. In 1974 he published in *Nature* his seminal paper on Lynch's Crater, demonstrating a radical vegetation change in northeast Australia around 38,000 years ago, at about the time when people were then thought to have first arrived on the continent (see Turney *et al.* 2006 for a subsequent redating of this vegetation change at Lynch's Crater, keeping in line also with subsequent redating of the first evidence of people in the landscape). This was also the year when the influential 'Sunda and Sahul' symposium was first planned (Allen *et al.* 1977), and the foundation year of the Australian Archaeological Association where, for the first time, researchers of Australian archaeology could assemble at an annual forum to share findings and discuss intellectual developments. This was a period of burgeoning interest in Australian Aboriginal archaeology and landscape history, where systematic connections were being forged between these two disciplines and interests (Aboriginal and landscape history). While discussions had by then already emerged on the role of people in shaping the environment (e.g. Jones 1968), especially through Rhys Jones' notion of 'firestick farming' (Jones 1969), and cause(s) of megafaunal extinctions had already long been debated since the 19th century (see Johnson 2007 for a review), what Peter brought to the equation were long environmental records and a conceptual shift signalling the necessary and obligatory incorporation of people into interpretations of landscapes, as managed and dynamic social spaces. From Peter's early works at Lynch's Crater, it was realised that it was simply not legitimate to interpret palaeoenvironments of the last *c.*40,000 years other than as *peopled* landscapes. This was not a matter of debating whether or not a particular archaeological or palaeontological assemblage showed evidence of human intervention, but rather a paradigm shift that newly saw palaeoecological sequences as only interpretable through consideration of human presence, given that people lived in the landscape and thus affected it. The question had become not so much whether or not people had a role to play in the evolution of the Australian landscape, nor whether they were palaeontologically or palynologically visible in that landscape, as a determination of the nature and scale of such interventions and their palaeoecological visibility.

From the outset Peter has been a key figure that led to a culture of research that closely enmeshed Australian archaeological with palaeoecological research during that period of the

Figure 1. Coring at Lynch's Crater in 1971.

1970s when environmental understanding proved critical to archaeological paradigms. While today connections between culture and environment have largely been reframed to incorporate new nuances of 'dwelling', 'inhabitation' and the like (e.g. Thomas 2008), the kinds of research connections that Peter helped to frame between environmental and cultural sequences have resulted in strong research bonds between the disciplines of palaeoecology and archaeology in Australia and beyond.

Peter Kershaw

To properly understand, and appreciate, Peter's impact on the study of the Australian landscape we begin where he himself began. Peter grew up in Littleborough in the north of England where the industrial edge of Lancashire nosed up the valley into the peat-covered Pennine hills. He attended the local schools then went to the University of Wales at Aberystwyth where, as a student of geography, he first became aware of Quaternary pollen analysis. After acquiring an Honours degree, he moved to Durham University to do a Masters in ecology supervised by Judith Turner. Soon afterward he successfully applied for a research assistantship to Donald Walker, in the still infant laboratory at the ANU, to help extend the then dominating interest in Papua New Guinea to continental Australia.

Arriving with his wife Susan in the summer of 1967, Peter was given the task of observing the drilling in the Lake George basin by the Bureau of Mineral Resources and taking samples of any deposit there which might have contained preserved pollen. Such material as he was able to obtain, with much scepticism, in the field he found to be virtually barren in the laboratory; a very disappointing start. So a shift of emphasis was called for.

Given the laboratory's existing tropical interests, it made sense to go north, specifically to the biogeographic boundary zone of the Atherton Tableland where, in 1962, Walker had established that pollen was plentifully preserved in a 'grab sample' of sediment from Lake Euramoo. Accordingly, the Walkers and the Kershaws entrained for Far North Queensland in June 1968 and set themselves up in half a house hired at the waterworks settlement of Tinaroo Falls where Peter voluntarily took on the additional evening duty of wheeling a two year-old Kate Walker around the streets to encourage her to sleep.

Fieldwork was not without its problems but, despite a raft that floated just below water level and what, in 2011, would be regarded as impossibly primitive coring equipment, a good sequence of samples was obtained and carried back to Canberra.

Months later, Peter presented Walker with an outline pollen diagram, undated, which undoubtedly showed sclerophyll woodland to have predated the existing rainforest of the Euramoo basin. Peter attributed this to a climatic change but Walker was not so sure, suggesting that it may have been a seral phenomenon immediately following the eruption by which the crater had been formed. The only way to solve the question was to sample more crater sediments of differing morphologies, and perhaps differing ages, in the region and date the vegetation change. It was also evident that Peter had a mind of his own in recognition of which he was awarded a scholarship to continue the work for a PhD and so threw off the shackles of assistantship. Thus was laid the foundation of a remarkable achievement in which Peter proved himself right (to Walker's delight) and established a basis from which he and others have made the Atherton Tableland the most concentrated source of palynological and related data in the tropical Quaternary world.

Figure 2. A biogeography field trip to Ironbark Basin, Angelsea, in the late 1970s prior to the establishment of a post-1983 fire regeneration project that has continued with second-year students to the present day.

Peter subsequently moved from the ANU to the School of Geography and Environmental Science at Monash University, where he continues to foster environmental education and research (Figure 2). One of Peter's great strengths in teaching and research is to gather around him people with exceptional skills and dedication to the research tasks at hand. His long-term collaboration with his student and then colleague Merna McKenzie continues to produce invaluable insights into the nature of glacial cycles and tree-line fluctuations in southeast Australia. The arrival of Sander Van der Kaars, one of the first Logan Fellows at Monash University, ushered in a period of prolific pollen counting in the department, focussing on the potential for marine records from the west and north coasts of Australia to unlock our understanding of the influence of the monsoon on Australian landscapes. And Peter long promoted the incorporation of Indigenous archaeology in teaching and research of environmental science, a logical outcome of his early views of peopled landscapes; indeed it was his presence that caused Bruno David to come to the School of Geography and Environmental Science rather than the anthropology department in 1997 and thereby establish Monash University's first Indigenous archaeology programme (further developed a few years later by the arrival of Ian McNiven). These collaborations, and those with his many successful honours, post-graduate and post-doctoral fellows led to the School being regarded as the pre-eminant department for undergraduate and graduate training in biogeography and palaeoecology.

After some 44 years of academic research and teaching in Quaternary ecology and biogeography, Professor Peter Kershaw retired in October 2010. A meeting was held on 1st November 2010 at the Royal Society of Victoria in Melbourne, for a day honouring Peter's contributions to palaeoecology, biogeography and archaeology in the Australian region (Figure 3). The presentations listed below represent a snapshot of the legacy of Peter's endeavours, collaborations and inspiration that will no doubt resonate into the future of Quaternary research in our region. He is a valued colleague driven by his commitment to the discipline and graduate students, and in each of these areas he has achieved outstanding results:

- Matt McGlone, Keynote, "Separated at birth: physical, biological and social aspects of the trans-Tasman relationship"
- David Mercer & Homer Le Grand, "Peter Kershaw's career at Monash"
- Martin Williams, "Did the 73 ka Toba super-eruption have an enduring effect? Insights from archaeology, genetics, palynology, stable isotope geochemistry and climate models"
- Jim Bowler, "Fishing at the LGM: A day in the life of early boat people"
- Patrick De Deckker, "Multidisciplinary studies applied to core Fr10/95-GC17 offshore Northwest Cape, Western Australia"
- John Dodson, "Paradise Lost: tools and lessons on how human-kind shaped the world"
- Richard Cosgrove, "The Archaeologists Palynologist: the connection between archaeology and palaeoecology in Australia"
- Lesley Head, "Tomorrow is a long time: palaeoecology and contested landscapes in Sweden and Australia"
- Peter Gell, "Palaeoecology as a means of auditing wetland condition"
- Patrick Moss, "Holocene Landscape Change in the Humid Tropics of Northeastern Australia"
- John Tibby, "Palaeolimnological evidence for European impact in Australia"
- Kale Sniderman, "New insights from the fossil record into the history of Australia's sclerophyllous vegetation"

Figure 3. Attendants at the one-day meeting held at the Royal Society of Victoria in Melbourne, 1st November 2010, in honour of Peter Kershaw (centre, back row, whitest hair).

Discussion

By those who shared the laboratory with Peter at the ANU, he is perhaps best remembered for his goodwill, hard work, Dylan-style songs and his forcefully enunciated views on anything from pollen morphology to the dangers of religious bigotry. He was also the only member of those early years of ANU pollen research to count pollen while smoking a cigarette, particularly in the evenings when nobody else was working, as evidenced by the accumulation of ash around his microscope each following morning.

Perhaps fittingly, then, it was the application of charcoal analysis alongside pollen counts to explore the role of fire in vegetation change that showed an unprecedented change in fire regimes accompanying the arrival of people into Australia and that led to fundamental changes in the extent and composition of rainforests in the Atherton Tableland region. Such an approach was new and innovative at the time, being utilised in the Department of Biogeography and Geomorphology at the ANU to address questions of landscape change. Whether or not the timing and extent of these transformations reflected the wider tempo of change in the Australian landscape is yet to be fully resolved and is likely to occupy the lives of many Quaternary researchers for years to come.

Peter's concern with the Australian landscape as a peopled landscape meant, and means, that archaeologists need to consider landscape processes in the interpretation of excavated sequences, while geomorphologists and biogeographers need to consider people in their own interpretations of landscape processes. In this context and for the Australian region in particular, Peter's contribution to these disciplines have revolved around a number of major themes:

1. For Australia's long history, the visibility of people in the landscape through the effects of anthropogenic landscape burning.

2. As best shown at Lynch's Crater, but evident at numerous other sites also, the ability of individual pollen *site* sequences to implicate *landscape* histories. That is, the ability to transcend different spatial scales of interpretation, as evident by the ability of palynological research within individual sites to implicate whole landscape histories, an

interpretative leap rarely legitimately achievable in purely archaeological research (Figure 4).

3. Peter was also the first person to securely date several glacial cycles in Australia and through the pollen records he showed that these cycles were different in their vegetation composition. This work came out of what was then seen as a pressing need to demonstrate synchronous orchestration of Australia's past climate with global climate signatures. He grasped the opportunity to open up the rich palynological fields of the western plains of Victoria, where maar deposits not unfamiliar to him from his earlier Atherton Tableland work yielded windows of opportunity to investigate the long Quaternary record of climate and vegetation change in southeast Australia.

4. The development of new and novel approaches to palaeoenvironmental reconstruction, particularly pioneering the use of bioclimatic profiles of extant taxa to generate quantitative palaeoclimatic estimates from pollen data (e.g. Kershaw and Nix 1988).

5. A preparedness to recognise that we do not know everything about the past, with present understandings sometimes turning to blind prejudice informed by the limitations of our data, meaning that we need to question conventional wisdom. Peter has thus been willing to play devil's advocate when new data hinted at the arrival of people in Australia 150,000-100,000 years ago, as evidenced by significant changes in pollen and carbonised particle frequencies in offshore sediments, northeast Australia (Kershaw *et al.* 1993); and a subsequent preparedness to reverse his own views in light of subsequent findings (e.g. Moss and Kershaw 2000). While these new interpretations flew in the face of conventional wisdom – and were subsequently shown to be wrong – Peter was prepared to shake the discipline(s) in light of evidence that required explanation.

6. Expanding our understanding of the deep-time biogeography of Australian rainforests and sclerophyll plants, most notably the Araucariaceae, an iconic Gondwanan family (Kershaw and Wagstaff 2001).

7. The encouragement of cross-disciplinary research and use of multi-proxy evidence to strengthen the reliability and interpretability of research findings.

These themes have each significantly contributed to how we now come to read, and understand, the Australian landscape as historically created from some 50,000 years of Aboriginal engagements with their surroundings. In this spirit of investigation, where understanding landscape history requires a joining rather than separation of parts (e.g. plants vs sediments vs people), in this volume we present a set of papers by scientists who have each been directly influenced by Peter's work. The case studies presented each consider the landscape as one that has developed with people in its midst. These are not prefigured landscapes as stages for people to subsequently act upon, but rather engaged landscapes at their very core: landscapes that are defined by such engagements. In this sense a peopled landscape is one that would not exist in that form without those who gave it its particular characteristics. One of our roles, as archaeologists, geomorphologists, palaeoecologists and biogeographers, is to determine the nature of those historical engagements that enable us to define how in history people have come to influence and shape the world in which we live today. This is the ongoing legacy of Peter Kershaw's ongoing contributions to the study of landscape history (Figure 5).

Figure 4. Peter Kershaw (right) celebrating another successful coring expedition at Lynch's Crater in 2004 with Damien Kelleher (left) and Chris Turney (middle).

Figure 5. Peter Kershaw at the Bromfield Swamp lookout in 2004.

References

Allen, J., Golson, J. and Jones R. (eds) 1977. *Sunda and Sahul: Prehistoric Studies in Southeast Asia, Melanesia and Australia.* Academic Press, London.

Hale, P. and Lamb, D. (eds) 1997. *Conservation Outside Nature Reserves.* Centre for Conservation

Biology, University of Queensland, Brisbane.

Johnson, C. 2007. *Australia's Mammal Extinctions: A 50,000-Year History*. Cambridge University Press, Cambridge.

Jones, R. 1968. The geographical background to the arrival of man in Australia and Tasmania. *Archaeology and Physical Anthropology in Oceania* 3:186-215.

Jones, R. 1969. Fire-stick Farming. *Australian Natural History* 16:224-228.

Kershaw, A.P. 1973. Late quaternary vegetation of the Atherton Tableland, north-east Queensland, Australia. PhD Thesis, Australian National University, Canberra.

Kershaw, A.P. 1974. A long continuous pollen sequence from northeastern Australia. *Nature* 251:222-223.

Kershaw, A.P., McKenzie, G.M. and McMinn, A. 1993. A Quaternary vegetation history of northeast Queensland from pollen analysis of ODP Site 820. *Proceedings of the Ocean Drilling Program* 133:107-114.

Kershaw, A.P and Nix, H.A. 1988. Quantitative palaeoclimatic estimates from pollen data using bioclimatic profiles of extant taxa. *Journal of Biogeography* 15:589-602.

Kershaw, A.P. and Wagstaff, B.E. 2001. The southern conifer family Araucariaceae: history, status, and value for palaeoenvironmental reconstruction. *Annual Review of Ecology and Systematics* 32:397-414.

Moss, P.T. and Kershaw, A.P. 2000. The last glacial cycle from the humid tropics of northeastern Australia: Comparison of a terrestrial and a marine record. *Palaeogeography, Palaeoclimatology, Palaeoecology* 155(1-2):155-176.

Thomas, J. 2008. Archaeology, landscape, and dwelling. In: David, B. and Thomas, J. (eds), *Handbook of Landscape Archaeology*, pp. 300-306. Left Coast Press, Walnut Creek.

Turney, C.S.M., Kershaw, A.P., James, S., Branch, N., Cowley, J., Fifield, L.K., Jacobsen, G. and Moss, P. 2006. Geochemical changes recorded in Lynch's Crater, northeastern Australia, over the past 50 ka. *Palaeogeography, Palaeoclimatology, Palaeoecology* 233:187-203.

Publications – A. Peter Kershaw, 1970-2011

Books and special jounal issues

Kershaw, A.P., Haberle, S.G, Turney, C.S.M. and Bretherton, S.C. (eds) 2007. *Environmental history of the humid tropics region of north-east Australia.* Special issue *Palaeogeography, Palaeoclimatology, Palaeoecology* 251 (1):1-173.

Kershaw, P., Chappellaz, J., Newman, L. and Kiefer, T. (eds) 2007. *Past Climate Dynamics: A Southern Perspective. Past Global Changes (PAGES) News* 15(2):1-27.

Turney, C.S.M., Kershaw, A.P. and Lynch, A. (eds) 2006. *Integrating High Resolution Past Climate Records for Future Prediction in the Australasian Region.* Special issue *Journal of Quaternary Science* 21 (7):679-801.

Bottjer, D.J., Correge, T., Kershaw, A.P. and Surlyk, F. (eds) 2006. *Exploring Life and Environments Through Time: Celebrating the 40th Anniversary of Palaeo-3.* Special issue *Palaeogeography, Palaeoclimatology, Palaeoecology* 232 (2-4):97-458.

Kershaw, A.P. and Orr, M.L. (eds) 2004. *Environmental History of the Newer Volcanic Province of Victoria.* Thematic issue, *Proceedings of the Royal Society of Victoria* 116 (1):1-182.

Kershaw, A.P., David, B., Tapper, N.J., Penny, D. and Brown, J. (eds) 2002. *Bridging Wallace's Line: The Environmental and Cultural History and Dynamics of the Southeast Asian – Australian Region.* Catena Verlag., Reiskirchen, Germany, 360 pp.

Dam, R.A.C., van der Kaars, S. and Kershaw, A.P. (eds) 2001. *Quaternary Environmental Change in the Indonesian Region.* Special issue, *Palaeogeography, Palaeoclimatology, Palaeoecology* 171 (3-4):421pp.

Kershaw, A.P. and Whitlock, C. (eds) 2000. *Last Glacial-Interglacial Cycle: Patterns and Causes of Change.* Special issue, *Palaeogeography, Palaeoclimatology, Palaeoecology* 155 (1-2):1-209.

Partridge, T., Kershaw, A.P. and Iriondo, M. (eds) 1999. *Palaeoclimates of the Southern Hemisphere During the Last 200,000 Years: Data, Models and Regional Syntheses.* Special issue, *Quaternary International* 58/58:1-235.

Kershaw, A.P., Pittock, B. and Simmonds, I. (eds) 1998. *Climate Change in the Australian Region: Quantifying the Past to Understand the Future.* Special Issue, *Palaeoclimates: Data and Modelling* 3:1-238.

Williams, M.A.J., Dunkerley, D.L., De Deckker, P., Kershaw, A.P. and Chappell, J. 1998. *Quaternary Environments.* 2nd Edition. Edward Arnold, 329pp.

Williams, M.A.J., Dunkerley, D.L., De Deckker, P., Kershaw, A.P. and Stokes, T. 1997. *Quaternary Environments.* Chinese Edition, 329pp.

Williams, M.A.J., Dunkerley, D.L., De Deckker, P., Kershaw, A.P. and Stokes, T. 1993. *Quaternary Environments.* Edward Arnold, 329pp.

Williams, M.A.J., De Deckker, P. and Kershaw, A.P. 1991. *The Cainozoic in Australia: A Re-appraisal of the Evidence.* Geological Society of Australia, Special Publication No. 18, 346 pp.

Werren, G.L. and Kershaw, A.P. (eds) 1991. *The Rainforest Legacy: Australian National Rainforests Study Vol. 2 - Flora and Fauna of the Rainforests.* Special Australian Heritage Publication Series No. 7 (2), Australian Heritage Commission, 414 pp.

Werren, G.L. and Kershaw, A.P. (eds) 1991. *The Rainforest Legacy: Australian National Rainforests Study Vol. 3 - Rainforest History, Dynamics and Management.* Special Australian Heritage Publication Series No. 7 (3), Australian Heritage Commission, 309 pp.

Werren, G.L. and Kershaw, A.P. (eds) 1987. *The Rainforest Legacy: Australian National Rainforests Study Vol 1 - The Nature, Distribution and Status of Rainforest Types.* Special Australian Heritage Publication Series No. 7 (1), Australian Heritage Commission, 344 pp.

Luly, J., Sluiter, I.R. and Kershaw, A.P. 1980. Pollen studies of Tertiary brown coals: preliminary analyses of lithotypes within the Latrobe Valley, Victoria. *Monash Publications in Geography No. 23,* 78pp.

Chapters in books

Kershaw, A.P. and van der Kaars, S. In Metcalfe, S.E. and Nash, D.J. (eds) (in press) Tropical Quaternary climates in Australia and the south-west Pacific. *Quaternary Environmental Change in the Tropics.* Blackwell Scientific Publishers, Oxford.

Porch, N. and Kershaw, A.P. 2010. Comparative AMS 14C dating of plant macrofossils, beetles and pollen preparations from two late Pleistocene sites in southeastern Australia. In: Haberle, S.G., Stevenson, J. and Prebble, M. (eds), *Altered Ecologies: Fire, Climate and Human Influence on Terrestrial Landscapes, Terra Australis 32,* pp. 395-403. ANU E-Press, Canberra, Australia.

Kershaw, A.P., G.M. McKenzie, J. Brown, R.G. Roberts, and S. van der Kaars 2010. Beneath the peat: A refined pollen record from an interstadial at Caledonia Fen, highland eastern Victoria, Australia. In: Haberle, S.G., Stevenson, J. and Prebble, M. (eds), *Altered Ecologies: Fire, Climate and Human Influence on Terrestrial Landscapes, Terra Australis 32,* pp. 33-48. ANU E-Press, Canberra, Australia.

Rowe, C. and Kershaw, A.P. 2008. Microbotanical remains in landscape archaeology. In: David, B. (ed) *Handbook of Landscape Archaeology,* pp. 430-441. Left Coast Press, California.

Kershaw, A.P. and van der Kaars, S. 2007. Pollen records, Late Pleistocene, Australia and New Zealand. In: Elias, S.A. (ed), *Encyclopedia of Quaternary Science, Vol. 4,* pp. 2613-2622. Elsevier, Amsterdam.

Kershaw, A.P., van der Kaars, S. and Flenley, J.F. 2007. The Quaternary history of far eastern rainforests. In: Bush, M.B. and Flenley, J.F. (eds), *Tropical Rainforest Responses to Climate Change,* pp. 77-115. Springer-Praxis, Berlin.

Tibby, J., Kershaw, A.P., Builth, H., Philibert, A. and White, C. 2006. Environmental change and variability in south-western Victoria: changing constraints and opportunities for occupation and land use. In: David, B., Bryson, B. and McNiven, I. (eds), *The Social Archaeology of Indigenous Societies,* pp. 254-269. Aboriginal Studies Press, Canberra.

Kershaw, A.P., Moss, P.T. and Wild, R. 2005. Patterns and causes of vegetation change in the Australian Wet Tropics region over the last 10 million years. In: Bermingham, E., Dick, C. and Moritz, C. (eds), *Tropical Rainforests: Past Present and Future,* pp. 374-400. The University of Chicago Press, Chicago.

Kershaw, A.P., van der Kaars, S., Moss, P.T. and Wang, X. 2002. Palynological evidence for environmental change in the Indonesian-northern Australian region over the last 140,000 to 300,000 years. In: Kershaw, A.P., David, B., Tapper, N.J., Penny, D. and Brown, J. (eds), *Bridging Wallace's Line: The Environmental and Cultural History and Dynamics of the Southeast Asian – Australian Region,* pp. 97-118. Catena Verlag, Reiskirchen, Germany.

David, B., Kershaw, A.P. and Tapper, N. 2002. Bridging Wallace's Line: Bringing home the Antipodes. In: Kershaw, A.P., David, B., Tapper, N.J., Penny, D. and Brown, J. (eds), *Bridging Wallace's Line: The Environmental and Cultural History and Dynamics of the Southeast Asian – Australian Region,* pp. 1-4. Catena Verlag, Reiskirchen, Germany.

Kershaw, A.P., Clark, J.S. and Gill, A.M. 2002. A history of fire in Australia. In: Bradstock, R., Williams, J. and Gill, A.M. (eds), *Flammable Australia: the Fire Regimes and Biodiversity of a Continent,* pp. 3-25. Cambridge University Press, Cambridge.

Clark, J.S., Gill, A.M. and Kershaw, A.P. 2002. Spatial variability in fire regimes: its effects on recent and past vegetation. In: Bradstock, R., Williams, J. and Gill, A.M. (eds), *Flammable Australia: the Fire Regimes and Biodiversity of a Continent,* pp. 125-141. Cambridge University Press, Cambridge.

Kershaw, A.P., Penny, D., van der Kaars, S., Anshari, G. and Thamotherampillai, A. 2001. Evidence for vegetation and climate in lowland southeast Asia at the Last Glacial Maximum. In: Melcalfe, I., Smith, J.M.B., Morwood, M. and Davidson, I. (eds), *Floral and*

Faunal Migrations and Evolution in Southeast Asia-Australasia, pp. 227-236. A.A. Balkema, Lisse.

Moss, P.T. and Kershaw, A.P. 1999. Evidence from marine ODP Site 820 of fire/vegetation/climate patterns in the humid tropics of Australia over the last 250,000 years. *Proceedings, Australian Bushfire Conference: Bushfire 99,* pp. 269-279. Charles Sturt University, Albury, NSW.

Kershaw, A.P., Bush, M., Hope, G.S., Weiss, K., Goldammer, J.G. and Sanford Jr., R. 1997. The contribution of humans to past biomass burning in the tropics. In: Clark, J., Cachier, H., Goldammer, J.G. and Stocks, B. (eds), *Sediment Records of Biomass Burning and Global Change,* pp. 413-442. Springer, Berlin.

Kershaw, A.P. 1997. Environments of mainland southeastern Australia at the climatic extremes of the last glacial cycle: evidence from pollen. In: Mecco, J. and Petit-Maire, N. (eds), *Climates of the Past,* pp. 115-122. Servicio de Publicaciones, Universidad de Las Palmas de Gran Canaria: Canary Islands.

Kershaw, A.P. and Bohte, A. 1997. The impact of prehistoric fires on tropical peatland forests. In: Rieley, J.O. and Page, S.E. (eds), *Biodiversity and Sustainability of Tropical Peatlands,* pp. 73-80. Samara Press, Tresaith, Cardigan.

Kershaw, A.P., Reid, M. and Bulman, D. 1997. The nature and development of peatlands in Victoria. In: Rieley, J.O. and Page, S.E. (eds), *Biodiversity and Sustainability of Tropical Peatlands,* pp. 81-91. Samara Press, Tresaith, Cardigan.

Kershaw, A.P. 1997. A history of biomass burning in the tropics: relative contributions of climate and people. In: Sirinanda, K.U. (ed), *Climate and Life in the Asia-Pacific,* pp. 130-142. Department of Geography, Universiti Brunei Darussalam, Brunei.

Allen, J. and Kershaw, A.P. 1996. The Pleistocene-Holocene transition in Greater Australia. In: Straus, L.G., Eriksen, B.V., Erlandson, J.M. and Yesner, D.R. (eds), *Humans at the End of the Ice Age. The Archaeology of the Pleistocene-Holocene Transition,* pp.171-199. Plenum Press, New York.

Balme, B.E., Kershaw, A.P. and Webb, J. 1995. Floras of Australian coal measures. In: Ward, C.R., Harrington, H.J., Mallett, C.W. and Beeston, J.W. (eds), *Geology of Australian Coalfields,* pp. 41-62. Geological Society of Australia, Sydney.

Kershaw, A.P. 1995. Pollen representation of peatland vegetation, Victoria, Australia. In: Dixon, G. and Aitken, D. (eds), *Institute of Australian Geographers: Conference Proceedings, 1993,* Monash Publications in Geography No. 45, pp. 162-167. Dept. of Geography and Environmental Science, Monash University, Melbourne.

Kershaw, A.P. and McGlone, M. 1995. The Quaternary history of the southern conifers. In: Enright, N. and Hill, R.S. (eds), *The Ecology of the Southern Conifers,* pp. 30-63. Melbourne University Press, Melbourne.

Dodson, J. and Kershaw, A.P. 1994. Evolution and history of Mediterranean vegetation types in Australia. In: Kalin Arroyo, M.T., Zedler, P.H. and Fox, M.D. (eds), *Ecology and Biogeography of Mediterranean Ecosystems in Chile, Califormia and Australia,* pp. 21-40, Springer-Verlag, New York.

Kershaw, A.P. 1994. Historical development of the vegetation. In: Bambrick, S. (ed) *The Cambridge Encyclopedia of Australia,* pp. 22-24. Cambridge University Press, Cambridge.

Ladiges, P., Kershaw, A.P. and Rich, P. 1994. Australian Environments and Biota. In: Knox, B., Ladiges, P. and Evans, B. (eds) *Biology,* pp. 904-933. McGraw-Hill, Sydney.

Kershaw, A.P., Martin, H.A. and McEwen Mason, J. 1994. The Neogene - a period of transition. In: Hill, R. (ed), *Australian Vegetation History. Cretaceous to Present,* pp. 435-462, Cambridge University Press, Cambridge.

McGlone, M., Kershaw, A.P. and Markgraf, V. 1992. El Nino-Southern Oscillation and climatic variability in Australasian and South American palaeoenvironmental records. In: Diaz, H.F. and Markgraf, V. (eds), *El Nino: Historical and Palaeoclimatic Aspects of the Southern Oscillation*, pp. 435-462. Cambridge, University Press Cambridge.

Kershaw, A.P. 1992. The development of rain-forest-savanna boundaries in tropical Australia. In: Furley, P.A., Proctor, J. and Ratter, J.A. (eds), *Nature and Dynamics of Forest-Savanna boundaries*, pp. 255-271. Chapman Hall, London.

Kershaw, A.P. 1992. The development and history of temperate zone rainforests in Australia. In: Gell, P. and Mercer, D. (eds), *Victoria's Rainforests: Perspectives on Definition, Classification and Management*. Monash Publications in Geography No. 41, pp. 107-115. Dept. of Geography and Environmental Science, Monash University, Melbourne.

Kershaw, A.P. 1992. Past vegetational and climatic change in Victoria: What can it show? In: Pittock, J. (ed), *Victoria's Flora and Fauna: Can it Survive the Greenhouse Effect?*, pp. 13-15. Victorian National Parks Association, Melbourne.

Hiscock, P. and Kershaw, A.P. 1992. Palaeoenvironments and prehistory of Australia's tropical top end. In J.R. Dodson (ed.) *The Naive Lands: Human/Environment Interactions in Australia and Oceania*, pp. 43-75. Longman Cheshire, Melbourne.

Kershaw, A.P., Sluiter, I.R., McEwen Mason, J., Wagstaff, B.E. and Whitelaw, M. 1991. The history of rainforest in Australia: evidence from pollen. In: Werren, G.L. and Kershaw, A.P. (eds), *The Rainforest Legacy: Australian National Rainforests Study Vol. 3 - Rainforest History, Dynamics and Management*, pp. 1-15. Special Australian Heritage Publication Series, Canberra.

Kershaw, A.P., Baird, J., D'Costa, D., Edney, P., Peterson, J.A. and Strickland, K.M. 1991. A comparison of long Quaternary pollen records from the Atherton and Western Plains volcanic provinces. In: Williams, M.A.J., De Deckker, P. and Kershaw, A.P. (eds), *The Cainozoic in Australia: a re-appraisal of the evidence*, pp. 288-301. Geological Society of Australia: Sydney.

Kershaw, A.P. and Strickland, K.M. 1990. The development of alpine vegetation on the Australian mainland. In: Good, R. (ed) *Proceedings of the First Fenner Conference: the Scientific Significance of the Australian Alps*, pp. 113-126. Australian Academy of Science, Canberra

Kershaw, A.P. and Gell, P.A. 1990. Quaternary vegetation and the future of the forests. In: Bishop, P. (ed), *Lessons for human survival: nature's record from the Quaternary*, pp. 11-20. Geological Society of Australia Symposium Proceedings 1, Sydney.

Kershaw, A.P. and Nix, H.A. 1989. Quantitative palaeoclimatic estimates from pollen data. In: Donnelly, T.H. and Wasson, R. (eds), *CLIMANZ 3: Proceedings of Symposium, Melbourne 1987*, pp. 78-85. Division of Water Resources, CSIRO, Canberra.

Kershaw, A.P. and Whiffin, T. 1989. Australia. In: Campbell, D.G. and Hammond, H.D. (eds), *Floristic Inventory of Tropical Countries: the Status of Plant Systematics, Collections and Vegetation, Plus Recommendations for the Future*, pp. 149-165. The New York Botanical Garden, New York.

Kershaw, A.P. 1988. Australasia. In: Huntley, B. and Webb lll, T. (eds), *Vegetation History*, pp. 237-306. Kluwer Academic Publishers, Dordrecht.

Williams, M.A.J., De Deckker, P. and Kershaw, A.P. 1988. Past environmental analogues. In: Pearman, G. (ed) *Greenhouse: planning for climate change*, pp. 473-488. CSIRO, Melbourne.

Truswell, E.M., Kershaw, A.P. and Sluiter, I.R. 1987. The Australian/Malaysian connection: evidence from the palaeobotanical record. In: Whitmore, T.C. (ed) *Biogeographic Evolution of the Malay Archipeligo*, pp. 32-49. Clarendon Press, Oxford.

Raby, G., van Djik, M. and Kershaw, A.P. 1987. Climate. In: Vampley, W. (ed), *Australians: Historical Statistics*, pp. 62-68. Fairfax, Syme and Weldon, Broadway, NSW.

Kershaw, A.P. 1987 A comparative vegetation history of southeastern Australia and New Zealand. In: Conacher, A. (ed), *Readings in Australian Geography*, pp. 433-445. Institute of Australian Geographers (W.A. Branch) and Department of Geography, University of Western Australia, Perth.

Kershaw, A.P., McEwen Mason, J.R., McKenzie, G.M., Strickland, K.M. and Wagstaff, B.E. 1986. Aspects of the development of cold-adapted flora and fauna in the Cenozoic of southeastern mainland Australia. In: Barlow, B.A. (ed), *Flora and Fauna of Alpine Australasia, ages and origins*, pp. 147-160. CSIRO, Canberra.

Kershaw, A.P. 1984 Some applications of studies on vegetation and fire history to forest management. In: Ealey, E.H.M. (ed), *Fighting fire with fire: A symposium on fuel reduction burning in forests*, pp. 55-69. Graduate School of Environmental Science, Monash University, Melbourne.

Kershaw, A.P., Sluiter, I.R., Dawson, J., Wagstaff, B.E. and Whitelaw, M. 1984. The history of rainforest in Australia. In: Werren, G.L. and Kershaw, A.P. (eds), *The Rainforest Legacy: Australian National Rainforests Study Vol 1 - The Nature, Distribution and Status of Rainforest Types*, Special Australian Heritage Publication Series No. 7 (1), pp. 462-77. Australian Heritage Commission, Canberra.

Kershaw, A.P. 1984. Late Cenozoic plant extinctions in Australia. In: Martin, P.S. and Klein, R.G. (eds), *Quaternary Extinctions, a Prehistoric Revolution*, pp. 691-707. University of Arizona Press, Tucson.

Kershaw, A.P. 1984. Review of Quaternary studies in Australia - plant and invertebrate palaeoecology. In: *Quaternary Studies in Australia: future directions*, pp. 67-82. Bureau of Mineral Resources, Geology and Geophysics Record 1984/14, Canberra.

Kershaw, A.P. 1983. The vegetation record from northeastern Australia 40,000 - 3000 B.P.. In: Chappell, J.M.A. and Grindrod, A. (eds), *CLIMANZ: Quaternary Climatic History of Australia*, pp. 25, 37, 61-2, 79-80, 100-1. Dept. of Biogeography and Geomorphology, Australian National University, Canberra.

Kershaw, A.P., Southern, W., Williams, J.M. and Joyce, L.J. 1983. The vegetation record from the southeastern highlands of mainland Australia 40,000 - 3000 B.P. In: Chappell, J.M.A. and Grindrod, A. (eds), *CLIMANZ: Quaternary Climatic History of Australia*, pp. 16, 37-8, 62, 80-1, 101, Dept. of Biogeography and Geomorphology, Australian National University, Canberra.

Kershaw, A.P. 1982. Holocene Palaeoecology. In: Thom, B.G. and Wasson, R. (eds), *Holocene Research in Australia 1978-1982*, pp. 78-110. Occasional Paper No. 33, Dept. of Geography, Royal Military College, Duntroon, Canberra.

Kershaw, A.P. 1981. Quaternary vegetation and environments. In: Keast, A. (ed), *Ecological Biogeography in Australia*, pp. 83-101. Dr. W. Junk, La Hague.

Singh, G., Kershaw, A.P. and Clark, R. 1981. Quaternary vegetation and fire history in Australia. In: Gill, A.M., Groves, R.A. and Noble, I.R. (eds), *Fire and Australian Biota*, pp. 23-54. Australian Academy of Science, Canberra.

Kershaw, A.P. 1980. Long term changes in north-east Queensland rainforest. In: Wright, J., Mitchell, N. and Watling, P. (eds), *Reef, Rainforest, Mangrove, Man*, pp. 32-38. Wildlife Preservation Society of Queensland, Brisbane.

Kershaw, A.P. 1980. Evidence for vegetation and climatic change during the Quaternary. In: Henderson, R.A. and Stephenson, P.J. (eds), *The Geology and Geophysics of northestern Australia*, pp. 398-402. Geological Society of Australia (Qld. Div.), Brisbane.

Kershaw, A.P. 1975. Late Quaternary vegetation and climate in north-eastern Australia. In: Suggate, R.P. and Creswell, M.M. (eds), *Quaternary Studies*, pp. 181-187. Royal Society of New Zealand, Wellington.

Kershaw, A.P. 1973. The numerical analysis of modern pollen spectra from north-east Queensland rainforests. In: Glover, J.E. and Playford, G. (eds), *Mesozoic and Cainozoic Palynology: Essays in Honour of Isabell Cookson*, pp. 191-9. Special publication, Geological Society of Australia, No. 4, Sydney.

Journal articles

Mooney, S.D., Harrison, S.P., Bartlein, P.J., Daniau, A.-L., Stevenson, J., Brownlie, K.C., Buckman, S., Cupper, M., Luly, J., Black, M., Colhoun, E., D'Costa, D., Dodson, J., Haberle, S.G., Hope, G.S., Kershaw, P., Kenyon, C., McKenzie, M. and Williams, N. 2011. Late Quaternary fire regimes of Australasia. *Quaternary Science Reviews* 30:28-46.

Haberle, S.G., Rule, S., Roberts, P., Heijnis, H., Jacobsen, G., Turney, C., Cosgrove, R., Ferrier, A., Moss, P., Mooney, S., and Kershaw, P. 2010. Paleofire in the wet tropics of northeast Queensland, Australia. *PAGES News* 18(2):78-80.

Walker, M., Johnsen, S., Rasmussen, S.O., Steffensen, J.-P., Popp, T., Gibbard, P., Hoek, W., Lowe, J., Andrews, J., Björck, S., Cwynar, L., Hughen, K., Kershaw, P., Kromer, B., Litt, T., Lowe, D.L., Nakagawa, T., Newnham, R. and Schwander, J. 2009. Formal definition and dating of the GSSP (Global Stratotype Section and Point) for the base of the Holocene using the Greenland NGRIP ice core, and selected auxiliary records. *Journal of Quaternary Science* 24:3-17.

Williams, M., Cook, E., van der Kaars, S., Barrows, T., Shulmeister, J. and Kershaw, P. 2009. Glacial and deglacial climatic patterns in Australia and surrounding regions from 35 000 to 10 000 years ago reconstructed from terrestrial and near-shore proxy data. *Quaternary Science Reviews* 28:2398-2419.

Goodall, R.A., David, B., Kershaw, P. and Fredricks, P.M. 2009. Prehistoric hand stensils at fern Cave, North Queensland, Australia: environmental and chronological implications of Raman spectroscopy and FT-IR imaging results. *Journal of Archaeological Science* 36:2617-2624.

Coulter, S.E., Turney, C.S.M., Kershaw, P. and Rule, S. 2009. The characterisation and significance of a MIS 5a distal tephra on mainland Australia. *Quaternary Science Reviews* 28:1825-1830.

Walker, M., Johnsen, S., Rasmussen, S.O., Steffensen, J.-P., Popp, T., Gibbard, P., Hoek, W., Lowe, J., Andrews, J., Björck, S., Cwynar, L., Hughen, K., Kershaw, P., Kromer, B., Litt, T., Lowe, D.L., Nakagawa, T., Newnham, R. and Schwander, J. 2008. The Global Stratotype Section and Point (GSSP) for the base of the Holocene Series/Epoch (Quaternary System/Period) in the NGRIP ice core. *Episodes* 31:264-267.

Builth, H., Kershaw, A.P., White, C., Roach, A., Hartney, L., McKenzie, M., Lewis, T. and Jacobsen, G. 2008. Environmental and cultural change on the Mt Eccles lava flow landscapes of south-west Victoria, Australia. *The Holocene* 18:421-432.

Brooks, B.W., Bowman, D.M.J.S., Burney, D.A., Flannery, T.F., Gagan, M.K., Gillespie, R., Johnson, C.N., Kershaw, A.P., Magee, J.W., Martin, P.S., Miller, G.H., Peiser, B. and Roberts, R.G. 2007. Would the Australian megafauna have become extinct if people had never colonized the continent? Comments on 'A review of the evidence for a human role in the extinction of Australian megafauna and an alternative explanation' by S. Wroe and J. Field. *Quaternary Science Reviews* 26:560-564.

Lynch, A.H., Beringer, J., Kershaw, P., Marshall, A., Mooney, S., Tapper, N., Turney, C. and van Der Kaars, S. 2007. Using the paleorecord to evaluate climate and fire interactions in Australia. *Annual Review of Earth and Planetary Sciences* 35:215-239.

Moss, P.T. and Kershaw, A.P. 2007. A late Quaternary marine palynological record (Oxygen isotope stages 1-7) for the Humid Tropics of northeastern Australia based on ODP Site

820). *Palaeogeography, Palaeoclimatology, Palaeoecology* 251:4-22.

Kershaw, A.P., Bretherton, S.C. and van der Kaars, S. 2007. A complete pollen record of the last 230 ka from Lynch's Crater, northeastern Australia. *Palaeogeography, Palaeoclimatology, Palaeoecology* 151:23-45.

Kershaw, A.P., McKenzie, G.M., Porch, N., Roberts, R.G., Brown, J., Heijnis, H., Orr, L.M., Jacobsen, G. and Newall, P.R. 2007. A high resolution record of vegetation and climate through the last glacial cycle from Caledonia Fen, south-eastern highlands of Australia. *Journal of Quaternary Science* 22:481-500.

Reid, M.A., Sayer, C.D., Kershaw, A.P. and Heijnis, H. 2007. Palaeolimnological evidence for submerged plant loss in a floodplain lake associated with accelerated catchment soil erosion (Murray River, Australia). *Journal of Paleolimnology* 38:191-208.

Sniderman, J.M.K., Pillans, B., O'Sullivan, P.B. and Kershaw, A.P. 2007. Climate and vegetation in southeastern Australia respond to Southern Hemisphere insolation forcing in the late Pliocene – early Pleistocene. *Geology* 35:41-44.

Turney, C.S.M., Haberle, S.G., Fink, D., Kershaw, A.P., Barbetti, M., Barrows, T.T., Black, M., Cohen, T.J., Correge, T., Hesse, P.P., Qua, Q., Johnston, R., Morgan, V., Moss, P., Nanson, G., Van Ommen, T., Rule, S., Williams, N.J., Zhao, J-X., D'Costa, D., Feng, Y-X., Gagan, M., Mooney, S. and Xia, Q. 2006. Integration of ice-core, marine and terrestrial records for the Australian Last Glacial Maximum and Termination: a contribution from the OZ INTIMATE group. *Journal of Quaternary Science* 21:751-761.

Kershaw, P., van der Kaars, S., Moss, P., Opdyke, B., Guichard, F., Rule, S. and Turney, C. 2006. Environmental change and the arrival of people in the Australian region. *Before Farming* [Online] http://www.waspress.co.uk/journals/beforefarming/journal_20061/ 2006/1 article 2.

Turney, C.S.M., Kershaw, A.P., James, S., Branch, N., Cowley, J., Fifield, L.K., Jacobsen, G. and Moss, P. 2006. Geochemical changes recorded in Lynch's Crater, northeastern Australia, over the past 50 ka. *Palaeogeography, Palaeoclimatology, Palaeoecology* 233:187-203.

Turney C.S.M., Kershaw, A.P., Lowe, J.J., van der Kaars, S., Johnston, R., Rule, S., Moss, P., Radke, L., Tibby, J., McGlone, M.S., Wilmshurst, J.M., Vandergoes, M.J., Fitzsimons, S.J., Bryant, C., Branch, N.P., Jacobsen, G. and Fifield, L.K. 2006. Climatic variability in the southwest Pacific during the Last Termination (20–10ka BP). *Quaternary Science Reviews* 25:886–903.

Moss, P.T., Kershaw, A.P. and Grindrod, J. 2005. Pollen transport and deposition in riverine and marine environments within the humid tropics of northeastern Australia. *Review of Palaeobotany and Palynology* 134:55-69.

Barry, M.J., Tibby, J., Tsitsilas, A., Mason, B., Kershaw, A.P. and Heijnis, H. 2005. A long term lake salinity record and its relationships to *Daphnia* populations. *Archiv. fur Hydrobiologie* 163:1-23.

Leahy, P.G., Tibby, J., Kershaw, A.P., Heijnis, H. and Kershaw, J.S. 2005. A palaeolimnological reconstruction of the impact of European settlement on the Yarra River floodplain, Victoria, Australia. *River research and Management* 21:131-149.

Wang, P., Clemens, S., Beaufort, L., Bracannot, P., Ganssen, G., Jian, Z., Kershaw, A.P. and Sarnthein, M. 2005. Evolution and variability of the Asian monsoon system: state of the art and outstanding issues. *Quaternary Science Reviews* 24:595-629.

Turney, C.S.M., Kershaw, A.P., Clemens, S., Branch, N., Moss, P.T. and Fifield, L.K. 2004. Millennial and orbital variations in El Niño/Southern Oscillation and high latitude climate in the last glacial period. *Nature* 428:306-310.

Sniderman, J.M.K., O'Sullivan, P.B., Hollis, J. and Kershaw, A.P. 2004. Late Pliocene vegetation and climate change in the Western Uplands of Victoria, Australia. *Proceedings, Royal Society of Victoria* 116:79-94.

Pickett, E.J., Harrison, S.P., Hope, G., Harle, K., Dodson, J.R., Kershaw, A.P., Prentice, I.C., Backhouse, J., Colhoun, E.A., D'Costa, D., Flenley, J., Grindrod, J., Haberle, S.G., Hassell, C., Kenyon, C., Macphail, M., Martin, H., Martin, A.H., McKenzie, M., Newsome, J.C., Penny, D., Powell, J., Raine, J.I., Southern, W., Sutra, J.-P., Thomas, I., van der Kaars, S. and Ward, J. 2004. Pollen-based reconstructions of biome distributions for Australia, South East Asia and the Pacific (SEAPAC region) at 0, 6000 and 18,000 14C yr B.P. *Journal of Biogeography* 30:1381-1444.

Hope, G.S., Kershaw, A.P., van der Kaars, S., Sun, X., Liew, P-M., Heusser, L.E., Takahara, H., McGlone, M., Myoshi, N. and Moss, P.T. 2004. History of vegetation and habitat change in the Austral-Asian region. *Quaternary International* 118-119:103-126.

Kershaw, A.P., D'Costa, D.M., Tibby, J., Wagstaff, B.E. and Heijnis, H. 2004. The last million years around Lake Keilambete, western Victoria. *Proceedings, Royal Society of Victoria* 116:95-106.

Sherwood, J, Oyston, B. and Kershaw, A.P. 2004. The age and contemporary environments of Tower Hill volcano, southwest Victoria, Australia, *Proceedings, Royal Society of Victoria* 116:71-78.

Harle, K.J., Kershaw, A.P. and Clayton, E. 2004. Patterns of vegetation change in southwest Victoria (Australia) over the last two glacial/interglacial cycles. *Proceedings, Royal Society of Victoria* 116:107-139.

Kershaw, A.P., Tibby, J., Penny, D., Yesdani, H., Walkley, R., Cook, E. and Johnston, R. 2004. Latest Pleistocene and Holocene vegetation and environmental history of the western plains of Victoria. *Proceedings, Royal Society of Victoria* 116:141-163.

Anshari, G., Kershaw, A.P. and van der Kaars, S. 2004. Environmental Change and peatland forest dynamics in the Lake Sentarum Area, West Kalimantan, Indonesia. *Journal of Quaternary Science* 19:637-655.

Kershaw, A.P and Turney, C. 2004. The day after tomorrow. *Australasian Science* 25(7):29-31.

McKenzie, G.M. and Kershaw, A.P. 2004. A Holocene pollen record from cool temperate rainforest, Aire Crossing, the Otway region of Victoria, Australia. *Review of Palaeobotany and Palynology* 132:281-290.

Kershaw, A.P., van der Kaars, S. and Moss, P.T. 2003. Late Quaternary Milankovitch-scale climate change and variability and its impact on monsoonal Australia. *Marine Geology* 201:81-95.

Kershaw, A.P., Moss, P.T. and van der Kaars, S. 2003. Causes and consequences of long-term climatic variability on the Australian continent. *Freshwater Biology* 48:1274-1283.

Tibby, J., Reid, M.A., Fluin, J., Hart, B.T. and Kershaw, A.P. 2003. Assessing long-term pH change in an Australian river catchment using monitoring and palaeolimnological data. *Environmental Science and Technology* 37:3250-3255.

Reid, M.A., Fluin, J., Ogden, R.W., Tibby, J. and Kershaw, A.P. 2002. Long-term perspectives on human impacts on floodplain-river ecosystems, Murray-Darling Basin, Australia. *Proceedings of the International Association of Theoretical and Applied Limnology* 28:710-716.

Harle, K.J., Heijnis, H., Chisari, R., Kershaw, A.P., Zoppi, U. and Jacobsen, G. 2002. A chronology for the long pollen record from Lake Wangoom, western Victoria (Australia) as derived from uranium/thorium disequilibrium dating. *Journal of Quaternary Science* 17:707-720.

Turney, C.S.M., Bird, M.I., Fifield, L.K., Kershaw, A.P., Cresswell, R.G., Santos, G.M., di Tada, M.L., Hausladen, P.A. and Youping, Z. 2001. Development of a robust 14C chronology for Lynch's Crater (North Queensland, Australia) using different pretreatment strategies. *Radiocarbon* 43:45-54.

Turney, C.S.M., Kershaw, A.P., Moss, P., Bird, M.I., Fifield, L.K., Cresswell, R.G., Santos, G.M., di Tada, M.L., Hausladen, P.A. and Youping, Z. 2001. Redating the onset of burning at Lynch's Crater (North Queensland): implications for human settlement in Australia. *Journal of Quaternary Science* 16:767-771.

Kershaw, A.P. and Wagstaff B.E. 2001. The southern conifer family Araucariaceae: history, status, and value for palaeoenvironmental reconstruction. *Annual Review of Ecology and Systematics* 32:397-414.

Anshari, G., Kershaw, A.P. and van der Kaars, S. 2001. A Late Pleistocene and Holocene pollen and charcoal record from peat swamp forest, Lake Sentarum Wildlife Reserve, West Kalimantan, Indonesia. *Palaeogeography, Palaeoclimatology, Palaeoecology* 171:213-228.

van der Kaars, S., Kershaw, A.P., Tapper, N., Moss, P. and Turney, C. 2001. Pollen records of the last glacial cycle in the Southern Hemisphere tropics of the PEPII transect. *PAGES News* 9(2):11-12.

Wagstaff, B.E., Kershaw, A.P., O'Sullivan, P.B., Harle, K.J. and Edwards, J. 2001. An Early to Middle Pleistocene palynological record from the volcanic crater of Pejark Marsh, Western Plains of Victoria, southeastern Australia *Quaternary International* 83-85:211-232.

Kershaw, A.P. 2001. The history, palaeoclimatic significance and present day status of the southern conifer families Araucariaceae and Podocarpaceae, with special reference to Australia. *Revista Universidade Guarulhos Geociencias* 6:5-21.

Kershaw, A.P., Quilty, P.G., van Huet, S., David, B. and McMinn, A. 2000. The Quaternary. In *Contributions to Phanerozoic Biogeography of Australasian Faunas and Floras. Australasian Association of Palaeontologists Memoirs* 23:461-506.

Bohte, A. and Kershaw, A.P. 1999. Taphonomic influences on the interpretation of the palaeoecological records from Lynch's Crater, northeastern Australia. *Quaternary International* 57/58:49-59.

Harle, K.J., Kershaw, A.P. and Heijnis, H. 1999. The contributions of uranium/thorium and marine palynology to the dating of the Lake Wangoom pollen record, Western Plains of Victoria, Australia. *Quaternary International* 57/58:25-35.

McKenzie, G.M. and Kershaw, A.P. 2000. The last glacial cycle from Wyelangta, the Otway Region of Victoria, Australia. *Palaeogeography, Palaeoclimatology, Palaeoecology* 155:177-193.

Moss, P.T. and Kershaw, A.P. 2000. The last glacial cycle from the humid tropics of northeastern Australia: comparison of a terrestrial and a marine record. *Palaeogeography, Palaeoclimatology, Palaeoecology* 155:155-176.

van der Kaars, S., Wang, X., Kershaw, A.P., Guichard, F. and Setiabudi, D.A. 2000. Late Quaternary palaeoecological record from the Banda Sea, Indonesia: patterns of vegetation, climate and biomass burning in Indonesia and northern Australia. *Palaeogeography, Palaeoclimatology, Palaeoecology* 155:135-153.

Petit-Maire, N., Bouysse, P., de Beaulieu, J.-L., Boulton, G., Iriondo, M., Kershaw, P., Lisitsyna, O., Partridge, T., Pflaumann, U., Sarnthein, M., Schulz, H., Soons, J., van Vliet-Lanö, J., Yuan, B., Guo, Z. and van der Zijp, M. 2000. Geological records of the recent past, a key to the near future world environments. *Episodes* 23:230-246.

Wang, X., van der Kaars, S., Kershaw, A.P., Bird, M. and Jansen, F. 1999. A record of vegetation and climate through the last three glacial cycles from Lombok Ridge core G6-4, eastern Indian Ocean, Indonesia. *Palaeogeography, Palaeoclimatology, Palaeoecology* 147:241-256.

Kershaw, A.P. 1998. Estimates of regional climatic variation within southeastern Australia since the Last Glacial Maximum from pollen data. *Palaeoclimates - Data and Modelling* 3:107-134.

Kershaw, A.P., Moss, P.T. and van der Kaars, S. 1997. Environmental Change and the human occupation of Australia. *Anthropologie* 35/2:35-43.

Kershaw, A.P. 1997. A modification of the Troels-Smith system of sediment description and portrayal. *Quaternary Australasia* 15/2:63-68.

Lloyd, P.J. and Kershaw, A.P. 1997. Late Quaternary vegetation and early Holocene quantitative climatic estimates from Morwell Swamp, Latrobe Valley, southeastern Australia. *Australian Journal of Botany* 45:549-563.

McKenzie, G.M. and Kershaw, A.P. 1997. A vegetation history and quantitative estimate of Holocene climate from Chapple Vale, the Otway region of Victoria, Australia. *Australian Journal of Botany* 45:565-581.

Jenkins, M.A. and Kershaw, A.P. 1997. A mid-late Holocene vegetation record from an interdunal swamp, Mornington Peninsula, Victoria. *Proceedings of the Royal Society of Victoria* 109:133-148.

Kershaw, A.P. 1997. A bioclimatic analysis of Early to Middle Miocene brown coal floras, Latrobe Valley, southeastern Australia. *Australian Journal of Botany* 45:373-387.

D'Costa, D.M. and Kershaw, A.P. 1997. An expanded recent pollen database from southeastern Australia and its potential for refinement of palaeoclimatic estimates. *Australian Journal of Botany* 45:583-605.

Kershaw, A.P. and Bulman, D. 1996. A preliminary application of the analogue approach to the interpretation of late Quaternary pollen spectra from southeastern Australia. *Quaternary International* 33:61-71.

Holdgate, G.R., Kershaw, A.P. and Sluiter, I.R.K. 1995. Sequence stratigraphic analysis and the origins of Tertiary brown coal lithotypes, Latrobe Valley, Gippsland Basin, Australia. *International Journal of Coal Geology* 28:249-275.

Sluiter, I.R.K., Kershaw, A.P., Holdgate, G.R. and Bulman, D. 1995. Biogeographic, ecological and stratigraphic relationships of the Miocene brown coal floras, Latrobe Valley, Victoria, Australia. *International Journal of Coal Geology* 28:277-302.

D'Costa, D. and Kershaw, A.P. 1995. A late Quaternary pollen record from Lake Terang, Western Plains of Victoria, Australia. *Palaeogeography, Palaeoclimatology, Palaeoecology* 113:57-67.

Kershaw, A.P. 1995. Environmental change in Greater Australia. *Antiquity* 69:656-675.

Kershaw, A.P. 1994. Pleistocene vegetation of the humid tropics of northeastern Queensland, Australia. *Palaeogeography, Palaeoclimatology, Palaeoecology* 109:399-412.

Crowley, G.M., Grindrod, J. and Kershaw, A.P. 1994. Modern pollen deposition in the tropical lowlands of northeast Queensland, Australia. *Review of Palaeobotany and Palynology* 83:299-327.

Kershaw, A.P. and Bulman, D. 1994. The relationship between modern pollen samples and environment in the humid tropics region of northeastern Australia. *Review of Palaeobotany and Palynology* 83:83-96.

Kershaw, A.P., Bulman, D. and Busby, J.R. 1994. An examination of modern and pre-European settlement pollen samples from southeastern Australia: assessment of their application to quantitative reconstruction of past vegetation and climate. *Review of Palaeobotany and Palynology* 82:83-96.

Crowley, G.M. and Kershaw, A.P. 1994. Late Quaternary environmental change and human impact around Lake Bolac, western Victoria, Australia. *Journal of Quaternary Science* 9:367-377.

Kershaw, A.P. and Nanson, G.C. 1993. The last full glacial cycle in the Australian region. *Global and Planetary Change* 7:1-9.

Kershaw, A.P., McKenzie, G.M. and McMinn, A. 1993. A Quaternary vegetation history of northeast Queensland from pollen analysis of ODP Site 820. *Proceedings of the Ocean*

Drilling Program 133:107-114.

Kershaw, A.P. 1993. Palynology, biostratigraphy, and human impact. *The Artifact* 16:12-18.

Harle, K.J., Kershaw, A.P., Macphail, M.K. and Neyland, M.G. 1993. Palaeoecological analysis of an isolated stand of *Nothofagus cunninghamii* (Hook.) Oerst. in eastern Tasmania. *Australian Journal of Ecology* 18:161-170.

Kershaw, A.P. 1993. Quantitative palaeoclimatic estimates from bioclimatic analyses of taxa recorded in pollen diagrams. *Quaternary Australasia* 11(1):61-64.

Aitken, D.L. and Kershaw, A.P. 1993. Holocene vegetation and environmental history of Cranbourne Botanic Garden. *Proceedings of the Royal Society of Victoria* 105:67-80.

Markgraf, V., Dodson, J.R., Kershaw, A.P., McGlone, M.S. and Nicholls, N. 1992. Evolution of late Pleistocene and Holocene climates in the circum-South Pacific land areas. *Climate Dynamics* 6:193-211.

Kershaw, A.P., Bolger, P., Sluiter, I.R., Baird, J. and Whitelaw, M. 1991. The origin and evolution of brown coal lithotypes in the Latrobe Valley, Victoria, Australia. *International Journal of Coal Geology* 18:233-249.

Kershaw, A.P., D'Costa, D.M., McEwen Mason, J. and Wagstaff, B.E. 1991. Quaternary vegetation of mainland southeastern Australia. *Quaternary Science Reviews* 10:391-404.

Edney, P.A., Kershaw, A.P. and De Deckker, P. 1990. A late Pleistocene and Holocene vegetation and environmental record from Lake Wangoom, Western Plains of Victoria, Australia. *Palaeogeography, Palaeoclimatology, Palaeoecology* 80:325-343.

Kershaw, A.P. and Strickland, K. 1990. A 10 year pollen trapping record from northeastern Australia. *Review of Palaeobotany and Palynology* 64:281-288.

Crowley, G.M., Anderson, P., Kershaw, A.P. and Grindrod, J. 1990. Palynology of a Holocene marine transgressive sequence, lower Mulgrave River Valley, north-east Queensland. *Australian Journal of Ecology* 15:231-240.

D'Costa, D.M., Edney, P.A., Kershaw, A.P. and De Deckker, P. 1989. The late Quaternary palaeoecology of Tower Hill, Victoria, Australia. *Journal of Biogeography* 16:461-482.

Bell, C.J.E., Finlayson, B.L. and Kershaw, A.P. 1989. Pollen analysis and dynamics of a peat deposit in Carnarvon National Park, central Queensland. *Australian Journal of Ecology* 14:449-456.

Kershaw, A.P. 1989. Was there a 'Great Australian Arid Period'? *Search* 20:89-92.

Kershaw, A.P and Nix, H.A. 1988. Quantitative palaeoclimatic estimates from pollen data using bioclimatic profiles of extant taxa. *Journal of Biogeography* 15:589-602.

Kershaw, A.P and Strickland, K.M. 1988. A Holocene pollen diagram from Northland, New Zealand. *New Zealand Journal of Botany* 26:145-152.

Kershaw, A.P. 1986. Climate change and Aboriginal burning in north-east Australia during the last two glacial/interglacial cycles. *Nature* 322:47-49.

Kershaw, A.P. 1985. An extended late Quaternary vegetation record from northeastern Queensland and its implications for the seasonal tropics of Australia, *Proceedings, Ecological Society of Australia* 13:179-189.

Kershaw, A.P., Edney, P., Peterson, J.A. and Coutts, P.J.F. 1985. Evidence of a Pleistocene age for Tower Hill, western Victoria, *Search* 16:302-303.

Kershaw, A.P. 1983. A Holocene pollen diagram from Lynch's Crater, northeastern Queensland, Australia *New Phytologist* 94:669-682.

Kershaw, A.P. and Green, J.E. 1983. Tawonga Bog revisited: the history of a low altitude peat deposit. *Victorian Naturalist* 100:256-259.

Kershaw, A.P. 1983. Considerations nouvelles sur la flore et la vegetation Australienne. *L'Espace Geographique* 12:185-194.

Kershaw, A.P. 1983. Huon Pine, Australia's longest living tree, tells an epic story of survival. *Habitat* 11:32-34.

Sluiter, I.R. and Kershaw, A.P. 1982. The nature of late Tertiary vegetation in Australia. *Alcheringa* 6:211-222.

Kershaw, A.P. and Sluiter, I.R. 1982. Late Cainozoic pollen spectra from the Atherton Tableland, northeastern Australia. *Australian Journal of Botany* 30:279-295.

Kershaw, A.P. and Sluiter, I.R. 1982. The application of pollen analysis to the elucidation of Latrobe Valley brown coal deposits and stratigraphy. *Australian Coal Geology* 4:169-186.

Kershaw, A.P. 1981. Climate and Australian flora. *Australian Natural History* 20:231-234.

Kershaw, A.P. 1980. An extension of the late Quaternary vegetation record from northeastern Australia. *Fourth International Palynological Conference, Lucknow (1976-77)* 3:28-35.

Hooley, A.D., Southern, W. and Kershaw, A.P. 1980. Holocene vegetation and environments of Sperm Whale Head, Victoria. *Journal of Biogeography* 7:349-362.

Kershaw, A.P. 1979. Local pollen deposition in aquatic sediments on the Atherton Tableland, north-eastern Australia. *Australian Journal of Ecology* 4:253-263.

Kershaw, A.P. 1979. The changing vegetation of northeastern Queensland. *See Australia* 2:152-154.

Kershaw, A.P. 1978. The analysis of aquatic vegetation on the Atherton Tableland, north-east Queensland, Australia. *Australian Journal of Ecology* 3:23-42.

Kershaw, A.P. 1978. Record of last glacial-interglacial cycle from northeastern Queensland. *Nature* 272:159-161.

Binder, R. and Kershaw, A.P. 1978. A late Quaternary pollen diagram from the southeastern highlands of Australia. *Search* 9:44-45.

Kershaw, A.P. 1977. The state of biogeographical research in Australia. *Geography Teacher* 17:5-11.

Kershaw, A.P. 1976. A late Pleistocene and Holocene pollen diagram from Lynch's Crater, northeastern Queensland, Australia. *New Phytologist* 77:469-498.

Kershaw, A.P. 1975. Stratigraphy and pollen analysis of Bromfield Swamp, northeastern Queensland, Australia. *New Phytologist* 75:173-191.

Kershaw, A.P. and Hyland, B.P.M. 1975. Pollen transfer and periodicity in a rainforest situation. *Review of Palaeobotany and Palynology* 19:129-138.

Kershaw, A.P. 1974. A long continuous pollen sequence from northeastern Australia. *Nature* 251:222-223.

Turner, J. and Kershaw, A.P. 1973. A late- and post-glacial pollen diagram from Cranberry Bog, near Beamish, County Durham. *New Phytologist* 72:915-928.

Kershaw, A.P. 1973. Quaternary history of rainforests in Australia. *Wildlife in Australia* 10:82-83.

Kershaw, A.P. 1971. A pollen diagram from Quincan Crater, north-east Queensland, Australia. *New Phytologist* 70:669-681.

Kershaw, A.P. 1970. Pollen morphological variation in the Casuarinaceae. *Pollen et Spores* 12:145-61.

Kershaw, A.P 1970. A pollen diagram from Lake Euramoo, north-east Queensland, Australia. *New Phytologist* 69:785-805.

I. Archaeology and Perceptions of Landscape

2

Hay Cave: A 30,000-year cultural sequence from the Mitchell-Palmer limestone zone, north Queensland, Australia

Harry Lourandos
Department of Anthropology, Archaeology and Sociology, School of Arts and Social Sciences,
James Cook University, Cairns, Queensland
harry.lourandos@jcu.edu.au

Bruno David
Monash University, Clayton, Victoria

Nicola Roche
Umwelt (Australia) Pty Ltd, Toronto, NSW

Cassandra Rowe
Monash University, Clayton, Victoria

Angela Holden
University of Queensland, St Lucia, Queensland

Simon J. Clarke
Charles Sturt University, Wagga Wagga, New South Wales

Introduction

Hay Cave is one of many limestone caves in the tropical Mitchell-Palmer area of north Queensland. Archaeologically, its major significance is a lengthy, more than 30,000 year-long, cultural sequence, with good preservation of faunal remains as well as stone artefacts and an abundance of rock art. Thus, it offers the opportunity to investigate long-term *local* archaeological trends in one site and to compare these with *regional* trends obtained from a wider range of sites throughout this archaeologically rich area (David and Lourandos 1997). How can these long-term cultural trends be characterised from an individual site? In what ways do they reflect wider regional trends and patterns? How do they compare with palaeoenvironmental trends? And, at a more general level, how can we connect different spatial scales of investigation (the local or site-specific and the regional) when seeking to explore long-term cultural trends? These were the questions guiding the research.

As a limestone cave with alkaline soils and good preservation, Hay Cave is well endowed in different kinds of archaeological materials, raising also the question of the relationship between different lines of archaeological evidence when exploring cultural trends through time. To what degree does each category of archaeological material represent independent sets of evidence, and to what degree can they be related inter-textually? With such questions in mind, the stone artefacts, animal bone, land-snail shell, mussel shell, brush-turkey egg shell, charcoal and hearths of Hay Cave are examined here in relation to wider regional chronological patterns for Cape York Peninsula (see David and Lourandos 1998). A large number of AMS radiocarbon determinations were obtained to investigate these data in adequate chrono-stratigraphic detail.

The Mitchell-Palmer limestone zone

The Mitchell-Palmer limestone zone is part of the Chillagoe Formation, located 180 km northwest of Cairns, north Queensland, Australia (Figure 1). The individual limestone towers ('bluffs') often exceed 1 km in length and 500 m in width, jutting up to 150 m above the surrounding landscape (Figure 2). These are impressive rock formations, possessing regionally distinctive lithologies and sediments (e.g. Galloway et al. 1970; Day et al. 1983:85), vegetation communities (e.g. Galloway et al. 1970) and fauna (for preliminary results, see Hall et al. 1996; Macrokanis 1996; see also Stanisic and Ingram 1998) in an otherwise dry sclerophyll landscape.

The Chillagoe Formation outcrops towards the western margin of the Hodgkinson Province as a steeply dipping, discontinuous belt running parallel to the Palmerville Fault line. The unit extends over a distance of approximately 150 km and varies in width from 10 km to a few hundred metres. Regional magnetic imagery indicates that the Chillagoe Formation continues beneath ground cover to the north in a north-northeasterly direction (Domagala and Fordham 1997).

The Chillagoe Formation consists of varying proportions of limestone, chert, basalt, arenite and mudstone, conglomerate and breccia. Limestone is the characteristic lithology. Given the geological components and an estimated early Silurian age (about 428 million years), the limestone belt probably was deposited as calcareous muds and coral reefs; earth movements since deposition have folded and tilted the limestone into its current vertical position. Where erosion has removed the surrounding material, the limestone is now exposed as numerous towers of variable height and size (Domagala and Fordham 1997; Stanisic 1997).

Vegetation on the Chillagoe Formation limestone outcrops is deciduous microphyll vine thicket. This is a stunted vine forest in which the canopy closes at 3 m to 9 m above the ground, with the majority of emergents deciduous, together with many understorey species existing as deciduous or semi-evergreen plants. The vine thickets lose their leaves in response to a drop in moisture availability, and the ready supply of leaf litter provided by the vegetation fills the rock crevices to further enhance the water-conserving properties of the limestone outcrops (Kahn and Lawie 1987). Fensham (1995, 1996) denoted these vine thickets as a floristically distinct group restricted mainly to limestone karst; the limestone substrate forms an additional natural fire barrier. Within the Chillagoe Formation, individual limestone towers are isolated by open woodland eucalypt communities devoid of limestone (Rowe et al. 2001). Climate is semi-arid, with a highly seasonal rainfall regime, falling predominately in the months of December to March. The enhanced humidity-preserving qualities of the limestone outcrops, particularly in caves surrounded by vine thickets, are especially relevant to long-term cultural trends spanning drier and wetter climatic phases, as at Hay Cave (see below).

The limestone outcrops are typically surrounded by alkaline rocky pediments conducive to the preservation of organic materials such as bone, egg shell and land-snail shell. In some rare instances, archaeological excavations have recovered well-preserved late-Holocene wooden digging sticks and other 'soft' organic items in buried cave sediments, including ancient fungus, fig fruit and *Pandanus* nuts (David and Dagg 1993). Hundreds of shallow rockshelters and also deep caves occur near the junction of the towers and their pediments. Some of these caves extend hundreds of metres into the rock, most often in pitch-black conditions, but also at times illuminated by roof collapses and sinkholes creating skylit chambers. While no evidence of people has ever been found in these deeper recesses, archaeological evidence of Aboriginal occupation is abundant in more open rockshelters and sunlit cave entrances. There is also ample evidence of Aboriginal presence on the plains surrounding the towers, particularly in the form of partly buried and surface stone artefact scatters.

Hay Cave

Hay Cave is a large limestone cave positioned between the Mitchell and Palmer rivers, towards the northern limits of the Mitchell-Palmer limestone belt and approximately 140 km west of the eastern Australian coastline, at 16°6'S latitude and 144°7'E longitude (Figures 1 and 3). The site was recorded during Bruno David's Mitchell-Palmer archaeological site surveys as site number PM18 (being the 18th archaeological site recorded in this programme), and also by the Chillagoe Caving Club under its speleological site number MP118. Hay Cave is situated at the southern end of Wilson's Tower, accessible from the east, with a large entrance (16 m wide, 3 m high and 10 m deep) (Figure 3). It sits at the base of the limestone tower, atop a small scree covered in vine thicket that slopes down to a surrounding plain of open sclerophyll forest. The cave entrance itself is positioned slightly above the level of the flat, soft cave floor within the overhang. Sunlight is subdued immediately inside the overhang, towards the centre of the entrance chamber. Here, where the floor is flat and sediments are soft, archaeological excavations were undertaken by Bruno David, Harry Lourandos and Chris Clarkson in 1996 (Figures 3 and 4). Four juxtaposed 50 x 50 cm squares were excavated, the deepest of which (M30) was removed in 63 Excavation Units (XU) of mean 2.0 cm thickness and following Stratigraphic Units (SU) where visible (Figure 5). Twenty-two distinct SUs were identified in the deepest excavation square of 133 cm depth. Sediments containing cultural materials were found at all depths except within the thick clays of SU3 at the very base of the excavations (Table 1).

The four Hay Cave squares were excavated independently of each other. Cultural items >2 cm maximum length were plotted in three dimensions and drawn on individual XU recording forms during the course of the excavation. All other sediments were sieved in 3 mm mesh sieves and bagged for later sorting in the laboratory. Sediment samples were collected from the <3 mm residue from each XU of each square.

At the University of Queensland and Monash University, over the ensuing years Angela Holden (1999) worked on the stone artefacts; Nicola Roche (1999) on the animal bone following preliminary work by Neville Terlich (1998); Cassandra Rowe (1998) on the land-snail shell; Simon Clarke (2005) on the egg shell; and Bruno David (e.g. David et al. 1997) on the rock art and chrono-stratigraphy. This, however, is the first full site report to appear and to assemble the different lines of previously unpublished evidence to explore long-term temporal trends.

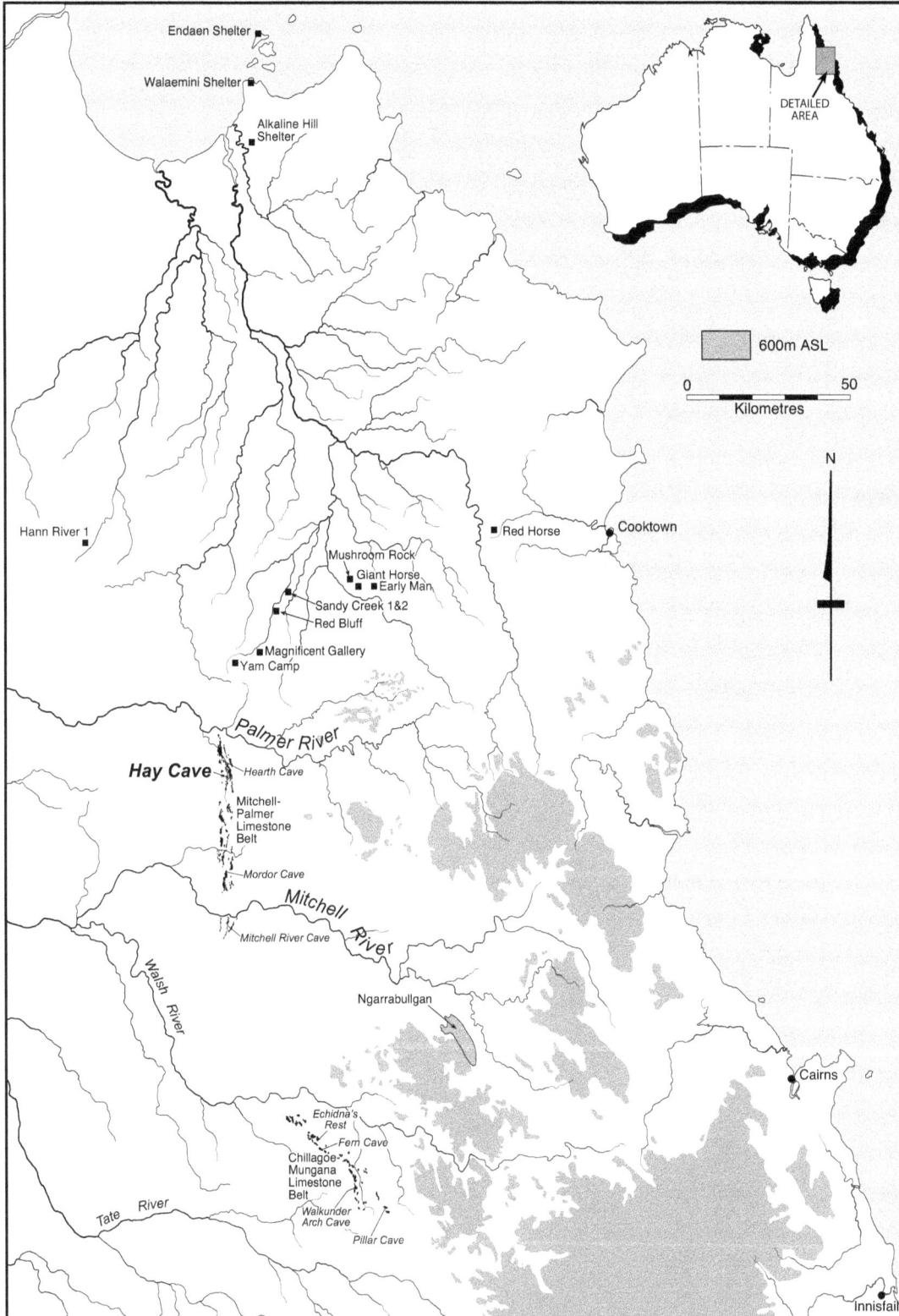

Figure 1. Map of southeast Cape York Peninsula showing location of the individual Mitchell-Palmer and Chillagoe-Mungana limestone towers and the excavated archaeological sites.

Figure 2. Limestone towers of the Mitchell-Palmer limestone zone.

Figure 3. Hay Cave, excavation in progress.

The sediments

Three major SU were revealed during the excavation (Figure 5). SU1, the uppermost, is subdivided into 18 subunits (SU1a-SU1r). All subunits of SU1 are ashy, and at least seven well-demarcated hearths are represented. The latter appear in SU1b/1c; SU1d/1f; SU1e/1h/1i/1k; SU1j; SU1m; SU1n; SU1q; and also probably SU1o. The other SU1 subunits are either of ill-defined hearths or contain mixed ashy sediments from hearths lying above or from those extending beyond the excavated deposits (SU1a, the ashy treadage zone; SU1l; SU1p; SU1r). Thus, the entire period spanned by SU1, the uppermost unit, represents the intensive construction of hearths within this part of the site.

SU2, lying immediately below SU1, is divided into three subunits (SU2a, 2b and 2c), each predominantly composed of silt and clay and variably rich in calcium carbonate concretions. No in situ evidence of hearths has been recovered from this unit.

SU3, the lowermost unit, is highly compact and clayey. Again, no stratigraphic evidence of hearths has been recovered from this unit.

Each SU is described in greater detail in Table 2.

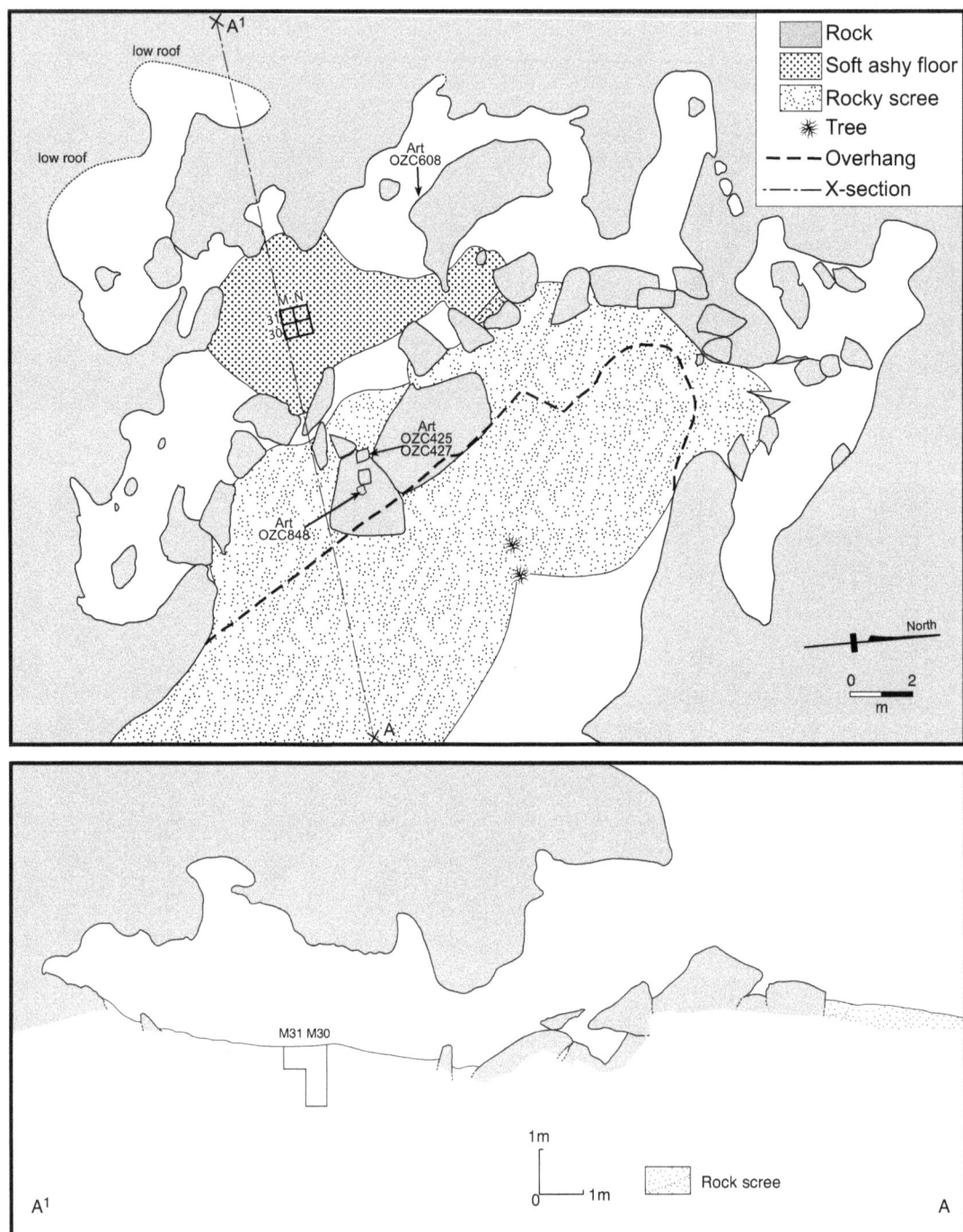

Figure 4. Map (top) and cross-section (bottom) of Hay Cave showing location of excavation squares and radiocarbon-dated rock art.

Radiocarbon dates

Twenty-six radiocarbon determinations have been obtained from Hay Cave. All are AMS dates, most on single pieces of charcoal (N=15), the others from single pieces of *Alectura lathami* egg shell (N=3), freshwater mussel shell (N=4), or from scrapings from charcoal wall drawings (N=4). The radiocarbon dates were obtained from three laboratories, ANSTO (OZ numbers, 24 dates), University of Colorado INSTAAR (NSRL- number, 1 date) and Waikato (Wk-number, 1 date). All the excavated radiocarbon samples were obtained from Square M30.

From the deepest excavated levels, a near-basal date was obtained on a riverine bivalve shell recovered from XU60 (122-125 cm below the surface). The resulting date of 29,700 ± 1100 BP (OZD948) indicates that cultural sediments towards the base of the excavation were

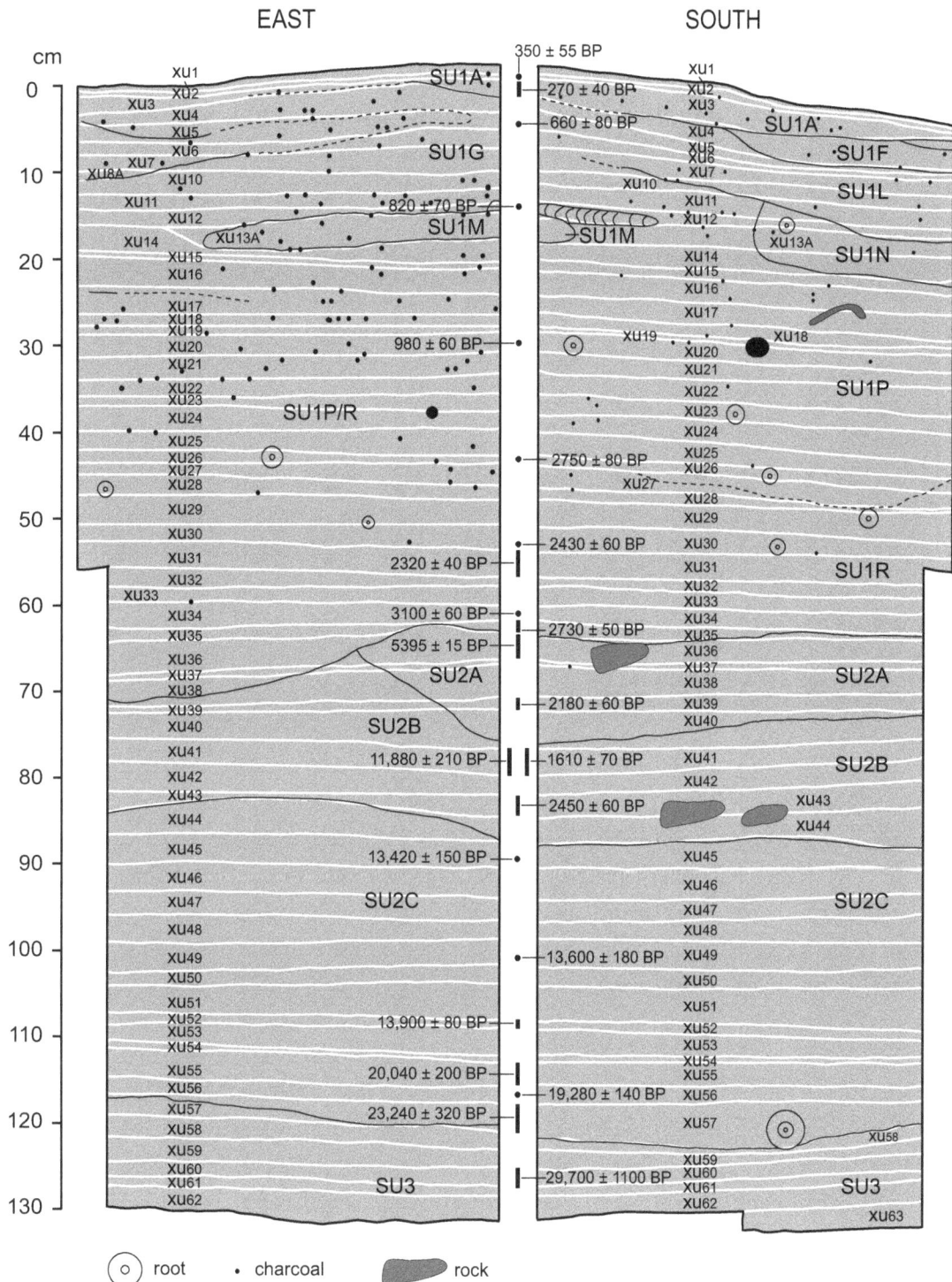

Figure 5. East and south sections, Square M30.

deposited around 30,000 years ago. A small number of artefacts also occur at the SU2-SU3 interface. Below this lowermost radiocarbon date, however, the SU3 sediments themselves are of very compact mottled clays, with small manganese nodules that appear to pre-date human occupation.

Radiocarbon dates were obtained from every three or so XU above XU60, sometimes more frequently. They appear to progress at a fairly regular sedimentation rate of about 0.10 cm/100 years from ca. 30,000 BP in XU60 to 13,900 BP in XU52 (see Table 3). After this time, sedimentation rates increase two- to three-fold, to 0.26 cm/100 years, until ca. 3000 BP

Table 1. Excavation details, Hay Cave Square M30.

XU	SU	Volume (litres)	Weight (kg)	Area (m²)	Mean thickness (cm)
1	1a	0.5	0.25	0.25	0.1
2	1a	4.5	4	0.25	1.3
3	1a	6	6	0.25	1.8
4	1a+1d+1l	5	5	0.25	1.7
5	1f+1l	4.5	5	0.25	1.4
6	1f+1l	5	5	0.25	1.8
7	1l+1p	5.5	5.5	0.25	2.2
8a	1l	1	1.25	0.12	0.9
9a	1l	<0.5	0.25	0.005	0.7
10	1l+1p	5.5	5.5	0.25	1.7
11	1l+1n+1p	6	5.25	0.25	2.0
12	1l+1n+1p	5	5	0.25	1.5
13a	1m+1n+1p	3	3	0.15	1.8
14	1n+1p	5	4.75	0.25	2.3
15	1p	5	5	0.25	1.7
16	1p	6	6	0.25	2.6
17	1p	5.5	5.5	0.25	2.2
18	1p	4	4.5	0.25	1.4
19	1p	1	1.5	0.25	0.7
20	1p	6		0.25	2.8
21	1p	6	6	0.25	2.2
22	1p	5.5	6	0.25	2.5
23	1p	5	4.5	0.25	1.9
24	1p	6	6.5	0.25	2.8
25	1p	6	6	0.25	2.1
26	1p	5	5	0.25	1.6
27	1p+1r	7	6.5	0.25	2.0
28	1r	4.5	4	0.25	1.2
29	1r	7	7	0.25	3.3
30	1r	6.5	7	0.25	2.7
31	1r	6.5	7	0.25	2.7
32	1r	5	5	0.25	1.6
33	1r	5	5.5	0.25	2.0
34	1r	5	6.5	0.25	2.3
35	2a	3.5	4	0.25	1.4
36	2a	6.5	7.5	0.25	2.5
37	2a	4	5.5	0.25	1.1
38	2a	6	8	0.25	3.2
39	2a	4	4.5	0.25	1.6
40	2a+2b	7.5	9	0.25	2.9
41	2b	8	9	0.25	2.9
42	2b	7.5	8.5	0.25	2.7
43	2b	7	9.5	0.25	3.0
44	2b	6	8	0.25	2.7
45	2c	8	9.5	0.25	3.2
46	2c	8.5	10	0.25	3.9
47	2c	6.5	8	0.25	2.2
48	2c	6.5	8	0.25	2.7
49	2c	9	10	0.25	3.3

Table 1. *Continued*

XU	SU	Volume (litres)	Weight (kg)	Area (m²)	Mean thickness (cm)
50	2c	6.5	8	0.25	2.4
51	2c	9	11	0.25	3.2
52	2c	3.5	5	0.25	1.6
53	2c	4.5	6	0.25	2.0
54	2c	4	5.5	0.25	1.4
55	2c	6	7.5	0.25	2.7
56	2c	4	5.5	0.25	2.1
57	2c	4.5	5.5	0.25	2.3
58	3	5	5.5	0.25	1.8
59	3	6	8	0.25	1.9
60	3	6	7.5	0.25	2.5
61	3	5	5.5	0.25	1.1
62	3	4	4.5	0.25	1.6
63	3	3	3	0.18	1.4

Table 2. Description of Stratigraphic Units, Hay Cave.

SU	Description	pH	Dry Munsell
1a	Very fine-textured surface sediments. Dry and ashy, fairly compact but unconsolidated and very easy to dig. Treadage zone of surface-disturbed sediments are extremely shallow, restricted to the uppermost 1-2 mm. This disturbed veneer consists of sediments indistinguishable in colour, content and texture from the rest of this SU. Cultural materials noticed during excavation.	9	10YR6/2 to 10YR6/3 Light brownish grey to pale brown
1b	Thin lens of light brownish brown heat-baked sediment. Its boundary with overlying SU1a is sharp, as it is also with underlying SU1c. This SU is not present across all the excavated squares, being of limited spatial extent: it is only present in the NW corner of the four-square pit. Sediments are homogeneous throughout.	8.5	10YR6/4 Light brownish brown
1c	Sediments are very ashy and cultural in origin. They are discontinuous across the four squares, being present only in the NW part of the excavation, immediately below SU1b but with a slightly greater spatial spread. Where SU1b does not occur, SU1c merges with SU1a. The interface between SU1c and SU1b is marked in most parts. The interface between SU1c and SU1e below it occurs over a vertical distance of ca. 1 cm. Sediments are very dry.	8.5	10YR6/2 Light brownish grey
1d	A thin lens of ash towards the southern wall near the SW corner of Square M30. Sediments are very fine in texture, dry, easy to dig but somewhat compact. The boundaries between SU1d and surrounding sediments are marked.	8.5	10YR8/1 White
1e	Very ashy lens. Sediments vary in colour from very pale brown in the SW parts of the pit to light brownish grey in the NW parts. This change in colour is gradual; a more reddish appearance of sediments towards the southern side of the pit is consistent with similar patterning in the immediately underlying SUs. SU1e is very fine in texture, dry, somewhat compact but easy to dig. Boundaries between SU1e and surrounding SUs are gradual but easily recognisable in situ and/or on the exposed sections, occurring over a vertical distance of ca. 2 cm.	8.5	10YR7/3 (SW parts of pit) to 10YR6/2 (NW parts of pit) Very pale brown to light brownish grey

Table 2. *Continued*

SU	Description	pH	Dry Munsell
1f	A small lens of colour-stained sediments immediately underlying SU1d. SU1f should not be interpreted as a distinct depositional unit, but rather the heat-stained sediments located immediately below the hearth represented by SU1d. Boundaries between SU1f and SU1a, and those between SU1f and SU1d, are marked. Boundaries between SU1f and SU1g are gradual, taking place over a vertical distance of 4-6 cm.	8.5	10YR5/3 Brown
1g	Dry and ashy sediments. While sediments are generally light brownish grey, towards the SW corner of the pit they become more reddish in colour (pink to light brown in Munsell terms). This more reddish towards the SW corner of the pit appears to be the result of the overlying hearth SU1e, which is most pronounced in this area. In this SW corner, the boundary between SU1e and SU1g is marked. The boundaries between SU1g and underlying SU1h and SU1i are also marked. SU1g was noticed to contain cultural materials during excavation.	8.5	10YR6/2 to 7.5YR6/4-7.5YR7/4 Light brownish grey to light brown to pink
1h	A small, localised lens of dry sediments immediately above SU1i. Like SU1f and SU1k, SU1h is interpreted as a reddish, heat-stained sediment, rather than as a distinct depositional unit. It is interpreted as a localised, heat-stained lower part of the SU1g hearth. The boundary between SU1g and SU1h is gradual, occurring over a vertical distance of 2 cm. The boundary with overlying SU1i is marked.	8.5	10YR6/1 Pale brown
1i	A very clearly demarcated hearth, including a fire pit intruding into underlying SU1k, SU1l and SU1p. SU1i is very well-defined throughout, and contains hearth stones. Sediments are white to light grey in colour, interspersed with light grey ash. SU1i contains very large pieces of charcoal randomly distributed throughout. This hearth is most evident in the western part of the pit where the western section dissects it in half.	8.5	10YR7/1-10YR8/1 with patches of 10YR7/2 White to light grey
1j	A localised lens of ash. This unit is a hearth whose boundaries are for the most part well-defined. Sediments are dry ash and very fine in texture.	8.5	2.5YR8/0 to 10YR8/1 white
1k	A localised patch of heat-stained sediments immediately below SU1i. It should not be interpreted as a distinct depositional unit, but rather as part of the SU1l sediments that were fire-stained by the overlying SU1i. Boundaries with SU1l and SU1p are gradual; the SU1k-SU1p interface is ca. 2 cm thick. Sediments are dry and ashy with a reddish tinge.	8.5	10YR6/3-10YR6/4 Light yellowish brown to pale brown
1l	Sediments are dry, very fine, ashy and homogeneous throughout. Boundary with underlying SU1p is extremely gradual, occurring over a vertical distance of ca. 3 cm. Sediments were noticed to possess cultural materials during excavation.	8.5	10YR6/2 Light brownish grey
1m	Apart from a difference in colour, the descriptions of SU1j apply here also.	8.5	10YR7/1 Light grey
1n	Relatively unconsolidated ash. Boundaries with surrounding sediments are fairly well-defined in most places, but not sharp, occurring over a distance of 1-2 cm. Numerous roots and rootlets occur in this SU. Sediments are dry. SU1n is interpreted as a hearth.	8.5	10YR6/2 Light brownish grey
1o	Sediments are fine and ashy. They are very similar to those of SU1g, but here they occur below SU1j and above SU1p. SU1o is indistinct and appears to be another ash-rich unit which only occurs in the extreme NE corner of the pit. It is probably the very edge of a hearth which remains in situ beyond the excavated pit. The boundaries of this SU are ill-defined and merge with surrounding units over a distance of 2 cm.	8.5	10YR6/2 Light brownish grey

Table 2. *Continued*

SU	Description	pH	Dry Munsell
1p	This unit is relatively homogeneous and thick compared with those above it. Its boundaries with surrounding SUs are always gradual, taking place over 2-4 cm, with the noticeable exception its boundary with SU1i, which is relatively marked. SU1p is very ashy and fine-grained in texture.	8.5	10YR6/3 Pale brown
1q	An uneven unit consisting of well-defined patches of white ash interspersed among brown sediments reminiscent of SU1p and SU1r. SU1q is here treated as a single SU with patches of ash constrained within the limits of this SU. SU1q is interpreted as a hearth whose patches of ash have been intermixed (probably largely during hearth construction and use), with sediments of SU1p and SU1r character. Note that both the white and brown components of SU1q are ashy. While the boundaries of SU1q are ill-defined, in the sense that the patches of ash appear interspersed among the brown sediments, the overall area where the ash occurs is very well-defined.	8.5	10YR5/3 to 10YR8/1 Brown to white
1r	Ashy silt whose boundary with overlying SU1p is gradual and occurs over ca. 5 cm. The boundary with underlying SU2a occurs over 2 cm. Sediments in SU1r are noticeably less ashy and more brown than those above. They are also noticeably more compact and slightly more humid.	8.5	10YR5/3-10YR5/4 Brown to yellowish brown
2a	Slightly moist clayey silt, rich in calcium carbonate concretions. This SU represents a major stratigraphic break from overlying SUs. The boundary with underlying SU2b is ill-defined.	8.5	10YR3/4-10YR4/4 Dark yellowish brown
2b	Sediments are similar to those of SU2a, but boundaries with surrounding SUs are less well-defined but still occur over some 2 cm of depth. The major difference from SU2a is that SU2b contains lower concentrations of calcium carbonate concretions. Sediments are fine, clayey silt, somewhat moist.	8.5	10YR4/4 Dark yellowish brown
2c	Silty clay. Sediments are moist and become more clayey with depth. They are relatively homogeneous throughout the pit, and some calcium carbonate concretions occur throughout. The boundary with the overlying SU2b is fairly distinct. Sediments are compact throughout. There is a large root which grows at the interface of SU2c and SU3.	8.5	10YR3/6-10YR4/6 Dark yellowish brown
3	This is the lowermost SU excavated, and appeared to be culturally sterile at the time of excavation (i.e. no cultural materials were noticed in situ). It consists of clayey matrix with large amounts of small manganese nodules (<1 cm in diameter). Sediments are mottled and extremely compact. In the western part of the pit, sediments are mottled whitish, but this is less apparent in the eastern part of the pit. Sediments in SU3 were very difficult to dig due to their high moisture content and their very compact nature. The interface between SU3 and overlying SU2c is marked.	8.5	7.5YR5/6-10YR4/6 Strong brown to dark yellowish brown

in XU41. Considerable mixing of sediments, however, is apparent from XU43 to XU35 (as is evident from dating reversals), making it more difficult to interpret average sedimentation rates on this part of the curve. The past 3000 years BP witnessed a further sustained nine- to 10-fold increase in sedimentation rates, to an average of 2.53 cm/100 years. The uppermost radiocarbon dates are consistent with the site finally being used by people around the time of European arrival in the region during the late 19th century AD.

We interpret the zone of dating reversals, most pronounced from 83 cm to 63 cm below ground (XU43-XU35), in two ways. This zone represents the boundary between SU1 and SU2 into SU2a and SU2b. Firstly, these levels include the early-Holocene period, a time when other archaeological sequences in north Queensland are also characterised by dating reversals and coarser sediments that have been argued by various authors (e.g. David and Chant 1995)

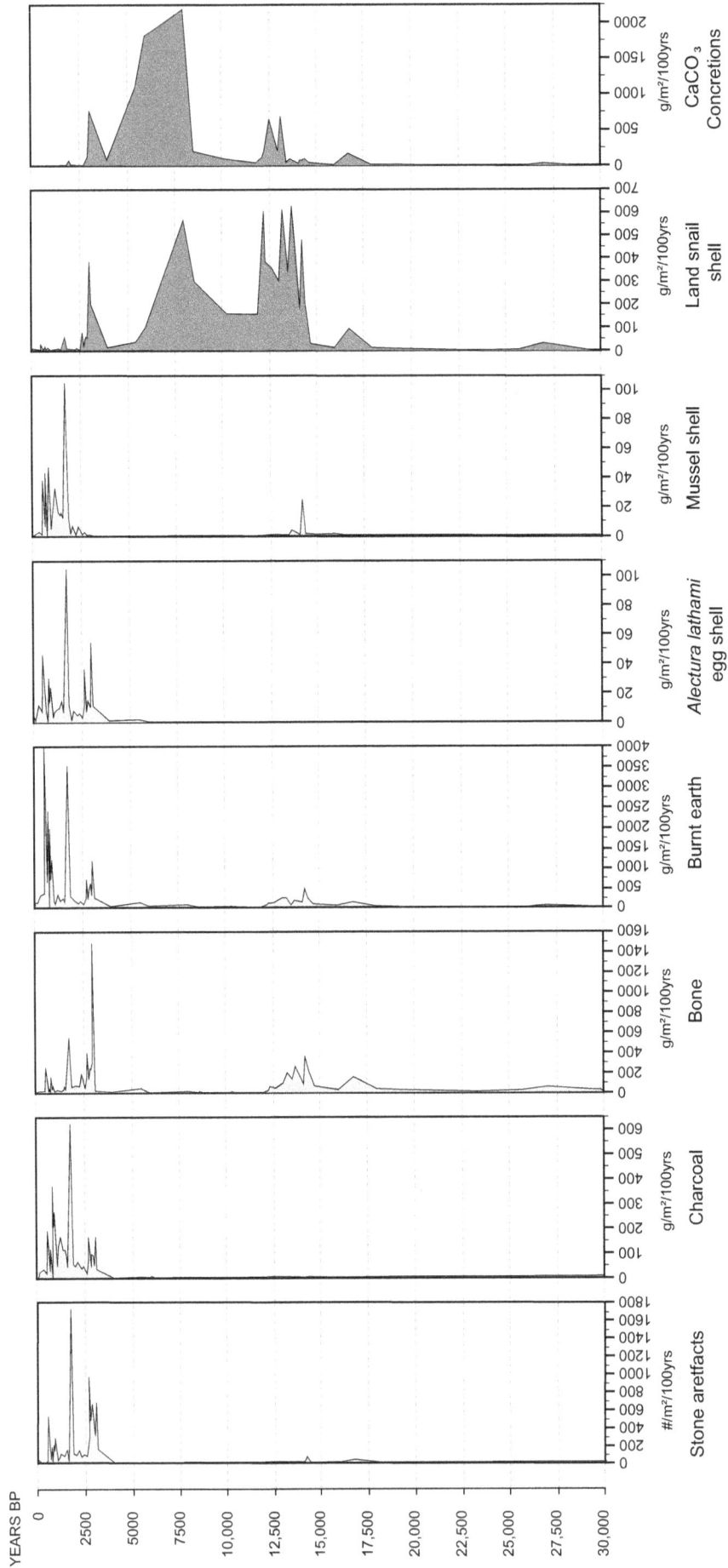

Figure 6. Deposition rates of excavated materials, Square M30.

Table 3. Radiocarbon dates from Hay Cave Square M30. Calibrations made on Calib 6.0 (Stuiver and Reimer 1993), using the SH Atmosphere option (McCormac et al. 2004); calibrated years in bold represent highest age probability range at 2 sigma.

XU	Depth (cm)	Material	Sample code	δ¹³C (‰)	Conventional radiocarbon age (years BP, ± 1σ)	Calibrated radiocarbon age in cal BP (1σ range [68.3% probability])	Calibrated radiocarbon age in cal BP (2σ range [95.4% probability])
2	1.4	charcoal	Wk-6053	-26.4 ± 0.2	350 ± 55	311-338 (20.9%) 355-449 (79.1%)	**285-495** (100%)
3	1.3-3.1	egg shell	OZF619	-8.5	270 ± 40	153-173 (20.9%) 177-209 (27.0%) 277-316 (52.0%)	145-223 (41.8%) **260-332** (44.6%) 364-443 (13.6%)
6	8.1	charcoal	OZD006	-25.14	660 ± 80	552-650 (100%)	**505-685** (99.4%) 709-714 (0.6%)
12	17.1	charcoal	OZD007	-30.22	820 ± 70	660-762 (100%)	563-602 (6.8%) **628-807** (89.9%) 879-901 (3.3%)
20	30.1	charcoal	OZD939	-26.77	980 ± 60	794-915 (100%)	**733-935** (98.4%) 943-955 (1.6%)
26	45.7	charcoal	OZD008	-27.18	2570 ± 80	2488-2644 (66.1%) 2654-2738 (33.9%)	**2362-2748** (100%)
30	53.0	charcoal	OZD940	-26.49	2430 ± 60	2337-2490 (89.4%) 2643-2672 (10.6%)	2209-2223 (0.9%) **2308-2619** (84.1%) 2631-2709 (15.1%)
31	53.2-55.9	egg shell	OZF620	-7.2	2320 ± 40	2180-2240 (55.9%) 2303-2342 (44.1%)	**2153-2276** (63.0%) 2290-2349 (37.0%)
34	63.5	charcoal	OZD009	-25.26	3100 ± 60	3164-3185 (9.5%) 3205-3355 (90.5%)	3043-3046 (0.2%) **3063-3393** (99.8%)
35	61.8-63.2	charcoal	OZD941	-26.96	2730 ± 50	2748-2801 (73.1%) 2817-2844 (26.9%)	**2721-2882** (99.0%) 2910-2919 (1.0%)
36	63.2-65.7	egg shell	NSRL-14448	-5.0	5395 ± 15	6020-6079 (47.1%) 6111-6155 (39.8%) 6174-6189 (13.1%)	**6000-6208** (98.5%) 6253-6261 (1.5%)

Table 3. *Continued*

XU	Depth (cm)	Material	Sample code	δ¹³C (‰)	Conventional radiocarbon age (years BP, ± 1σ)	Calibrated radiocarbon age in cal BP (1σ range [68.3% probability])	Calibrated radiocarbon age in cal BP (2σ range [95.4% probability])
39	70.0-71.6	charcoal	OZD942	-25.61	2180 ± 60	2000-2155 (88.2%) 2270-2295 (11.8%)	1949-1963 (1.7%) **1967-2213** (77.6%) 2218-2310 (20.7%)
41	74.5-77.4	charcoal	OZD010	-24.82	1610 ± 70	1379-1524 (100%)	**1305-1570** (98.5%) 1583-1600 (1.5%)
41	74.5-77.4	charcoal	OZD943	-25.0	11,880 ± 210		
43	80.1-83.1	charcoal	OZD944	-26.92	2450 ± 60	2345-2492 (83.1%) 2602-2608 (1.9%) 2641-2678 (15.0%)	**2333-2619** (82.2%) 2632-2708 (17.8%)
45	90.7	charcoal	OZD011	-26.37	13,420 ± 150		
49	102.8	charcoal	OZC422	assumed -25.00	13,600 ± 180		
52	106.7-108.2	charcoal	OZD945	-27.11	13,900 ± 80		
55	111.6-114.3	freshwater mussel shell	OZD946	-5.68	20,040 ± 200		
56	118.2	freshwater mussel shell	OZD012	-5.35	19,280 ± 140		
57	116.4-118.7	freshwater mussel shell	OZD947	-9.41	23,240 ± 320		
60	122.4-125.0	freshwater mussel shell	OZD948	-9.09	29,700 ± 1100		
Rock-art dates							
Radiating lines		charcoal	OZC608	assumed -25.0	1010 ± 60	801-888 (74.1%) 897-927 (25.9%)	**746-963** (100%)
Anthropo-morph		charcoal	OZD427	assumed -25.0	1480 ± 50	1292-1360 (100%)	**1263-1414** (100%)
Anthropo-morph		charcoal	OZC848	assumed -25.0	1570 ± 110	1316-1518 (100%)	1186-1202 (0.8%) 1244-1247 (0.1%) **1256-1635** (96.3%) 1647-1694 (2.8%)
Anthropo-morph		charcoal	OZD425	assumed -25.0	1700 ± 90	1415-1620 (94.8%) 1673-1687 (5.2%)	**1343-1740** (98.8%) 1757-1778 (1.2%)

to have been caused by higher rainfall levels resulting in the disturbance of sediments and lag deposits. Increased humid conditions appear to have damaged this part of the archaeological assemblage at Hay Cave through migrating driplines and lag or mixed deposits whereby fine particles, but not the coarse sediment fraction, have been removed (see below). Between XU36 and XU46 sediments are clayey but incorporate large amounts of coarser-grained components, including peak quantities of land-snail shell and calcium carbonate concretions (Figure 6). These XUs represent the final period of the Pleistocene to the mid Holocene when rainfall levels are known to have peaked. The large numbers of land-snail shell and peak frequencies of calcium-carbonate concretions in these levels include the early Holocene period of heightened humidity, consistent with both items themselves signalling wetter conditions.

Secondly, the complex of hearths from SU1 (cf. Figure 5) appear to have intruded into and disturbed underlying stratigraphic levels from earlier periods, such as those of SU2a and SU2b. SU1 is composed of numerous in situ hearths and hearth material, as well as representing the period of time when the cave was most intensively used by people and thus suffered the greatest amount of treadage (see below). Intrusive hearths and related activities, therefore, can be expected to have caused some disturbance to earlier sediments, including mixing of small charcoal particles such as those used here in our AMS determinations, particularly at stratigraphic interfaces representing ancient surfaces.

Palaeoenvironments: Land-snail shells

Information on palaeoenvironmental dynamics for the Chillagoe Formation has been obtained through an examination of the geographical distribution, and ecological and biological characteristics, of the helicinid terrestrial mollusc shell *Pleuropoma extincta* (Odhner 1917). The use of *P. extincta* remains for palaeoenvironmental reconstruction at Hay Cave is dependent on two main factors: firstly, the capacity of the species to show a measurable response to environmental change; and secondly, the ability to recover sufficient undamaged shell material from the archaeological excavation.

Pleuropoma extincta's current restriction to the vine-thicket communities of the Chillagoe Formation is indicative of an acute sensitivity to regional and local moisture supply. The species occurs abundantly both as living populations on the limestone rock and associated vegetation, and as dead shells in the litter deposits. It has not been recorded from the more expansive eucalypt forest/woodland communities which dominate much of the region. An analysis of 1140 modern *P. extincta* shells from 23 collection localities within the Chillagoe Formation confirmed a sensitivity in shell growth (measured as size-correlated differences in shell whorl count) to environmental moisture availability. Differences in shell size and pattern between populations of *P. extincta* correlate positively with a declining regional north-south gradient in annual average rainfall and mean number of rain days. Measurements made on shells collected from the northern Mitchell-Palmer limestone belt are consistently more than those gathered from the Chillagoe-Mungana limestone belt further south, where the former recorded higher rainfall statistics than the latter. This apparent variation in the shell size of modern *P. extincta* is consistent with findings of environmental moisture-related variation in the shell size of terrestrial molluscs from the Kimberley in northwestern Australia (Solem and Christensen 1984) and several other detailed studies (e.g. Tillier 1981; Gould 1984) that suggest moisture is the critical factor in determining size and shape in mollusc species and between populations, more so than temperature, insolation, population density or the availability of calcium. Full details of these modern *P. extincta* investigations have been presented in Rowe (1998) and Rowe et al. (2001).

A total of 295 sub-fossil specimens of *P. extincta* were recovered from archaeological excavations at Hay Cave. A high incidence of terrestrial mollusc shell (of all species) by weight occurred after ca. 13,600 BP to around or shortly before 3100 BP (XU49-XU38), incorporating a peak (688.0 g) at XU40. Before this time, total shell incidence is comparatively low. Above XU38, a pronounced decrease to almost negligible occurrence is evident, encompassing the period ca. 3100 BP to present. In the excavated *P. extincta* shells, mean whorl counts, heights and diameters peak between 13,600 BP and 13,420 BP (XU49-XU45) and all shell parameters have high values between 13,600 BP and 3100 BP (XU49-XU34), declining after ca. 3100 BP before a secondary peak through the period ca. 820-660 BP (XU12-XU6). Low values continue to the present.

Palaeoenvironmentally, both the incidence of terrestrial mollusc shell and the variation in *P. extincta* shell size with depth indicate dry conditions during the late Pleistocene leading into the terminal Pleistocene/Holocene boundary. This is followed by a period of increased (to peak) precipitation encompassing the early and mid Holocene, until approximately 3100 BP, in turn followed by an effective drying. Although dry, the last ca. 3100 years appear to have incorporated a wetter phase around 820-660 BP, suggesting a degree of local precipitation and general climatic variability. These major changes in moisture regimes are thought to have affected the spread, floristic structure and complexity of the vine-thicket communities throughout the Chillagoe Formation. Vine-thicket communities would have had a maximum distribution beyond the current confines of the limestone outcrops, displacing areas of sclerophyll vegetation between ca. 13,600 BP and 3100 BP. A late-Holocene drier climate allowed for a reinvasion of the sclerophyll vegetation, reducing the *P. extincta* habitat. The Chillagoe Formation vine thickets are therefore remnants of a more extensive late-Pleistocene to mid-Holocene flora, now represented as permanent refugia. The local Chillagoe Formation palaeoenvironmental results are, overall, consistent with higher rainfall regimes and associated maximum extent of humid rainforests during the early and mid Holocene on the Queensland east coast, as informed by palynological data (Kershaw et al. 1993; Kershaw 1994). However, the *P. extincta* shell data suggestion of an earlier increase in moisture availability, close to 13,600 BP west of the Great Dividing Range, warrants further investigation.

Palaeotemperatures: Alectura lathami *egg shells*

The chronology of Hay Cave has also been assisted by examination of the remains of brush-turkey egg shell from its deposits. The extent of isoleucine epimerisation in Australian brush-turkey (*Alectura lathami*) egg shell from Hay Cave was reported by Clarke 2005 and Clarke et al. 2007. Isoleucine epimerisation (a diagenetic reaction analogous to amino acid racemisation) is the interconversion of the epimers L-isoleucine (Ile) and D-alloisoleucine (aIle), and is expressed as the ratio of these two molecules (aIle/Ile). In modern *Alectura* egg shell the aIle/Ile value is approximately 0.02 and increases towards a value of 1.30 at dynamic equilibrium.

A total of 99 *Alectura* egg-shell fragments excavated from Square M30 of Hay Cave were subjected to isoleucine epimerisation analyses. These egg shells primarily come from the upper 75 cm of the sequence; below these depths, egg shell was rare or absent. The values fall into three main clusters: 0.02 < aIle/Ile < 0.05, 0.05 < aIle/Ile < 0.08, and 0.13 < aIle/Ile < 0.16. These clusters are identified as A, B and C, respectively. Not incorporated in these three clusters are the aIle/Ile values of three egg shells, one of which falls between clusters B and C. The other two have aIle/Ile values greater than those observed in cluster C. The two egg shells with aIle/Ile values exceeding these ranges are not considered further as they are likely to reflect the influence of campfire heating on the rate of isoleucine epimerisation (Clarke et al. 2007).

The correlation between *Alectura* egg shell aIle/Ile values and calibrated radiocarbon ages (median of the 1σ range) obtained on the same samples can be approximated using a simple linear model that estimates that the extent of isoleucine epimerisation increases by 0.020 aIle/

Ile units every 1000 years. The root mean square error of this regression predicts that the model can be used to interpret aIle/Ile in terms of sample age with an uncertainty of approximately ± 470 years (when the average analytical uncertainty of ± 0.005 aIle/Ile units is included).

Three groups corresponding to clusters A, B and C are identified in the depth-age profile of the isoleucine epimerisation ages in the Hay Cave sequence. The implication is that three major chrono-stratigraphic phases can be identified from these data. The egg shells of cluster A were recovered from depths close to the surface down to approximately 40 cm and have age estimates of ca. 0-1500 years. The cluster B egg shells occupied depths between 40 cm and 65 cm and have ages of ca. 2000-3000 years. The egg shells of cluster C came from sediments at about 65-70 cm depth and have ages of approximately ca. 6000-7000 years. The apparently young egg shells from clusters A and B found at depths below the above-mentioned respective depths are explained, along with other radiocarbon dates, as most likely due to one of two possibilities: either post-depositional contamination through treadage and related activities, or the intrusion of hearths in the suite of Excavation Units spanning XU35-44, at the stratigraphically indistinct Pleistocene-Holocene interface (see above).

Archaeological materials: Temporal trends

To address questions of temporal trends in cultural activity and intensity of site occupation, studies have been carried out on the rock art and the main archaeological materials from Square M30: stone artefacts, animal bone, land-snail shell and brush-turkey egg shell (see Table 4 for a list of excavated materials by XU). The results of these studies are presented below.

Stone artefacts

The analysis of the stone artefacts focuses on the question of 'intensity of site use' through time. Holden (1999:31-46) followed the methods of Hiscock (1988:163-228) at Colless Creek, Mitchell (1988) at Seal Point, and Lamb (1994) at Fern Cave to demonstrate that discard rates of stone artefacts alone are insufficient measures of 'intensities of site use', as other factors may also critically affect the number of artefacts produced and discarded at a site. These factors include the manufacturing process (e.g. the use of blade versus non-blade technologies; bipolar versus freehand flaking techniques; heat treatment), lithic resource manipulation (e.g. decortication at the quarry versus all stages of core reduction on-site) and taphonomic processes (e.g. post-depositional heat shatter from adjacent hearths; removal of small artefacts by water action). A set of appropriate multiple attributes was thus used to investigate these processes at Hay Cave (see Holden 1999 for a detailed analysis of the Hay Cave stone artefacts). Here we present a summary of these results.

Quantities of stone artefacts from SU2 and SU3 at Hay Cave were too low to conclusively assess changes in human activity during those periods. Suggestive trends, however, were apparent. In general, deposition rates of the weight of stone material through the deposit strongly correlate with those of burnt earth, mussel shell and egg shell (Figure 6). The few stone artefacts in the lower levels suggest overall low levels of site use between ca. 30,000 BP and 13,600 BP, followed by much reduced numbers or a possible hiatus in occupation during the early Holocene when the cave might have been too humid for regular occupation, with peak levels of human activity taking place after ca. 3000 BP and continuing until the near-present. Holden argued that after ca. 3000 BP, increases in the frequency, length of duration of stay or number of people using the site (all of which potentially contribute to increased intensities of site use) boosted the demand on stone, leading to practices aimed at conserving the resource. For example, during the late Holocene, an emphasis was placed on later stages of reduction, with high levels of controlled flaking, along with increased lithic discard rates. Fluctuations in lithic deposition rates appear to be unrelated to differential activity locations, raw-material selection or taphonomic processes, and are thus largely explained by changing intensities of site use by people (Figure 6; see Holden 1999 for specific details).

Table 4. Excavated materials, Hay Cave square M30.

XU	Flaked stone artefacts (#)	Flaked stone artefacts (g)	Mean weight of flaked stone artefacts (g)	Bone (g)	Mussell shell (g)	Egg shell (g)	Charcoal (g)	Burnt earth (g)	Hearth stones (#)	Hearth stones (g)	Bone points (#)	Burnt seeds (#)	Burnt seeds (g)	Land-snail shells (g)	CaCO₃ concretions (g)	Roots (g)
1	1	0.17	0.17			<0.1	<0.1	2.8						0.1		16.0
2	4	1.05	0.26	4.0	0.2		5.2	27.5						1.5		16.0
3	2	0.03	0.02	11.7	1.2	6.2	17.6	156.8						3.1		11.0
4	13	0.90	0.07	12.0	0.4	3.2	9.0	162.5						1.6		6.7
5	26	6.07	0.23	12.9	1.9	2.3	9.5	197.9						1.3		4.6
6	17	2.46	0.14	7.8	2.5	2.3	9.1	217.1						0.8		1.9
7	18	3.38	0.19	4.4	4.3	1.5	11.6	236.3						1.5		2.5
8a	1	0.25	0.25	0.8	<0.1	0.2	1.3	34.2						0.9		0.7
9a								2.2								
10	10	3.88	0.39	8.1	0.9	1.5	18.4	96.4						0.5		2.4
11	14	5.62	0.40	4.4	4.7	1.3	20.6	69.0						0.9		7.0
12	21	2.05	0.10	5.0	3.3	1.8	20.1	87.6						0.8		3.6
13a	12	1.04	0.09	2.1	1.5	1.0	15.9	43.9						0.3		2.1
14	15	18.33	1.22	4.5	2.6	1.7	30.9	17.9				1	0.4	0.7		11.0
15	25	9.80	0.39	9.4	8.1	2.2	41.3	80.5						1.5		1.3
16	31	4.70	0.15	8.0	6.1	3.7	41.6	55.7						2.8	3	13.0
17	31	16.18	0.52	7.9	3.6	3.4	28.1	38.4						1.9		7.5
18	26	2.05	0.08	11.7	2.8	2.5	16.6	38.4						0.1	1	4.1
19	4	3.25	0.81	5.0	1.6	0.8	5.5	14.8						0.5		0.5
20	43	6.74	0.16	13.9	2.6	2.6	15.5	86.5						1.4		7.5
21	33	4.46	0.14	22.2	4.0	3.8	17.5	85.5						2.4		3.3
22	28	2.35	0.08	24.5	0.4	0.2	15.2	65.2						2.1	22	2.4
23	33	3.33	0.18	18.6	1.6	1.8	14.5	34.9						1.5		4.1
24	34	12.20	0.36	38.5	<0.1	2.2	18.0	45.1			1			2.6	7	17.0
25	25	15.62	0.02	53.3	1.8	1.7	13.1	40.5	1	71.2				2.5	3	35.0
26	38	6.66	0.18	23.5	0.3	1.3	7.4	30.1						2.3		38.0
27	59	30.16	0.51	37.0	0.1	2.2	15.0	46.4						4.4		31.0
28	48	11.46	0.24	20.2	0.1	1.8	8.2	33.2						3.9		4.2
29	118	29.72	0.25	35.8	0.3	1.9	11.5	51.1						4.6	7	22.0

30	97	32.32	0.33	36.2	<0.1	2.2	13.8	64.4					7.4	3	5.4
31	82	13.46	0.17	36.3	0.2	2.1	13.0	87.2					8.7	2	4.7
32	62	22.87	0.37	61.9		2.1	10.2	59.3					11.7	25	3.0
33	34	26.04	0.77	74.5		2.7	8.3	57.2					19.4	20	5.0
34	35	6.45	0.18	5.8		2.5	7.3	54.4					45.9	174	4.1
35	20	2.51	0.13	34.0		1.8	2.2	38.9					31.3	174	1.6
36	12	4.14	0.35	58.2	0.1	2.1	2.4	130.7					49.2	1368	2.8
37	4	6.64	2.21	6.0		0.2	0.2	43.4					126.3	2255	1.0
38	4	2.25	0.56	11.7		0.4	0.6	51.2					427.1	1634	1.4
39	3	0.12	0.04	12.4		0.3	0.2	21.3					452.9	301	1.4
40	8	10.07	1.26	21.5		0.4	0.5	43.5					688.0	406	3.7
41	2	10.06	5.03	18.1			0.4	40.7					634.2	171	4.9
42	8	5.56	0.70	20.5		0.1	0.1	37.6					457.2	87	0.9
43	6	8.94	1.49	61.0		<0.1	0.4	87.2					388.6	202	3.7
44	1	0.07	0.07	34.5	0.8	<0.1	0.6	77.7					274.7	481	2.0
45	1	1.85	1.85	97.3		<0.1	1.5	225.7					306.2	209	3.3
46	1	0.03	0.03	99.0			0.3	109.2					307.8	344	1.0
47	8	7.79	1.03	100.9	0.2		0.2	41.5					259.4	39	1.6
48	3	0.25	0.07	96.5	1.4	<0.1	0.2	57.6					237.0	37	0.9
49	3	66.89	22.30	98.8	0.9		0.3	124.1					210.4	38	2.3
50	10	6.15	0.62	90.0	6.1		<0.1	112.0					121.4	20	5.3
51				104.2	0.3		0.5	100.2					104.9	49	3.2
52	5	5.96	1.19	50.6	0.3		0.2	54.5					24.5	34	0.6
53	18	11.66	0.65	95.8	1.1		0.1	126.2					43.2	53	1.1
54	12	15.13	1.26	78.1	0.1		0.1	58.4					48.6	89	1.2
55	6	12.83	2.14	113.8	0.4		0.1	61.7					42.9	73	1.6
56	15	17.81	1.19	95.7	0.2		0.1	51.9					33.8	55	1.3
57	17	6.16	0.36	101.9	0.1		0.1	62.6					43.7	21	0.9
58	7	14.51	2.10	113.9				57.0					51.7		0.8
59	22	12.46	0.57	160.0				127.0					116.9		0.5
60	9	3.61	0.40	103.1	<0.1		<0.1	48.0					53.0	130	0.5
61	6	0.38	0.06	22.2				8.4					5.9	4	0.1
62	1	2.11	2.11	8.1		<0.1		0.5					0.4		0.4
63				2.0									0.2	4	0.2

Table 5. Teeth and jaw fragment NISP, Hay Cave Square M30 (after Roche 1999:82, table VI).

Taxon	Excavation Unit																									Total
	1	4	7	10	13	16	19	22	25	28	31	34	37	40	43	46	49	52	55	58	59	60	61	62	63	
Macropus agilis									2		1	1				2	1	2				1				10
Macropus giganteus		1																								1
Macropus parryi										1						1		1	2	5						10
Macropus robustus		1	1								1						2					1				6
Petrogale sp.					1				7	5	7	5		2	4	5	3	3	12	18	16	12	3			103
Macropodidae (sp. indeterminate)			1							1		3	1		4	3	4	1	1	6	8	8	1	2		44
Trichosurus vulpecula		1							1	1		2														5
Petaurus norfolcensis																					2					2
Isoodon obesulus		1															1									2
Isoodon macrourus				1																	1					2
Perameles nasuta										1	1															2
Bettongia tropica			1																		1					2
Dasyurus hallacatus				1	1																					2
Dasyuridae (sp. indeterminate)																							1	1		2
Phascogale tapoatafa																						1				1
Phascogale (sp. indeterminate)																					1					1
Sminthopsis murina																	2					1				3
Conilurus penicillatus														1								2	3	1		7
Mesembriyos gouldii																					1					1
Uromys caudimaculatus																					1					1
Rattus tunneyi																	1				2					3
Rattus (sp. indeterminate)																				1	8	1	1	2		13

Table 5. *Continued*

Taxon	Excavation Unit																									
	1	4	7	10	13	16	19	22	25	28	31	34	37	40	43	46	49	52	55	58	59	60	61	62	63	Total
Melomys burtoni																					1	1		1		3
Zyzomys argurus																							1			1
Pseudomys gracilicau-datus														2	2	1	4			2	10	9	3	1	1	35
Pseudomys patrius																						1				1
Pseudomys sp. 1														1		1	1				1			1		5
Muridae (sp. indetermi-nate)					1							3		1	2	3	4	3	4	7	39	72	24	13	1	177
Tiliqua scincoides																1										1
Total		4	2	3	3	0	0	0	10	9	10	14	1	7	12	17	23	10	19	39	88	110	37	22	6	443

Vertebrate faunal remains

Roche (1999) extended the above study by examining the animal bone from Square M30. Because of the large quantities of bone present, the bones were identified from every third XU down to XU57, below which every XU was analysed for taxonomic identifications. For analytical purposes, Roche divided the excavated sequence into arbitrary units, each representing a period of approximately 3000 years (Analytical Units 1-5, excluding Analytical Unit 2a, which is added here for the purposes of this paper). The Analytical Units followed the timeframes below, and were based on extrapolation from the depth-age curve (Hughes and Djohadze 1980):

Analytical Unit 1: [XU1-34, representing 150 BP to 3000 BP]

Analytical Unit 2a: [XU35 to 40, representing 3000 BP to 12,000 BP]

Analytical Unit 2b: [XU41-52, representing 12,000 BP to 15,000 BP]

Analytical Unit 3: [XU53-56, representing 16,000 BP to 19,500 BP]

Analytical Unit 4: [XU57-59, representing 23,500 BP to 27,000 BP]

Analytical Unit 5: [XU60-63, representing 29,500 BP to 32,000 BP]

These Analytical Units are distinct from the SUs and XUs that structured the excavation itself, and were employed to facilitate examination of temporal trends through the use of chronological units of roughly equal, and therefore more comparable, duration. The identifiable bone was analysed by Minimum Number of Individuals (MNI), and the unidentified, fragmented bone by the Number of Identified Specimens (NISP) (cf. Grayson 1984:28, 30; Lyman 1994:98, 100). Bone deposition rates (Figure 6) conform to the general trends observable in other cultural indices, except that high levels also occur between ca. 32,000 BP and 18,000 BP when other indices are low; and levels increase markedly between ca. 15,000 BP and 12,500 BP. Bone, therefore, appears prominent in the lower and middle units (Anaytical Units 5, 4, 2).

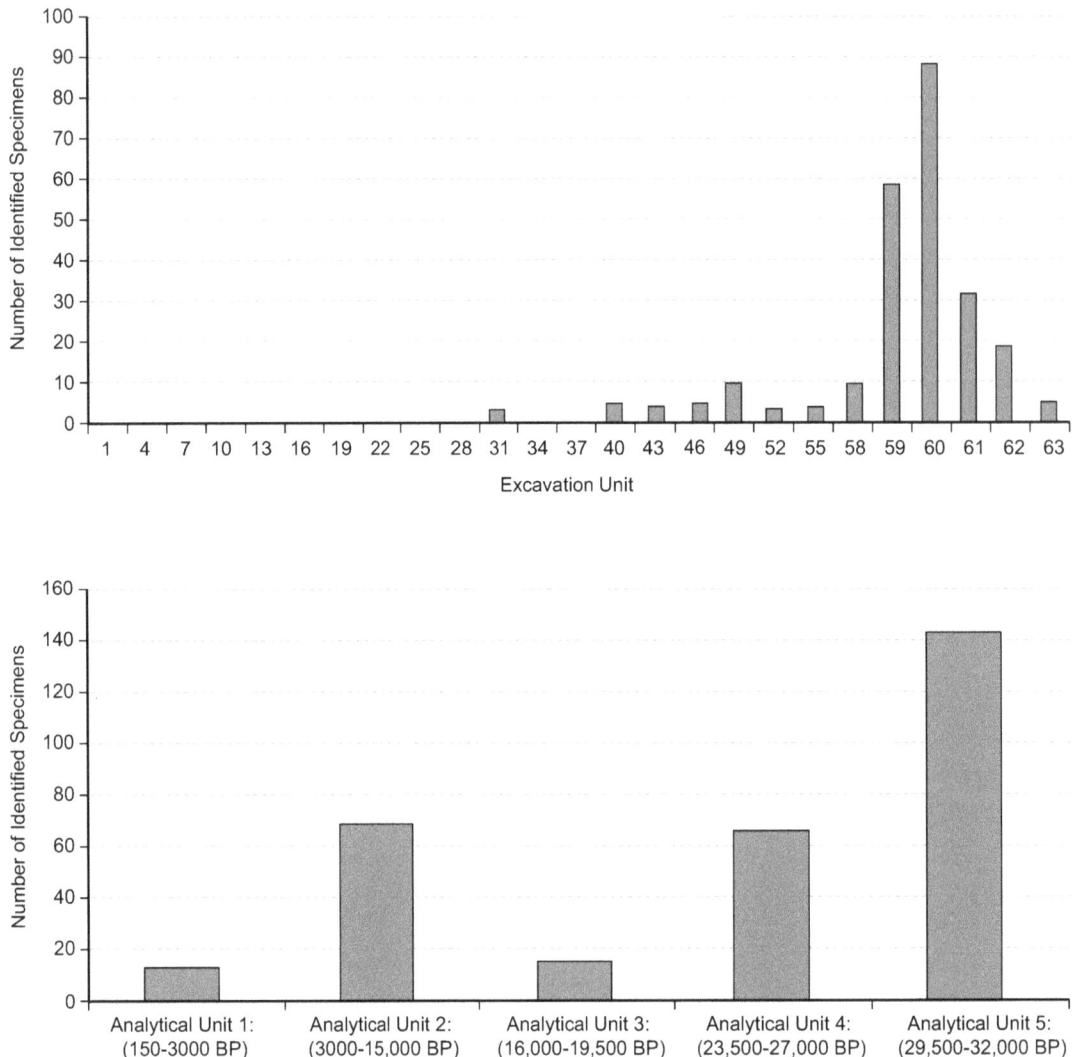

Figure 7. Distribution of murid bones by XU (top) and by Analytical Unit (bottom), Square M30.

As soil pH remained constant throughout the deposit, general variations in quantities of bone over time are not related to changes in the acidity of the matrix. There is a generally high rate of post-cranial fragmentation throughout (more than 90% of the fragments are smaller than 15 mm); frequencies of skeletal elements revealed little variation between XU, with limb bones being the most common post-cranial bones. All calculations in this analysis were adjusted for temporal variation between XUs. Calcined and carbonised bones are present in almost all XUs and highest between XU34-28 and XU52-46. That is, burned bone is most evident between ca. 3000-2500 BP and ca. 15,000-13,500 BP, which we argue represents those levels when the site was most heavily used by people. The presence of burnt bone in Hay Cave cannot be attributed to natural (bush) fires due to its protected location inside a cave which is itself surrounded by fire-protected limestone pediments; furthermore, the presence of calcined bone is generally understood to be closely related to human activity (e.g. Shipman et al. 1984:323; David 1990:75).

Taxonomic identifications were undertaken on teeth and jaws, indicating the presence of 24 distinct species plus five indeterminate family- or genus-level taxa which could not be attributed to species (Table 5). Macropod bones are present in almost all XUs, with NISP peak deposition periods in XU34-25 (the lower part of Analytical Unit 1) and XU52-43 (Analytical Unit 2b) – that is, between ca. 3000 BP and 2500 BP and between ca. 15,000 BP and 13,500 BP. Macropod MNI are highest in Analytical Unit 1 (ca. 3000 BP and 150 BP)(Table 6).

Table 6. Description of Analytical Units (see Roche 1999:101-11 for further details).

Analytical Unit	Years BP	Description
1	150-3000	This unit contains the fastest sedimentation rates, peak proportions of macropods, and the second-highest value of calcined bone. There is also evidence for the rationing and conservation of stone (Holden 1999:96), indicating that demands on lithic raw materials were high. All of the dated rock art belongs to this time. The complete absence of grassland faunal taxa (with murid NISP at their lowest) cannot be related to differential environmental or climatic conditions. Rather, people, with their high incidence of hearths, are inversely correlated with the frequency of murids at this time (and, by implication, probably also the use of the cave by predatory birds). Forest faunal taxa also show a marginal decrease from the prior wetter period (of the early to mid Holocene represented by Analytical Unit 2a). Lower precipitation during Analytical Unit 1 is also reflected in low deposition rates of land-snail shell and calcium-carbonate concretions (cf. Rowe 1998:78-79).
2a	3000-12,000	Between ca. 3000 BP and 13,500 BP, land-snail shell and calcium-carbonate deposits increased noticeably, indicating peak levels of precipitation (Rowe 1998:78-79). The period between 3000 BP and 12,000 BP is not clearly differentiated at Hay Cave, partly due to the presence of mixed sediments during this period, the coarser nature of those sediments (indicating that finer sediments may have been washed away through percolating water action), and the cementing of sediments in calcium-carbonate concretions.
2b	12,000-15,000	Human use of the site is less than during the late Holocene, but higher than preceding Analytical Units, with high weights of calcined bone and other indices. Murids and macropods are also abundant, with peak incidence of open forest/woodland taxa. Grassland taxa, however, are little different in proportion to those of Analytical Units 4 and 5, indicating that grasslands were not expansive at this time. During this time, the increased use of the site by people and other fauna, including murid predators, is related to higher levels of precipitation than previously.
3	16,000-19,500	Human use was marginally more than during the preceding Analytical Unit 4. Here, there is also a low level of murids, and the second-lowest weight of calcined bone, murid and macropod NISP. Climatic conditions were dry, as indicated by low levels of land-snail shell and calcium-carbonate concretions. Low levels also of open forest/woodland animals are consistent with drier environmental conditions and a reduction in forests.
4	23,500-27,000	A relatively low level of human use is shown by average levels of calcined bone and macropods, and low levels of flaked stone and burnt earth. There is a high proportion of murids, indicating that animals and probably predatory birds were frequent users of the cave.
5	29,500-32,000	A low level of human use is indicated by the low amount of calcined bone, the slow rate of sedimentation, and the lowest values of macropods contrasting with the highest proportions of murids.

Greater amounts of macropod than murid bone at a site are usually suggestive of people rather than other predators, even though the diets of people and animal predators may overlap to some extent (Balme et al. 1978:52; Bowdler 1984:63; David 1984b:40). Macropods form a key element of human diets, while murids are more prominent in the diets of other predators, such as birds using caves as roosts. More intensive human use of a cave or rockshelter, therefore, often results in a decline in murid bones due to the presence of hearths and other disturbances to the lair or roost of non-human predators (David 1984a:49). As rates of bone fragmentation remained constant throughout the Hay Cave sequence, differential bone preservation does not appear to account for the changes in the absolute or relative representation of macropods and murids. Frequencies of murid bones can thus be contrasted with frequencies of culturally deposited bones (in particular, macropod) within the deposit. Accordingly, and in contrast with the pattern of macropod bone at Hay Cave, murid NISP are almost totally absent in Analytical Unit 1 when human activity is most prominent (murid bones are absent in XU31-1, except for XU13). They are present in all other Analytical Units, with a peak presence in Analytical Units 4 and 5 (Figure 7). Comparison between Analytical Units 1 and 5, representing the two extremes of highest and lowest human occupation respectively, indicates a marked negative correlation between murid and macropod bone concentrations; when murid concentrations are high, macropod are low, and vice versa. In line with previous studies, such as at Walkunder

Figure 8. Panel of pigment anthropomorphs, including the two radiocarbon-dated pictures, Hay Cave.

Arch Cave also in the Chillagoe Formation, the implication is that the human use of Hay Cave is significantly greatest during the last ca. 3000 years in Analytical Unit 1 (Roche 1999:80-92).

The frequency of open forest/woodland taxa, as measured by NISP/1000 years, indicates their presence throughout the deposit, with very low frequencies in Analytical Unit 3 (the Last Glacial Maximum) and peak frequencies in Analytical Unit 2 (terminal Pleistocene). This may be an indication of a climatic gradient from maximum aridity to increasing humidity, and therefore a decline in open forest/woodland during the drier Last Glacial Maximum followed by an increase during the more humid terminal Pleistocene into the early Holocene. However, the complete absence of grassland taxa (for example, murids) in Analytical Units 1 and 3 may be related to the increased presence of humans during the late Holocene and possibly the Last Glacial Maximum also. As Analytical Unit 3 represents the glacial maximum – a time of peak aridity when grasslands were more dominant – the results in this case are somewhat ambiguous and may be interpreted either as evidence of changed environmental conditions and/or as a period of increased human presence (Roche 1999:92-100).

If we summarise the data regarding people's activities at Hay Cave, comparing vertebrate faunal evidence with the deposition rates of other cultural materials and with the speed of sediment accumulation through time, peak rates of use by people consistently occur during the late Holocene (especially the past ca. 3000 years in Analytical Unit 1), followed by Analytical Unit 2b (the terminal Pleistocene, ca. 15,000 BP to 12,000 BP), with Analytical Units 3, 4 and 5 showing sequentially decreasing levels.

The rock art

Hay Cave contains 92 rock pictures, all but four of which are either drawn (mainly in black charcoal) or painted. The drawn and painted art is from here on referred to as 'pigment art'. The other four rock pictures are moderately patinated engravings pecked into the rock wall. Appendix 1 lists each of these rock pictures.

Four radiocarbon determinations were obtained from small scrapings of charcoal pigment art. Two of these are from generalised anthropomorphs located on a small panel near the cave entrance, but well inside the dripline (Figure 8). This decorated panel is found high up on the overhanging rock wall, in a part of the cave that could only be accessed by artists when standing on a large boulder on the scree slope and reaching up. This localised panel is segregated spatially from the main body of artwork in the cave, marking the cave entrance, and the charcoal drawings found here would appear, on both spatial and stylistic grounds, to represent a single artistic event. The two radiocarbon-dated charcoal drawings from this panel are part of a spatially and stylistically close-knit set of 'Generalised Anthropomorphs' and 'Anthropomorphs with Inverted V-Headdresses' (see David et al. 1997 for details of the Mitchell-Palmer region's rock art). About 1 m away around the corner of the same rock that protrudes down from the ceiling, an 'Anthropomorph with Knobbed Headdress', drawn in a similar manner to the other two radiocarbon-dated anthropomorphs, was also radiocarbon dated. Each of these three radiocarbon determinations overlaps at two standard deviations, and indicates that this panel and the nearby paintings were probably all drawn around 1500 BP.

The fourth radiocarbon-dated rock drawing is located in a protected part of the site in semi-dark conditions on the vertical wall of a large limestone pillar close to the back wall. It consists of a set of 'Radiating Lines' (see David et al. 2001:112 for a tracing of this drawing). The radiocarbon date for this image is 1010 ± 60 BP (OZC608).

The two major implications from the radiocarbon dates at Hay Cave are that, firstly, artistic activity took place late during the late Holocene, during a span of time when cultural activity (including the application of pigment art on rock) and demographic conditions appear to have reached a peak across much of north Queensland (see David and Chant 1995). And, secondly, that the artistic activity at Hay Cave took place at a point in time with no in-situ evidence of human presence in that part of the site we have excavated. For example, even when the out-of-sequence ^{14}C sample OZD010 is taken into account, there appears to be an absence of sedimentation and cultural deposition between ca. 1500 BP and 1000 BP. Therefore, it could be argued that at that time the cultural significance of the site shifted, the cave being no longer suitable for habitation, rather being reserved for artistic practices and their symbolic and cosmological associations.

Discussion

Palaeoenvironmental trends at Hay Cave

Based on the incidence of grassland and open forest/woodland animal taxa, Roche (1999:111-13) concluded that gross vegetation patterns in the Hay Cave area remained generally stable through time (relative to more coastal parts of southeast Cape York Peninsula). That is, throughout the past 30,000 years, the Hay Cave environment retained its patchwork of open forest/woodland and grassland with small pockets of wet forest vegetation amongst the karst towers.

Nevertheless, some changes did take place. Between ca. 19,000 BP and 16,000 BP, low proportions of open forest/woodland animal taxa indicate more arid conditions than previously, but less dramatic changes than have been described for the Atherton Tableland some 150 km to the southeast (cf. Butler 1998:5; Kershaw 1986:48). This drier period is followed by an increase in open forest/woodland animal taxa between ca. 15,000 BP and 12,000 BP, suggesting increasing precipitation (Roche 1999). Rowe's (1998) complementary land-snail shell and calcium-carbonate concretion data also indicate dry conditions leading into the terminal Pleistocene/ Holocene boundary, followed by a period of increased-to-peak precipitation encompassing the early and mid Holocene ca. 13,600-3100 BP. Peak humid conditions during this time began shortly after ca. 13,600-13,420 BP. The sediments of the mid-Holocene period at Hay Cave include the phase ca. 7000-6000 BP, as indicated by the Australian brush-turkey (*Alectura lathami*) egg-shell cluster C.

At Hay Cave, the onset of increased precipitation (and end of peak aridity) in the terminal Pleistocene (between ca. 15,000 BP and 12,000 BP) generally overlaps in timing with that of other parts of southeast Cape York Peninsula ca. 13,000-10,000 BP (Torgensen et al. 1988:259), although its commencement at Hay Cave may be slightly earlier. The analysis of land-snail shell (Rowe 1998) present throughout the Hay Cave sequence, indicates an unbroken association with vine thicket. While the karst formations may have acted as refugia for such relatively humid forest vegetation, low quantities of rainforest/wet forest vertebrate taxa at the site suggest that humid forest or vine thicket itself was never expansive across the region during the past 30,000 years. Here, relatively wet forest/vine thicket may have been restricted mainly to small refugia like the karsts and creek edges, even during the wetter terminal-Pleistocene to mid-Holocene period.

The past ca. 3000 years, by way of contrast, were generally drier, with Rowe's land-snail shell data indicating that a short wetter phase occurred also between ca. 900 BP and 600 BP, suggesting overall general climatic variability. Open forest/woodland vertebrate taxa accordingly show a marginal decrease from the prior wetter period of the early to mid Holocene.

Cultural trends at Hay Cave

Drawing all the material together, we have the following sequence. People first started using Hay Cave around 30,000 BP or shortly beforehand, but at first, human occupation was low. The following, increasingly drier, periods that led into the glacial maximum (ca. 19,500-16,000 BP), continued to experience low levels of site use by people. Murids are absent during this phase of peak aridity. Throughout this long period of time, the cave was occupied more frequently by animals, including predatory birds. These trends were measured by the various indices – deposition rates of sediment, animal bone, mussel shell and egg shell; and construction rates of hearths, incorporating charcoal, burnt earth and structural features. These indicate that low levels of site use continued until around 15,000 BP when evidence of human presence begins to increase. This was a time of open forest/woodland and grassland communities.

The first noticeable long-term rise in occupation by people coincided with increasing rates of precipitation between ca. 15,000 BP and 12,000 BP that continued into the early to mid Holocene until around 3100 BP. With the onset of increasing precipitation at this time, people began to use the cave more often and there is a peak in the incidence of open forest/woodland animal taxa. Murids also are present, indicating a continuation of mixed human and non-human use of the cave.

Patterns of human occupation at the site are unclear to some extent during the early to mid Holocene due to stratigraphic disturbances. The significant rise in levels of precipitation between ca. 13,600 BP and 3100 BP (Rowe 1998:78-79) may have resulted in disturbance to the cave's deposits, as indicated by larger sediment particle sizes and high quantities of calcium-carbonate concretions, together with increases in land-snail shell. To what extent people continued using the cave throughout this lengthy period is uncertain, although the presence of cultural remains in those deposits (often encrusted in thick calcium-carbonate concretions) testifies against total abandonment. During this period, radiocarbon dates on charcoal and egg shell at Hay Cave, and chronological estimates based on degrees of isoleucine epimerisation among Australian brush-turkey egg shells, point to depositional events around 12,000 BP, between 7000-6000 BP and around 5400 BP.

The most intensive period of occupation of the cave was during the late Holocene, between ca. 3100 BP until the near-present, a period of drier, changeable climate. This peak period of human use of Hay Cave between ca. 3000 BP and 150 BP is shown by all indices, including the fastest deposition rates of sediments, and also of a broad class of cultural materials, including peak proportions of macropod bone and high calcined bone frequencies. Attempts were made during this time to ration and conserve stone, indicating that demands for lithic raw materials were high relative to availability. The decline in murids may also be linked to increased human presence. Indeed, murids are only present in XU13, coinciding with a short wet phase dated to within ca. 900-600 BP, suggesting that wetter conditions may have then limited people's use of the damp cave, leaving it to other predators including birds. There is no strong evidence here to link murids predominantly to human diet.

It is during this recent period also that rock art – in particular pigment art – makes its appearance. Around 1500 BP, artists began to use the site in a new way, and in doing so, ceased to live in it – or at least in the area we excavated, using it instead in a specialised way. The symbolic marking of the cave with charcoal anthropomorphs at its entrance, and a broader range of designs within its chambers, signals the onset of a new system of territorial place marking and cosmological referencing that point to a changing social world.

Regional trends and patterns of site use

General comparisons can be drawn between the chronological trends observable at Hay Cave and trends derived from regional archaeological data. Long-term trends for southeast Cape York Peninsula have been obtained in three ways by analysing temporal patterning in (a) all radiocarbon dates, (b) the number of occupied sites, and (c) the rates of site establishment, at a regional scale and over a 30,000-year period (David and Lourandos 1997; Lourandos and David 1998, 2002). While accepting the generalised nature of all these trends and their methodological limitations, they nevertheless offer the opportunity to draw comparisons between different kinds of data sets and between data derived at different analytical scales: in this case, comparing site trends with regional trends.

A general comparison between the Hay Cave trends and the regional trends shows some strong similarities. In both, rates of site and regional land use by people are low from before 30,000 BP until after the Last Glacial Maximum, with noticeable increases in all indices during the terminal Pleistocene after ca. 15,000 BP, and with even higher rises in the late Holocene after ca. 3000 BP until recent times. These trends apply both to highly durable materials such as stone artefacts and to hard organic items such as animal bones and shell at sites with good preservation. Similar trends are observed also in other site data, including the incidence of hearths, evidence for treadage-induced disturbance, and overall sedimentation rates. They also apply to rock art, which evidences major quantitative increases in pigment art creation (see below). The *general* trend, therefore, is consistent across the varied lines of evidence (e.g. stone artefacts, faunal remains, rock art) and across analytical scales (site and regional).

In this context, and at both site and regional scales, the glacial maximum is clearly associated with changes to environment and site use. There are, however, no signs of disuse or abandonment of Hay Cave at this time, as occur at some other sites in the region. During the Last Glacial Maximum, the karst towers may have served as more humid locations, to which people periodically retreated, within a generally drier landscape. This pattern of increased use during the Last Glacial Maximum is observable at the sites of Fern Cave, Hearth Cave and Sandy Creek 1, all located near permanent sources of water (Morwood and Hobbs 1995:180; David and Lourandos 1997:6). Likewise, Hay Cave lies close to Limestone Creek, assumed to have been active during the Last Glacial Maximum.

The early Holocene period remains less clear to some extent, at both local (Hay Cave) and regional scales, due to sediments being affected by post-depositional taphonomic factors such as lag deposits caused by increased precipitation levels in most sites excavated so far. In addition to Hay Cave, this problem has been documented also at Walkunder Arch Cave (Campbell 1982:65), Green Ant Shelter (Flood and Horsfall 1986:39), Sandy Creek 1 (Morwood and Hobbs 1995:78) and Fern Cave (David and Chant 1995:401). In contrast to this issue, however, the *general regional* trends for this same early-mid Holocene period indicate significant *rises* in site establishment and use (David and Lourandos 1997).

The Hay Cave sequence, therefore, appears to reflect the site's particular location, and individual characteristics, within the environmentally varied region of southeast Cape York Peninsula. Throughout its occupational history, the Hay Cave environment appears to have remained largely a mixture of open forest/woodland and grassland, with small pockets of dry rainforest vegetation in the immediate vicinity of karst towers. Overall environmental change during the Last Glacial Maximum, for example, was less marked than has been described for the Atherton Tableland on the eastern plateau (Kershaw et al. 2007). In general, Hay Cave reflects a more stable and drier climate when compared with the more humid areas some 150 km further to the southeast, towards the higher parts of the Great Dividing Range and the coast. Lying within a rain shadow, the Hay Cave area appears to have been shielded from the more dramatic climatic oscillations experienced in more coastal regions nearby (see also Rowe 1998:78; Roche 1999:115-25). Nevertheless, the Mitchell-Palmer region (including Hay Cave) witnessed similar directions of environmental change to those further in the east.

Regional trends and patterns: Rock art

We have elsewhere argued that the past 2000 years in particular saw an increased regionalisation of social practice across southeast Cape York Peninsula (David and Chant 1995; David and Lourandos 1998, 1999; David 2002). In Princess Charlotte Bay and its neighbouring Flinders Island group, the late-Holocene art includes numerous figurative paintings of moths/butterflies and zoomorphs with crescent heads (David and Chant 1995). Close by to the south, in the Koolburra Plateau, large numbers of paintings of echidna-human therianthropes predominate (Flood 1987). Across the Kennedy River, just off the eastern edge of the Koolburra Plateau, the Laura sandstones are richly decorated with a broad range of anthropomorphs and, to a lesser degree, zoomorphs, many of which are life-sized and contain elaborate decoration in bichrome or polychrome (e.g. Cole 1992; Morwood and Hobbs 1995). To the immediate south of the Laura sandstones, in the Mitchell-Palmer limestone belt, generalised anthropomorphs abound, many painted upside-down (David et al. 1997). To the south, neighbouring the Mitchell-Palmer limestones, are the Bonney Glen granite boulders, where again, generalised anthropomorphs and zoomorphs occur (David 1998a). Immediately to the south of the Mitchell and Walsh Rivers is Ngarrabullgan, a large, cliff-lined mountain where the art suddenly and dramatically contrasts with that of all its northern neighbours (David 1998b): here, no more do we find a predominance of figurative paintings of human and animal shapes. Instead, we find linear abstract designs, such as single and composite lines, circles and their variants, radiating and grid designs. In the Mungana, Chillagoe, and Almaden areas on the other side of the Featherbed Ranges and to the immediate southwest of Ngarrabullgan – in a similar geological setting to the Mitchell-Palmer limestone belt further to the north – we again find a predominance of abstract linear paintings (e.g. David and David 1988). In the Davies Creek region on the edge of the rainforest further to the east, there is a predominance of localised figurative forms in the shape of generalised anthropomorphs with upraised arms and down-turned legs, in what some authors have described as a 'dancing' style (Clegg 1978; see David and Chant 1995, and David and Lourandos 1998 for reviews of regional rock-art conventions). There is, therefore, significant late-Holocene regionalism in southeast Cape York Peninsula's rock art. During that time, individual sites began to be used in new ways, adding to our understanding of social process. Socio-territorial changes documented for the *region* thus now also appear to be associated with changes in the *meaning* of specific places. We suggest that the increased regionalism of the past two millennia took place as territorial constrictions, and were inscribed by local artistic practices and legitimated by increasingly localised cosmological referents to local ancestral spirit beings that came to be known ethnographically as 'Dreamings'. These Dreamings became referents to 'country'; to landscapes populated not only by people but also ancestors and spirit-forces that gave the land its particular character in language and culture, as was apparent during the early European contact period. The portrayal of Dreaming beings and spirit-forces in rock art during the past 1500 years at Hay Cave, and the past 3000-2000 years across southeast Cape York Peninsula, at a time coincident with significant increases in site and regional land use, suggests the onset of population increases and social and territorial restructuring (David and Lourandos 1997, 1998, 1999), in association with a new or transforming cosmology; akin to that which through time became the ethnographic Dreaming system of this region. The predominance of anthropomorphs in the rock art of the entire Mitchell-Palmer region is a graphic articulation of this process of territorial and cosmological regionalisation expressed in site-specific depictions and use.

Conclusions

At Hay Cave, long-term archaeological trends have been characterised in a number of ways by employing a multifarious approach, composed of a range of separate indices and trends. The overall *general* archaeological trend is consistent across the varied lines of evidence and across different analytical scales, both site-specific and regional. At Hay Cave, generalised, long-term archaeological trends compare favourably with general, regional temporal trends derived from separate data sets. In both, rates of site and regional land use by people are low from before ca. 30,000 BP until after the Last Glacial Maximum, with noticeable increases in all indices during the terminal Pleistocene after ca. 15,000 BP, and with even higher rises in the late Holocene after ca. 3000 BP until recent times.

At Hay Cave, palaeoenvironmental trends were analysed by use of the land-snail shells and vertebrate faunal material and these also reflected the wider, general regional trends, while at the same time presenting a more localised focus. Lying further inland, Hay Cave falls outside the more humid coastal belt and its more dramatic climatic oscillations. The broader relationships between the long-term, regional palaeoenvironmental trends and cultural patterns, as well as human demographic trends, have been discussed elsewhere (cf. David and Lourandos 1997, 1998, 1999; Lourandos and David 2002). At Hay Cave, the rock art, for example, is clearly a local manifestation of the regionalised rock-art patterns of the past 2000 years found across the wider Cape York Peninsula area that have been viewed in terms of increased and denser Aboriginal populations (David and Lourandos 1998). In all, these spatio-temporal patterns of human behaviour observed through the analysis of individual sites and their broader regional patterning – incorporating emplaced art and deposition rates of varied cultural materials – amount to a *spatial history* of Aboriginal Australia.

Acknowledgements

We thank Kuku Yalanji representative Qawanji and his family; Bob Bultitude, Dave Currie, Les Hall, Lana Little and Andy Spate for fieldwork assistance; Mrs Wilson and George Wilson for permission to excavate on their land; Earthwatch for funding the fieldwork; Bob Bultitude for his generous time and advice with stone artefact raw material identifications; AINSE for funding the radiocarbon determinations; Toby Wood (Monash University) for drafting the figures. BD thanks the School of Geography and Environmental Science at Monash University for support, and the Australian Research Council for a QEII Fellowship to enable the writing up of this paper. Thanks also to Simon Haberle and anonymous referees for comments on an earlier version of this paper.

References

Balme, J., Merilees, D. and Porter, J.K. 1978. Late Quaternary mammal remains spanning about 30,000 years from excavations in Devil's Lair, Western Australia. *Journal of the Royal Society of Western Australia* 61(2):33-65.

Bowdler, S. 1984. *Hunter Hill, Hunter Island*. Canberra: The Australian National University.

Butler, D. 1998. Environmental change in the Quaternary. In David B. (ed), *Ngarrabullgan: Geographical Investigations in Djungan Country, Cape York Peninsula*. Monash Publications in Geography and Environmental Science 51:78-97. Monash University, Clayton.

Campbell, J.B. 1982. New radiocarbon results for north Queensland prehistory. *Australian Archaeology* 14:62-66.

Clarke S.J. 2005. Isoleucine epimerisation and stable isotope ratio studies of cassowary, megapode and Aepyornis egg shells: biogeochemical and palaeoenvironmental implications. Unpublished PhD thesis, University of Wollongong. Available online at http://www.library.uow.edu.au/theses/

Clarke, S.J., Miller, G.H., Murray-Wallace, C.V., David, B. and Pasveer, J.M. 2007. The geochronological potential of isoleucine epimerisation in cassowary and megapode eggshells from archaeological sites. *Journal of Archaeological Science* 34:1051-63.

Clegg, J. 1978. Mathesis words, mathesis pictures. Unpublished MA (Honours) thesis, University of Sydney, Sydney.

Cole, N. 1992. 'Human' motifs in the rock paintings of Jowalbinna, Laura. In: McDonald, J. and Haskovec, I.P. (eds), *State of the Art: Regional Rock Art Studies in Australia and Melanesia*, pp. 164-73. Australian Rock Art Research Association, Melbourne.

David, B. 1984a. Walkunder Arch Cave: a faunal report. *Australian Archaeology* 18:40-54.

David, B. 1984b. *Man vs Dingo: the identification of bone remains from archaeological sites with specific reference to Walkunder Arch Cave, Chillagoe, north-east Queensland*. Brisbane: Department of Aboriginal and Islander Affairs Archaeological Branch.

David, B. 1990. How was this bone burnt? *Tempus* 2:65-79.

David, B. 1998a. Rock art of southeast Cape York Peninsula: Bonney Glen Station. *Memoirs of the Queensland Museum* 1:127-36.

David, B. 1998b. The rock art. In: David, B. (ed), *Ngarrabullgan: Geographical Investigations in Djungan Country, Cape York Peninsula. Monash Publications in Geography and Environmental Science* 51, pp. 143-56. Monash University, Clayton.

David, B. 2002. *Landscapes, Rock-art and the Dreaming: An Archaeology of Preunderstanding*. Leicester University Press, London.

David, B., Armitage, R.A., Rowe, M.W. and Lawson, E. 2001. Landscapes in transition? New radiocarbon dates on cave paintings from the Mitchell-Palmer limestone belt (northeastern Australia). *American Indian Rock Art* 27:107-116.

David, B, and Chant, D. 1995. Rock art and regionalization in north Queensland prehistory. *Memoirs of the Queensland Museum* 37, 2:357-528.

David, B. and Dagg, L. 1993. Two Caves. *Memoirs of the Queensland Museum* 33:143-62.

David, B. and David, M. 1988. Rock pictures of the Chillagoe-Mungana limestone belt, north Queensland. *Rock Art Research* 5:147-56.

David, B. and Lourandos, H. 1997. 37,000 years and more in tropical Australia: investigating long-term archaeological trends in Cape York Peninsula. *Proceedings of the Prehistoric Society* 63:1-24.

David, B. and Lourandos, H. 1998. Rock art and socio-demography in north-eastern Australian prehistory. *World Archaeology* 30:193-219.

David, B. and Lourandos, H. 1999. Landscape as mind: land use, cultural space and change in north Queensland prehistory. *Quaternary International* 59:107-123.

David, B., Walt, H., Lourandos, H., Rowe, M., Brayer, J. and Tuniz, C. 1997. Ordering the rock paintings of the Mitchell-Palmer limestone zone (Australia) for AMS dating. *The Artefact* 20:57-72.

Day, R.W., Whitaker, W.G., Murray, C.G., Wilson I.H. and Grimes, K.G. 1983. *Queensland Geology: A companion volume to the 1:2,500,000 scale geological map (1975)*. Geological Survey of Queensland Publication 383, Brisbane.

Dogmagala, J. and Fordham, B.G. 1997. Silurian-Early Devonian: Chillagoe Formation (Hodgkinson Province). In: Bain, J.H.C. and Drapper, J.J. (eds), *North Queensland Geology*. AGSO Bulletin 240/Queensland Geology, Brisbane.

Fensham, R.J. 1995. Floristics and environmental relations of inland dry rainforests in north Queensland, Australia. *Journal of Biogeography* 22:1047-1063.

Fensham, R.J. 1996. Land clearance and conservation of inland dry rainforest in north Queensland, Australia. *Biological Conservation* 75:289-298.

Flood, J. 1987. Rock art of the Koolburra Plateau, north Queensland. *Rock Art Research* 4(2):91-126.

Flood, J. and Horsfall, N. 1986. Excavation of Green Ant and Echidna shelters, Cape York Peninsula. *Queensland Archaeological Research* 3:4-64.

Galloway, R.W., Gunn, R.H. and Stony, R. 1970. *Lands of the Mitchell-Normanby Area*. CSIRO Land Research Series 26, CSIRO, Canberra.

Gould, S.J. 1984. Covariance sets and ordered geographic variation in Cerion from Aruba, Bonaire and Curacao: a way of studying nonadaptation. *Systematic Zoology* 33:217-237.

Grayson, D.K. 1984. *Quantitative Zooarchaeology*. New York: Academic Press.

Hall, L., Whittier, J. and Macrokanis, C. 1996. The fauna surveys. Unpublished report to Bruno David for the Earthwatch Institute, Melbourne.

Hiscock, P. 1988. Prehistoric settlement patterns and artefact manufacture at Lawn Hill, north-west Queensland. PhD thesis, Department of Anthropology and Sociology, University of Queensland, St. Lucia.

Holden, A. 1999. A technological analysis of the lithic assemblage from Hay Cave, SE Cape York Peninsula: considering diachronic variations in patterns of 'intensity of site use'. Unpublished B.A. (Hons) thesis, University of Queensland, St Lucia.

Hughes, P.J. and Djohadze, V. 1980. *Radiocarbon dates from archaeological sites on the South Coast of New South Wales and the use of depth/age curves*. Occasional papers in prehistory 1, Australian National University, Canberra.

Kahn, T.P. and Lawrie, B.C. 1987. Vine thickets of the inland Townsville region. In: *Australian Heritage Commission (ed), The Rainforest Legacy: Australian National Rainforest Study*. Australian Government Publishing Service, Canberra. pp. 159-201.

Kershaw, A.P. 1986. Climate change and Aboriginal burning in north-east Australia during the last two glacial/interglacial cycles. *Nature* 322(3):47-49.

Kershaw, A.P. 1994. Pleistocene vegetation of the humid tropics of northeastern Queensland, Australia. *Palaeogeography, Palaeoclimatology, Palaeoecology* 109:339-412.

Kershaw, A.P., McKenzie, G.M. and McMinn, A. 1993. A Quaternary vegetation of northeastern Queensland from the pollen analysis of ODP site 820. In: McKenzie, J.A., Davie, P.J. and Palmer-Johnston, A. (eds), *Proceedings of the Ocean Drilling Program, Scientific Results* 133:107-114.

Kershaw, A.P., Haberle, S.J., Turney, CS.M. and Bretherton, S.C. (eds), 2007. *Environmental history of the humid tropics region of north-east Australia*. Special Issue *Palaeogeography, Palaeoclimatology, Palaeoecology* 251(1):1-173.

Lamb, L. 1994. A technological analysis of lithic material from Fern Cave, Queensland. B.A. (Hons) thesis, Department of Anthropology and Sociology, University of Queensland, St. Lucia.

Lourandos, H. and David B. 2002. Long-term archaeological and environmental trends: a comparison from late-Pleistocene-Holocene Australia. In: Kershaw, P., David, B., Tapper, N., Penny, D. and Brown, J. (eds), *Bridging Wallace's line: the Environmental and Cultural History and Dynamics of the SE-Asian-Australian Region*, pp. 307-338. Advances in Geoecology 34, Catena Verlag GMBH, Reiskirchen.

Lourandos, H. and David, B. 1998. Comparing long-term archaeological and environmental trends: north Queensland, arid and semiarid Australia. *The Artefact* 21:105-114.

Lyman, R.L. 1994. Vertebrate Taphonomy. Cambridge: Cambridge University Press.

McCormac, F.G., Hogg, A.G., Blackwell, P.G., Buck, C.E., Higham, T.F.G., and Reimer, P.J. 2004. SHCal04 Southern Hemisphere Calibration 0-11.0 cal Kyr BP. *Radiocarbon* 46, 1087-1092.

Macrokanis, C. 1996. Journal of the University of Queensland and Earthwatch expedition to Mitchell-Palmer valley from a zoological perspective. Unpublished report to Bruno David for the Earthwatch Institute, Melbourne.

Mitchell, S. 1988. Chronological change in intensity of site use: a technological analysis. B.A. (Hons) thesis, Department of Anthropology and Sociology, University of Queensland, St. Lucia.

Morwood, M.J. and Hobbs, D.R. (eds), 1995. Quinkan prehistory: the archaeology of Aboriginal art in S.E. Cape York Peninsula. *Tempus* 3. University of Queensland, Brisbane.

Odhner, N.H. 1917. Results of Dr. E. Mjöberg's Swedish scientific expeditions to Australia 1910-1913. XVII. Mollusca. *Kungliga Svenska Vetens Kapsakademiens Handlingar* 52(16):1-115.

Roche, N. 1999. Reading the bones: an analysis of cultural and palaeoenvironmental trends at Hay Cave, S.E. Cape York. Unpublished B.A. (Hons) thesis, University of Queensland, St Lucia.

Rowe, C. 1998. The value of land snails for reconstruction of palaeoenvironments in northern Queensland. Unpublished B.A. (Hons) thesis, Monash University, Clayton.

Rowe, C., Stanisic, J., David, B. and Lourandos, H. 2001. The helicinid land snail *Pleuropoma extincta* (Odhner, 1917) as an environmental indicator in archaeology. *Memoirs of the Queensland Museum* 46(2):741-70.

Shipman, P., Foster, G. and Schoeninger, M. 1984. Burnt bones and teeth: an experimental study of colour, morphology, crystal structure and shrinkage. *Journal of Archaeological Science* 11:307-325.

Solem, A. and Christensen, C.C. 1984. Camaenid land snail reproductive cycle and growth patterns in semi-arid areas of northwest Australia. *Australian Journal of Zoology* 32:471-91.

Stanisic, J. 1997. Land snails of the Chillagoe Limestones. *Australian Shell News*, 96(Oct):3-6.

Stanisic, J. and Ingram, G. 1998. Sandstone, snails and slaters: invertebrate fauna. In: David, B. (ed), *Ngarrabullgan: Geographical investigations in Djungan country, Cape York Peninsula*, pp. 112-18. Monash Publications in Geography and Environmental Science, Monash University, Clayton.

Stuiver, M. and Reimer, P.J. 1993. Extended 14C database and revised CALIB radiocarbon calibration program. *Radiocarbon* 35:215-230.

Terlich, N.J. 1998. Teeth, bones and other indices. Unpublished B.A. (Hons) thesis, Department of Anthropology and Sociology, University of Queensland, St. Lucia.

Tillier, S. 1981. Clines, convergence and character displacement in New Caledonian diplommatinids (land prosobranchs). *Malacologia* 21:177-208.

Torgerson, T., Luly, J., De Deckker, P., Jones, M.R., Searle, D.E., Chivas, A.R. and Ullman, W.J. 1988. Late Quaternary environments of the Carpentaria Basin, Australia. *Palaeogeography, Palaeoclimatology, Palaeoecology* 67:245-61.

Appendix 1. The rock pictures of Hay Cave; motif types follow David et al. (1997).

Picture #	Motif type	Engraving	Drawing or painting colour	Maximum length (cm)	^{14}C date (BP)	Comments
1	generalised anthropomorph		black			
2	generalised anthropomorph		black			male
3	generalised anthropomorph		black			
4	generalised anthropomorph		black			
5	anthropomorph with inverted V-headdress		black			male
6	anthropomorph with inverted V-headdress		black			male
7	generalised anthropomorph		black			
8	anthropomorph with inverted V-headdress		black			male
9	aeneralised anthropomorph		black			
10	anthropomorph with inverted V-headdress		black		1480 ± 50 (OZD427)	male
11	linear non-figurative		black			
12	linear non-figurative		black			
13	generalised anthropomorph		black			
14	anthropomorph with inverted V-headdress		black	33	1700 ± 90 (OZD425)	
15	generalised anthropomorph		black			
16	generalised zoomorph	x		68		lizard with crescent-head shape
17	linear non-figurative	x		25		
18	generalised anthropomorph		red	31		male
19	generalised anthropomorph		red	47		
20	generalised anthropomorph		orange	48		male

Appendix 1. *Continued*

Picture #	Motif type	Engraving	Drawing or painting colour	Maximum length (cm)	¹⁴C date (BP)	Comments
21	drawn area		black	9		his motif type was not described in David et al. 1997. It consists of numerous drawn lines covering a small area a few centimetres square
22	drawn area		black	10		See Picture 21
23	generalised anthropomorph with fingers and/or toes		red	56		
24	generalised anthropomorph		red	54		male
25	generalised anthropomorph with fingers and/or toes		white outline-red infill	31		upside-down anthropomorph
26	generalised zoomorph		black			lizard-shape
27	complex anthropomorph		white outline-red infill-white internal decoration	49		upside-down male
28	generalised anthropomorph with fingers and/or toes		red	54		male
29	generalised anthropomorph		red	35		
30	generalised anthropomorph		red	ca. 32		male
31	generalised anthropomorph with fingers and/or toes		red	44		female
32	generalised anthropomorph		red	53		
33	generalised anthropomorph	·	red	44		female
34	anthropomorph with knobbed headdress		red	80		
35	anthropomorph with knobbed headdress		red	73		
36	generalised anthropomorph		red	72		male

Appendix 1. *Continued*

Picture #	Motif type	Engraving	Drawing or painting colour	Maximum length (cm)	¹⁴C date (BP)	Comments
37	generalised anthropomorph with fingers and/or toes		red	84		upside-down female
38	generalised anthropomorph		black	31		male
39	grid		black	40		
40	radiating lines		black	31	1010 ± 60 (OZC608)	
41	bird track	x		16		trident with heel
42	bird track	x		15		
43	generalised anthropomorph with fingers and/or toes		red	20		upside-down anthropomorph
44	anthropomorph with knobbed headdress		black	30	1570 ± 110 (OZC848)	male
45	anthropomorph with inverted V-headdress		black	20		
46	linear non-figurative		red	20		
47	indeterminate		black			
48	indeterminate		red			
49	generalised anthropomorph		black			male
50	indeterminate		black			
51	anthropomorph with inverted V-headdress		black			
52	generalised anthropomorph		black			
53	generalised anthropomorph		black			male
54	generalised anthropomorph		black			
55	generalised anthropomorph		black			
56	generalised anthropomorph		black			male
57	generalised anthropomorph		black			
58	generalised anthropomorph		black			
59	linear non-figurative		red	29		
60	generalised anthropomorph		black			male
61	linear non-figurative		black			
62	linear non-figurative		red			

Appendix 1. *Continued*

Picture #	Motif type	Engraving	Drawing or painting colour	Maximum length (cm)	¹⁴C date (BP)	Comments
63	linear non-figurative		red			
64	linear non-figurative		red			
65	linear non-figurative		red			
66	linear non-figurative		red			
67	generalised anthropomorph		black			
68	generalised anthropomorph		black			male
69	generalised anthropomorph		black			male
70	generalised anthropomorph		black			male
71	generalised anthropomorph		black			male
72	generalised anthropomorph		black			male
73	generalised anthropomorph		black			male
74	linear non-figurative		black			
75	linear non-figurative		black			
76	indeterminate		black			
77	generalised anthropomorph		black			male
78	indeterminate		black			
79	linear non-figurative		red			
80	radiating lines		red			
81	generalised anthropomorph		black			
82	generalised anthropomorph		black			
83	generalised anthropomorph		black			
84	indeterminate		black			
85	linear non-figurative		black			
86	generalised anthropomorph		black			
87	linear non-figurative		black			
88	generalised anthropomorph		black			male
89	linear non-figurative		black			
90	indeterminate		black			
91	linear non-figurative		black			

Appendix 1. *Continued*

Picture #	Motif type	Engraving	Drawing or painting colour	Maximum length (cm)	¹⁴C date (BP)	Comments
92	indeterminate		black			

3

An early-Holocene Aboriginal coastal landscape at Cape Duquesne, southwest Victoria, Australia

Thomas Richards

School of Geography and Environmental Science, Monash University, Clayton, Victoria
thomas.richards@monash.edu

Introduction

Peter Kershaw has contributed substantially to the understanding of palaeoenviromental change in Australia, particularly in relation to the timing of Aboriginal colonisation and anthropomorphic alterations of vegetation communities. More recently, Kershaw and colleagues have studied the palaeoenvironment of southwestern Victorian landscapes, with emphasis on the palaeoecology of lakes and swamps, especially in regard to the appearance of Aboriginal water management and fish-trapping systems on the Mt Eccles lava flow and the relationship of these systems to socioeconomic complexity of Aboriginal groups in the southwest (e.g. Kershaw 2004; Tibby et al. 2006; Builth et al. 2008; Kershaw and Lewis 2011). This chapter addresses similar issues of Aboriginal social complexity in southwest Victoria, but examines them from a nearby coastal landscape perspective.

There has been controversy regarding the complexity of Aboriginal societies in southwest Victoria since the 1880s, with disagreement focused on the nature of leadership in the ethnographic period (e.g. Dawson 1881, 1887; Curr 1886; Howitt 1887, 1904; Corris 1968; Lourandos 1977, 1980a, b, 1983, 1984, 1987, 1997; Barwick 1984; Williams 1985, 1987; Critchett 1990, 1998; Edwards 1987; Hiatt 1996; Keen 2006; Hayden 2006) and expanding a century later into debates regarding Lourandos's socioeconomic intensification modelling for the late-Holocene (e.g. Lourandos 1980a, 1983, 1985a, b, 1988, 1993, 1996, 1997; Beaton 1983, 1985; McBryde 1984; Williams 1985, 1987, 1988; Godfrey 1989; Bird and Frankel 1991a,b, 1998, 2005; Lourandos and Ross 1994; Bird et al. 1998, 1999; McNiven et al. 1999; David et al. 2006; Keen 2006; Hiscock 2008). 'Complex', 'affluent', or 'transegalitarian' foragers or hunter-gatherers are societies thought to exhibit characteristics of cultural and social complexity that contrast with an idealised view of egalitarian, highly mobile hunter-gatherers (Koyama and

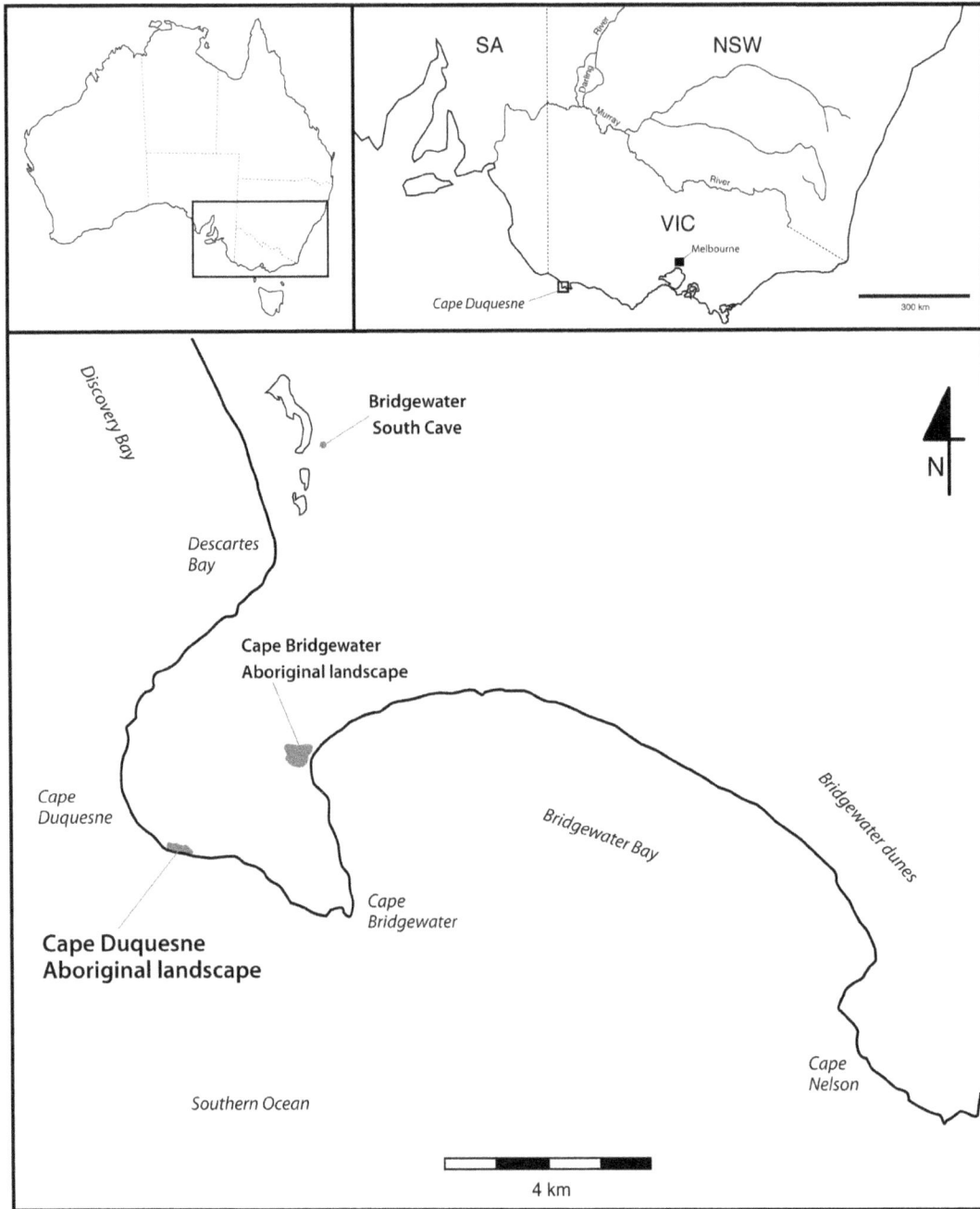

Figure 1. Location of Cape Duquesne Aboriginal landscape in southwest Victoria, Australia.

Thomas 1981; Price and Brown 1985; Hayden 1995; Grier et al. 2006):

> Transegalitarian societies are societies that are neither egalitarian nor politically stratified; they are thus intermediate between generalized hunter-gatherers and chiefdoms in terms of the social and economic inequalities that characterize them. (Owens and Hayden 1997:121)

While much of the debate regarding transegalitarian features in the archaeological record of southwest Victoria has revolved around Aboriginal water control and eel management infrastructure, and earth mounds, Aboriginal marine shell middens are a major source of contention in the region, with Lourandos (1983, 1993, 1997:224-227; Lourandos and Ross 1994:58-59) documenting the increasing use and establishment of middens from ca. 3500 years ago as evidence in support of his intensification model, although this is disputed by some other researchers (e.g. Godfrey 1989; Bird and Frankel 1991a, b; Hiscock 2008:190-191). Missing from these discussions have been comparisons between the structure and contents of late-Holocene middens and earlier middens, due to the absence of data from formally excavated early-Holocene middens in the region.

Excavations on an Aboriginal landscape at Cape Duquesne provide crucial evidence for early-Holocene coastal occupation in this region and this paper will document the chronology, contents and structure of several excavated middens from this period and characterise midden deposition rates and littoral resource exploitation patterns (Figure 1). This baseline data set for the early-Holocene will be compared with data from the late-Holocene as an additional means of evaluating possible late-Holocene changes in coastal resource use related to or reflecting increased complexity in regional societies.

Before the Cape Duquesne Aboriginal landscape data are presented, the regional archaeological context and local environmental setting are briefly reviewed.

Previous archaeology

Over the past four decades, numerous researchers have focused on the southwest Victoria and adjacent southeast South Australia coastal region (e.g. Lourandos 1976, 1980a, 1983, 1997; Witter 1977; Clark 1979; Godfrey 1980, 1984, 1989, 1994, 1996, 2000; Godwin 1980; Wesson and Clark 1980; Simmons and Djekic 1981; Head 1985; Frankel 1986, 1991; Cann et al. 1991; Webb 1995; Richards and Jordan 1996; Everett 1998; Schell 2000a, b; Bird and Frankel 2001; Debney and Cekalovik 2001; Richards and Johnston 2004; Richards and Webber 2004). Yet excavation data for the early-Holocene is limited to two well-researched sites – Bridgewater South Cave and Koongine Cave.

Bridgewater South Cave, only 8 km north of Cape Duquesne, was excavated by Lourandos in the mid 1970s (1976, 1980a, 1983, 1997). The site consists of superficially disturbed stratified deposits both inside and in front of a medium-sized limestone rockshelter that had excellent preservation of organic material (see cover photograph of this volume). Of interest here is the late-Pleistocene-early-Holocene occupation evidence within stratigraphic Phase A, dating between ca. 13,250 cal BP and ca. 9350 cal BP (Table 1) (Lourandos 1983:83; Head 1985:5):

> During this phase there is an over-riding emphasis on land mammals that consisted of a substantial proportion of macropods (including the grey kangaroo) and wombat. There was correspondingly very little representation of marine foods. Apart from scattered pieces of shell of both sandy beach and ocean rock platform species there was evidence of one seal and one fish. Flaked stone was in low frequency... (Lourandos 1980a:348)

Lourandos (1980a:349-350) characterises occupation during Phase A as likely to have occurred during autumn-winter, with the evidence suggesting '...an ephemeral use of the site as a hunting bivouac...' (Lourandos 1997:201-202).

Although currently 1.5 km from the sea in a straight line, Bridgewater South Cave would have been ca. 3.75 km from the Discovery Bay coastline at 11,000 cal BP, ca. 3.0 km at 10,000 cal BP and only ca. 2.25 km by 9000 cal BP, thus within easy reach of the shoreline; however, it is apparent that decisions were made not to exploit coastal resources from this camp, with the major exception of a high-value resource such as seals.

In several ways, Koongine Cave is a twin of Bridgewater South Cave, located at the opposite end of Discovery Bay, some 85 km to the northwest (Bird and Frankel 2001:74). It is also a

Table 1. Calibrated radiocarbon age determinations for southwest Victorian and southeast South Australian late-Pleistocene and early-Holocene Aboriginal coastal sites discussed in paper (other than Cape Duquesne).

Site	Area	Material	Lab code	Dating method	[14]C age (years BP)	Calibrated age BP 95.4% prob.	Cal BP med. prob.	Reference
Bridgewater South Cave	Pit C	charcoal	Beta-3923	Conventional	11390 ± 310	12644-13845	13250	Lourandos 1983
Bridgewater South Cave	Pit B	charcoal	SUA-2175	Conventional	10900 ± 90	12600-13064	12800	Head 1985
Bridgewater South Cave	Pit C	charcoal	Beta-8465	Conventional	10760 ± 10	12575-12737	12650	Head 1985
Bridgewater South Cave	Pit I	charcoal	Beta-8464	Conventional	8350 ± 130	9027-9536	9350	Head 1985
East Monbong	Site 4, Midden 1	marine shell	Wk-1105	Conventional	8410 ± 90	8715-9273	9000	Godfrey 1989
East Swan Lake	Site 2, Midden 1	marine shell	LTU-18	First Order	9450 ± 1250	7509-13304	10400	Frankel 1991
Koongine	J9/59	charcoal	Beta-14862	Conventional	9590 ± 140	10523-11243	10900	Bird and Frankel 2001
Koongine	J9/44	charcoal	Beta-14861	Conventional	9710 ±180	10560-11708	11050	Bird and Frankel 2001
Koongine	J9/26	charcoal	Beta-15996	Conventional	9240 ± 100	10229-10664	10400	Bird and Frankel 2001
Koongine	J9/20	charcoal	Beta-21541	Conventional	8900 ± 110	9631-10240	10000	Bird and Frankel 2001
Koongine	J9/14	charcoal	Beta-14859	Conventional	8270 ± 400	8345-10224	9200	Bird and Frankel 2001
Noble's Rocks East	Site 1, Midden 1	charcoal	Wk-410	Conventional	8340 ± 110	9033-9528	9350	Godfrey 1989
Noble's Rocks East	Site 1, Midden 1	marine shell	Wk-605	Conventional	8840 ± 80	9318-9705	9500	Godfrey 1989
Noble's Rocks Survey Area	Site 45, Midden 1	marine shell	Wk-1262	Conventional	8940 ± 70	9465-9835	9600	Godfrey 1989
Sutton's Rocks	Site 13, Midden 2	marine shell	Wk-1263	Conventional	8680 ±60	9194-9482	9350	Godfrey 1989

Radiocarbon determinations used the Libby ^{14}C half life (5568 years).

Marine shell determinations were calibrated with the Marine09 radiocarbon age calibration curve; no Delta R correction was employed due to the age of the samples (Reimer et al. 2009).

Charcoal determinations were calibrated using the IntCal09 radiocarbon age calibration curve as some of the samples were too old for SHcal04 (Reimer et al. 2009; McCormac et al. 2004).

All calibrations were undertaken with the CALIB Radiocarbon Calibration Program rev. 6.0.1. (Stuiver and Reimer 1993; Stuiver et al. 2011).

Median probability of calibrated dates is rounded to the nearest 50 years.

substantial limestone cave set in a scarp on the edge of the coastal plain, containing significant early-Holocene occupation deposits, and has a well-preserved faunal assemblage.

Both the early and middle phases of occupation (ca. 11,000-9000 cal BP) are interpreted as representing repeated camping events during which a range of medium and small land mammals, including possums, bandicoots, potoroo, wallabies and wombats, and large mammals such as grey kangaroo, were hunted (Bird and Frankel 2001:71). In this sense, the period of occupation and the faunal assemblage is remarkably similar to that documented at Bridgewater South Cave (Lourandos 1980a: Table 13:2, 1983:83; Head 1985:5). The middle phase is essentially a less intensive version of the early phase, characterised by shorter and less frequent occupation episodes (Bird and Frankel 2001:74).

Although the shoreline would have been 10-15 km distant during the early and middle phases, a small amount of marine mollusc shell was recovered from Koongine Cave (Bird and Frankel 2001:73), as at Bridgewater South Cave. Finally, it is inferred from the presence of emu eggshell that occupation occurred in winter, although additional seasonal usage is possible (Bird and Frankel 2001:74) – yet another similarity to Bridgewater South Cave (Lourandos 1980a:352).

For Discovery Bay and Cape Bridgewater coastal sites, Godfrey (1989, 1994) and Frankel (1991) have used spot samples of marine shell from deflated middens for radiocarbon dating, producing several early-Holocene age determinations (Table 1). Unfortunately, this type of uncontrolled sampling does not provide a reliable basis for characterising the composition of faunal assemblages, as surface shell species proportions at Cape Duquesne are typically not representative of nearby deposits as revealed through excavation (see surface vs. excavated shell species proportions for shell midden investigations reported in this chapter). The itinerant dating approach of Godfrey and Frankel tells us only that early-Holocene Aboriginal coastal occupation of this region probably did occur, but little about the nature of this occupation.

Landscape description and environment

The study area is located in a very exposed position along the top of steep cliffs at Cape Duquesne, a headland at the southwestern tip of Cape Bridgewater on the Portland Peninsula (Figures 1, 2). This is a high-energy coast dominated by swell waves and winds from the west through south-southwest (Short 1988:125; Buckley 1992:13).

Geomorphology, geology and soils
The Portland Peninsula is a large promontory jutting southward into the Southern Ocean, tipped by a sequence of protruding headlands and indented bays. Cape Duquesne is the westernmost of the headlands and is bounded by the extensive northwest-southeast trending Discovery Bay to the west, and the much smaller, protected Bridgewater Bay to the east. The juncture of Cape Duquesne with Discovery Bay is known as Descartes Bay.

Coastal geomorphology is dominated by three major geological formations exposed in cliff faces and on the surface – Plio-Pleistocene basalts and tuffs of the Newer Volcanic Formation, overlain by Pleistocene beach and dune calcarenites of the Bridgewater Formation and capped by terminal Pleistocene-Holocene dune sands (Boutakoff 1963; Bird 1993:24) (Figures 2-4). Much of this sequence is exposed in 25-120 m high vertical cliffs extending from Descartes Bay in the northwest, around the Cape Duquesne and Cape Bridgewater headlands to the south, and east and northeast into Bridgewater Bay. The present shoreline is mainly sheer basalt cliff, with the occasional small sandy beach (e.g. White's Beach to the northwest), narrow shingle, boulder talus or sea caves (e.g. Seal Cave to the east). There are also frequent off-shore intertidal basalt platforms.

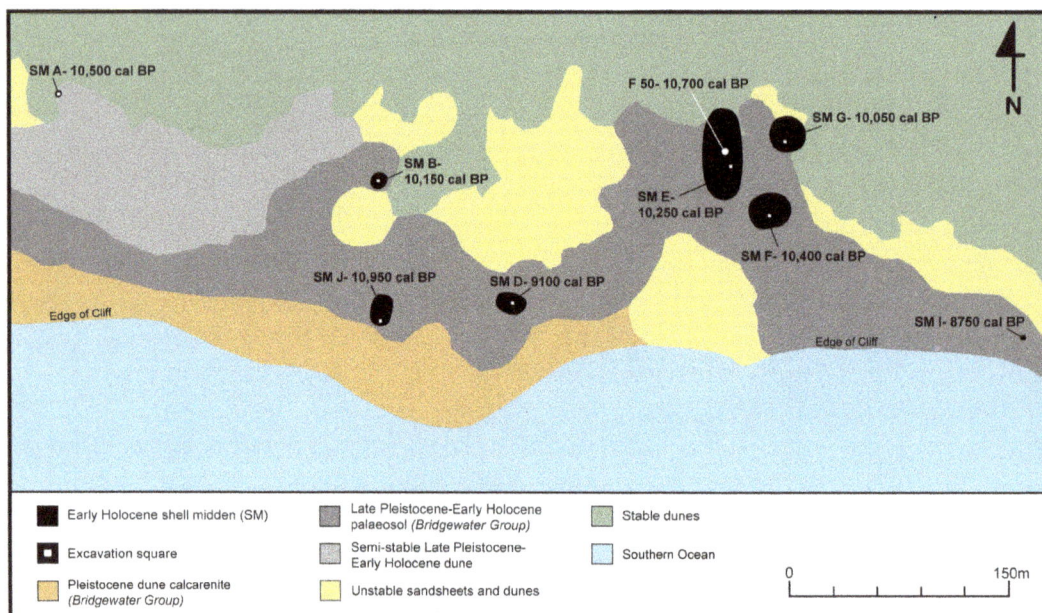

Figure 2. Exposed strata on Cape Duquesne Aboriginal landscape, showing the earliest calibrated radiocarbon age determination (median probability) for each dated midden and feature and the location of excavation squares.

Boutakoff (1963:48-51) defined the Bridgewater Formation, which consisted of a series of lithified calcareous sand dunes ('limestone dunes'), formed on sand derived from weathering of Tertiary dunes exposed by regressing seas during Pleistocene glacial periods, and also included palaeosols capping the calcarenite dunes. The latter, which were classed as rendzina, *terra rossa* and laterite fossil soils, were considered to have formed following calcification of the underlying dunes during interstadial/interglacial pluvial conditions (Boutakoff 1963:49). They were described as calcareous, sandy and red, reddish-brown or reddish-pink soils (Boutakoff 1963:51). Subsequent research in the region confirmed Boutakoff's sequence (e.g. Kenley 1976; Douglas 1979; Land Conservation Council 1981), until Cupper et al. (2003:343-344) upgraded the Bridgewater Formation to the Bridgewater Group and expanded its distribution more broadly across southwestern Victoria.

Shortly after Boutakoff's (1963) geological study appeared, Gibbons and Downes (1964) published a study of soils along the southwest coast of Victoria. They described a *terra rossa* soil, the Nelson Sandy Loam, and a rendzina soil, the Bridgewater Sandy Loam, both with parent materials attributed to Pleistocene dune limestone or aeolianite of the Bridgewater Formation (Gibbons and Downes 1964: Appendix 1). Nelson Sandy Loam is a dark reddish-brown (2.5YR 2/4-3/4 wet) sandy loam, with a weak fine sub-angular blocky and fine crumb structure, loose to very friable, very porous and with a pH of 6.6 to 7.8 (Gibbons and Downes 1964: Appendix 1). Bridgewater Sandy Loam is a very dark brown (5 to 7.5YR 2/2) sandy loam, with a moderate medium crumb structure to sub-angular blocky structure, friable, and with a pH of 8.5 to 8.9 (Gibbons and Downes 1964:Appendix 1). They described one further soil of relevance, Discovery Bay Sand, found in high and unstable sand dunes, as a light yellowish brown (7.5YR to 10YR 7/4 dry, 6/4 moist) coarse sand composed mostly of calcium carbonate (finely broken seashells), with massive structure, loose, no organic matter, and a pH of 8.9 to 9.4 (Gibbons and Downes 1964:Appendix 1).

The Land Conservation Council (1981:14-15) described soils developed on dune limestone in the region as 'Red-black, uniform-gradational, sandy loams (terra rossa) ... generally shallow, with a crumb structure, friable throughout, and ... little horizon differentiation'. The Land Conservation Council (1981:14) study also describes 'Undifferentiated Calcareous Sands' generally found in extensive unconsolidated to mobile dunes, formed on sands moved onshore

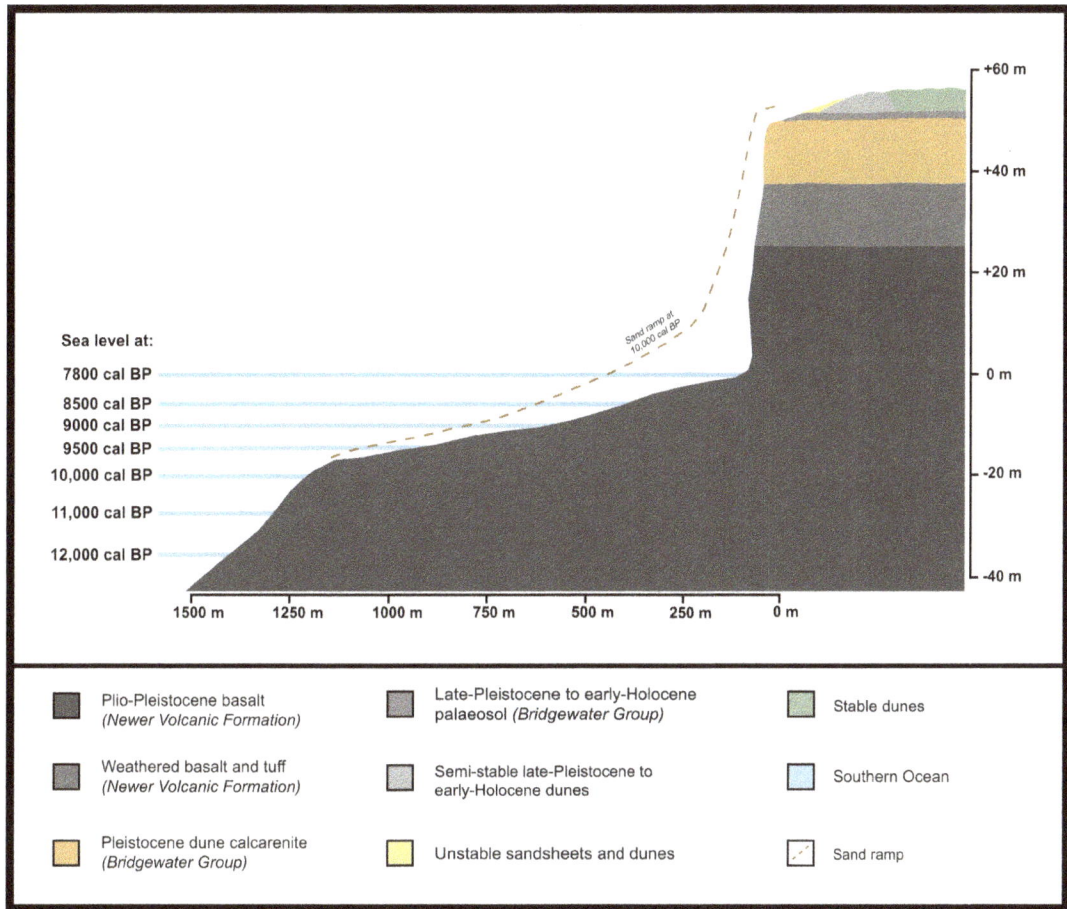

Figure 3. Cross-section through Cape Duquesne Aboriginal landscape and sea bed, showing sea levels (Sloss et al. 2007), geology (Boutakoff 1963) and the early-Holocene sand ramp (Short 1988). Vertical exaggeration 12.5 x.

from the continental shelf and reworked by wind action. These sands were characterised as yellow-brown, with little clay or organics, loose and structureless and extremely alkaline (Land Conservation Council 1981:14).

The Coastal Management and Co-ordination Committee report (1981:29) described the Holocene Bridgewater Bay mobile to stabilised dunes as overlying Pleistocene calcarenites; they comprise friable, uniform calcareous sand with little or no soil formation and a pH of up to 9.5.

More recently, Buckley (1992) studied coastal beach and dune sediments from southwest Victoria in detail. He mentioned unconsolidated Holocene dune sands overlying Pleistocene calcarenite, notably the stranded clifftop dunes at Cape Bridgewater, and his analysis of Discovery Bay beach sands is very relevant to understanding of the origin of the clifftop sediments (Buckley 1992:13-20). He concluded that the majority of sand found on the Discovery Bay beaches was derived from calcareous continental shelf deposits moved landward during Pleistocene glacial phases (Buckley 1992:20). The sands were described as grey-brown, well sorted (0.44 phi), fine-medium calcareous sand, with a mean grain size of 0.26 mm and a carbonate content of 75% (Buckley 1992:15-20).

In the recent synthetic overview by Cupper et al. (2003:343), the Bridgewater Group included the palaeosol horizons derived from pedogenesis and weathering of aeolian calcarenite surfaces first identified by Boutakoff. Overlying the Bridgewater Group are the Holocene Discovery Bay and Bridgewater Bay Sands, a composite of earlier work (e.g. Gibbons and Downes 1964; Kenley 1976; Land Conservation Council 1981). These sands were found in unconsolidated calcareous sandy beach, foredune and dune complexes, and were summarised as fine-grained, cream to white calcareous sand (Cupper et al. 2003:344).

Rosengren (2001a, b) carried out an applied geological study of Cape Bridgewater and Cape Duquesne for a wind-farm development. He reviewed the previously reported geology, and reconfirmed Boutakoff's Pleistocene dune limestone Bridgewater Formation. He attributed the limestone calcrete to Oxygen Isotope Stages (OIS) 5 to 2, and noted that dunes formed during OIS 2 to 1 had much slower rates of cementation than in previous Stages (Rosengren 2001a:213-215). Reddish-brown silty and clayey sands overlying calcrete, characterised as *terra rossa* and rendzina palaeosols, were again observed, as were poorly consolidated calcareous sands and clifftop dunes (Rosengren 2001a:213-215). The calcareous sands, of estimated OIS 2 to 1 age, were described as poorly differentiated to gradational, yellowish-brown sands, with only minor surface organic matter accumulation (Rosengren 2001a:214-215).

Palaeoenvironment

According to Kershaw et al. (2004:158), a decline in effective moisture in the terminal Pleistocene, largely due to rising temperatures, culminated in a period of maximum aridity from ca. 17,000-14,000 cal BP, during which woody plants were uncommon. In the succeeding period, ca. 14,000-11,500 cal BP, increases in both temperature and rainfall resulted in an expansion of the distribution of trees, accompanied by a change from steppe grassland to grassland (Kershaw et al. 2004:158). By the beginning of the Holocene, 11,500 cal BP, tree cover had reached early 19th century levels (pre-European clearing) (Kershaw et al. 2004:158). Vegetation community composition continued to change, with a sustained increase in *Eucalyptus* relative to *Casuarinacea* trees around 8900-7800 cal BP, so that essentially the vegetation cover present in the early 19th century, dry sclerophyll forest/woodland, was established by the end of this period (Kershaw et al. 2004:139,158-159).

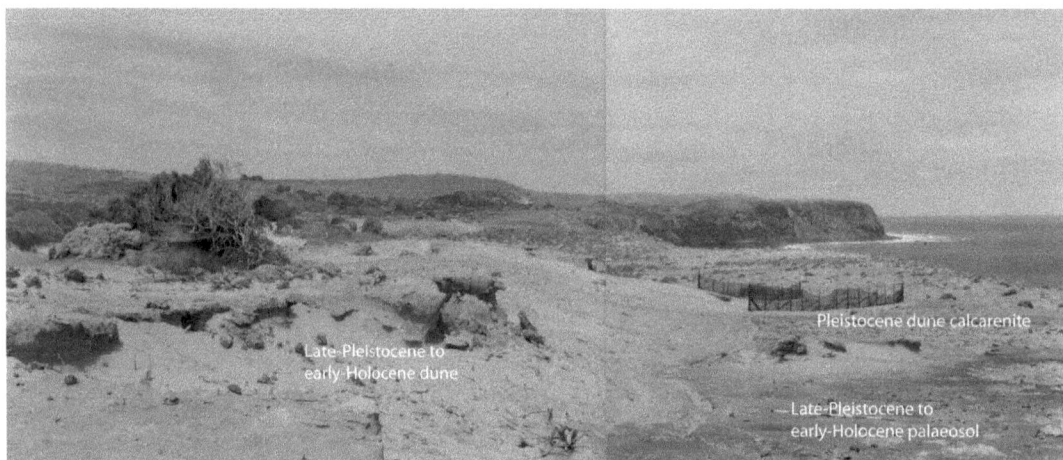

Figure 4. View east from western end of landscape, showing surface exposure of strata, including the actively eroding dune formed in the late-Pleistocene and early-Holocene, the exposed late-Pleistocene early-Holocene palaeosol and Pleistocene dune calcarenite. A remnant sediment pedestal can be seen on the left top of the dune and numerous carbonate root casts are exposed all over the eroding face of the dune.

Cape Duquesne investigations

Investigations at Cape Duquesne (AAV 7121/233) have revealed a series of open-air Aboriginal shell midden deposits at the top of a 50 m high cliff, with the ocean currently washing against its base (Figures 2-4). As noted in the previous section, vast areas of carbonate sands were exposed on the continental shelf along southern Australia during the late Pleistocene and much of this sand was mobilised as the high-energy sea transgressed and strong southwesterly winds prevailed (Short 1988:121). These sands, mainly locally sourced from the exposed sea bed, piled up against obstructions such as the cliffs at Cape Duquesne, and occasionally overtopped these cliffs (Short 1988:138). During the terminal Pleistocene through early-Holocene, the resulting sand ramp would have extended from the exposed sea bed near the shoreline up and over the top of the Cape Duquesne cliffs, providing easy human access between the two locations. This explains the presence of marine shell middens at the top of the seemingly impassable barrier of the cliffs (Figure 3, Table 2). Later, the transgressing sea would have directly eroded the base of the ramp, removing it entirely as the current sea level was attained (and then exceeded), leaving the stranded clifftop dunes as the last remnant of the former ramp (Short 1988:138-139; Rosengren 2001a:215).

Table 2. Distance to shoreline from Cape Duquesne early-Holocene Aboriginal landscape (sea level curve of Sloss et al. 2007).

Years cal BP	Sea level (m) (relative to present)	C. Duquesne occupation	Horizontal distance from C. Duq. to shore (m)	Vertical distance from C. Duq. to shore (m)
12,000	-35		1384	90
11,500	-31	pre-SM A	1312	86
11,000	-27		1275	82
10,500	-23	F 50, SM J	1236	78
10,000	-19	SM A, SM B, SM E, SM F, SM G	1184	74
9500	-14	SM A, SM G	888	69
9000	-10	SM D	632	66
8500	-6	SM I	400	61
8000	-3		240	58
7800	0		50	55

Wind erosion has significantly shaped the Cape Duquesne landscape, so that the present surface is a patchwork of differentially exposed Pleistocene and Holocene landforms and sediments (Figures 2, 3). There are extensive exposures of lithified dune deposits (calcarenite) along the southern edge of the landscape (the clifftop) in the centre and west of the landscape, where wind action has scoured away all overlying sediment. These are calcarenites of the Pleistocene Bridgewater Group.

Running approximately east-west across the landscape is an exposed palaeosol, comprising greyish sandy loam, most similar to the Bridgewater Sandy Loam, a rendzina soil (Gibbons and Downes 1964: Appendix 1). The sandy loam directly overlies the calcarenite, varying from mere remnant patches a few centimetres thick filling in hollows on uneven, calcarenite surfaces, to deposits more than 1 m thick. Such palaeosols are commonly associated with calcarenites of the Bridgewater Group and are considered part of this unit. In the present erosional situation, the palaeosols are again active soils, although generally very poorly vegetated.

Dominating the western end of the landscape is a large dune, for the most part actively eroding with little vegetation cover, but which is vegetated and stable to the north (Figure 4). Remnant sediment pedestals and exposed carbonate root casts indicate that more than 1 m of sand has been removed from the southern surface of this dune by wind action. This dune is a remnant of the clifftop dunes deposited since the Last Glacial Maximum (LGM), and much of the present eroding greyish-brown sandy surface is at least early-Holocene in age. It overlies both the sandy loam palaeosol and the calcarenites of the Bridgewater Group (Figures 2-4).

The vegetated area 50-150 m to the north of the cliff edge all along the landscape comprises the remaining stabilised clifftop dune field (Figure 2). Poorly vegetated sandsheets are present south of the dunefield in patches on the landscape, representing destabilised clifftop dunes reworked by wind action during the Holocene (Figure 2).

Archaeological survey and the surface record

The area subject to detailed systematic archaeological survey, much of it deflated to some extent, extended 1 km along the clifftops and up to 400 m inland. The goal of the survey was to identify the extent of the Aboriginal landscape, focusing on the distribution of flaked and ground stone artefacts, shell middens, hearth features and other evidence of human occupation. The initial pedestrian survey identified the gross extent of the distribution of Aboriginal cultural material exposed on the surface – 58 hearth features, eight shell middens and thousands of flaked stone artefacts were found over an area of approximately 500 m east-west and 125 m north-south (62,500 m²). The landscape was mapped, hearth features and shell middens were recorded in detail, three of the exposed hearth features were excavated and eight shell midden deposits were tested. None of the great numbers of lithic artefacts observed on the surface were in situ, so they were generally not mapped individually, although samples of stone artefacts, animal bones, marine shell, hearth stones, charcoal and sediments were mapped and collected for identification and analysis.

In situ shell midden deposits are only located on early-Holocene and late-Pleistocene features, namely clifftop dunes and sandy loam palaeosols capping calcarenite, while stone artefacts and deflated hearth features are distributed more widely across the landscape, suggesting they mostly derive from more recent deposits that have largely eroded away. Technology of flaked stone artefacts and radiocarbon age determinations on two remnant hearth features indicate a substantial late-Holocene occupation across this landscape, but the focus of the present study is on the early-Holocene shell middens and hearth features.

Surface exposures of middens displayed obvious loose shells that were not in their primary context, but other shells appeared to be eroding out of the surface, due to wind removal of surrounding soil particles, and seemed to be in situ. Shell Midden A (SM A) was excavated first to determine whether in situ deposits were indeed present, to identify the subsurface structure and characteristics of the midden and to obtain samples for identification, analysis and chronometric dating.

Excavation of Shell Midden A

SM A is located at the northwestern corner of the landscape (Figure 2). It manifested on the surface as a dense concentration of *Turbo undulata* and *Cellana tramoserica* shells and a small amount of charcoal exposed by wind erosion (Figure 5). Some of the shells were loose, but what appeared to be an in situ shell deposit was eroding out of sediments below the loose shell and recently blown-in sand. The area of midden exposed on the ground surface measured approximately 3 m in diameter; however, it extended under partially consolidated dune deposits to the northwest and is probably many times larger in area than the exposed, eroding portion.

Methods

Square 1, measuring 1 m x 1 m was staked out over the area, with the highest concentration of eroding midden material on the surface (Figure 5). All sediment was excavated with trowels and dry sieved through 3 mm mesh screens in the field (with the exception of bulk soil samples); all sieve residue was retained for cataloguing and analysis. The first step in the investigation involved the sweeping up and collecting of all loose material on the surface of the square. Although sorted and catalogued, this material has been excluded from the present analysis and discussion because it was not *in situ*. Two excavation units (XUs), each 5 cm deep, were dug with trowels across Square A. A third XU, also 5 cm deep, was dug in the northeast quadrant of the square.

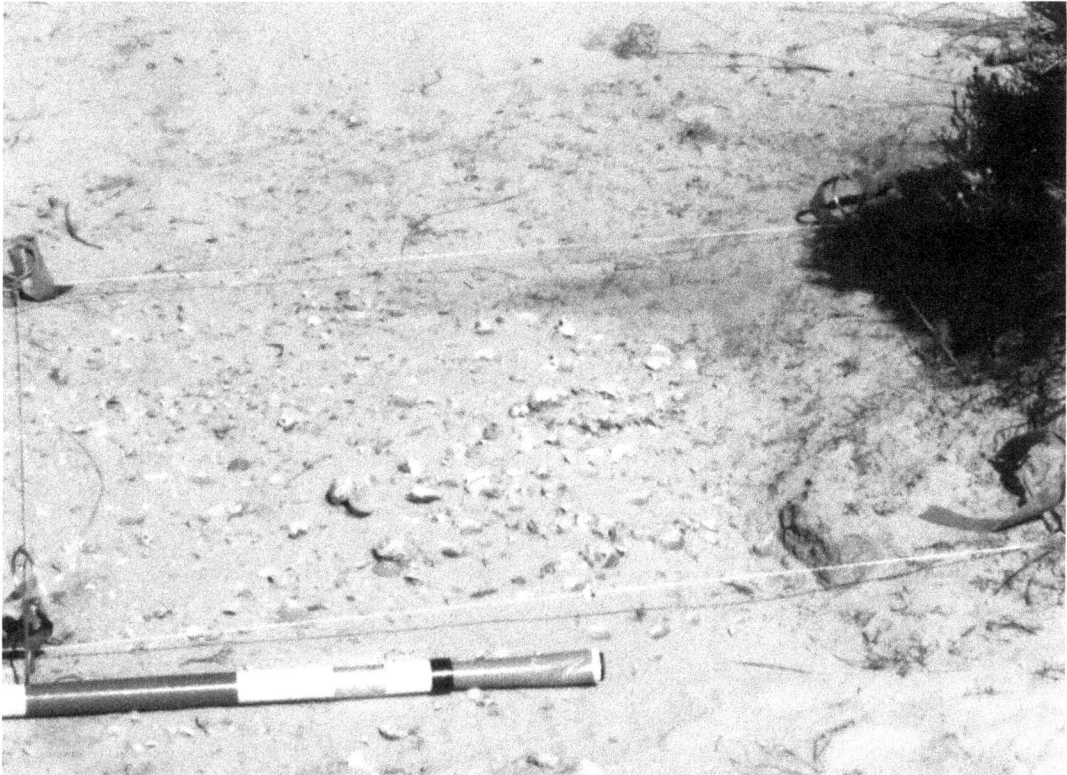

Figure 5. Shell Midden A, Square 1, pre-excavation view showing dense midden material on surface. Ranging pole bars are 20 cm long.

Stratigraphy

Excavation revealed a very dense deposit of whole and fragmented marine shellfish shell that extended across the square, with associated carbonate-coated chunks of charcoal and flaked stone, to an average depth of 4 cm below surface. Small amounts of shell and charcoal continued for another 2 cm in the unconsolidated sand (Figure 6). Below the base of the midden, a patch of burnt sediment and charcoal with an associated large flat rock (manuport) was uncovered in the northeast of the square and extended to 9 cm below surface. This feature is stratigraphically earlier than SM A.

Sediment in the midden deposit was an unconsolidated, dark greyish-brown (10YR4/2 wet) to light-grey (10YR7/2 dry) calcareous sand containing carbonate clods and sediment aggregates, as well as land snails and rootlets providing a non-cultural minor organic component (Johnston 1996). Soil pH values varied from 8 (field), a highly favourable environment for the preservation of bone, to 10 (lab), which is less favourable (Reitz and Wing 2008:140-141). Detailed analysis revealed that the sand was medium to fine grained (Wentworth 1933), and composed of 45% quartzite, 40% carbonate and 15% quartz. Negative skewness in the particle size distribution, grain shape and surface texture, as well as the moderate to well-sorted nature of the deposit, indicated that this was an aeolian deposit originating from beach sands (Johnston 1996). Below the midden, the sediment appeared virtually identical in the field, although slightly darker (10YR3/2).

Finally, the fact that the loose surface shell appeared to be present only in a small area directly over in situ deposits suggests that the top of the midden was only exposed a short time before the investigations occurred and that not much had been lost to wind action (i.e. no downwind surface trail of smaller items was present).

Marine shellfish

The Minimum Number of Individuals (MNI) identified to species in SM A comprised 650

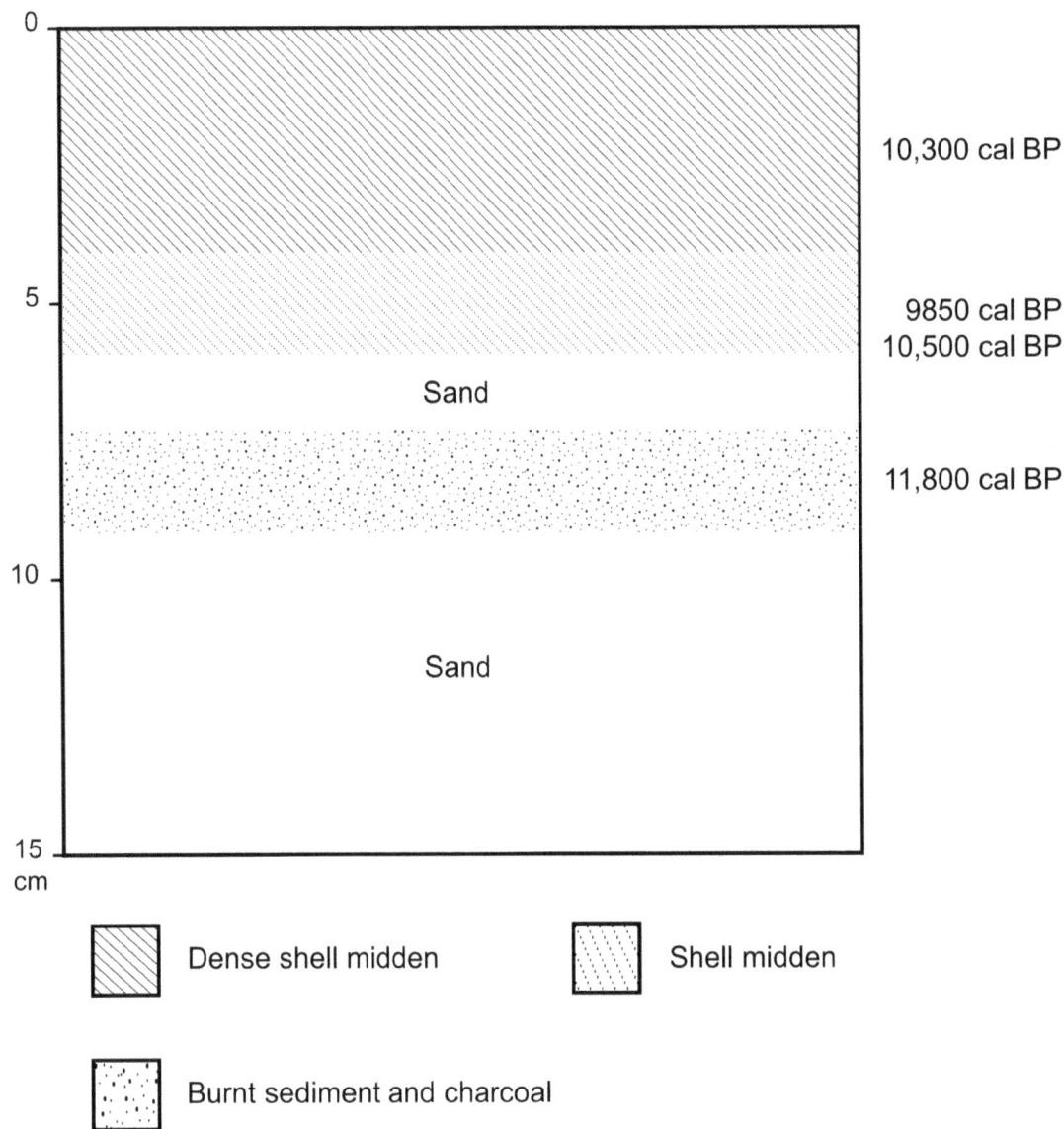

Figure 6. Shell Midden A northeast corner stratigraphy.

individuals from eight taxa of marine mollusc identified in more than 7.5 kg of shell recovered (Tables 3-5). The most common shellfish is *Turbo undulata*, at just under 70% of the total MNI, with *Cellana tramoserica* contributing 18%, *Austrocochlea concemerata* 9%, and five other taxa each representing 1% or less (Table 5).

Radiocarbon dating

Four radiocarbon age determinations are pertinent to the dating of SM A, three of them on *Turbo undulata* shell from the midden itself, and one on wood charcoal located below the midden. Initially, one shell sample from XU 1 yielded an age determination of ca. 10,300 cal BP (Beta-93569) and another from XU 2 dated to ca. 9850 cal BP (Beta-96584) (Table 6). The determination for XU 1 was slightly older than that from XU 2, although they overlap at two standard deviations. A second shell determination was run for XU 2, ca. 10,500 cal BP (UB-4369), with the resulting date older than both of the previous determinations and in sequence (Figure 7).

A further determination on wood charcoal from the charcoal-rich layer underlying the midden deposit yielded a terminal-Pleistocene radiocarbon age determination of ca. 11,800 cal BP (UB-4370), indicating a distinct occupation horizon pre-dating midden formation (Table 6).

Other excavations and investigations

Following excavation of SM A, with its evident in situ midden deposits, test pits were excavated at six other surface shell scatters (see Table 7). The purpose of these small-scale excavations was to evaluate whether intact midden deposits were also present, to characterise the nature of any midden layers and to obtain samples of shell and other materials for identification and dating. Similar excavation and recording methods to those employed at SM A were used on these smaller test pits.

Shell Midden B

SM B is located in a flat, nearly vegetation-free expanse of dark grey, very consolidated sandy loam, located 140 m to the east of Shell Midden A (Figure 2). *Turbo undulata* shell and a flint flake appeared to be eroding out of sandy loam palaeosol in an area measuring 7.0 m east-west by 6.1 m north-south.

SM B was tested with a single 25 cm square placed in the approximate centre of the exposed midden in an area of high surface shell density. Almost all shell was concentrated in the uppermost 3 cm of the first XU, although a small amount extended to a maximum depth of 7 cm in the second XU.

Two species were identified in the 70 g of shell recovered (Tables 3, 4). *Turbo undulata*

Table 3. Finds by weight (g) for Cape Duquesne investigations.

Area	Excavation Unit	Flaked stone	Shell	Bone	Charcoal	Ochre
SM A	surface		816.49			
SM A	1	1.03	7414.81		41.54	
SM A	2		122.83		64.24	
SM A	3		3.24		4.54	
SM B	1		60.30		2.91	
SM B	2		9.57		13.82	
SM D	1		199.28		0.01	
SM E	surface		271.51			
SM E	1		60.42		0.20	
SM E, F 50	surface		65.70			
SM F	1		1030.94	0.09	3.39	
SM F	2		17.68		0.53	
SM G	surface		61.05			
SM G	1		126.49		0.80	
SM G	2		5.30		6.70	
SM G	3		0.04		0.25	
SM I	surface	0.75	74.90			
SM J	1		11.56		0.10	2.15
SM J	2		1.90		0.80	0.10
SM J	3		0.05		0.05	
Sub Total	surface	0.75	1289.65			
Sub Total	excavated	1.03	9064.41	0.09	139.88	2.25
Totals		**1.78**	**10354.06**	**0.09**	**139.88**	**2.25**

Table 4. Shellfish taxa weight (g) distribution by excavated shell midden.

Shell taxa	SM A (g)	%	SM B (g)	%	SM D (g)	%	SM E (g)	%	SM F (g)	%	SM G (g)	%	SM J (g)	%	Total (g)	Total % weight
Turbo undulata	3706.10	72.89	55.81	91.09	2.50	1.28	32.61	64.68	406.80	44.33	123.04	100.00			4326.86	67.13
Cellana tramoserica	1257.69	24.74					2.20	4.36	420.08	45.78					1679.97	26.06
Donax deltoides					193.28	98.72									193.28	3.00
Thais orbita	84.92	1.67							27.30	2.97					112.22	1.74
Austromytilus rostratus			5.46	8.91			0.11	0.22	0.22	0.02			13.37	100.00	19.16	0.30
Polyplacophora	6.77	0.13							63.25	6.89					70.02	1.09
Ostrea angasi							15.50	30.74							15.50	0.24
Austrocochlea concemerata	14.72	0.29													14.72	0.23
Melanerita melanotragus	12.50	0.25													12.50	0.19
Dicathais baileyana	1.01	0.02													1.01	0.02
Cominella lineolata	0.69	0.01													0.69	0.01
Totals	**5084.40**	**100.00**	**61.27**	**100.00**	**195.78**	**100.00**	**50.42**	**100.00**	**917.65**	**100.00**	**123.04**	**100.00**	**13.37**	**100.00**	**6445.93**	**100.00**

Table 5. Shellfish taxa MNI percentages for each excavated shell midden.

Shell taxa	SM A MNI	SM A % MNI	SM B MNI	SM B % MNI	SM D MNI	SM D % MNI	SM E MNI	SM E % MNI	SM F MNI	SM F % MNI	SM G MNI	SM G % MNI	SM J MNI	SM J % MNI	Total MNI	Total % MNI
Turbo undulata	454	69.85	7	53.85	1	5.00	7	70.00	41	29.50	10	100.00		0.00	520	61.39
Cellana tramoserica	120	18.46		0.00		0.00	1	10.00	81	58.27		0.00		0.00	202	23.85
Austrocochlea concemerata	59	9.08		0.00		0.00		0.00		0.00		0.00		0.00	59	6.97
Donax deltoides		0.00		0.00	19	95.00		0.00		0.00		0.00		0.00	19	2.24
Thais orbita	7	1.08		0.00		0.00		0.00	11	7.91		0.00		0.00	18	2.13
Austromytilus rostratus		0.00	6	46.15		0.00	1	10.00	1	0.72		0.00	5	100.00	13	1.53
Polyplacophora	3	0.46		0.00		0.00		0.00	5	3.60		0.00		0.00	8	0.94
Melanerita melanotragus	5	0.77		0.00		0.00		0.00		0.00		0.00		0.00	5	0.59
Ostrea angasi		0.00		0.00		0.00	1	10.00		0.00		0.00		0.00	1	0.12
Dicathais baileyana	1	0.15		0.00		0.00		0.00		0.00		0.00		0.00	1	0.12
Cominella lineolata	1	0.15		0.00		0.00		0.00		0.00		0.00		0.00	1	0.12
Totals	650	100.00	13	100.00	20	100.00	10	100.00	139	100.00	10	100.00	5	100.00	847	100.00

Table 6. Calibrated radiocarbon age determinations from the Cape Duquesne Aboriginal landscape investigations.

Area	Depth (cm)	Material	Lab code	Dating method	δ13C ‰	14C age (years BP)	Calibrated age BP 95.4% prob.	Rel. prob.	Cal BP med. prob.
pre-SM A	7-9	wood charcoal	UB-4370	Conventional	*	10,207 ± 55	11,704-12,108	0.993	11800
							11,649-11,663	0.007	
SM J	0-3	Austromytilus rostratus	Wk-9532	AMS	2.0	9971 ± 58	10,708-11,121	1.000	10950
F 50	0-1	Ostrea angasi	Wk-9563	Conventional	-0.6	9814 ± 71	10,533-10,955	0.973	10700
							10,965-11,009	0.027	
SM A	5-6	Turbo undulata	UB-4369	Conventional	*	9619 ± 52	10,345-10,584	1.000	10500
SM F	5-6	Turbo undulata	Wk-29818	AMS	2.5	9525 ± 33	10,263-10,487	1.000	10400
SM A	0-5	Turbo undulata	Beta-93569	Conventional	1.6	9430 ± 100	10,074-10,533	1.000	10300
SM E	0-5	Turbo, Ostrea	Beta-93568	Conventional	1.0	9390 ± 80	10,069-10,476	0.996	10250
							10,039-10,050	0.004	
SM B	0-3	Turbo undulata	Beta-93567	Conventional	1.2	9300 ± 70	9893-10,288	1.000	10150
SM F	0-5	Cellana tramoserica	Wk-9604	Conventional	1.4	9290 ± 56	9923-10,234	1.000	10150
SM G	0-3	Turbo undulata	Wk-9564	Conventional	1.0	9213 ± 56	9833-10,184	1.000	10050
SM B	5-7	Turbo undulata	Wk-29816	AMS	1.8	9183 ± 31	9868-10,135	1.000	10000
SM A	5-6	Turbo undulata	Beta-96584	Conventional	1.5	9110 ± 70	9622-10,120	1.000	9850
SM G	5-10	Turbo undulata	Wk-29817	AMS	2.4	9068 ± 32	9630-9920	1.000	9800
SM D	0-3	Donax deltoides	Beta-93566	Conventional	0.1	8490 ± 70	8954-9323	1.000	9100
SM I	0-1	Donax deltoides	Wk-9562	Conventional	1.0	8238 ± 57	8593-8957	1.000	8750

Radiocarbon determinations used the Libby 14C half life (5568 years).

Marine shell determinations were calibrated with the Marine09 radiocarbon age calibration curve; no Delta R correction was employed due to the age of the samples (Reimer et al. 2009).

The charcoal determination was calibrated using the IntCal09 radiocarbon age calibration curve as the sample was too old for SHcal04 (Reimer et al. 2009; McCormac et al. 2004).

All calibrations were undertaken with the CALIB Radiocarbon Calibration Program rev. 6.0.1. (Stuiver and Reimer 1993; Stuiver et al. 2011).

* δ 13C ‰ value measured, but not reported.

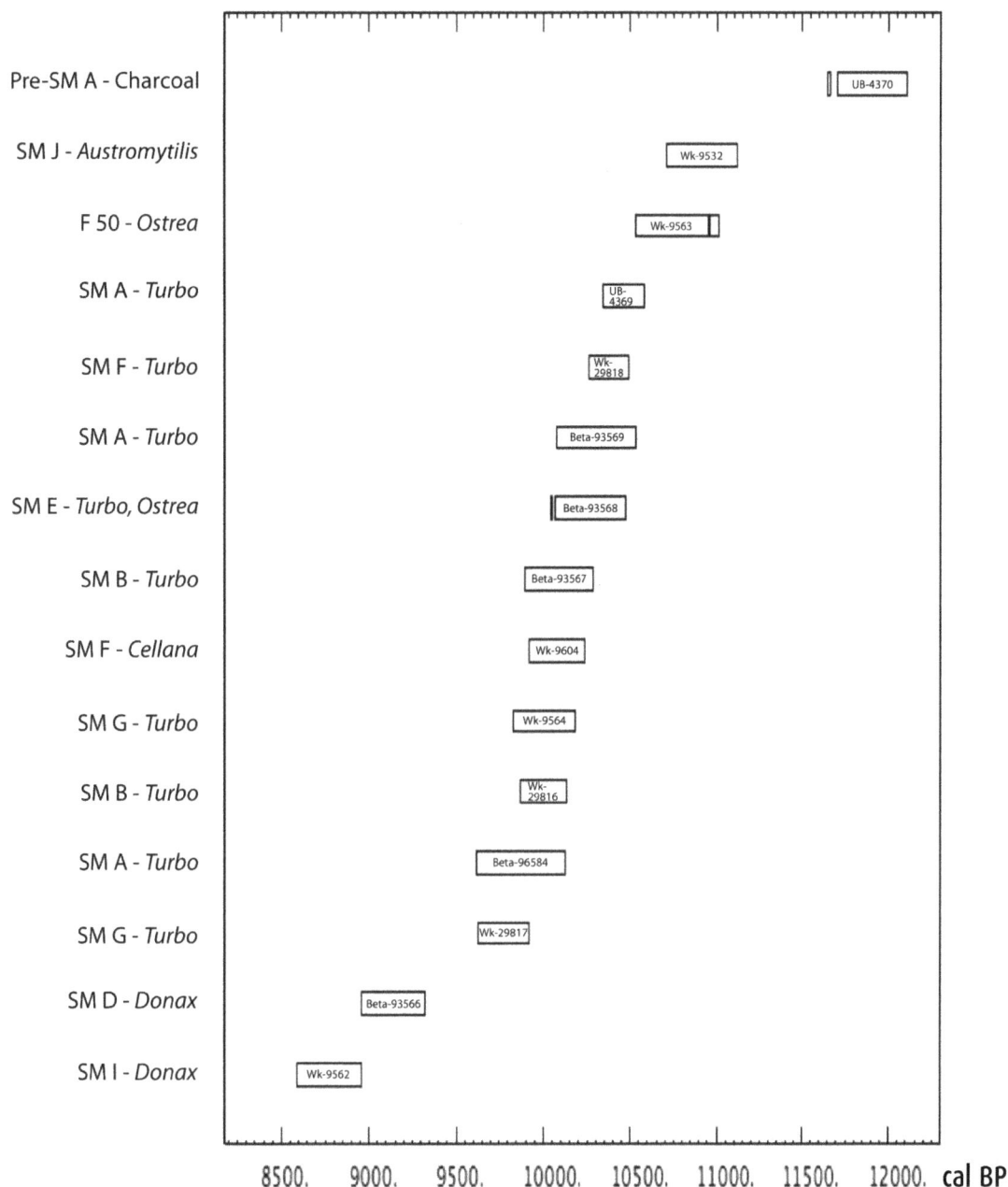

Figure 7. Cape Duquesne Aboriginal landscape radiocarbon age determinations arranged in chronological order (see Table 6 for details of age determinations). Dates are presented as bars spanning the two sigma range.

again dominated the small assemblage, at 54% of the MNI, with another rock-platform species, *Austromytilus rostratus*, contributing 46% (Table 5).

A sample of *Turbo undulata* fragments from XU 1 was radiocarbon dated to ca. 10,150 cal BP (Beta-93567) and another on a *Turbo undulata* shell from XU 2 yielded an AMS determination of ca. 10,000 cal BP (Wk-29816) (Table 6, Figure 7). The samples are statistically the same age at the 95% level (T statistic) (Stuiver et al. 2011).

Shell Midden D

SM D is located in the south-central portion of the landscape, in a flat, sparsely vegetated area of exposed dark grey palaeosol some 50 m from the cliff edge (Figure 2). Flint artefacts and *Donax deltoides*, *Cellana tramoserica*, *Turbo undulata* and Polyplacophora shells were sitting on the surface, with some of the *Donax deltoides* appearing to be in situ and eroding out of the sediment over an expanse measuring 13.6 m east-west and 8.9 m north-south within a much larger area of the exposed dark-grey sediment.

SM D was tested with a 25 cm square placed in the centre of the exposed deposits in a surface shell concentration. Excavation of a single 5 cm XU in dark-grey, consolidated sandy loam revealed that dense in situ midden deposits extended from the surface to 2-3 cm, with a sharply defined termination.

The SM D assemblage totals 199 g of shell, with a trace amount of charcoal (Table 3). *Donax deltoides* comprises 95% of the MNI, with minor representation by *Turbo undulata* (5%) (Table 5). A sample of whole *Donax deltoides* shells was radiocarbon dated to ca. 9100 cal BP (Beta-93566) (Table 6, Figure 7).

Shell Midden E

Ancient sediments up to 1 m deep are exposed around the periphery of a large blowout towards the northeastern periphery of the landscape. Cultural material comprising SM E was exposed in section on the steep-edged eastern side of the blowout and on the surface beyond the blowout (Figure 2). In situ items in section and partially exposed on the eroding surface (not the blowout) included shell, notably *Ostrea angasi*, flint artefacts including a flake core, and charcoal, as well as a hearth feature (Feature 50, discussed below). At more than 750 m², SM E is the largest midden on the landscape.[1]

A 25 cm square test pit was placed on the eastern edge of the eroded area where in situ shell, charcoal and flint artefacts were observed both in section and on the surface. Time constraints did not allow the excavation of more than one XU, which revealed a moderately dense midden to at least 6 cm depth (supported by material visible in the eroded section to a similar depth). Excavation was very difficult, especially within the first few centimetres, due to the sediments being cemented with calcium carbonate. Beneath this crust was a consolidated light-grey sandy loam.

The excavated sample of 60 g was dominated by *Turbo undulata*, which comprised 70% of the MNI (Tables 3-5). Three other species, *Cellana tramoserica*, *Austromytilus rostratus* and *Ostrea angasi*, were each represented by 10% of the MNI. A sample of *Ostrea* and *Turbo undulata* shell from the excavation provided a radiocarbon age determination of ca. 10,250 cal BP (Beta-93568) (Table 6, Figure 7).

Feature 50

Feature 50 (F 50) is located 10 m northeast of the blowout on SM E. F 50 was a cluster of burnt calcarenite cobbles measuring 4.6 m east-west by 4.2 m north-south, some in situ and some simply resting on the greyish brown, compact sandy loam surface, the surrounding sediments having been removed by wind action (Figures 2, 8). These hearthstones are the only ones observed in situ in early-Holocene sediments on the Cape Duquesne landscape. Several *Ostrea angasi* shells were observed among the hearthstones, including two in situ shells, each approximately 50% exposed on the surface. One in situ *Ostrea* shell was removed for radiocarbon dating, providing a determination of ca. 10,700 cal BP (Wk-9563) (Table 6, Figure 7).

Shell Midden F

Located on the eastern side of the landscape, SM F is a largely unvegetated surface exposure of eroding shells, measuring 18.8 m east-west by 14.6 m north-south (Figure 2). A single pass

---

1   Detailed examination of the area around Shell Midden E, Shell Midden H and Feature 50 during contour mapping operations revealed that the distribution of surface cultural material showed no gaps between Shell Middens E and H. As a result, SM H is now incorporated within SM E and is no longer regarded as a discrete midden. Feature 50 is within the boundaries of SM E and is now regarded as part of that midden, although it still retains its designation as a feature.

Figure 8. Feature 50 at Shell Midden E, showing burnt calcarenite rocks and *Ostrea angasi* shell in compact sandy loam palaeosol. Ranging pole bars are 20 cm long; small scales are 10 cm long.

by plough by the land management authority for the purpose of creating furrows for replanting to stabilise the area exposed significant densities of *Cellana tramoserica, Turbo undulata, Thais orbita* and *Donax deltoides* shell.

A 40 cm square was placed over an average surface expression of *Cellana tramoserica* and *Turbo undulata* shell located between furrows (Figure 9:top). The loose shells, some nearly whole, some fragmented, were swept up and bagged before the excavation of XU 1.

Once the excavation started, *Cellana tramoserica* and *Turbo undulata* shells were found to be very dense, comparable to SM A and SM D (Figure 9:bottom). A few small chunks of dubiously cultural charcoal were found near the surface in the northwest corner and also deeper in the southwest. Pointing trowels and dental picks were used to remove the compact sediment from around the tightly packed shells so they could be removed without damaging them. The first XU was excavated to the bottom of the dense concentration of shell, at 5 cm below the surface.

XU 2 was excavated to establish the depth of sterile deposits. Only small fragments of shell were recovered and they clearly came from the upper 1 cm of the 5 cm deep XU. Small pieces of charcoal were found throughout the XU.

The surface consisted of a greyish-brown (10 YR 5/2) fine sand with almost no vegetation cover. The sediment in both excavated XUs was identical to that on the surface, except it was lightly consolidated in comparison to the very consolidated surface crust.

SM F is the only midden dominated by *Cellana tramoserica*; out of 1049 g of shell, *Cellana tramoserica* comprised 58% of the total MNI, *Turbo undulata* 29%, *Thais orbita* 7%, Polyplacophora 3% and *Austromytilus rostratus* <1% (Tables 3-5).

A sample of whole *Cellana tramoserica* shells from XU 1 was submitted for radiocarbon dating, yielding an age estimate of ca. 10,150 cal BP (Wk-9604) (Table 6, Figure 7). An AMS age determination of ca. 10,400 cal BP (Wk-29818) was obtained on a *Turbo undulata* shell from XU 2. The samples are significantly different at the 95% level (T statistic) (Stuiver et al. 2011).

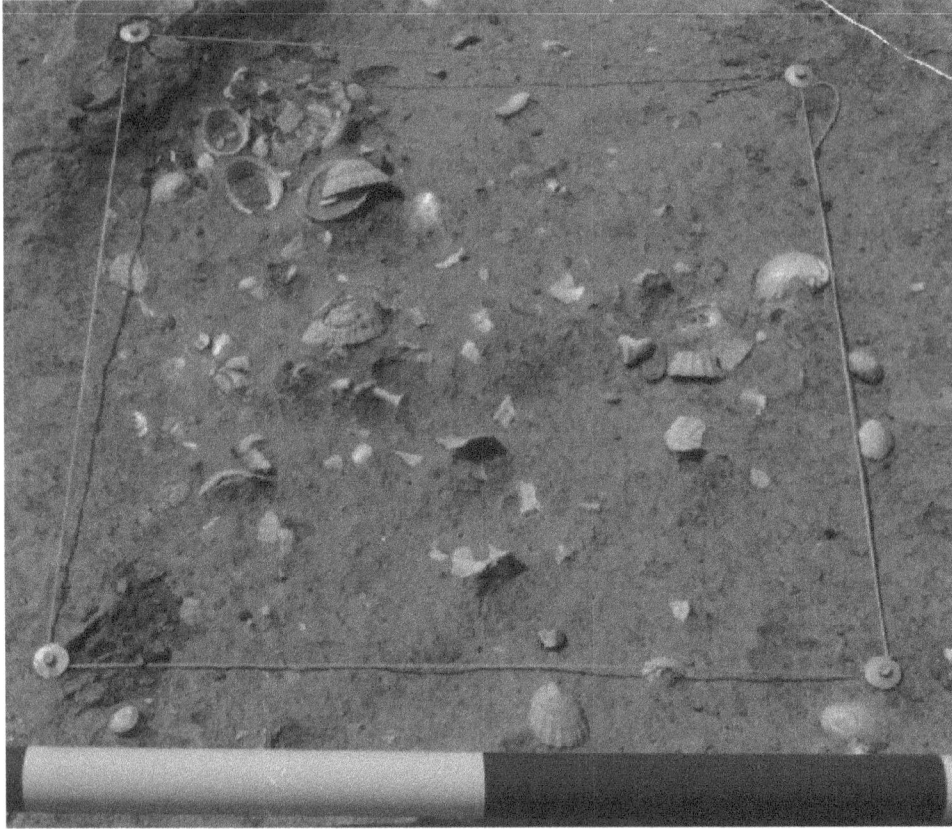

Figure 9. Shell Midden F, Square 1: (top) pre-excavation view showing midden material on surface of palaeosol; (bottom) initial stage of excavation following removal of loose surface material, exposing dense in situ shell. Ranging pole bars are 20 cm long.

Shell Midden G

SM G, located in the northeast of the landscape, was eroded by wind action and largely devoid of vegetation (Figure 2). Flint artefacts, *Turbo undulata*, *Cellana tramoserica* and *Donax deltoides* shells and fragmented animal bone were exposed on the surface over an area approximately 16 m in diameter. SM G was investigated with a 40 cm square test pit excavated in an area of average surface shell density. Surface sediments were a loose, light-grey (10YR 7/1) sand. Sediment in XUs 1 and 2 consisted of lightly consolidated, light-grey (10YR 5/2), fine sand with few rootlets. The only change in the XU 3 sediment was the addition of small calcarenite pebbles to the sand.

Turbo undulata shell was concentrated in the upper few centimetres of the first XU, with very little in the two XUs excavated below that (Table 3). A sample of 132 g of shell was recovered, with *Turbo undulata* comprising 100% of the shellfish MNI (Tables 3-5).

Large *Turbo undulata* fragments from XU 1 provided a radiocarbon determination of ca. 10,050 cal BP (Wk-9564) (Table 6, Figure 7). A single *Turbo* operculum from XU 2 yielded an AMS age determination of ca. 9800 cal BP (Wk-29817). The samples are significantly different at the 95% level (T statistic) (Stuiver et al. 2011), although they overlap at 2 s.d.

Shell Midden I

SM I, located at the eastern end of the landscape, is approximately 3.2 m in diameter and consists of a cluster of *Donax deltoides* shell and a few flint artefacts on the surface (Figure 2). No excavations were undertaken at this midden, but a *Donax deltoides* sample consisting of some partially exposed, in situ shells, but mostly loose surface shells, was collected from an approximately 1 m diameter area (Table 3). A radiocarbon age determination of ca. 8750 cal BP (Wk-9562) was obtained on whole *Donax deltoides* shell (Table 6, Figure 7).

Shell Midden J

Shell Midden J, located in the south-central part of the landscape, measures 13.5 m north-south, by 9.1 m east-west on a much larger exposure of sandy loam sediment almost devoid of vegetation cover and containing numerous surface cracks (Figure 2). SM J had one dense in situ exposure of mainly *Turbo undulata* shell covering ca. 5 m², with scattered *Turbo undulata*, *Cellana tramoserica*, *Donax deltoides* and *Austromytilus rostratus* shell evident over a larger area. Notably, two *Austromytilus rostratus* shells were present on the surface on the southern edge of the midden. As the other middens had only yielded a small amount of shell of this species, it was decided to recover some from this area, both to obtain basic information on subsistence and to provide material for radiocarbon dating. A 50 x 50 cm square was laid out on the southern edge of the midden to include the partially exposed *Austromytilus rostratus* shells.

The *Austromytilus rostratus* shells, which were in a crumbly condition and only held together by the highly consolidated sediment, fell apart as they were gently exposed with pointing trowels. Excavation generally proved to be very difficult and progressed very slowly despite the use of highly sharpened tools. Very little additional shell was found in the first XU, and what was there was fragmentary and friable. Almost no shell was apparent near the bottom of the level, the last 1 cm of which was completed by careful shovel shaving. Hand picks were cautiously used to excavate the second, 5 cm-thick XU. Very little shell was found, and what was there was fragmented. A third XU was excavated only in the southeast quadrant (a 25 cm square). Only one tiny shell fragment and a few pieces of charcoal were present (Table 3).

There was no vegetation on the surface of the square, which was a greyish-brown (10 YR 5/2), very consolidated sandy loam. Sediment in XUs 1 and 2 is a greyish-brown (10 YR 3/2), very consolidated sandy loam. Cracks visible on the surface were still present down to XU 2. Sediment in XU 3 is a dark-brown (10 YR 3/3), consolidated sandy loam with small, rounded, calcarenite pebbles.

Very little shell was recovered from the test excavation in SM J, totalling only 13 g (Table 3). *Austromytilus rostratus* was the only species represented, with a MNI of 5 (Table 5). A trace of charcoal was present in each of the excavated XUs and a few grams of red ochre were present in XU 1, with a tiny amount in XU 2 (Table 3).

An AMS radiocarbon age determination of ca. 10,950 cal BP (Wk-9532) was obtained on *Austromytilus rostratus* shell from XU 1 (Table 6). This is the earliest age determination on shell from the landscape (Figure 7).

Sediment discussion

SM A sediment is similar to the late-Pleistocene-to-Holocene Discovery Bay Sand (Gibbons and Downes 1964), Undifferentiated Calcareous Sands (Land Conservation Council 1981), Discovery Bay and Bridgewater Sands (Cupper et al. 2003) and Poorly Consolidated Calcareous Sand (Rosengren 2001a) previously described for the region. It is clearly not a Pleistocene *terra rossa* or rendzina palaeosol, typically associated with the Bridgewater Group in this region. The location of the excavation, on a large sand dune, well above contact with the underlying Bridgewater Group calcarenite, further reinforces this interpretation.

In fact, the sediment in SM A closely matches Discovery Bay beach and dune sediments that also consist of grey-brown, fine-medium calcareous sand, with a virtually identical mean grain size (Buckley 1992:18, Table 2). In addition, there is a highly similar carbonate content to samples from Descartes Bay (Buckley 1992: Appendix 2). The data strongly indicate that the SM A sediment and the Discovery Bay-Descartes Bay beach and dune deposits mainly derive from the same calcareous continental shelf deposits exposed from the LGM (ca. 20,000 cal BP) well into the Holocene (Land Conservation Council 1981:14; Short 1988:121; Buckley 1992:20; Bird 1993:25; Rosengren 2001a:213-215; Sloss et al. 2007).

None of the other excavations had sediment analysis undertaken as with SM A, but the field observations from SM J are sufficient to indicate that this is a substantially different sediment. Firstly, SM J is present in a shallow, very consolidated, greyish-brown sediment, with surface cracks evident, directly overlying calcarenite. Excavation was very difficult at this site due to the very consolidated nature of the soil, which had a sandy loam texture and contained small calcarenite pebbles. This sediment does not compare well with SM A or the nearby beach and dune sands, and is rather more like the Bridgewater Group palaeosols that commonly cap calcarenite. It is especially similar to the Bridgewater Sandy Loam, a rendzina soil (Gibbons and Downes 1964: Appendix 1). Radiocarbon dating of a shell sample from the top of this deposit yielded the oldest shell determination for the site, an early-Holocene date of ca. 10,950 cal BP, further supporting this interpretation. The matrix of F50 was a similar dark-grey, very consolidated sandy loam, overlying calcarenite, a shell from which produced a radiocarbon age determination of ca. 10,700 cal BP.

Of the other shell middens, the sediment textures, pH values, consolidation and colour varied somewhere between the extremes of the loose, sandy SM A and the very consolidated sandy loam of SM J. Texture was typically finer and more compacted than SM A – varying from lightly consolidated fine sand (SM F, SM G), to consolidated sandy loam (SM B, SM D) and very consolidated sandy loam (SM E, SM J).

Table 7. Summary of excavated early-Holocene shell midden characteristics at Cape Duquesne.

Midden	Surface area (m²)	Excavation pit size (m)	Depth excav (m)	Mean midden thickness (m)	Shellfish (g)	Shellfish g/m²	Shellfish g/m³	Excavated midden density	Comments
SM A	9	1.0 x 1.0	0.15	0.060	7537.64	7537.64	125,627.33	very dense	most dense area of surface shell exposure sampled
SM B	43	.25 x .25	0.10	0.030	60.30	964.80	32,160.00	dense	most dense area of surface shell exposure sampled
SM D	121	.25 x .25	0.05	0.025	199.28	3188.48	127,539.20	very dense	most dense area of surface shell exposure sampled
SM E	750	.25 x .25	0.06	0.060	60.42	966.72	16,112.00	moderately dense	area with statified cultural material visible in section sampled
SM F	274	.40 x .40	0.10	0.050	1030.94	6701.11	134,022.20	very dense	average surface density area sampled
SM G	256	.40 x .40	0.15	0.030	126.49	822.185	26,352.08	dense	most dense area of surface shell exposure sampled
SM J	123	.50 x .50	0.15	0.030	11.56	46.24	1,541.33	diffuse	low-density surface shell exposure sampled

Variability in midden density

Intact shell midden deposits were encountered in all but one of the excavations.[2] There was some variability in the structure of the shell midden deposits and they ranged from diffuse through very dense. In all cases, a single shell layer was present, and these varied from tightly defined 2-3 cm thick middens (SM B, SM D, SM G) to well defined 5-6 cm thick SMs A and F, through to much less well defined deposits, such as the tested areas in SM E and SM J (Table 7). For comparative purposes, shell weight per cubic metre has been extrapolated from the testing, which reveals that SM A, SM D and SM F had the highest ratio of shell weight to excavated volume, expressed as kilograms of shell per cubic metre, at 125-134 kg/m^3, and these three middens are characterised as very dense (Table 7). SM B and SM G are described as dense middens, at 26-32 kg/m^3, SM E is moderately dense, with 16 kg/m^3, and SM J can be considered diffuse, at 1.5 kg/m^3. These values are best regarded as an indicator of the range of midden densities to be found on the landscape, rather than an average density for each midden.

Size and extent of middens on the landscape

The middens had a variety of contents, including hearth features, flaked stone, charcoal, red ochre and animal bone, but with a consistent domination by marine shellfish remains. In situ midden deposit was found in patches ranging from <10 to 752 m^2, with a total for the eight early-Holocene middens of just over 1575 m^2 (Table 7). This figure should be read as a minimum because deposits were covered by overlying sediments in some cases; for example, the observed portion of SM A was only about 9 m^2, yet it may well have covered an area in excess of 100 m^2, but this could not be confirmed without disturbing overlying semi-stable dune deposits that were protecting the midden from wind attack. The landscape has also suffered enormously from wind erosion and every midden observed has had an unknown quantity of material removed from above the observed in situ shell, indicating the possibility that the middens covered a larger area than presently preserved. It is also a near certainty that some early-Holocene middens have eroded away entirely, but of course this is something that cannot be proven. Finally, a reasonable estimate of the extent of early-Holocene middens on this landscape, taking into consideration those entirely removed by wind erosion and those probably still covered by overlying sediment, would be to double the observed extent of midden, to 3000 m^2 as a maximum figure.

Midden formation

Multiple radiocarbon age determinations are available for several middens that are each structurally different and which provide insights into midden formation. In the case of SM A, there is a dense shell concentration 4 cm thick, overlying 2 cm of less-dense shell deposit that yielded two age determinations that are statistically the same. A third determination, from the less-dense deposit, is ca. 500 years more recent than the other dates, out of sequence, and statistically different from them. A single occupational event is indicated in the midden deposit, and although the out-of-sequence date suggests later material is also present, this may be intrusive from eroded deposits overlying the stratified occupation of ca. 10,400 cal BP.

The structure of SM F is similar to SM A, with a concentration of shell in the upper 5

---

2   Shell Midden C was also tested, but is not shown on Figure 2 or discussed, because as Richards and Jordan (1996) observed: 'Excavation of one spit quickly revealed that all cultural materials were only superficially embedded in the upper few millimetres of the sediment and no in situ deposits were present. The materials observed here had clearly been redeposited, having in all likelihood been moved from a location to the north down to the 'midden' by slopewash.'

Table 8. Shellfish taxa description and ecology (after Macpherson and Gabriel 1962; Coutts 1967; Dakin 1976, 1987; Wilson and Gillett 1979, 1980; O'Sullivan 1980; Godfrey 1984; Davey 2000; Australian Museum 2010; Beechey 2009; Barwon Bluff Marine Sanctuary 2010; NSW Fisheries 2010; Species Bank 2010).

Shell taxa	Common name	Substrate	Preferred habitat	Tidal zone	Coast type	Feeding strategy	Frequency	Mobility	Average size mm
Turbo undulata	Wavy Turbo	rock	intertidal platform	mid to sub tidal	med-high energy	algae grazer	very common	mobile	70
Cellana tramoserica	Variegated Limpet	rock	intertidal platform	high to low tidal	med-high energy	algae grazer	abundant	mobile	60
Austrocochlea concemerata	Wavy Top Shell	rock	intertidal platform	high to mid tidal	med-high energy	algae grazer	common	mobile	25
Donax deltoides	Pipi	sand	sandy beach	high to mid tidal	low-high energy	filter feeder	very common	sedentary	60
Thais orbita	Dog Whelk	rock	intertidal platform	mid to sub tidal	med-high energy	carnivore	common	mobile	80
Austromytilus rostratus	Beaked Mussel	rock	intertidal platform	mid tidal	med-high energy	filter feeder (mainly plankton)	abundant	sedentary	40
Polyplacophora	Chiton	rock	intertidal platform	low to sub tidal	low-high energy	grazer (encrusted plants and animals)	very common	mobile	85
Melanerita melanotragus	Black Nerite	rock	intertidal platform	high to low tidal	low to medium energy	algae grazer	very common	mobile	30
Ostrea angasi	Common Mud Oyster	sand, silt	estuary, lagoon	low to sub tidal	low to medium energy	phyto-plankton filter feeder	uncommon to abundant	sedentary	120
Dicathais baileyana	Bailey's Dog Winkle	rock	intertidal platform	mid to sub tidal	med-high energy	carnivore	uncommon	mobile	35
Cominella lineolata	Lineated Whelk	rock	intertidal platform	high to sub tidal	med-high energy	carnivore/scavenger	very common	mobile	25

cm and sparse shell extending only 1 cm below that. The age determinations from each are statistically different, a few hundred years apart, and in sequence. In this instance, repeated occupation is indicated.

SM B has a concentration of shell in the upper 3 cm, with sparse shell below that to 7 cm depth, and dates from the dense and sparse shell deposits are statistically the same age. A single occupational event is indicated.

The structure of SM G is similar to SM B, with a concentration of shell in the upper 3 cm and sparse shell below that to 10 cm. In this case, the age determinations from the dense and sparse deposits are statistically different, but not in sequence, with the underlying date a few hundred years more recent than the overlying one. On the other hand, there is still a 100-year overlap between the two dates at two standard deviations and the difference in median ages is only 250 years on ca. 10,000 year dates. Repeated occupation is indicated.

To summarise, these open-air middens show a strong integrity considering that they are thin and shallow deposits (around 5 cm in depth) that have survived for 10,000 years or longer. Treadage alone is enough to explain the minor mixing revealed by the dating in some of these deposits. Some middens represent single, short-term occupation events and others represent a few short-term occupations over a few hundred years. The general chronological signature is of repeated, short-term occupation of this landscape.

Shellfish species frequency

The majority of the 11 shellfish taxa recorded in the shell midden deposits are rock-platform species (Tables 5, 8). *Turbo undulata* dominates the overall excavated shellfish assemblage, at nearly 62% of the total MNI of 847, and is the most common species in four of the seven tested middens (SMs A, B, E and G), as well as being represented in all of the other middens. It may well be the dominant species in SM J as well, because the testing there deliberately avoided the most dense in situ deposits exposed on the surface, which were strongly dominated by *Turbo undulata*. The next most important species in the total assemblage is *Cellana tramoserica*, which represents nearly 24% of the MNI. This species dominates one of the middens (SM F) and is prominent in another (SM A).

Austrocochlea concemerata comprises close to 7% of the total MNI, but this species is only represented in SM A (Table 5). *Thais orbita* comprises just over 2% of the MNI. Sandy shore species *Donax deltoides* make up slightly less than 2% of the excavated MNI, although it is the dominant species in SM D and appears to be dominant at the unexcavated SM I. *Austromytilus rostratus* comprises just over 1.5% of the assemblage, and although it is absent in SM A, it is represented in four other middens (SM D, E, F and J). Of the five other species present in the excavated assemblage, all comprise <1% of the total MNI, and they are mostly only present in SM A. The estuarine/muddy shore *Ostrea angasi* is only represented at SM E and F 50, and although its large size makes it appear prominent, there are only around five examples exposed on the surface over a very large area. Looked at another way, rock-platform species dominate six middens (SM A, B, E, F, G and J), while sandy-shore species dominate two middens (SM D and I).

Shellfishing chronology

The overall chronological range of shell middens on this landscape, considering the two sigma range of calibrated radiocarbon age determinations, indicates middens were deposited as late as ca. 8600 cal BP and as early as ca. 11,100 cal BP, a maximum span of 2500 years (Table 6, Figure 7). If the median probability of each calibrated date is used, the midden-related occupation span is 2200 years (8750-10,950 cal BP).

There is some chronological patterning in shellfish species presence in the middens. The earliest shell dates are on *Austromytilus rostratus* and *Ostrea angasi*, from the period ca. 10,950-10,700 cal BP from SM J and F 50 (Table 6). The *Austromytilus rostratus* and *Ostrea angasi* dates may indicate an early period of landscape occupation when these species were important economic contributors, but *Turbo undulata* is also common in deposits exposed near the dated locations.

The main period of landscape occupation is defined by eight dates on middens (SM A, B, E, F and G), with a strong focus on *Turbo undulata*, which range from ca. 10,500 to ca.10,000 cal BP (which may be expanded slightly to include two out-of-sequence age determinations of ca. 9800 cal BP and ca. 9850 cal BP) (Figure 7).

Finally, there is a late-occupation pulse with a focus on *Donax deltoides* around 9000 cal BP at SM D and SM I (ca. 9100-8750 cal BP).

Nature and distribution of shellfish resources

Characteristics of the shellfish species found in the middens on the Cape Duquesne landscape are provided in Table 8. The majority of species found in the middens at Cape Duquesne are adapted to high-wave energy coasts. Further, all but two species are adapted to rock substrates. All tidal zones are represented in the above species, from high tidal to sub tidal (down to a depth of 10 m); almost all species can be found in either the high or mid-tidal zones, with only two exclusively in the low-tidal zone or deeper (and one of these is commonly found in shallow lagoons).

In terms of characteristics of the shellfish themselves, they range significantly in size, frequency, gregariousness, occurrence and mobility. Most are currently common to abundant along the Victorian coast, with one species, *Ostrea angasi*, rare generally, but abundant in its fairly limited estuarine or lagoon habitat. Most are gregarious to a greater or lesser degree, and occur in small groups to very dense beds. All but one of the rock substratum adapted species are mobile, while both soft substratum species are sessile. The largest species, *Ostrea angasi*, is also currently the most rare, and the rest of the species range from 25 mm to 85 mm for adults.

It is always risky to apply modern-day faunal distributional studies to Pleistocene situations and the present study has the usual difficulties in this regard. The coastline in the vicinity of Cape Duquesne was considerably different 10,000 years ago and it is difficult to accurately and independently model the nature of shellfish habitats in detail for that period. Nevertheless, on the basis of sea-bed mapping, geological studies and information from local crayfishermen, it is apparent that the shoreline off what is now Cape Duquesne would have been rocky, dominated by basalt boulders and platforms, with some sandy coves. This is currently a high-energy shoreline, bearing the brunt of high breaker waves, predominantly from the southwest (Short 1988:125); it was likely at least as energetic during the late-Pleistocene and early-Holocene. It is reasonable to conclude that the immediate early-Holocene coastal marine environment off what is now Cape Duquesne provided a favourable habitat for shellfish that prefer cold water, tolerate high energy waves and feed on algae that live on a rocky substratum.

Discovery Bay would have been likely dominated by high-energy sandy beaches as at present, with occasional rock outcrops and possibly lagoons, while Bridgewater Bay would not have existed as a water body; rather, a stretch of sandy beaches, possibly with backing lagoons, would have joined the southward-projecting Cape Bridgewater and Cape Nelson headlands. Patchy basalt outcrops and occasional offshore platforms would probably also have been present, and all would have been exposed to high-energy waves.

Subsistence

Shellfish are a predictable resource in that the density and location of species of interest varies little from year to year (e.g. Meehan 1982:160), making them an ideal regular staple, especially on a short-term or seasonal basis. Very little evidence for the presence of other types of food was found in the excavations at Cape Duquesne, despite alkaline pH values conducive to the preservation of bone (Reitz and Wing 2008:140-141). It must be concluded that the economic basis of occupation at Cape Duquesne was largely shellfish, although non-economic factors are likely to have also influenced selection of this locality. The sustained and repeated usage of the locality for essentially the same purpose strongly indicates that shellfishing was an ongoing component of the subsistence economy.

If the proportions of shellfish species found in the middens are considered, the top three species, *Turbo undulata*, *Cellana tramoserica* and *Austrocochlea concemerata*, account for 92% of the total assemblage MNI, spanning the entire 2200-year period that middens were deposited on this landscape. These species are very alike in terms of habitat, all being found on intertidal rock platforms on medium to high-energy coasts (Table 8). They are also all mobile algae grazers, are gregarious, and occur in small to large groups. One difference between them is that while *Austrochlea* and *Cellana tramoserica* can be found in the high-tidal zone, often exposed to the air, and all three species occur in the mid-tidal zone, only *Turbo undulata* is commonly found in the sub-tidal zone down to a depth of 10 m. The other notable difference is that while adult *Turbo undulata* average about 70 mm in diameter and *Cellana tramoserica* averages 60 mm, *Austrocochlea concemerata* only average 25 mm in size.

Looking at proportions of shellfish species in the excavated middens more closely, the most common species by a considerable margin is *Turbo undulata*, at 61% of the entire assemblage MNI. One interpretation of the excavation results, in light of the likely shoreline habitat characteristics during the main period of occupation at Cape Duquesne ca. 10,000 cal BP and the current understanding of shellfish ecology, is that the occupants primarily targeted *Turbo undulata*, and that the specific rock-platform locations where *Turbo undulata* was to be found also held significant numbers of *Cellana tramoserica* and *Austrocochlea concemerata*, usually at shallower depths, and that these were secondary targeted species. The other rock-platform species represented in the assemblage were likely collected as bycatch during the main activity of obtaining *Turbo undulata*. To obtain meaningful amounts of fully grown *Turbo undulata*, some may be collected by wading, but complete immersion in the ocean would likely be necessary. If there were no *Turbo undulata* to hand while submerged in the very cold Southern Ocean, there would be considerable incentive to collect anything edible otherwise encountered (hence, the bycatch species in the assemblage).

During the latter period of occupation, it appears that there was a significant change in focus from mobile, intertidal rock-platform shellfish species that had to be individually located and collected by hand, to a sessile, sandy-beach species that occurred in large, highly dense beds. *Donax deltoides* are typically found in water-saturated sand, around 10-20 cm below the surface, and using bare feet to locate them and hands to pick them up, collectors may expect to get arms wet up to the elbows and legs wet to the knees, which is a radically different proposition to the collection of *Turbo undulata* in sub-tidal waters while being battered against a rock platform by enormous waves.

Discussion

Relationship with contemporary early-Holocene inland sites

In terms of their place in a larger-scale landscape settlement pattern, the three early-

Holocene occupations at Cape Duquesne, Bridgewater South Cave and Koongine Cave represent repeated, short-term, small-group resource-extraction camps, with Bridgewater South Cave and Koongine Cave having a nearly exclusive focus on terrestrial mammals and Cape Duquesne on littoral invertebrates. The balance of the evidence suggests cold-season occupation for the inland caves and warm-season occupation for the open coastal landscape, although there is no absolute seasonal control on resource availability or access. In all three cases, food resources were killed or collected away from the camping area and brought back to the camps for processing and consumption. The cave sites, with their hunting emphasis, contain considerable evidence for stone working, which included wood working, tool manufacturing and maintenance. At the Cape Duquesne camps, stone working was a minimal activity. Viewed through general ethnographic patterns, the data could indicate a division of activity and site occupation by gender, with hunting and stone working providing a predominantly male emphasis at Bridgewater South Cave and Koongine Cave, while shellfishing and associated processing suggest a female activity focus at Cape Duquesne.

The distance from Bridgewater South Cave to the camping area at Cape Duquesne at 10,000 cal BP was 8 km, with less than 1.2 km further south to the littoral resources on the shore. Given the contemporaneity of occupation of the two locations, and the two to three hour walk between them, it is conceivable that the same people occupied both sequentially, or even at the same time. With little effort, scenarios could be constructed involving: (1) small family groups spending time shellfishing on the coast while camped at Cape Duquesne and then moving on to Bridgewater South Cave to hunt terrestrial mammals; (2) mainly women, children and older people camping at Cape Duquesne and collecting shellfish, while a related party of men were hunting terrestrial resources out of a field camp in the Bridgewater South Cave; (3) groups of mainly men or women, in different hunting and foraging parties from as-yet-unknown, larger base-camp locations, using Bridgewater South Cave and Cape Duquesne as temporary field camps, where activities concentrated on field processing food resources for transportation to the base camps.

All three sites should be viewed as indicative of early-Holocene short-term field camps in this region, along with even more ephemeral contemporary inland evidence of occupation at Blackfellows Waterhole (Richards 2004), Billimina Rockshelter and Drual Rockshelter (Bird and Frankel 2005), located to the northeast. This regional pattern of short-term, repeatedly occupied sites suggests a highly mobile, sparse human population in the early-Holocene.

Comparison with late-Holocene midden structure and contents

Recent research at a very large late-Holocene midden some 3 km to the northeast of Cape Duquesne, has revealed a distinct chronology of landscape usage and shellfishing patterns (Richards and Johnston 2004; Richards and Webber 2004). The Cape Bridgewater landscape contains Aboriginal midden deposits measuring up to 237,000 m² in area, and with an estimated average thickness of 10 cm there may have been up to 23,700 m³ of midden at this location (Richards and Johnston 2004). In contrast, the Cape Duquesne landscape, although around half to one quarter the size of Cape Bridgewater, only contains small patches of visible midden deposit, measured at just over 1500 m² and estimated to have been no more than 3000 m² (including buried and destroyed deposits). Testing of the Cape Duquesne middens indicates a mean thickness of 4 cm, providing an estimate of up to 120 m³ of early-Holocene midden deposit on this landscape, vs. 23,700 m³ at Cape Bridgewater, or 198 times as much midden at the latter.

Radiocarbon age determinations indicate an occupation span at Cape Bridgewater of some 3600 years, from ca. 4000 cal BP to ca. 400 cal BP, with an estimated average of up to 6.6 m³ of shell midden deposition per year. In contrast, occupation of Cape Duquesne occurred over a span of 2200 years, resulting in an estimated midden deposition rate of up to .06 m³ per year. The estimated average annualised rate of midden deposition on the late-Holocene Cape Bridgewater landscape is thus 110 times greater than the average annualised rate of midden deposition on the early-Holocene Cape Duquesne landscape.

Comparison with late-Holocene shellfishing patterns

At Cape Duquesne, there was a focus on one species, *Turbo undulata*, a relatively large sea snail that had to be collected by hand in the sub-tidal zone off wave-battered intertidal rock platforms. Two other species which had similar adaptations and habitat, although largely confined to the mid to high-tidal zones, were also collected, but to a lesser extent. Combined, the three top species in the assemblage, which were able to be collected at the same location, but at varying depths, comprised 92% of the MNI; almost all of the remaining species, which also could have been found in the same location, can be regarded as bycatch. The collection of sandy-shore species *Donax deltoides* occurred after the main period of landscape occupation and represents a switch to a new catchment, with *Donax* replacing *Turbo* as the dominant species.

The investigated sample from Cape Bridgewater totals 11,308 MNI, comprising 17 distinct taxa (vs. 11 taxa at Cape Duquesne); however, there was a concentration on mass collection and processing of one small species of shellfish found in sandy beaches, *Paphies angusta* (Narrow Wedge Shell). In the period 4000 to 1500 cal BP at Cape Bridgewater, 70% of MNI were *Paphies angusta*, while in the period 1400-400 cal BP this species increased to 90% of MNI (Richards and Johnston 2004:106). Significant numbers of both sandy and rocky substrate species were collected at Cape Bridgewater, so that while there was a major focus on one sandy-shore species, the fact that a distinct rocky-shore area was also accessed to collect rock-platform species indicates that these were deliberately targeted as supplemental species and not bycatch.

The economic focus on the Cape Bridgewater landscape was shellfish extraction, processing and consumption, which occurred on a large scale, repeatedly and over an extended duration (Richards and Johnston 2004:110). Aboriginal bands used this location as a base camp, undertaking a range of activities, probably on a semi-annual, seasonal (summer) basis for durations of a few weeks to a few months at a time (Richards and Johnston 2004:108-109). Cape Duquesne also had a focus on shellfish extraction, processing and consumption, and this also appeared to occur repeatedly over an extended duration and was probably semi-annual and seasonal in nature; however, this was a very small-scale activity that was targeted mainly towards a few species found in a few vertically stratified zones at intertidal rock platforms. In contrast to Cape Bridgewater, all of the evidence at Cape Duquesne indicates that small groups, on the scale of extended families, camped at this location for short periods of time, for durations of a few days to a few weeks at most.

Conclusions

Three natural phenomena allowed the formation and subsequent preservation of the Cape Duquesne Aboriginal landscape: (a) a cliff; (b) a steeply sloping sea bed; and (c) a sand ramp. The sand ramp joined the clifftop with the early-Holocene shoreline, providing an environmental setting that Aboriginal people turned into their own landscape, collecting marine littoral resources from the Southern Ocean and transporting them to their camping places at the top of Cape Duquesne for processing and consumption. Because of the steeply sloping sea bed, at 10,000 cal BP, the clifftop location would have been only just over 1 km to the north and about 75 m above the shoreline, providing a highly convenient camping location with a commanding view over an extensive coastal plain.

This is not an unusual occurrence of one isolated early midden; instead, the evidence from the Cape Duquesne landscape indicates that coastal occupation was a regular feature of the settlement-subsistence pattern in southwest Victoria from at least the onset of the Holocene.

The archaeological research at Cape Duquesne contributes to the study of socio-cultural complexity in southwest Victoria by providing a baseline dataset for comparisons with late-Holocene coastal Aboriginal landscape use, especially:

1. The physical nature of the middens, including size, composition and structure, indicating discrete patches of midden ranging up to 750 m² in area, containing 11 shellfish taxa, stone artefacts and hearth features.

2. A firm chronology based on 15 radiocarbon dates, demonstrating repeated marine shell midden occupations of the area over a maximum 2500-year span (ca. 11,100 cal BP to ca. 8600 cal BP) with a focus on the period ca. 10,500-10,000 cal BP.

3. Statistics derived from midden measurements and chronology, such as annualised midden deposition rate, estimated at up to .06 m³ per year.

4. The character of marine littoral resource exploitation, which was a highly patterned focus on three species that were available from high-tidal to sub-tidal zones at rock platforms. Other shellfish species were collected incidentally as they were encountered at these locations. The primary focus was on relatively K-selected shellfish that were individually collected and processed by hand prior to consumption.

5. The character of human occupation of the landscape; this was small scale and probably seasonal, indicating short-term reliance on shellfish as a food source by small groups of people.

In marked contrast, a nearby late-Holocene Aboriginal landscape at Cape Bridgewater, occupied some 5000 years after Cape Duquesne, contained a very large midden deposit with evidence for a concentration on mass collection and processing of one very small, sessile species of shellfish that was available in very large beds in the high-tidal zone. This late-Holocene midden had an annualised midden deposition rate that was estimated to be 110 times greater than at Cape Duquesne in the early-Holocene. The Cape Bridgewater midden itself is structurally different, being much larger and thicker, having huge areas of continuous midden deposit, while at Cape Duquesne the midden deposit is discontinuous, patchy and much thinner.

It must be concluded that there is no indication of complexity in the early-Holocene archaeological record at Cape Duquesne; in fact the evidence provides a textbook signature for generalised, highly mobile, egalitarian hunter-gatherers. This finding is consistent with other documented early-Holocene occupations in southwest Victoria, such as Bridgewater South Cave, Koongine Cave, Blackfellow's Waterhole, Billimina Rockshelter and Drual Rockshelter. It is also clear that something very different was happening in terms of Aboriginal coastal occupation and littoral resource use in this area during the late-Holocene in comparison with the early-Holocene. The late-Holocene occupations were more highly organised, and probably more tightly scheduled, by larger groups that stayed on the coast for longer periods, and who utilised a broader range of species while increasingly focusing over time on the mass collection and processing of one highly r-selected species. This pattern is entirely consistent with the late-Holocene appearance of semi-sedentary, high population density, complex, transegalitarian societies, as postulated by Lourandos (1980a, b, 1983, 1997) and Williams (1985, 1987, 1988).

It is also consistent with the results of Kershaw and his collaborators on the Mt Eccles lava flow.

Acknowledgements

The research reported in this paper resulted from fieldwork in 1995 and 2001 undertaken in a cultural heritage research, management and training partnership between the Gunditj Mirring Traditional Owners and Aboriginal Affairs Victoria.

I acknowledge with gratitude the Elders and members of the Gunditj Mirring Traditional Owners Aboriginal Corporation (and the Winda Mara Aboriginal Corporation, Gunditjmara Native Title Claimants and Kerrup Jmara Elders Corporation) who initiated the idea of archaeological investigations, protection and management for the Cape Duquesne Aboriginal landscape, and then followed through with strong support and action: Eileen Alberts, Damein Bell, Michael Bell, Denise Lovett, Mick Onus, Darryl Rose, Denis Rose, Ken Saunders and Theo Saunders. Special recognition is due to Denise Lovett and Darryl Rose for their superb job of organising and co-directing the 2001 field school. We thank the following 2001 field school participants from the Winda Mara Aboriginal Corporation who excavated at Cape Duquesne: John Bell, Darren Bell, Leonard Cooper, Keisha Day, John Kanoa, Jessica Lovett, Gordon Slade and Leon Walker.

I thank former and present senior managers of Aboriginal Affairs Victoria for funding and supporting the research and training under my direction: Terry Garwood, Jill Gallagher, Joy Elley, Tony Cahir, Ian Hamm and especially David Clark who originally guided me to Cape Duquesne. Many other present and former members of Aboriginal Affairs Victoria (including volunteers) provided important assistance and support during various phases of the research, including background research, excavations, mapping, labwork, map drafting and teaching students: Grant Cochrane, David Doyle, Joanna Freslov, Pam Gait, Simon Greenwood, Tania Hardy-Smith, Shawn Ilsley, Sharon Lane, Richard MacNeill, Rob McWilliams, Joanna Newby, Christina Pavlides, Jamie Reeves, Petra Schell, Harry Webber, Bianca Weir and Christine Williamson. I especially note the indispensable contributions from Joanne Jordan as assistant director of the 1995 investigations and Rochelle Johnston for her work on site in 1995 and 2001, for her Shell Midden A sediment analysis, and for detailed comments on a draft of this paper.

Parks Victoria staff generously supported this research, both through the timely provision of permits for conducting scientific research within the Discovery Bay Coastal Park and in numerous practical ways in the field.

Southwest TAFE provided an accredited teaching framework for the archaeological field school investigations in 2001, and 17 participants received Certificate II qualifications in *Aboriginal Material Culture – Care and Management*.

More recently, the School of Geography and Environmental Science, Monash University, has supported the final writing up of this chapter, especially my principal supervisor, Dr Ian McNiven, and Dr Bruno David who provided funding for several radiocarbon dates.

I thank an anonymous reviewer and the editors of this volume for their helpful comments on a draft of this chapter.

References

Australian Museum 2010. *Molluscs.* http://australianmuseum.net.au/Molluscs/, accessed 13 October 2010.

Barwick, D.E. 1984. Mapping the past: an atlas of Victorian clans 1835-1904. *Aboriginal*

History 8:100-131.

Barwon Bluff Marine Sanctuary 2010. *Life on the edge: a guide to the animals and plants of the Barwon Bluff Marine Sanctuary.* http://www.barwonbluff.com.au/barwon_index.htm, accessed 13 October 2010.

Beaton, J.M. 1983. Does intensification account for changes in the Australian Holocene archaeological record? *Archaeology in Oceania* 18:94-97.

Beaton, J.M. 1985. Evidence for a coastal occupation time-lag at Princess Charlotte Bay (North Queensland) and implications for coastal colonization and population growth theories for Aboriginal Australia. *Archaeology in Oceania* 20:1-20.

Beechey, D. 2009. *The Seashells of New South Wales. Release 14, 1 October 2009.* http://seashellsofnsw.org.au/index.htm, accessed 17 September 2010.

Bird, C.F.M. and Frankel, D. 1991a. Chronology and explanation in western Victoria and south-east South Australia. *Archaeology in Oceania* 16:1-16.

Bird, C.F.M. and Frankel, D. 1991b. Problems in constructing a prehistoric regional sequence: Holocene south-east Australia. *World Archaeology* 23:180-192.

Bird, C.F.M. and Frankel, D. 1998. Pleistocene and Early Holocene archaeology in Victoria. A view from Gariwerd. *The Artefact* 21:48-62.

Bird, C.F.M. and Frankel, D. 2001. Excavations at Koongine Cave: Lithics and land-use in the Terminal Pleistocene and Holocene of South Australia. *Proceedings of the Prehistoric Society* 67:49-83.

Bird, C.F.M. and Frankel, D. 2005. *An archaeology of Gariwerd from Pleistocene to Holocene in western Victoria.* Tempus 8. St. Lucia: Anthropology Museum, University of Queensland.

Bird, C.F.M., Frankel, D. and van Waarden, N. 1998. New radiocarbon determinations from the Grampians-Gariwerd region, western Victoria. *Archaeology in Oceania* 33:31-36.

Bird, C.F.M., Frankel, D. and van Waarden, N. 1999. Prokrustes in Gariwerd. *Archaeology in Oceania* 34:86.

Bird, E.C.F. 1993. *The Coast of Victoria. The Shaping of Scenery.* Melbourne: Melbourne University Press.

Boutakoff, N. 1963. The geology and geomorphology of the Portland area. *Geological Survey of Victoria*, Memoir 22. Melbourne: Government Printer.

Buckley, R. 1992. Victorian Coastal Sediment Survey, final report. Unpublished Report No. 92-03-15, Coastal Investigations Unit, Port of Melbourne Authority, Melbourne.

Builth, H., Kershaw, A.P., White, C., Roach, A., Hartney, L., McKenzie, M., Lewis, T. and Jacobsen, G. 2008. Environmental and cultural changes on the Mt Eccles lava-flow landscapes of southwest Victoria, Australia. *The Holocene* 18:413-424.

Cann, J.H., de Deckker, P. and Murray-Wallace, C.V. 1991. Coastal Aboriginal shell middens and their palaeoenvironmental significance, Robe Range, South Australia. *Transactions of the Royal Society of South Australia* 115:161-175.

Clark, D.J. 1979. The Gambieran Stone Tool Industry. Unpublished BA Honours thesis, Division of Prehistory, La Trobe University.

Coastal Management and Co-ordination Committee 1981. *Western Coastal Study. Volume One, Resource Document.* Melbourne: Department of Crown Lands and Survey.

Corris, P. 1968. *Aborigines and Europeans in Western Victoria.* Occasional Papers in Aboriginal Studies No 12, Ethnohistory Series No. 1. Canberra: Australian Institute of Aboriginal Studies.

Coutts, P.J.F. 1967. Wilson's Promontory, Victoria from prehistoric times to present. Unpublished PhD thesis, Australian National University.

Critchett, J. 1990. *A Distant Field of Murder. Western District Frontiers 1834-1848.* Melbourne: Melbourne University Press.

Critchett, J. 1998. *Untold Stories. Memories and Lives of Victorian Kooris.* Melbourne: Melbourne University Press.

Cupper, M.L., White, S. and Neilson, J.L. 2003. Quaternary: Ice Ages – Environments of Change. In: Birch, W.D. (ed), *Geology of Victoria* (3rd ed.), pp. 337-360. Geological Society of Australia Special Publication 23. Melbourne: Geological Society of Australia.

Curr, E.M. 1886-87. *The Australian Race: its Origins, Languages, Customs, Places of Landing in Australia and the Routes by Which it Spread Itself Over that Continent, Volumes 1-4.* Melbourne: Victorian Government Printer.

Dakin, W.J. 1976. *Australian Seashores.* Sydney: Angus and Robertson Publishers.

Dakin, W.J. 1987. *Australian Seashores.* Fully revised and updated by I. Bennett. Sydney: Angus and Robinson.

Davey, K. 2000. *Life on Australian seashores.* http://www.mesa.edu.au/friends/ seashores, accessed 10 September 2005.

David, B., Barker, B. and McNiven, I.J. (eds) 2006. *The Social Archaeology of Australian Indigenous Societies.* Canberra: Aboriginal Studies Press.

Dawson, J. 1881. *Australian Aborigines. The Languages and Customs of Several Tribes of Aborigines in the Western District of Victoria, Australia.* Melbourne: George Robertson.

Dawson, J. 1887. Mr. Curr's work on the Australian race. Letter to Editor of *The Argus*, 21 October.

Debney, T. and Cekalovik, H. 2001. Portland Wind Energy Project EES, cultural heritage study. In: *Portland Wind Energy Project, Environment Effects Statement and Planning Report, Supplemental Volume C*, pp. 165-239. Blue Wind Energy, Pacific Hydro and Sinclair Knight Merz.

Douglas, J.G. 1979. Explanatory notes on the Portland 1:250,000 Geological Map. *Geological Survey Report No. 62.* Melbourne: Geological Survey of Victoria.

Edwards, W.H. 1987. Leadership in Aboriginal society. In: Edwards, W.H. (ed), *Traditional Aboriginal Society: a Reader*, pp. 153-173. Melbourne: MacMillan.

Everett, C. 1998. An archaeological investigation of proposed windfarm sites, Portland, Victoria. Unpublished report, Aboriginal Affairs Victoria, Melbourne.

Frankel, D. 1986. Excavations in the lower southeast of South Australia: November 1985. *Australian Archaeology* 22:75-87.

Frankel, D. 1991. First-order radiocarbon dating of Australian shell-middens. *Antiquity* 65:571-574.

Gibbons, F.R. and Downes, R.G. 1964. *A Study of the Land in South Western Victoria.* Melbourne: Soil Conservation Authority Victoria.

Godfrey, M.C.S. 1980. An archaeological survey of the Discovery Bay Coastal Park, Volumes 1 and 2. Unpublished report for the National Parks Service, Melbourne.

Godfrey, M.C.S. 1984. Seasonality and shellfishing at Discovery Bay, Victoria. Unpublished MA thesis, Division of Prehistory, La Trobe University, Bundoora.

Godfrey, M.C.S. 1989. Shell midden chronology in south western Victoria: Reflections of change in prehistoric population and subsistence? *Archaeology in Oceania* 24(2):65-79.

Godfrey, M.C.S. 1994. The archaeology of the invisible. Seasonality and shellfishing at Discovery Bay Victoria: the application of oxygen isotope analysis. Unpublished PhD thesis, School of Chemistry, La Trobe University, Bundoora.

Godfrey, M.C.S. 1996. The Yellow Rock project. Unpublished report, Aboriginal Affairs Victoria, Melbourne.

Godfrey, M.C.S. 2000. Access and protection? An archaeological survey of the Bridgewater Bay Dune. Unpublished report, Aboriginal Affairs Victoria, Melbourne.

Godwin, L. 1980. What you can do with 27,000 pieces of bone: A taphonomic study of the vertebrate fauna of Bridgewater Cave South. Unpublished BA (Honours) thesis, University of New England, Armidale.

Grier, C., Kim, J. and Uchiyama, J. 2006. *Beyond Affluent Hunter-Gatherers: Rethinking Hunter-Gatherer Complexity.* Proceedings of the 9th Conference of the International Council of Archaeozoology, Durham, August 2002. Oxford: Oxbow Books.

Hayden, B. 1995. Pathways to power: Principles for creating socioeconomic inequalities. In: Price, T.D. and Feinman, G.M. (eds), *Foundations of Social Inequality*, pp. 15-86. New York: Plenum Press.

Hayden, B. 2006. Comment on Constraints on the development of enduring inequalities in late Holocene Australia by I. Keen. *Current Anthropology* 47(1):21-22.

Head, L. 1985. Pollen analysis of sediments from the Bridgewater Caves archaeological site, southwestern Victoria. *Australian Archaeology* 20:1-15.

Hiatt, L.R. 1996. *Arguments About Aborigines. Australia and the Evolution of Social Anthropology.* Cambridge: Cambridge University Press.

Hiscock, P. 2008. *The Archaeology of Ancient Australia.* London: Routledge.

Howitt, A.W. 1887. The Australian Aborigines. Letter to the editor of *The Argus*, 28 October.

Howitt, A.W. 1904. *The Native Tribes of South-East Australia.* London: Macmillan and Co. Ltd. (Facsimile edition, Aboriginal Studies Press, Canberra 1996.)

Johnston, R. 1996. Sediment report for the Moorabool Basin and Cape Duquesne Statewide Survey investigations. Unpublished report, Aboriginal Affairs Victoria, Melbourne.

Keen, I. 2006. Constraints on the development of enduring inequalities in late Holocene Australia. *Current Anthropology* 47(1):7-38.

Kenley, P.R. 1976. Southwestern Victoria. In: Douglas, J.G. and Ferguson, J.A. (eds), *Geology of Victoria*, pp. 290-298. Melbourne: Geological Society of Australia.

Kershaw, A.P. (ed) 2004. Environmental History of the Newer Volcanic Province of Victoria. *Proceedings of the Royal Society of Victoria*, Vol. 116, No. 1. Melbourne.

Kershaw, A.P. and Lewis, T. 2011. Environmental history of the Budj Bim landscape: implications for Aboriginal occupation and management and World Heritage nomination. Paper presented at the Budj Bim World Heritage Symposium, Heywood Victoria.

Kershaw, A.P., Tibby, J., Penny, D., Yezdani, H., Walkley, R., Cook, E.J. and Johnston, R. 2004. Latest Pleistocene and Holocene vegetation and environmental history of the Western Plains of Victoria, Australia. In: Kershaw, A.P. (ed), Environmental History of the Newer Volcanic Province of Victoria, pp. 139-161. *Proceedings of the Royal Society of Victoria*, Vol. 116, No. 1. Melbourne.

Koyama, S. and Thomas, D.H. (eds) 1981. *Affluent Foragers: Pacific Coasts East and West.* Senri Ethnological Series 9. Osaka: National Museum of Ethnology.

Land Conservation Council 1981. *Report on the South-Western Area, District 1 – Review.* Melbourne: Land Conservation Council.

Lourandos, H. 1976. Aboriginal settlement and land use in south western Victoria: a report on current field work. *The Artefact* 1:174-193.

Lourandos, H. 1977. Aboriginal spatial organization and population: South western Victoria reconsidered. *Archaeology and Physical Anthropology in Oceania* 12:202-225.

Lourandos, H. 1980a. Forces of change. Unpublished PhD thesis, Department of Anthropology, University of Sydney.

Lourandos, H. 1980b. Change or stability?: Hydraulics, hunter-gatherers and population in temperate Australia. *World Archaeology* 11:254-266.

Lourandos, H. 1983. Intensification: a Late Pleistocene-Holocene archaeological sequence from south western Victoria. *Archaeology in Oceania* 18:81-97.

Lourandos, H. 1984. Review of Australian Aborigines: the languages and customs of several tribes of Aborigines in the Western District of Victoria, Australia by James Dawson. *Aboriginal History* 8:215-219.

Lourandos, H. 1985a. Intensification and Australian prehistory. In: Price, T.D. and Brown, J.A. *Prehistoric Hunter-Gatherers – The Emergence of Cultural Complexity*, pp. 385-423. Academic Press, Inc.

Lourandos, H. 1985b. Problems with the interpretation of late Holocene changes in Australian prehistory. *Archaeology in Oceania* 20:37-39.

Lourandos, H. 1987. Swamp managers of southwestern Victoria. In: Mulvaney, D.J. and White, J.P. (eds), *Australians to 1788* (Vol. 1), pp. 292-307. Sydney: Fairfax, Syme and Weldon Associates.

Lourandos, H. 1988. Palaeopolitics: Resource intensification in Aboriginal Australia and Papua New Guinea. In: Ingold, T., Riches, D. and Woodburn, J. (eds), *Hunters and gatherers 1: History, evolution, and social change*, pp. 148-160. Oxford: Berg.

Lourandos, H. 1993. Hunter-gatherer cultural dynamics: long- and short-term trends in Australian prehistory. *Journal of Archaeological Research* 1(1):67-88.

Lourandos, H. 1996. Change in Australian prehistory: scale, trends and frameworks of interpretation. In: Lilley, I., Ross, A. and Ulm, S. (eds), Proceedings of the 1995 Australian Archaeological Association Conference, pp. 15-21. *Tempus* 6.

Lourandos, H. 1997. *Continent of Hunter-Gatherers. New Perspectives in Australian Prehistory.* Cambridge: Cambridge University Press.

Lourandos, H. and Ross, A. 1994. The great 'intensification debate': its history and place in Australian archaeology. *Australian Archaeology* 39:54-63.

Macpherson, J.H. and Gabriel, C.J. 1962. *Marine Molluscs of Victoria.* Melbourne: Melbourne University Press/Museum of Victoria.

McBryde, I. 1984. Exchange in south eastern Australia: an ethnohistorical perspective. *Aboriginal History* 8:132-153.

McCormac, F.G., Hogg, A.G., Blackwell, P.G., Buck, C.E., Higham, T.F.G. and Reimer, P.J. 2004. SHCal04 southern hemisphere calibration 0-11.0 cal kyr BP. *Radiocarbon* 46:1087-1092.

McNiven, I.J., David, B. and Lourandos, H. 1999. Long-term Aboriginal use of western Victoria: Reconsidering the significance of recent Pleistocene dates for the Grampians-Gariwerd region. *Archaeology in Oceania* 34:83-85.

Meehan, B. 1982. *Shell Bed to Shell Midden.* Canberra: Australian Institute of Aboriginal Studies.

NSW Fisheries 2010. *Native oyster Ostrea angasi.* NSW Department of Primary Industries, Fishing and Aquaculture. http://www.dpi.nsw.gov.au/fisheries/recreational/ saltwater/sw-species/flat-oyster, accessed 13 October 2010.

O'Sullivan, B.W. 1980. The fertility of the Port Lincoln Oyster (*Ostrea angasi sowerby*) from West Lakes, South Australia. *Aquaculture* 19:1-11.

Owens, D. and Hayden, B. 1997. Prehistoric rites of passage: a comparative study of transegalitarian hunter-gatherers. *Journal of Anthropological Archaeology* 16:121-161.

Price, T.D. and Brown, J.A. (eds) 1985. *Prehistoric Hunter-Gatherers – The Emergence of Cultural Complexity.* San Diego: Academic Press, Inc.

Reimer, P.J., Baillie, M.G.L., Bard, E., Bayliss, A., Beck, J.W., Blackwell, P.G., Bronk Ramsey, C., Buck, C.E., Burr, G.S., Edwards, R.L., Friedrich, M., Grootes, P.M., Guilderson, T.P., Hajdas, I., Heaton, T.J., Hogg, A.G., Hughen, K.A., Kaiser, K.F., Kromer, B., McCormac, F.G., Manning, S.W., Reimer, R.W., Richards, D.A., Southon, J.R., Talamo, S., Turney, C.S.M., van der Plicht, J., Weyhenmeyer, C.E. 2009. Intcal09 and Marine09 radiocarbon age calibration curves, 0-50,000 Years Cal BP. *Radiocarbon* 51:1111-1150.

Reitz, E.J. and Wing, E.S. 2008. *Zooarchaeology, Second Edition.* Cambridge Manuals in Archaeology. Cambridge: Cambridge University Press.

Richards, T. 2004. Blackfellows Waterhole 1 – Aboriginal Pre-Contact Site Excavation. In: Webber, H. and Richards, T. (eds), Barrabool Flora and Fauna Reserve Aboriginal Heritage Investigation and Training Project, pp. 23-31. Unpublished Report on Activities of the Aboriginal Community Heritage Investigations Program. Aboriginal Affairs Victoria, Melbourne.

Richards, T. and Johnston, R. 2004. Chronology and evolution of an Aboriginal landscape at Cape Bridgewater, south west Victoria. *The Artefact* 27:97-112.

Richards, T. and Jordan, J. 1996. Archaeological investigations at Cape Bridgewater, Victoria: management recommendations and preliminary excavation report. Unpublished report, Aboriginal Affairs Victoria, Melbourne.

Richards, T. and Webber, H. (eds) 2004. Cape Bridgewater Aboriginal Cultural Heritage Field School. Contributions by Thomas Richards, Michael Godfrey, Harry Webber, Rochelle Johnston, Richard MacNeill and Michael Westaway. Unpublished report, Aboriginal Affairs Victoria, Melbourne.

Rosengren, N. 2001a. Geotechnical and Geomorphological Issues. *Pacific Hydro Portland Wind Energy Project EES.* Sinclair Knight Merz.

Rosengren, N. 2001b. Supplementary Report – Geomorphological Indicators of Potential Archaeological Sites. Appendix 6, *Pacific Hydro Limited Portland Wind Energy Project EES,* Cultural Heritage Study Phase 2. Biosis Research.

Schell, P. 2000a. Cape Bridgewater and Bridgewater Bay Aboriginal cultural heritage review – Volume 1: heritage management. Unpublished report, Aboriginal Affairs Victoria, Melbourne.

Schell, P. 2000b. Cape Bridgewater and Bridgewater Bay Aboriginal cultural heritage review – Volume 2: Aboriginal Cultural Heritage Survey. Unpublished report, Aboriginal Affairs Victoria, Melbourne.

Short, A.D. 1988. Holocene coastal dune formation in Southern Australia: a case study. *Sedimentary Geology* 55:121-142.

Simmons, S. and Djekic, A. 1981. Alcoa Portland Aluminium Smelter Environmental Studies Report No. 2, supplementary archaeological survey. Unpublished report, Aboriginal Affairs Victoria, Melbourne.

Sloss, C.R., Murray-Wallace, C.V. and Jones, B.G. 2007. Holocene sea-level change on the southeast coast of Australia: a review. *The Holocene* 17:999-1014.

Species Bank 2010. *Ostrea angasi.* Australian Government, Department of the Environment, Water, Heritage and the Arts. http://www.environment.gov.au/cgi-bin/species-bank/sbank-treatment.pl?id=69271, accessed 27 May 2010.

Stuiver, M. and Reimer, P.J. 1993. Extended 14C database and revised CALIB radiocarbon calibration program. *Radiocarbon* 35:215-230.

Stuiver, M., Reimer P.J. and Reimer R. 2011. *CALIB Radiocarbon Calibration Program rev.6.0.1.* http://calib.qub.ac. uk/calib, accessed May 2011.

Tibby, J., Kershaw, A.P., Builth, H., Philibert, A. and White, C. 2006. Environmental change and variability in southwestern Victoria: changing constraints and opportunities for occupation and land use. In: David, B., Barker, B. and McNiven, I.J. (eds), *The Social Archaeology of Australian Indigenous Societies,* pp. 354-369. Canberra: Aboriginal Studies Press.

Webb, C. 1995. An evaluation of the archaeological resources of six lightstation reserves in Victoria: Cape Nelson, Cape Otway, Cape Schanck, Wilsons Promontory, Point Hicks, Gabo Island. Unpublished report, Aboriginal Affairs Victoria, Melbourne.

Wentworth, C.K. 1933. The shapes of rock particles: a discussion. *Journal of Geology* 41:306-

309.

Wesson, J.P and Clark, D. 1980. Alcoa Portland Aluminium Smelter Working Paper Number 2, archaeology. Unpublished report, Aboriginal Affairs Victoria, Melbourne.

Williams, E. 1985. Wet underfoot? Earth mound sites and the recent prehistory of southwestern Victoria. Unpublished PhD thesis, Australian National University.

Williams, E. 1987. Complex hunter-gatherers: a view from Australia. *Antiquity* 61:310-321.

Williams, E. 1988. *Complex Hunter-Gatherers: a Late Holocene Example from Temperate Australia.* British Archaeological Reports, International Series 423. Oxford.

Wilson, B.R. and Gillett, K. 1979. *A Field Guide to Australian Shells Prosobranch Gastropods.* Sydney: A.H. & A.W. Reed Pty Ltd.

Wilson, B.R. and Gillett, K. 1980. *Australian Shells.* Sydney: A.H. & A.W. Reed Pty Ltd.

Witter, D. 1977. The archaeology of the Discovery Bay area, Victoria. In: A collection of papers presented to ANZAAS 1977, Vol. 2, pp. 51-72. *Memoirs of the Victorian Archaeological Survey.* Melbourne: Ministry for Conservation.

4

Aboriginal exploitation of toxic nuts as a late-Holocene subsistence strategy in Australia's tropical rainforests

Åsa Ferrier
Archaeology Program, La Trobe University, Melbourne campus, Bundoora, Victoria
a.ferrier@latrobe.edu.au

Richard Cosgrove
La Trobe University, Bundoora, Victoria

Introduction

Human occupation of Sahul (Australia-Tasmania-New Guinea) began about 50,000 years ago, and by 40,000 BP most environments had been colonised (O'Connell and Allen 2008). By this time, people had adapted to the tropics, the arid centre and the glacial areas of Tasmania. In his groundbreaking palynological and palaeoecological studies, Peter Kershaw provided an essential framework within which to examine human responses to changing vegetation and climate variability in Australia. At Lynch's Crater on the Atherton Tableland, Kershaw demonstrated major climatic changes based on the study of pollen and charcoal concentrations fluctuating over the past ca. 120,000 years (Kershaw 1986; Moss and Kershaw 2000). He and his teams identified a date of ca. 45,000 BP for the onset of burning on the Atherton Tableland, through increases in carbon particles around this time (Turney et al. 2001). This evidence was said to support an argument for early peopling of the region, between 45,000 and 55,000 years ago, which also heralded the initial human impacts on Sahul's ecosystems, including the extinction of megafauna. However, despite the evidence for widespread human occupation of Sahul by 40,000 BP, no dates earlier than 8000 years ago have been found in north Queensland's tropical rainforest region. Thus, the present archaeological data run counter to suggestions for the presence of people on the Atherton Tableland ca. 40,000 years ago. However, it is possible there was some Aboriginal occupation on the edge of the rainforest ca. 30,000 years ago (Cosgrove et al. 2007) and further work may reveal late-Pleistocene human occupation in the core rainforest area, as has been detected to the north of the study area (Haberle and David 2004; Summerhayes et al. 2010).

Wide-ranging archaeological research in the Atherton-Evelyn Tableland region of north Queensland (Horsfall 1987, 1996; Cosgrove and Raymont 2002; Cosgrove 2005; Cosgrove et al. 2007; Ferrier 2010) has shown that the majority of the archaeological material dates to within the past 5000 years, with human occupation initially at very low levels, while it was only within the past 2000 years that occupation intensified. Here, we expand on the current understanding of what appears to be a late-Holocene intensification in site occupation and Aboriginal use of toxic food plants in north Queensland's Wet Tropics Bioregion, with an analysis of the macro-botanical remains excavated from Urumbal Pocket, an open archaeological site on the Evelyn Tableland (Figure 1). We use ethnographic analogy and a modern botanical reference collection to understand past techniques of toxic plant food processing and the identification of archaeological specimens.

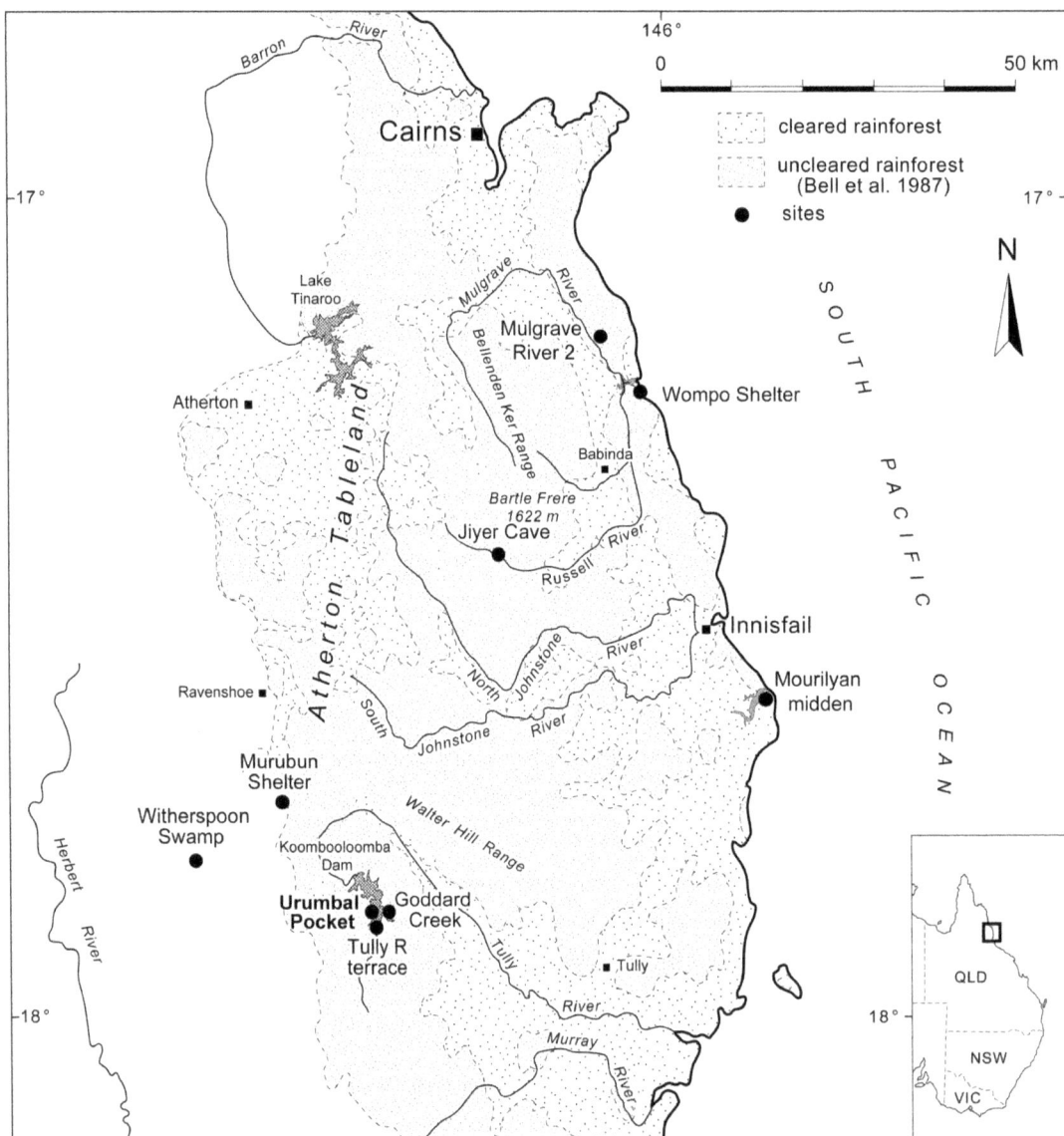

Figure 1. Location of Urumbal Pocket at Koombooloomba Dam and other sites investigated on the Atherton-Evelyn Tableland in Far North Queensland's rainforest region.

Aboriginal exploitation of toxic foods

A variety of explanations have been put forward for continent-wide changes in Aboriginal subsistence patterns and site occupation during the mid- to late-Holocene period (Denham et al. 2009). These include social intensification (Lourandos 1997:303, 305) and broad-spectrum resource use (Haberle and David 2004), population increase and ceremonial activities (Beaton 1983, 1985, 1990), large-scale climatic change (Morwood and Hobbs 1995:182) and high-intensity El Niño Southern Oscillation (ENSO) activity (Rowland 1999; Cosgrove 2005; Turney and Hobbs 2006; Cosgrove et al. 2007). Unravelling the causes for an increase in the use of poisonous food plants is dependent on a capacity to build solid regional frameworks of archaeological and palaeoecological data that include an understanding of (i) substantial archaeological evidence of ancient plant remains, (ii) site formation and taphonomic processes, (iii) an identification of associated changes in other cultural remains with matching palaeoenvironmental signals, and (iv) commensurate regional site comparisons that have stratigraphically intact dated deposits. Previous studies have argued that the development of toxic plant processing techniques occurred outside Australia and diffused into this country about 4000 to 3000 years ago (Beaton 1983). However, the significant inclusion of tree nuts into rainforest hunter-gatherer diets suggests it was part of a specific regional economic development occurring across Australasia in the Holocene (Denham et al. 2009). It has been argued that this development was linked to a much earlier generalist late-Pleistocene subsistence strategy of habitat modification by fire and broad-spectrum plant exploitation. The noted lack of evidence for early plant exploitation may have an underlying taphonomic explanation (Asmussen 2008, 2009, 2010), and unlike faunal remains in some late-Pleistocene sites (Cosgrove and Allen 2001), preservation of organic materials has perhaps hampered the investigations into early plant-food exploitation patterns. We see more consistent evidence in the Holocene from which to make predictions about Aboriginal dietary plant use.

Background

Aboriginal tropical-rainforest occupation and the use of plant foods by Aboriginal rainforest dwellers was extensively recorded in the early contact period by Europeans such as explorers, botanists, Aboriginal Protectors and naturalists (Lumholtz 1889; Meston 1889; Roth 1900, 1901-1910; Coyyan 1918; Mjöberg 1918). Much of the ethno-historical literature has been summarised by Harris (1975, 1978, 1987), Horsfall (1987, 1990) and Pedley (1992, 1993), who have discussed Aboriginal foods of the rainforest. From the historical literature, it is apparent that the rainforest provided a wealth of vegetable foods. Historical documents and Aboriginal oral histories demonstrate that plant foods comprised a significant proportion of the Aboriginal rainforest diet, which included the collection, processing and consumption of a large number of rainforest nuts (e.g. Mjöberg 1918:492-494; Pedley 1993; M. Barlow pers. comm. 2004). More than 112 plants have been identified as food sources consumed by Aboriginal rainforest dwellers. Of these, 10% to 13% are toxic and require extensive processing (Horsfall 1987; Pedley 1993). Most historical descriptions emphasise specific toxic tree nuts that apparently provided an important food source during the wet season (late November through to March). Experimental work by Pedley (1993:179-180) and Tuechler (2010) has shown that the contribution of toxic nuts to the Aboriginal rainforest diet was significant, being important sources of carbohydrates, protein and fats in various quantities. It has been estimated (Pedley 1993) that toxic nuts comprised around 10% to 14% of the diet of rainforest people at the time of Aboriginal-European contact. Their total contribution to the Aboriginal diet in prehistory is unknown, but considering the early ethnographic observations and estimated nutritional values, it was probably considerable.

Gold prospector Mick O'Leary made observations in the early 1880s of nut use in the upper Tully River area:

> The principal food trees are the koah [yellow walnut], burra [black walnut], bean tree, tchupella [black pine] and a number of smaller varieties; there are also a few vines or tree climbers that at times bear edible fruit. The bean tree is not often used by those people on account of its poisonous nature and the amount of work that is attached to preparing it. The nuts are pared into very thin slices using a piece of sharp quartz, then there is a considerable time that it has to go under the water process and fire before it is fit for consumption. The tchupella is a smaller nut and grows on the trees we know as black pine. When the season is on the food is eagerly sought for by those people and they will travel over miles of country to partake of it. They also grind those nuts, into fire, but it does not require the water treatment, baking in the hot ashes being sufficient. The tchupella is an annual bearer, but is not too plentiful and generally found on the high or tableland country. (Coyyan 1918)

Historical descriptions of Aboriginal toxic-nut exploitation in the rainforest mostly refer to two types of walnut, *Beilschmiedia bancroftii* (yellow walnut) and *Endiandra palmerstonii* (black walnut), the 'black pine' nut, *Sundacarpus amara*, and the black bean, *Castanospermum australie*. Many of the toxic species utilised by Aboriginal rainforest people are endemic to Australia's Wet Tropics Bioregion (Hyland et al. 2002). The yellow and black walnuts are available for around eight months of the year, mainly over the spring and summer months, and grow at altitudes ranging from 0 m to 1300 m. The black pine has a considerably shorter fruiting period and is available only between October and December and has a more limited distribution, growing between altitudes of 600 m and 1200 m (Cooper and Cooper 2004). The black-bean tree fruits between March and November and grows at altitudes between 0 m and 840 m (Cooper and Cooper 2004:204). These varieties of toxic nuts have a high food value, high seasonal abundance and storage potential (both above and below ground), and as a result are sought after by both people and rainforest animals. Yellow walnuts and black pine nuts are eaten by cassowaries and bush rats, while the black walnut is popular with white-tailed rats. A particular attraction of the yellow walnut for humans is that much of the fallen fruit can lie on the ground for a short period without attack from predators. Predators avoid the fallen black bean and black pine nuts, but for a more limited time (Pedley 1993:193), while the black walnut must be collected straight away to prevent competition from the white-tailed rat (Pedley 1993:193). These hard-shelled nuts could be stored for several months below ground for later consumption (Coyyan 1918; Mjöberg 1918; Harris 1975).

Based on historical accounts as well as ethnographic observations, it is possible to reconstruct the processes involved in detoxification. Elaborate lawyer-cane (*Calamus australis*) baskets were used for the collection of nuts on the ground and lawyer-cane ropes were used for climbing trees to collect fresh nuts (e.g. Roth 1901-1910; Mjöberg 1918). O'Leary observed Aboriginal rainforest people using a sharp piece of quartz to slice toxic nuts on the Tully River, and in other areas of the rainforest region, snail-shell graters were used (Roth 1900; Pedley 1992:51). Earth ovens were used to steam the toxic nuts and other foods, including meat and fish, sometimes lining the pit with river cobbles as well as ginger leaves, placing the nuts in the pit and covering them up with more leaves, and finally placing hot coals on top. Following this baking and steaming procedure, the nuts were cracked open using a small cobble as a nut-cracking stone. Lastly, the grated pulp was put in lawyer-cane dilly bags and leached for two to three days in a small running creek. Once leached of their toxins, the pulp was chewed and formed into a paste that was eaten raw, and later in the contact period, Aboriginal rainforest people made it into 'Johnny cakes', or flat cakes that were baked on hot coals (Coyyan 1918; Mjöberg 1918; Pedley 1992). What remains in the archaeological record are the carbonised fragments of the hard layer of endocarp that enclose a single seed (nut).

Archaeological investigations

The archaeological open site at Urumbal Pocket is located on a flat spur, high on a bank above the original course of the Tully River (Figure 2).

Urumbal Pocket is a series of *Eucalyptus* patches among rainforest, bordering a stretch of the upper Tully River. In total, six 1 m x 1 m pits were excavated at the site to investigate the

Figure 2.
The archaeological open site at Urumbal Pocket, located on a flat spur, high on a bank above the original course of the Tully River.

density of archaeological remains and the age and stratigraphy of the site, as well as to establish the spatial distribution of cultural materials. Two main stratigraphic units are distinguished on the basis of soil colour and structure – an upper dark humic layer, with an underlying lighter layer (Figure 3).

The soil type is common in the area and defined as Yellow Kandosol (McKenzie et al. 2004:246-247). Unit 1 consists of an artefact-rich black loam, with a Munsell ranging from 10YR1/7 in the top layers to 7.5YR4/6 in the bottom layers. The sediments in Unit 1 can be described as a homogeneous and unconsolidated sandy loam deposit with fine grit throughout, derived from the surrounding organic soil. A transition layer separates Unit 1 and 2 (7.5YR 3/2) and consists of a light brown gritty soil that is distinguished by some orange mottling. Its top starts at a depth of 40 cm and is approximately 10 cm thick. Sediments in Unit 2 (7.5YR 4/6) become increasingly gritty and are less homogeneous with depth, with particles from the decomposing granite bedrock becoming incorporated into the soil.

Chronology

Charcoal was recovered from all squares excavated. Seventeen radiocarbon dates were obtained from *in situ* samples of charcoal recovered during the excavations (Table 1). The dates show a good chronological order, suggesting that the site is relatively undisturbed. The earliest dates at Urumbal Pocket, 7445 ± 68 BP (8273 ± 67 cal BP [Wk-13578]) and 7212 ± 46 BP (8052 ± 65 cal

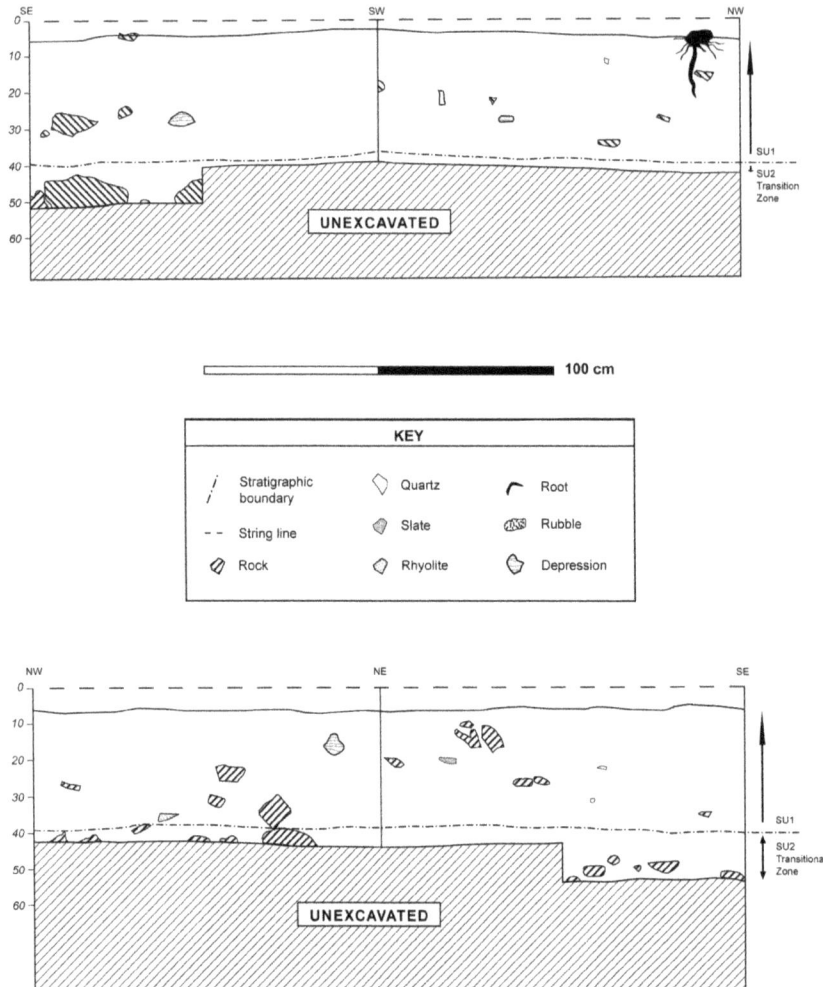

Figure 3. Stratigraphic sections in square Z3 at Urumbal Pocket.

BP [Wk-13571]) are from charcoal found in the lower spits of Unit 2 together with a small number of quartz artefacts. The dates coincide with an initial period of Holocene rainforest expansion, and correspond with increases in microscopic charcoal in the pollen record (Haberle 2005; Cosgrove et al. 2007). A further six AMS dates (bold in Table 1) were obtained from *in situ* diagnostic endocarp fragments. The dates on the nutshell fragments broadly correlate with the dates on the charcoal samples. One excavated Lauraceae endocarp fragment, a yellow walnut (*Beilschmiedia bancroftii*) or a black walnut (*Endiandra palmerstonii*), was submitted for ^{14}C analysis. The fragment returned a date of 1585 ± 40 BP (uncalibrated age; OZJ718) and is consistent with evidence for toxic plant use from other archaeological sites in the region (Horsfall 1987:268; Cosgrove et al. 2007). The dated Lauraceae fragment from Urumbal Pocket provides a minimum age for the appearance of toxic nut processing in the rainforest.

Table 2 below shows how the grouping of radiocarbon dates from Urumbal Pocket suggests three phases of Aboriginal occupation at Urumbal Pocket, where ca. 70% of the dates are younger than 1500 years. The three occupation phases are correlated with increasingly high numbers of cultural materials through time. Some mixing of the deposits in Unit 1 may be related to increased human activity at the site in the past 1000 years. The chronology points to very low intensity initial site use around 8000 years ago, with a hiatus in site use from around 7000 BP until 5000 BP. The evidence points to an occasional use of the site between 5000 BP and 2000 BP. At the end of Phase 2 (transition zone), cultural materials start to increase, and Phase 3, dating from around 2000 BP to the late 1800s, is rich in cultural materials. Thousands of stone artefacts, small amounts of ochre and plant remains were recovered from these upper layers.

Table 1. Conventional and calibrated radiocarbon dates for Urumbal Pocket.

Square	Spit	Material dated	Uncalibrated age	Calendric age cal BP	68% range cal BP	Code
A2	3	charcoal	514 ± 51 BP	569 ± 48	520-617	Wk-11341
A2	5	charcoal	1045 ± 51 BP	980 ± 51	929-1031	Wk-11342
A2	8	charcoal	3339 ± 66 BP	3579 ± 82	3497-3661	Wk-11343
A2	10	charcoal	4887 ± 93 BP	5627 ± 111	5516-5738	Wk-11344
O2	7	charcoal	422 ± 40 BP	449 ± 62	387-511	Wk-13566
O2	10	charcoal	2201 ± 46 BP	2229 ± 69	2160-2298	Wk-13567
S2	13	charcoal	1497 ± 34 BP	1381 ± 30	1351-1411	Wk-13568
S2	15	charcoal	1660 ± 44 BP	1573 ± 54	1581-1627	Wk-13569
V5	7	charcoal	1581 ± 41 BP	1472 ± 47	1424-1519	Wk-13570
V5	9	charcoal	7212 ± 46 BP	8052 ± 65	7987-8117	Wk-13571
V8	6	charcoal	1374 ± 39 BP	1304 ± 21	1283-1325	Wk-13572
V8	8	charcoal	2628 ± 51 BP	2762 ± 34	2728-2796	Wk-13573
V8	**9**	**endocarp**	**1585 ± 40 BP**	**1474 ± 47**	**1427-1521**	**OZJ718**
Z3	2	charcoal	190 ± 37 BP	156 ± 122	33-278	Wk-13574
Z3	**4**	**endocarp**	**470 ± 60 BP**	**502 ± 44**	**457-546**	**OZJ719**
Z3	**5**	**endocarp**	**850 ± 40 BP**	**778 ± 51**	**727-829**	**OZJ720**
Z3	**7**	**endocarp**	**720 ± 40 BP**	**676 ± 21**	**655-697**	**OZJ721**
Z3	8	charcoal	672 ± 39 BP	622 ± 45	577-667	Wk-13575
Z3	11	charcoal	1244 ± 40 BP	1181 ± 64	1117-1245	Wk-13576
Z3	**11**	**endocarp**	**1605 ± 40 BP**	**1486 ± 51**	**1434-1537**	**OZJ722**
Z3	**13**	**endocarp**	**1595 ± 40 BP**	**1480 ± 48**	**1431-1528**	**OZJ723**
Z3	14	charcoal	2143 ± 48 BP	2169 ± 104	2065-2273	Wk-13577
Z3	16	charcoal	7445 ± 68 BP	8273 ± 67	8205-8340	Wk-13578

The plant assemblage

The plant assemblage from Urumbal Pocket consists for the most part of robust endocarp fragments, generically referred to as nutshells. The plant remains are highly fragmented, both as a result of nut-cracking during processing and probably also from human activities at the site, such as trampling and cleaning activities. A small number of complete and incomplete seeds with diagnostic features were also recovered. All of the botanical remains are carbonised inert charcoal, which has allowed for their survival in the archaeological record (Horsfall 1987, 1990). It is not yet clear how the plant remains became burnt, but experiments show that nutshell fragments put on coals produced from a small log fire burn fast and disintegrate to ash. Nutshell charring is therefore most likely to have taken place in a low oxygen environment and their survival is possibly due to anaerobic carbonisation during the steaming and baking of the nuts. This step in the detoxification process has been described as 'nuts roasted in hot ash' (Mjöberg 1918:494), which suggests that the nuts came in direct contact with hot ash and as a result perhaps became burnt. A second explanation is that the nutshells were incorporated into coals in dying fires as waste products during cleaning up activities at the site. Further experimental work may reveal an explanation of their survival in the archaeological record.

Creating a modern reference collection

To facilitate identification of the excavated nutshell fragments, modern samples of yellow

Table 2. Chronologically ordered uncalibrated radiocarbon dates from Urumbal Pocket, grouped into three phases of site use; about 70% date to more recently than 2000 BP (bold dates indicate transition zone).

Stratigraphic Unit 2		Stratigraphic Unit 1
Phase 1 Low level use ca. 8000-7000 BP	Phase 2 Occasional use ca. 5000-2000 BP	Phase 3 Intensive use after ca. 2000 BP
7445 ± 68 BP		
7212 ± 46 BP		
	4887 ± 93 BP	
	3339 ± 66 BP	
	2628 ± 51 BP	
	2201 ± 46 BP	
	2143 ± 48 BP	
		1605 ± 40 BP
		1595 ± 40 BP
		1585 ± 40 BP
		1581 ± 41 BP
		1497 ± 34 BP
		1374 ± 39 BP
		1244 ± 40 BP
		1045 ± 51 BP
		850 ± 40 BP
		720 ± 40 BP
		672 ± 39 BP
		514 ± 51 BP
		470 ± 60 BP
		422 ± 40 BP
		190 ± 37 BP

and black walnuts, as well as black pine nuts, were collected and brought back to the laboratory for comparison. The black bean does not contain a thick endocarp layer and as a result was eliminated as a possible candidate. Samples of modern yellow and black walnuts and black pine nuts were collected from underneath trees located in the study area and in locations on the Atherton-Evelyn Tableland with a vegetation structure similar to pre-European times. This ensured that the modern reference collection was representative, taking into account any size variation within the modern and archaeological plant assemblages. The morphological features of 22 modern samples of yellow and black walnut and five modern samples of black pine nut were recorded. These attributes were then used as a guide to compare and contrast attributes recorded on the archaeological plant material.

Methods applied in the analysis

All nutshell fragments greater than 10 mm in maximum dimension and with distinctive surface features visible to the naked eye were selected for analysis. Surface structures were also compared using a standard binocular microscope (7-40 X). The significant morphological features of modern walnuts and black pine nuts recorded were (i) shape, (ii) size, (iii) thickness of the endocarp wall, and (iv) surface structures. The results from recording these morphological features on the modern samples were compared with results from analysis of the archaeological nutshell fragments (Table 3). In the modern samples, yellow and black walnuts are spherically shaped. The main surface feature on the black walnut is a pointed and sharp apex with an opposite blunt base, and on the yellow walnut the apex and base both have sharp protrusions (Figure 4).

Black pine nuts are also spherical, but are slightly more oblong, smaller and lack the pointed ends. In the first instance, large curved endocarp fragments and fragments with distinctly pointed or swollen ends were selected for further analysis and identification (Figure 5).

Some size variability in the modern walnut samples was observed, similar to that recorded in the botanical literature on tropical rainforest plants (Cooper and Cooper 2004:242). Modern black pine nut samples were all relatively small (20-25 mm) and fell outside the modern walnut size range (28-50 mm). A circle template was used to measure large, curved endocarp fragments in the archaeological assemblage so that the complete modern samples could be compared. This enabled an estimation of the size of the nut, including the complete seed surrounded by the endocarp layer. The modern sample sizes were used to assess potential endocarp shrinkage during the carbonising process that could result in erroneous identifications. Those samples that matched the modern walnut size were selected for further analysis. At this point, fragments significantly smaller than the modern range of walnuts and without any other morphological indicators such as endocarp thickness or surface ornamentations were eliminated from the analysis and recorded as unidentified fragments. Black pine nuts are outside the walnut size range and due to their lack of diagnostic morphological features other than size and shape, it was difficult to make a positive identification. The thickness of the endocarp wall was measured in millimetres using callipers, with the aim of investigating whether or not fragments shrunk in the process of carbonisation, potentially making identification ambiguous (Figure 6).

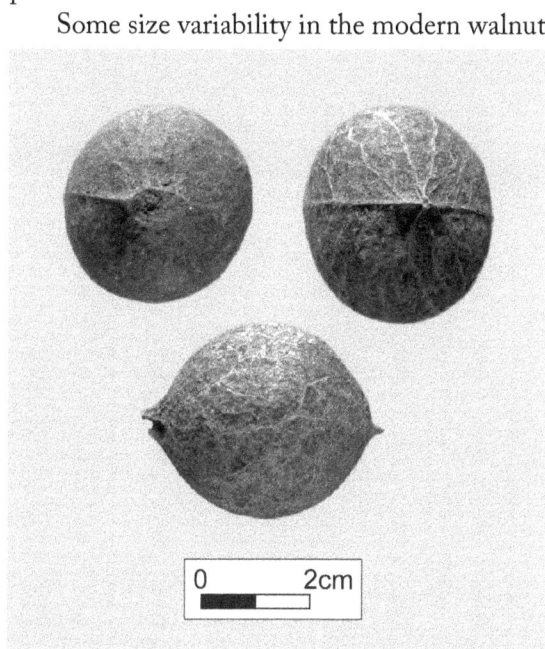

Figure 4. Distinctive features on a modern yellow walnut.

Figure 5. Excavated nutshell fragments showing preserved parts of pointed apex and surface ornamentations.

The endocarp walls of the modern black pine nut were consistently thinner than those of the walnuts, making them susceptible to breakage, and difficult to identify among the archaeological assemblage. Although some thin pieces of endocarp with a smaller circumference were tentatively attributed to black pine, further microscopic analysis of cell structure would be required for a conclusive identification. Surface features were carefully examined and recorded, following Anderberg (1994:9). Endocarp fragments preserving a complete or part of a pointed apex were easily identified to either black or yellow walnut. Other surface structures recorded on modern samples were their relative smoothness, the presence of veins, and surfaces with foveate (pitted), rough and ribbed ornamentations.

A number of complete and incomplete seeds were also recovered in the archaeological deposits. The excavated seeds were identified by comparison with modern reference material held at CSIRO's tropical herbarium in Atherton and with samples collected in the field.

Table 3. Morphological features recorded on modern walnuts and black pine nuts compared with results from analysis of the archaeological nutshell fragments.

Attribute	Modern yellow and black walnut	Excavated nutshell fragment	Modern black pine	Excavated nutshell fragment?
Shape	globose	curved	globose but slightly more oblong than walnuts	slightly curved
Size	28-50 mm	25-50 mm	20-25 mm	<25 mm
Endocarp thickness	1-3 mm	1-3 mm	0.5-1 mm	0.5 mm
Surface ornamentation	sharp protrusions on one or two ends, smooth surface with veins or ribbed structures	sharp protrusions on ends, smooth surface with veins, pitted or rough structures	lack protrusions on ends, overall smooth surface	smooth surface

A number of these were identified as Sapotaceae, genus *Pouteria* (B. Grey pers. comm. 2006). *Pouteria* spp. seeds have a number of key distinguishing features. They are ovate with one or two pointed ends and have a smooth surface with a groove running down the centre of the body (Figure 7).

The interpretation that *Pouteria* species were used and deposited in the sites by humans cannot be supported by reference to ethnographic analogy, although another variety, *Pouteria sericea*, is considered 'bush tucker' in the field guides to vegetation of dry tropical areas in Queensland (Brock 2005:287). In this case, it is the fleshy pericarp that is consumed by humans. Another type of seed found in the archaeological deposits is a small, round to slightly oval seed, between 10 mm and 14 mm in diameter, with distinct surface ornamentations. The seed is enclosed within a thin, wrinkled and woody endocarp (Figure 8). This type of seed has tentatively been identified as belonging to the Elaeocarpaceae and is probably one of the quandong species (B. Hyland pers. comm. 2006). The use of other *Elaeocarpus* species by rainforest dwellers was recorded in the contact period, for example *Elaeocarpus bancroftii*, the Johnstone River almond (Harris 1975:39-43).

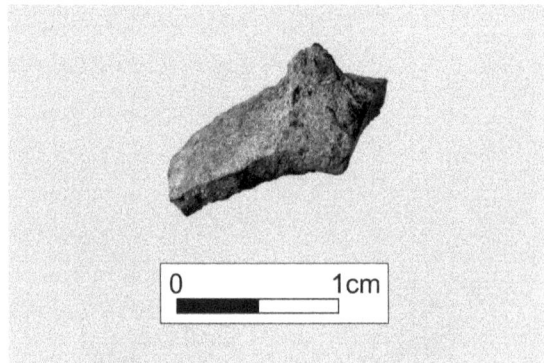

Figure 6. Walnut endocarp fragment showing characteristic morphological features discussed in the text.

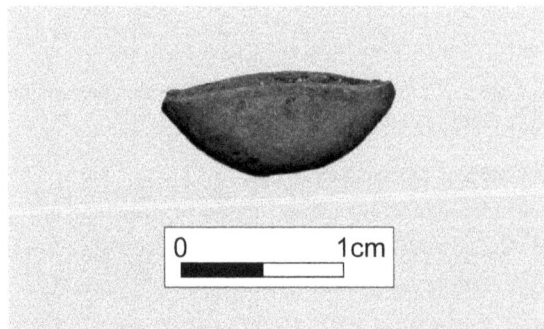

Figure 7. Excavated *Pouteria euphlebia* (fam. Sapotaceae) seed, showing characteristic groove running down the centre of its body.

Potential limitations of the data

Investigations into past plant use by humans must include examinations of taphonomy and site formation processes in order to understand their vectors of introduction (Smith 1982, 1996; Clarke 1985, 1988; McConnell and O'Connor 1997; Wallis 2001, 2003; Martinoli and Jacomet 2004). Thus, discriminating between cultural and natural plant accumulations is important to the interpretation of plant evidence found at archaeological sites in general (Beck et al. 1989). It has been suggested that good indications of the past cultural use of a plant are (i) its presence in high concentrations, (ii) a continuous presence through time in cultural deposits, (iii) similar patterns within a range of taxa, and (iv) the nature of the preservation of the remains (Minnis 1981). Criteria to consider when trying to discriminate between cultural and natural plant accumulations in the archaeological record include radiocarbon dating of plant remains, comparison of species represented in cultural soils with those in adjacent non-cultural soils, and a presence of high concentrations and good preservation of identifiable elements, particularly those that are burnt and charred (Keepax 1977:226-228; Minnis 1981). The plant remains excavated from Urumbal Pocket fulfil these criteria. However, to further assist in the identification of cultural and natural plant accumulations in the archaeological record at Urumbal Pocket, the spatial distribution of plant remains was tested for in a series of shovel tests dug in a 150 m forest transect perpendicular to the site. Some wood charcoal was collected, but no nutshell or other cultural materials were identified in the pits. This suggests that adjacent non-cultural soils contain no charred nutshells and reduces the possibility that the botanical remains were deposited at the site from natural fires and subsequent slope wash.

In addition, none of the identified species grow within the immediate vicinity of the site. The surrounding vegetation at Urumbal Pocket is currently dominated by eucalypts. Several dates on charcoal from soil pits near the excavation suggest fire has influenced the vegetation for the last 8000 years (7181 ± 30 BP = 7988 ± 23 cal BP [Wk-28722]), suggesting that the archaeological plant

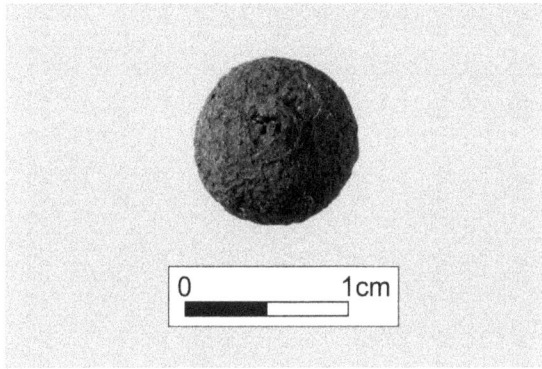

Figure 8. Excavated seed tentatively identified as belonging to the Elaeocarpaceae family.

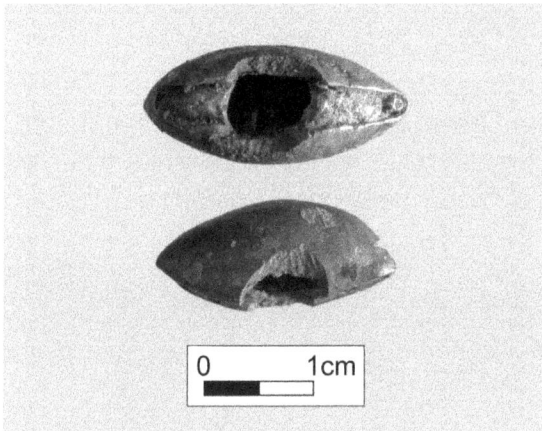

Figure 9. Rat gnaw marks on modern samples of *Pouteria* spp.

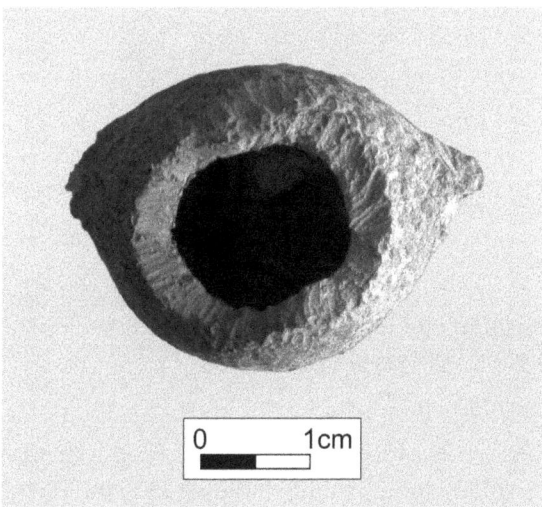

Figure 10. Rat gnaw marks on a modern yellow walnut, *Beilschmedia bancroftii.*

remains are not the result of nuts and seeds falling into the site from overhanging rainforest vegetation. A further indicator that strongly supports the hypothesis that humans discarded the plant remains is the lack of evidence for animal consumption on the archaeological plant remains. Rodents and cockatoos fed on nuts and seeds of the species identified archaeologically, leaving distinct gnaw marks such as those in Figures 9 and 10.

Thus, the evidence supports the notion that the carbonised plant remains found in the Urumbal Pocket excavations were brought to the site by Aboriginal people. To summarise, the evidence for this is:

1. Aboriginal people reportedly used the species represented archaeologically, according to oral traditions and ethno-historical documents.

2. There is a narrow range of species represented.

3. The plant remains are charred.

4. Pits dug outside the site contain no charred plant remains.

5. The remains have not been chewed by rats and other animals.

Results

More than 9100 pieces of plant remains, weighing a total of 422.7 g, were recovered from the Urumbal Pocket archaeological excavations. Of these, more than 90% are unidentified carbonised endocarp fragments fewer than 10 mm in size. The remainder consists for the most part of diagnostic endocarp fragments greater than 10 mm, in addition to a number of complete and partially complete seeds. All of the identified plant remains are from Unit 1, and account for 5.3% of the total plant assemblage. An MNI of 86 pieces of endocarp is derived from the two species of toxic walnut, either *Beilschmiedia bancroftii* or *Endiandra palmerstonii*. At least another 312 fragments are curved pieces of endocarp from a large type of fruit (greater than 25-30 mm in diameter). Given that the estimated size, endocarp wall thickness and surface ornamentations of the partial remains are consistent with features recorded on modern walnuts, there is a strong possibility that these are also walnut fragments. Modern and archaeological samples showed no significant difference in the size of the nuts or endocarp wall thickness and it is concluded that no

Figure 11. Fragment of incised slate grinding stone recovered from the Urumbal Pocket excavations.

major shrinkage occurred to the fragments at the time they were burnt, thus eliminating the possibility of misidentification. Starch grain analysis was carried out on an excavated fragment of an incised slate grinding stone (UP/A2/SP5/1022) from Urumbal Pocket (Figure 11) to provide evidence for what types of plants were being processed on a particular stone artefact (Cosgrove et al. 2007; Field et al. 2009).

These stone tools, sometimes referred to as graters, are restricted to Far North Queesnsland's rainforest region. The results showed that the grooves on incised grinding stones act as 'residue traps' and preserve micro-fossils such as starch and phytoliths. Starch recovered from the fragment excavated at Urumbal Pocket was identified to yellow walnut, indicating on-site processing (Cosgrove et al. 2007:164; Field et al. 2009). The results support an interpretation that toxic starchy seeds were being exploited at Urumbal Pocket, and are consistent with the identification of carbonised nutshells of the yellow walnut (*Beilschmiedia bancroftii*). The remaining 98 identified specimens, which consist of complete and partly complete seeds, have been identified to Sapotaceae, and more specifically to varieties of *Pouteria* spp. A small round seed is also represented at the two sites and has been tentatively identified to Elaeocarpaceae, probably one of the quandong species. Neither of these two species has been previously identified as economically important, or historically documented as a food source by Aboriginal rainforest people.

Results indicate that toxic walnuts were being processed at Urumbal Pocket, a detoxification process that takes several days. It is likely that Aboriginal people stayed at the site more than one night during each visit, assuming that the occupants had the same toxic-nut processing techniques as historically recorded and that they did the whole processing cycle on site.

Table 4. Distribution of carbonised plant remains in grams per square and per composite spit in Analytical Units 1 and 2.

Square	Analytical Unit 1 Weight (g)	Analytical Unit 2 Weight (g)
A2	75.8	1.5
Z3	130.5	10.1
V5	49.7	0.5
V8	93.4	0.9
S2	45.1	2.8
O2	12.0	0.4
Total	**406.5**	**16.2**

Despite the burning process associated with nut processing, our analysis has shown that it is possible to specifically identify a small proportion of the excavated macro-fossils from the archaeological record. We have shown that these nuts belong to the Lauraceae family; i.e. the yellow (*Beilschmedia bancroftii*) and black (*Endiandra palmerstonii*) walnuts. Two other types of seeds recovered in the excavations were identified to a variety of *Pouteria* spp. and a variety of Elaeocarpaceae. Neither is referred to in the ethno-historical literature or remembered by Aboriginal elders as a food source in the recent past (M. Barlow pers. comm.). In contrast, use of *Sundacarpus amara* (black pine) as a food source in the contact period was observed and documented (e.g. Mjöberg 1918). Black pine could not with any certainty be identified in the plant macrofossil assemblage. It is possible that black pine nuts are too fragile to survive site formation processes, although it is also possible that, since the trees only produce nuts during a short period between October and December, the site was occupied at other times.

Long-term Aboriginal plant use at Urumbal Pocket

The temporal distribution of plant remains was assessed in the two Analytical Units. Table 4 shows that most of the plant remains were excavated from Unit 1.

Nutshell fragments from Unit 2 are characteristically smaller in size and lack any clear diagnostic features. However, it demonstrates that carbonised nutshells can survive in older deposits and that Aboriginal people were most likely exploiting rainforest environments and bringing rainforest plant foods to the site before 2000 years ago, as unidentified nutshell fragments were recovered in association with stone artefacts and charcoal in layers radiocarbon dated to ca. 2500 years old. At Urumbal Pocket, numbers of plant remains peak in layers dated to between 800 BP and 400 BP, a time period also associated with high stone-artefact numbers and the presence of rich charcoal deposits (Cosgrove et al. 2007; Ferrier 2010). A number of identified endocarp fragments recovered from Urumbal Pocket were dated (refer to Table 1), suggesting a minimum age of ca. 1600 years BP for toxic-nut use at the site. Interpreting the long-term change and continuity in Aboriginal plant use at Urumbal Pocket is difficult, due to the fragmented plant macro-fossil record. It nevertheless appears that tree-nut exploitation has been an important and consistent component of Aboriginal rainforest diet for at least 1000 years. These trends are also reflected in other archaeological sites investigated in the rainforest region: Murubun Shelter, Goddard Creek (Cosgrove et al. 2007), Jiyer Cave and the Mulgrave River sites (Horsfall 1996) (Figure 1).

Discussion

Palaeoenvironmental reconstructions in the rainforest region demonstrate that in the mid Holocene an extended period of dry conditions and environmental pressures began. It has been suggested that as a response to these environmental changes, Aboriginal people living near or on the rainforest fringe were forced to undertake more frequent journeys into the rainforest as a survival strategy (Ferrier 2010). This interpretation is supported in the archaeological record from Urumbal Pocket and from two other sites investigated in this area – Murubun Shelter and Goddard Creek (Figure 1). The archaeological records from these sites suggest a change from the exploitation of the semi-dry landscape bordering the rainforest region to a more permanent life in the rainforest in the late Holocene (Cosgrove et al. 2007). The archaeological record from Urumbal Pocket indicates that increases in Aboriginal activity began to take place around 2000 years ago, accelerating and peaking in the past 1000 years. Before this time, a relatively low level of human occupation occurred at the site. The past 2000 years of Aboriginal occupation at Urumbal Pocket show major increases in cultural remains, including the appearance of incised grinding stones used to process toxic nuts, evidence that points to significant changes in the way Aboriginal people were now exploiting the rainforest environment and its resources. Numbers of plant remains increase dramatically in archaeological deposits dated to the past 1000 years of occupation, a pattern that is reported to be similar in archaeological sites investigated across the rainforest region (Horsfall 1996; Cosgrove et al. 2007). At the time of Aboriginal-European contact, historical accounts demonstrate that varieties of toxic nuts provided a reliable staple food source on the Atherton-Evelyn Tableland during the wet season. Thus, one possibility is that toxic nuts played a significant role in the development of complex semi-sedentary rainforest societies that were recorded at contact, and perhaps also provided the means for large ceremonial gatherings to be held on the Atherton-Evelyn Tablel and during the wet season when large quantities of toxic nuts were consumed (Coyyan 1918).

Conclusion

Peter Kershaw's pioneering research on the Atherton Tableland in Far North Queensland's tropical rainforest region over the past 40 years has established the important palaeoenvironmental backdrop to human-environment interactions on the Atherton Tableland. These include his palaeovegetation framework derived from pollen and charcoal studies from Lynch's Crater. We have benefited from this important work, allowing us to investigate the ecological relationships between people and changing subsistence patterns in a unique part of Australia. Results of the analysis on archaeobotanical macro-fossils excavated from the Urumbal Pocket open site in Far North Queensland's Wet Tropics World Heritage Bioregion demonstrate that the application of ethnographic analogy to the archaeological plant remains provides clues to past human subsistence behaviour. By applying this method, we have identified toxic food use that extends back to at least 1600 years ago. Based on the evidence, it appears that Aboriginal people repeatedly collected and processed rainforest walnuts at the Urumbal Pocket site over a period of ca. 1600-2000 years. The evidence also suggests that this subsistence strategy was already in place on a much smaller scale before the shift towards more permanent rainforest occupation sometime in the past 2000 years. This adaptive shift has been interpreted as the outcome of highly unstable ENSO activity beginning about 5000 years ago (Turney and Hobbs 2006; Cosgrove et al. 2007). Our research demonstrates that explanations of cultural change in the mid- to late-Holocene period can be linked to broad-scale environmental changes. Processes of social intensification probably led to a major re-orientation of social, political and economic structures. It also appears that two languages developed during this time, Yidinj in the north and Dyirbal in the south (Dixon 1991:4). This facilitated the emergence of a fully functioning rainforest society that was unique in Aboriginal Australia.

Acknowledgements

This research was funded by an ARC Discovery Grant DP0210363. La Trobe University provided Ferrier with a scholarship to carry out her PhD research on the Urumbal Pocket archaeological material. AINSE funded the radiocarbon dating of the nutshell fragments (Research Award 07/034). Radiocarbon determinations were conducted at the Waikato Radiocarbon Laboratory in New Zealand, and we thank Alan Hogg and Fiona Petchey for their assistance. The collection of modern plant specimens were conducted with permission from the Department of Environment and Resource Management Queensland (WISP07417410). All research procedures were approved by the relevant Ethics Committee (Reference Number: 04-44). We thank the Jirrbal Aboriginal people in Ravenshoe for granting us permission to excavate the Urumbal Pocket site, and particularly Maisie Barlow for sharing her childhood memories and knowledge with us. We would also like to thank Bruno David and Simon Haberle and two anonymous referees for constructive comments on our paper.

References

Anderberg, L. 1994. *Atlas of Seeds and small fruits of Northwest-European plant* species. Swedish Museum of Natural History, Stockholm.

Asmussen, B. 2008. Anything more than a picnic? Re-considering arguments for ceremonial Macrozamia use in mid-Holocene Australia. *Archaeology in Oceania* 43:93-103.

Asmussen, B. 2009. Another burning question: hunter-gatherer exploitation of Macrozamia spp. *Archaeology in Oceania* 44:142-9.

Asmussen, B. 2010. In a nutshell: the identification and archaeological application of experimentally defined correlates of macrozamia seed processing. *Journal of Archaeological Science* 37:2117-25.

Beaton, J. 1983. Does intensification account for change in the Australian Holocene Archaeological record? *Archaeology in Oceania* 18:94-97.

Beaton, J. 1985. Evidence for a coastal occupation time-lag at Princess Charlotte Bay (North

Queensland) and implications for coastal colonisation and population growth theories for Aboriginal Australia. *Archaeology in Oceania* 20:1-20.

Beaton, J. 1990. The importance of past population for prehistory. In: Meehan, B. and White, N. (eds), Hunter-Gatherer Demography: Past and Present. *Oceania Monograph* 39:23-40.

Beck, W.A., Clarke, A. and Head, L. 1989. *Plants in Australian Archaeology*, Tempus Vol. 1, Anthropology Museum, University of Queensland, St. Lucia.

Bell, F.C., Winter, J.W., Pahl L.I. and Atherton, R.B. 1987. Distribution, area and tenure of rainforest in northeastern Australia. *Proceedings of the Royal Society of Queensland* 98:27-39.

Brock, J. 2005. *Native Plants of Northern Australia*. Reeds Books, Chatswood, NSW.

Clarke, A. 1985. A preliminary archaeobotanical analysis of the Anbanagbanag 1 site. In: Jones, R. (ed), *Archaeological research in Kakadu National Park*, pp. 77-96. Australian National Parks and Wildlife Service, Special Publication 13.

Clarke, A. 1988. Archaeological and ethnobotanical interpretations of plant remains from Kakadu National Park, Northern Territory. In: Meehan, B. and Jones, R. (eds), *Archaeology with ethnography: An Australian perspective*, pp. 123-36. Department of Prehistory, Research School of Pacific Studies, Canberra.

Cooper, W. and Cooper, W.T. 2004. *Fruits of the Australian Tropical Rainforest*. Nokomis Editions Pty Ltd, Melbourne.

Cosgrove, R. 2005. Coping with noxious nuts. *Nature Australia* 28(6):47-53.

Cosgrove, R. and Allen, J. 2001. Prey Choice and Hunting Strategies in the Late Pleistocene: Evidence from Southwest Tasmania. In: Anderson, A., Lilley, I. and O'Connor, S. (eds), *Histories of old ages. Essays in honour of Rhys Jones*, pp. 397-429. Research School of Pacific and Asian Studies, ANU.

Cosgrove, R., Field, J. and Ferrier, Å. 2007. The archaeology of Australia's tropical rainforests. *Palaeogeography, Palaeoclimatology, Palaeoecology* 251:150-73.

Cosgrove, R. and Raymont, E. 2002. Jiyer Cave revisited: preliminary results from northeast Queensland rainforest. *Australian Archaeology* 54:29-36.

Coyyan (M. O'Leary) 1918. *The Aboriginals*. Columns I-X. *The Northern Herald* (*The Tablelander*).

Denham, T., Fullagar, R. and Head, L. 2009. Plant exploitation in Sahul: From colonisation to the emergence of regional specialization during the Holocene. *Quaternary International* 202:29-40.

Dixon, R.M.W. 1991. *Words of Our Country: Stories, Place Names and Vocabulary in Yidiny. The Aboriginal Language of the Cairns-Yarrabah Region*. Queensland University Press.

Ferrier, Å. 2010. Journeys into the rainforest. Long-term change and continuity in Aboriginal tropical rainforest occupation on the Evelyn Tableland in far north Queensland. Unpublished PhD thesis. La Trobe University, Melbourne.

Field, J., Cosgrove, R., Fullagar, R. and Lance, B. 2009. Survival of starch residues on grinding stones in private collections: a study of morahs from the tropical rainforests of NE Queensland, pp. 228-38. In: Haslam, M., Robertson, G., Crowther, A., Nugent, S. and Kirkwood, L. (eds), *Archaeological science under a microscope. Studies in Residue and ancient DNA analysis in Honour of Thomas H. Loy*. Terra Australis 30, ANU E Press, Canberra.

Haberle, S. 2005. A 22Ka pollen record from Lake Euramoo, Wet Tropics of NE Queensland, Australia. *Quaternary Research* 64(3):343-56.

Haberle, S.G. and David, B. 2004. Climates of change: human dimensions of Holocene environmental change in low latitudes of the PEPII transect. *Quaternary International* 118-9, 165-79.

Harris, D.R. 1975. Traditional patterns of plant-food procurement in the Cape York Peninsula and Torres Strait Islands: report on fieldwork carried out Aug-Nov 1974.

Harris, D. 1978. Adaptation to a tropical rain-forest environment: Aboriginal subsistence in northeastern Queensland. In: Blurton Jones, N.G. and Reynolds, V. (eds), *Human behaviour and adaptation*, pp. 113-34. Taylor & Francis, London.

Harris, D. 1987. Aboriginal subsistence in a tropical rain forest environment: food procurement, cannibalism, and population regulation in northeastern Australia. In: Harris, M. and Ross, E.B. (eds), *Food evolution: toward a theory of human food habits*, pp. 357-85. Temple University Press, Philadelphia.

Horsfall, N. 1987. Living in rainforest: the prehistoric occupation of North Queensland's humid tropics. Unpublished PhD thesis. James Cook University, Townsville.

Horsfall, N. 1990. People and the rainforest: an archaeological perspective. In: Webb, L.J. and Kikkawa, J. (eds), *Australian tropical rainforests: science-values-meaning*, pp. 33-9. CSIRO, Melbourne.

Horsfall, N. 1996. *Holocene occupation of the tropical rainforests of North Queensland.* Tempus Vol. 4, pp. 174-90.

Hyland, B.P.M., Wiffin, T., Christophel, D.C. and Elick, R.W. 2002. *Australian Tropical Rain Forest Plants: Trees, Shrubs and Vines*. CSIRO Publishing, Melbourne.

Keepax, C. 1977. Contamination of archaeological deposits by seeds of modern origin with particular reference to the use of flotation machines. *Journal of Archaeological Science* 4:221-9.

Kershaw, A.P. 1986. Climatic change and Aboriginal burning in northeast Australia during the last two glacial/interglacial cycles. *Nature* 322:47-9.

Lourandos, H. 1997. *Continent of hunter-gatherers: new perspectives in Australian prehistory.* Cambridge University Press, New York.

Lumholtz, C. 1889. *Among Cannibals: account of four years travels in Australia and of camp life with the Aborigines of Queensland*. Australian National University Press, Canberra (reprinted 1980).

Martinoli, D. and Jacomet, S. 2004. Identifying endocarp remains and exploring their use at Epipalaeolithic Öküzini in southwest Anatolia, Turkey. *Vegetation History and Archaeobotany* 13:45-54.

McConnell, K. and O'Connor, S. 1997. 40,000 year record of food plants in the southern Kimberley. *Australian Archaeology* 45:20-31.

McKenzie, N., Jacquier, D., Isbell, R. and Brown, K. 2004. *Australian Soils and Landscapes: An Illustrated Compendium*, Melbourne, CSIRO.

Meston, A. 1889. Report of the Government Scientific Expedition to Bellender-Ker Range. Government Printer, Brisbane.

Minnis, P.E. 1981. Seeds in archaeological sites: sources and some interpretive problems. *American Antiquity* 48(1):143-52.

Mjöberg, E. 1918. *Bland Stenåldersmänniskor i Queenslands Vildmarker (Amongst Stone Age People in the Queensland Wilderness)*. Albert Bonniers Boktryckeri, Stockholm.

Morwood, M.J. and Hobbs, D.R. 1995. Conclusions. In: Morwood, M.J. and Hobbs, D.R. (eds), *Quinkan Prehistory: The archaeology of Aboriginal Art in S.E. Cape York Peninsula, Australia*. Tempus Vol. 3, pp. 178-85. Archaeology and Material Culture Studies in Anthropology. The University of Queensland, Brisbane.

Moss, P.T. and Kershaw, A.P. 2000. The last glacial cycle from the humid tropics of northeastern Australia: comparison of a terrestrial and a marine record. *Palaeogeography, Palaeoclimatology, Palaeoecology* 155:155-76.

O'Connell, J. and Allen, J. 2008. Getting from Sunda to Sahul. In: Clark, G., Leach, F. and O'Connor, S. (eds), *Islands of Inquiry. Colonisation, seafaring and the archaeology of maritime landscapes*, pp. 31-46. Terra Australis 29, ANU E Press.

Pedley, H. 1992. *Aboriginal life in the rainforest*. Queensland Department of Education, Brisbane.

Pedley, H. 1993. Plant detoxification in the rainforest: The processing of poisonous plant foods by the Jirrbal-Girramay people. Unpublished MA thesis. Material Culture Unit, James Cook University, Townsville.

Roth, W.E. 1900. Scientific report [to the Under-Secretary, Brisbane] on the natives of the (lower) Tully River, Cooktown.

Roth, W.E. 1901-1910. *North Queensland ethnography*. Bulletins 1-8. Department of Home Secretary, Brisbane.

Rowland, M. 1999. Holocene environmental variability: have its impacts been underestimated in Australian pre-history? *The Artefact* 22:11-48.

Smith, M. 1982. Late Pleistocene zamia exploitation in southern Western Australia. *Archaeology in Oceania* 17:117-21.

Smith, M. 1996. Revisiting Pleistocene Macrozamia. *Australian Archaeology* 42:52-3.

Summerhayes, G.R., Leavesley, M., Fairbairn, A., Mandui, H., Field, J., Ford, A. and Fullagar, R. 2010. Human Adaptation and Plant Use in Highland New Guinea 49,000 to 44,000 Years Ago. *Science* 330:78-81.

Tuechler, A. 2010. Toxic plant food processing in north-east Queensland's rainforest region. A study in cost-benefit ratio. Unpublished Honours dissertation. La Trobe University.

Turney, C.S.M., Kershaw, A.P., Moss, P., Bird, M.I., Fifield, L.K., Cresswell, R.G., Santos, G.M., Di Tada, M.L., Hausladen, P.A. and Zhou, Y. 2001. Redating the onset of burning at Lynch's Crater (North Queensland): implications for human settlement in Australia. *Journal of Quaternary Science* 16 (8):767-71.

Turney, C.S.M. and Hobbs, D.R. 2006. ENSO influence on Holocene Aboriginal populations in Queensland, Australia. *Journal of Archaeological Science* 33:1744-48.

Wallis, L. 2001. Environmental history of northwest Australia based on phytolith analysis at Carpenter's Gap 1. *Quaternary International* 83-85:103-17.

Wallis, L. 2003. An overview of leaf phytolith production patterns in selected northwest Australian flora. *Review of Palaeobotany and Palynology* 125:201-48.

5

Terrestrial engagements by terminal Lapita maritime specialists on the southern Papuan coast

Ian J. McNiven
School of Geography and Environmental Science, Monash University, Clayton, Victoria
ian.mcniven@monash.edu

Bruno David
Monash University, Clayton, Victoria

Ken Aplin
National Museum of Natural History, Smithsonian Institution, Washington D.C., United States of America

Jerome Mialanes
Monash University, Clayton, Victoria

Brit Asmussen
Queensland Museum, South Brisbane, Queensland

Sean Ulm
James Cook University, Cairns, Queensland

Patrick Faulkner
University of Queensland, Brisbane, Queensland

Cassandra Rowe
Monash University, Clayton, Victoria

Thomas Richards
Monash University, Clayton, Victoria

Introduction

In 1974, Peter Kershaw published a paper in *Nature* outlining a remarkable pollen core sequence from Lynch's Crater in tropical northeast Queensland (Kershaw 1974). From an archaeological perspective, the most interesting dimension to this work was the novel and provocative suggestion that the transition from rainforest to sclerophyll forest beginning around 38,000 BP may have been related to anthropogenic burning of the landscape since Aboriginal colonisation of the continent (Kershaw 1974:222). In a sense, Kershaw was giving empirical veracity to Rhys Jones's (1969) paradigmatic notion of 'fire-stick farming' and the proposition that 'the arrival of Aboriginal man [to Australia] increased the fire frequency by an enormous amount'. In 1981, Kershaw and colleagues gave further support to the anthropogenic burning interpretation by showing increases in charcoal counts within the Lynch's Crater core coincident with the rainforest-sclerophyll forest transition (Singh et al. 1981). While initially archaeologists were wary of this new and alternative approach to the human past in Australia, by 1993 Kershaw could rightly claim that 'information from some pollen records has been important to the debate on the time of arrival of Aboriginal people' (Kershaw 1993:14).

A key concern of archaeologists over the use of pollen and charcoal records to date human colonisation was the controversial claim that charcoal in some cores pointed to an Aboriginal presence well over 100,000 years ago and well before the earliest archaeological evidence of less than 50,000 years ago (Singh et al. 1981; White and O'Connell 1982:42; Kershaw 1993). Yet few disagreed with the general thrust of Singh et al.'s (1981:45) proposition that 'the impact of the increase in the frequency of fires through the early activities of Man may have been marked, so that evidence from the vegetation history of an area might be used to supplement, or foreshadow, the archaeological record'. More recently, Hiscock (2008:37) has suggested that fire records at sites such as Lynch's Crater may underestimate human presence, as 'intensified fire frequencies might signal a time when humans began regularly using fire to burn their ecosystem, which could be long after the colonisation of Australia'. Despite these useful caveats, the re-dating of anthropogenic firing at Lynch's Crater to 45,000 years ago by Kershaw and colleagues (Turney et al. 2001) now comes close to current archaeological evidence for the antiquity of the human colonisation of Australia (O'Connell and Allen 2004; see also Mooney et al. 2011).

To the immediate north of Australia in New Guinea, palaeoenvironmental approaches to human colonisation and use of landscapes have similarly complemented archaeological approaches to the human past (Hope and Haberle 2005). Here, one question that has long intrigued archaeologists and ecologists alike is the history and formation of grasslands across various parts of the highlands and the role of human vegetation clearance and landscape firing in this process, particularly over the past 4000 years (e.g. Hope 1976, 1983, 2009; Haberle 1998; Haberle et al. 2001; Swadling and Hope 1992). Indeed, such is the scale and significance of these anthropogenic transformations that Hope et al. (1983:41) suggest that 'Man-made landscapes occur over about 200,000 km² [25%] of the 800,000 km² of New Guinea'.

Intense research into the impacts of human colonisation and long-term occupation of the Papua New Guinea highlands has not been matched across the surrounding coastal lowlands. In terms of Pleistocene occupation, this paucity of research has led to a situation where 'there are no coastal sites from mainland New Guinea that can provide data with which to assess human impacts on the coastal resources for this crucial early period of time' (Summerhayes et al. 2009:730). A similar situation exists for insular eastern Papua New Guinea, where evidence for 'an environmental impact signature for these early peoples' during the Pleistocene is generally lacking (Summerhayes et al. 2009:731). Yet the lowlands (or at least islands) were the setting for the second major colonisation process across Melanesia – Austronesian expansions, particularly those associated with Lapita peoples of the past ca. 3500 years. As Enright and Gosden (1992) and Summerhayes et al. (2009) point out, here evidence for human environmental impacts becomes much more apparent in terms of:

- introduction of new animals (e.g. pigs, dogs, rats);
- introduction of horticultural plants (e.g. bananas);
- extinctions (e.g. birds);
- firing of vegetation (associated with agriculture and hunting);
- accelerated erosion of hillsides; and
- lowland sedimentation and coastal progradation.

Insights into Austronesian environmental impacts on the New Guinea mainland are negligible, as until now no conclusive evidence for Lapita settlement of mainland New Guinea had been found (e.g. Lilley 2008:79) and the period of concern reveals 'very little correlation with [anthropogenic] environmental change' (Hope and Haberle 2005:548). The recent excavation of Lapita and post-Lapita sites dating between 2900 and 2000 years ago at Caution Bay immediately northwest of Port Moresby, southern Papua New Guinea, negates the first conclusion of a purported absence of Lapita, and provides scope to challenge the second (McNiven et al. 2011). For 40 years, the accepted view has been that Austronesian colonisation of the southern Papuan coast took place around 2000 years ago by maritime peoples possessing a pottery tradition similar to but post-dating Lapita (David et al. in press). Ethnographically, the descendants of these peoples in the Port Moresby region practised a mixed economy of marine fishing and shellfishing, along with wallaby hunting across grasslands and savannah vegetation maintained by firing. While the antiquity of this process of anthropogenic landscape modification is unknown, it is considered to be no more than 2000 years ago. Here, we extend the known antiquity of mixed economic practices and possible anthropogenic landscape modification in the Port Moresby region using recent excavation results from Caution Bay, focusing on Edubu 1 site, dating from <2350 to 2650 cal BP. Our key aim is to establish a historical framework for pre-2000 cal BP human landscape engagements and transformations along the southern coast of mainland Papua New Guinea.

Edubu 1

Edubu 1 (aka AH15) is located 1 km inland from the southern end of Caution Bay and 20 km northwest of Port Moresby (Figure 1). The site name refers to the local place name for a nearby creek. The site falls within the Fairfax Land System, with characteristic gently undulating terrain, brown clay soils and savannah and grassland dominated by *Themeda australis* grass and *Eucalyptus* trees (Mabbutt et al. 1965). The site is 19 m above sea level within the Vaihua River drainage catchment and is positioned on flat to gently sloping ground elevated above the southern margins of Moiapu creek, which supports *Pandanus* trees. It is on the ecotone of grasslands (that extend inland from the coast) and the start of open woodlands and eucalypt savannah (Paijmans 1975). Narrow zones of riparian or gallery forest vegetation can be found along some waterways and along the frontal dune flanking the nearby shoreline. Extensive mangrove forests up to 600 m wide front Caution Bay opposite the site. The Owen Stanley Range provides a visually impressive backdrop to the area and commences its steep rise around 40 km inland. As part of the broader Port Moresby region, Caution Bay experiences a tropical monsoonal climate with a characteristic wet season (December to April), which accounts for 80% of precipitation, and a dry season (April to November) (McAlpine et al. 1983). Annual rainfall averages 1000 mm on the coast and increases nearly four-fold, moving inland over a distance of 50 km (Mabbutt et al. 1965:89).

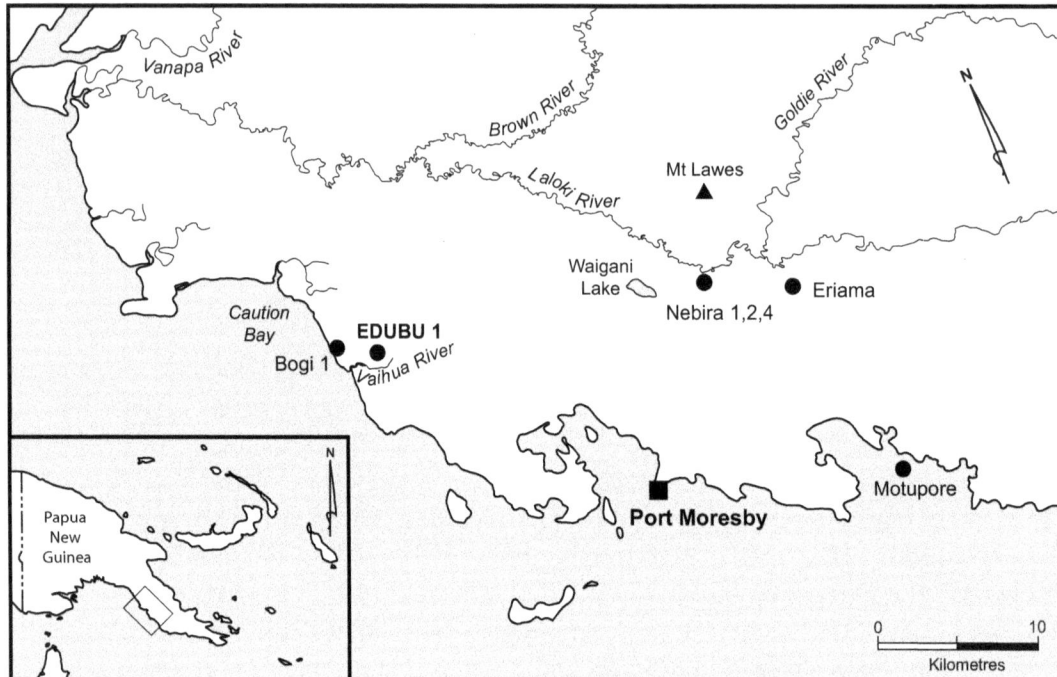

Figure 1. Study area.

Site description

Edubu 1 is located on a flat area that drops away on the northwest side down to an ephemeral waterway known locally as Moiapu creek (Figure 2). The site is covered in grass, with scattered and isolated low shrubs. Sediments are silty clays. The adjacent waterway is also covered in grass but includes scattered pandanus trees. The archaeological site extends along the edge of the flat land and down the slope towards the waterway. The highest concentration of cultural material is located between the top of the slope and back across the flat for some 10 m. The length of the site is at least 30 m. Cultural material extends down the slope towards the waterway for a distance of 10 m. Most if not all of the slope cultural materials derive from in-situ cultural deposits up on the flat area. Cultural materials across the site surface are mainly marine shells, flaked stone artefacts, fire-cracked rocks and pottery sherds. A low termite mound was located in the middle of the site up on the flat area. Squares A to C were located across the main concentration of surface cultural materials on the flat area. However, Square C was located on the edge of the flat area and presumably the edge of the main settlement area.

Excavation and stratigraphy

Three 1 m x 1 m pits (Squares A, B and C) were excavated across Edubu 1. The three squares were aligned in an approximate southwest-northeast direction, with Square A in the southwest and Square C in the northeast. Centrally located Square B was located 7 m from Square A and 6 m from Square C, the distance between Squares A and C being 13 m. Similar but not identical cultural deposit was encountered in all squares. For the purposes of this chapter, analyses focus on Square A, which revealed less post-depositional disturbance and greater stratigraphic integrity than Squares B and C. Square A was dug to a maximum depth of 90 cm (southeast corner) in 40 Excavation Units (XUs), with a mean thickness of 2.2 cm (Figure 3). The weight and volume of each XU was measured to the nearest 0.5 kg and 0.5 litres respectively. Elevations were taken at the start and end of each XU and bulk sediment samples were taken for each XU. All excavated sediments were wet sieved through 2.1 mm mesh.

Figure 2. Edubu 1 during excavation, with Moiapu creek in the foreground (looking southeast), 28 September 2009. Squares A to C located right to left are positioned under each shade tent. Photo: Ian J. McNiven.

Figure 3. Edubu 1 Square A, east and south sections, at end of excavation (scale in 10 cm units). Photo: Ian J. McNiven.

Diagnostic and fragile artefacts (e.g. obsidian, decorated ceramic sherds, adzes) and charcoal fragments for AMS dating were plotted in 3D and bagged separately.

The deposit was divided into four major Stratigraphic Units (SUs) (Figure 4). SUs 1 to 3 account for the upper three-quarters of the deposit and are the main layers bearing cultural materials (e.g. shells, bones, stone artefacts and pottery). SU1 is consolidated very dark brown to black silty clay and takes in the upper 13 cm of the deposit. SU2 is consolidated dark-brown silty clay that varies in depth from 16 cm to 29 cm below the surface. SU3 is consolidated brown clay with a maximum depth of 68 cm. SU3a was isolated due to infiltration of SU2 sediments. SU3b is slightly more consolidated and mottled compared with other SU3 sediments. SU3c sediments contain few cultural materials. SUs 3d to 3f represent burrows. SU4 features consolidated clays that take in the lower 27 cm of deposit. SU4a is consolidated brown clay, with mottling and some cultural materials from infiltration of SU3c sediments. SUs 4b to 4c are culturally sterile and mottled sediments, with layer orientation tilting downwards and very different to that seen for SUs 1 to 3. Sediments vary from mottled orange and brown clays (SU4b), to mottled white, grey and brown clays with limestone inclusions (SU4c), and mottled grey and dark-brown clays (SU4d).

Cultural materials

Square A contains a diverse range of cultural materials. They include: marine shell (24,090.1 g), sea-urchin exoskeleton (1757.5 g), crustacean exoskeleton (49.8 g), bone (283.3 g), egg shell (<0.1 g), ceramic sherds (6552.2 g), stone artefacts (1776.7 g) and charcoal (0.9 g). Vertical changes in the distribution of these cultural materials reveal three major concentrations of activity – an upper concentration between XU1 and XU16 (0 cm to 38 cm below the surface) and a middle concentration between XU16 and XU23 (38 cm to 52 cm below the surface), which merge at their boundary, and a more isolated lower concentration between XU27 and XU32 (60 cm to 73 cm below the surface) (Figure 5).

Radiocarbon dating and chronology

Eight AMS radiocarbon dates are available for Edubu 1 (Table 1). All dates were obtained on single fragments of wood charcoal and determined by the University of Waikato Radiocarbon Dating Laboratory in New Zealand. Radiocarbon dates were calibrated into calendar years using the online calibration program Calib 6.0 (Stuiver and Reimer 1993) and the IntCal09 dataset (Reimer et al. 2009). Due to a paucity of charcoal in the site, a more comprehensive and additional series of AMS dates is planned using marine shell, applying species-specific Delta-R marine reservoir correction values for the Caution Bay area as determined by a comprehensive program of paired shell-charcoal dates from the area.

The available charcoal dates range from 2339 ± 30 BP (ca. 2350 cal BP) to 2546 ± 30 BP (ca. 2650 cal BP). The two dates for Square A come from near the base of the upper (ca. 2350 cal BP) and middle (ca. 2500 cal BP) concentrations of cultural materials. Insight into the age of the lower concentration of cultural materials is provided by the six AMS dates in Squares B and C. The lowest and oldest dates in Square B (XU29a) and Square C (XU22) are 2520 ± 30 BP (ca. 2600 cal BP) and 2546 ± 30 BP (ca. 2650 cal BP) respectively. These dates suggest strongly that the lowest cultural materials in Square A date to around 2600-2650 cal BP, or more broadly to within the period 2500-2750 cal BP at 2 sigma probability.

Subsistence (marine)

Shellfish. Marine mollusc shells were recovered from throughout the 90 cm-deep Square A sequence, with nearly all (99.9% by weight) recovered from XUs 1 to 32, taking in the three concentrations of cultural materials down to a depth of 73 cm. The assemblage comprised

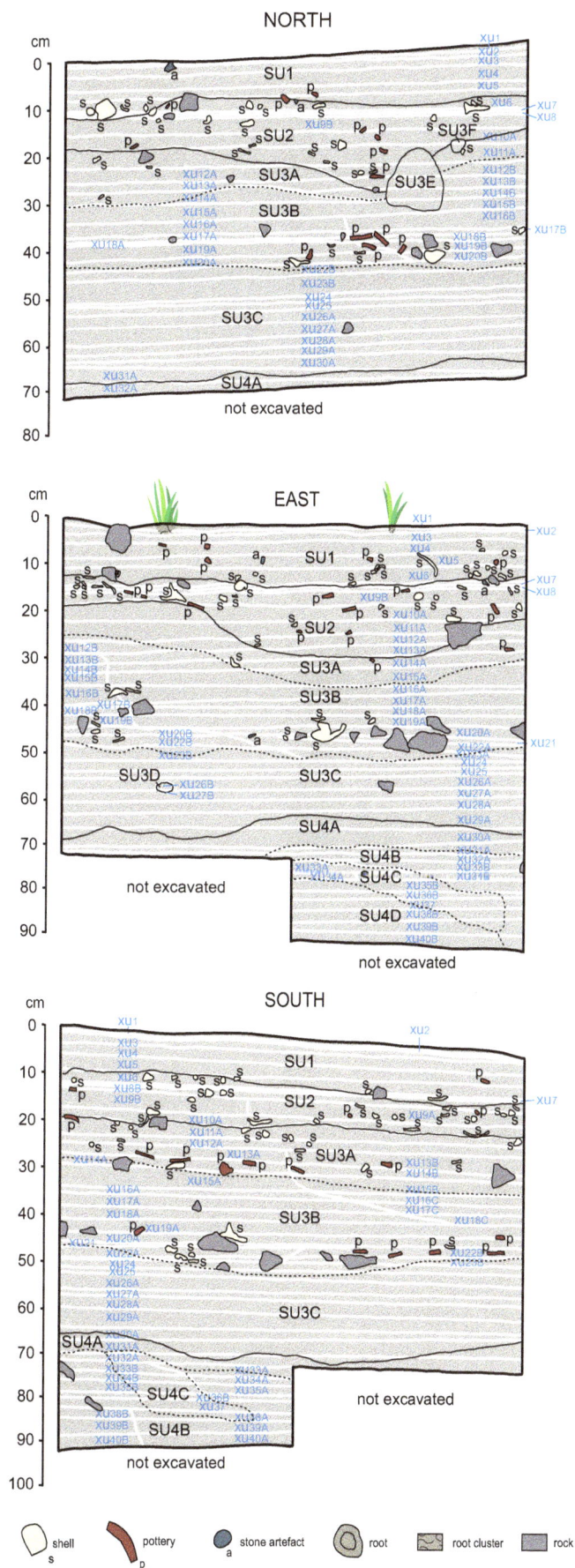

Figure 4. Stratigraphy, Square A, Edubu 1.

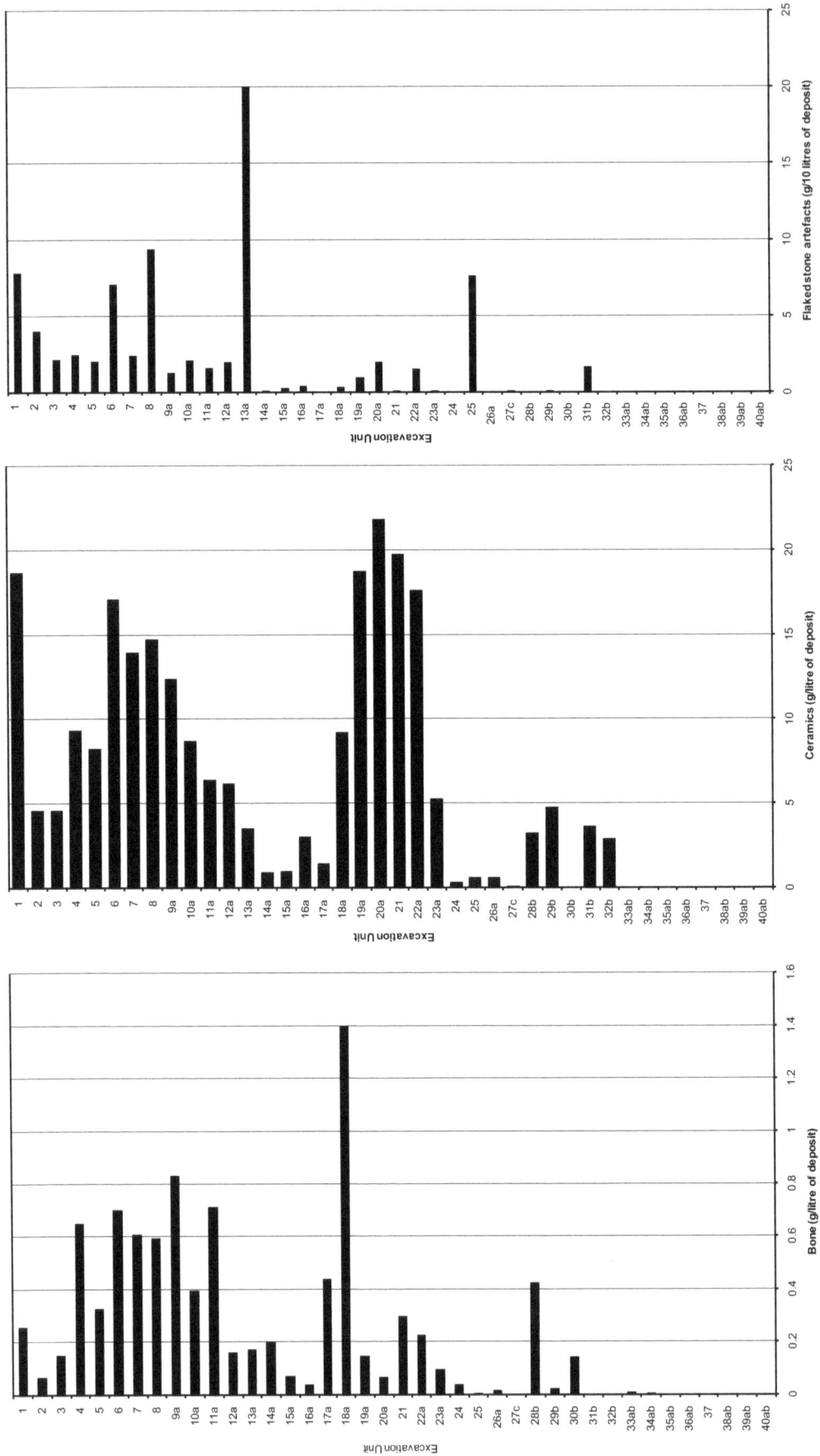

Figure 5. Vertical changes in cultural materials, Square A, Edubu 1.

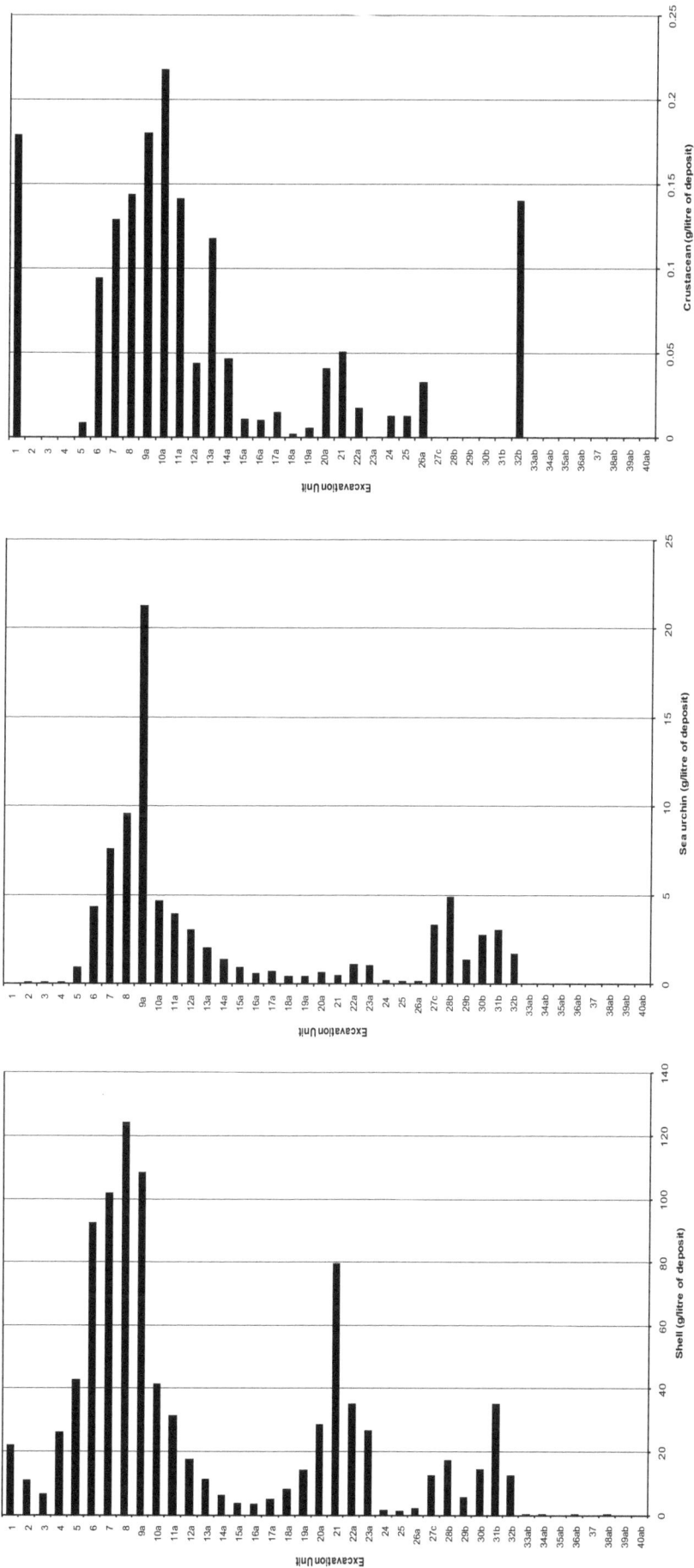

Figure 5. *Continued*

Table 1. AMS radiocarbon charcoal dates, Square A, Edubu 1. *Median probability of calibrated dates rounded to the nearest 50 years.

Lab. code	Square	XU (find #)	Depth below surface (cm)	δ¹³C‰ ± 0.2	¹⁴C Age (years BP)	Calibrated age BP 68.3% 1 σ (probabilities)	Calibrated age BP 95.4% 2 σ (probabilities)	cal BP circa*
Wk-27302	A	13b(3)	28.5	74.7	2339 ± 30	2335-2358 (1.000)	2318-2376 (0.835) 2384-2459 (0.165)	2350
Wk-27301	A	22a(4)	46.2	73.8	2440 ± 30	2363-2419 (0.338) 2428-2492 (0.402) 2601-2608 (0.034) 2641-2678 (0.227)	2355-2544 (0.677) 2563-2571 (0.010) 2586-2616 (0.080) 2635-2700 (0.233)	2500
Wk-27510	B	18(2)	36.6	73.0	2531 ± 30	2541-2590 (0.383) 2615-2636 (0.220) 2698-2738 (0.397)	2492-2601 (0.470) 2608-2641 (0.186) 2678-2744 (0.344)	2600
Wk-27511	B	27(1)	56.7	72.9	2537 ± 30	2543-2564 (0.175) 2568-2587 (0.131) 2616-2635 (0.226) 2700-2740 (0.467)	2494-2599 (0.427) 2610-2640 (0.185) 2681-2746 (0.388)	2650
Wk-27512	B	29a(1)	71.1	73.1	2520 ± 30	2507-2528 (0.136) 2539-2593 (0.425) 2614-2637 (0.200) 2697-2725 (0.239)	2488-2644 (0.711) 2665-2741 (0.289)	2600
Wk-27514	C	12a(4)	29.0	73.2	2502 ± 30	2498-2596 (0.694) 2612-2623 (0.074) 2628-2638 (0.074) 2691-2714 (0.158)	2470-2730 (1.000)	2600
Wk-27515	C	17a(2)	40.3	73.1	2514 ± 30	2503-2530 (0.172) 2537-2594 (0.442) 2614-2637 (0.188) 2696-2720 (0.197)	2488-2644 (0.722) 2653-2739 (0.278)	2600
Wk-27516	C	22(1)	58.3	72.8	2546 ± 30	2545-2561 (0.129) 2573-2582 (0.056) 2616-2635 (0.221) 2701-2744 (0.594)	2497-2597 (0.350) 2612-2639 (0.182) 2689-2748 (0.469)	2650

Table 2. Marine shellfish MNI, Square A, Edubu 1.

XU	1	2	3	4	5	6	7	8	9	10	11	12	13	14	15	16	17	18	19	20	21	22	23	24	25	26	27	28	29	30	31	32	33	34	35	36	37	38	39
Bivalves																																							
Acrosterigma reeveanum						2		2	2	1																													
Anadara antiquata					9	16	6	6	2	5	3	1	2			1	1	1																					
Anadara granosa					1			1		1	1					1				2	3	1		1															
Anadara sp.			2	2	2	1	1	4	2				1		1						1																		
Anodontia edentula					2	1	1				3	2		1			2			1	1																		
Anomalocardia squamosa															1		1			1																			
Antigona puerpera									1																														
Arca avellana											1																												
Arca cf. *avellana*				1																																			
Arca sp.					1																																		
Arcidae																															1								
Astriella corrugata					2																											1							
Atactodea striata									1		1	1								1									1										
Barbatia amygdalumtostum								1	2			1																											
Barbatia foliata											1				1												1												
Barbatia sp.					1	2			1																														
Batissa violacea					3	1					2																												
Cardiiae											2	2								1																			
Chama sp.																						1																	
Gafrarium pectinatum					1	1				2		1																											
Gafrarium tumidum					3	1																																	
Gafrarium sp.				2	1	4		2	2	1																	1												
Isognomon sp.		1			2			4	5	5		2	2			1										1													
Mactra cuneata					1																																		
Mactra sp.						1																																	
Malleus sp.						1		2	1	1																													

Table 2. *Continued*

XU	1	2	3	4	5	6	7	8	9	10	11	12	13	14	15	16	17	18	19	20	21	22	23	24	25	26	27	28	29	30	31	32	33	34	35	36	37	38	39
Marcia hiantina								1										1																					
Mytilidae							1	2																															
Ostreidae				7	15	74	32	29	29	21	21	12	8	8	4	3	6		1	10	2	14	3						1	1									
Pinctada sp.				2	2	2	2	1	4	3	1	1	3	1	1				1	1		1	1						1										
Pinna bicolour																	1	1	1																				
Polymesoda erosa					3	10	3	1	6	3	4	1	2	1	1	1	1		1			1										1							
Ruditapes variegatus						1																																	
Semele sinensis				1	1	5		1																															
Tellina palatum	1			2	2	2			1	1	1				1																								
Tellina remies				4	4	4	3	3	4	1																													
Tellina sp.							3	1	1	2	3			2	1	1	3	1	4	1		1	5	1	1						1								
Trachycardium cf. *rugosum*																		1																					
Veneridae														2																									
Bivalve													1																										
Bivalve C						2																																	
Bivalve D					1	5																																	
Bivalve K						2			1																														
Bivalve L					1	1	1				1																												
Bivalve O	1																1						1		2														
Gastropods																																							
Angaria delphinus			1	1		12	4	1	3	5	1	1												1															
Bulla ampulla					2	7	6	6	6	2	2	2	3	2	1	1	3		3	1	1		1	2	1		1	1	1	1	1								
Calliostoma sp.							1			1				1																									
Cerithidea cingulata		1		2																																			
Cerithidea largillierti	4	3	4	71	268	709	202	312	105	110	115	80	51	41	30	4	15	7	4	1		2	5	3	2	1	2	2	1		1	4							
Cerithium echinatum							1	2	1		1																												
Chicoreus brunneus					1	1	1	1	1	1	1																			1	1								
Chicoreus capucinus							1	1				1																			1								
Chicoreus sp.	1		1	1	1		1	4	1																														
Conus abbas						1																																	

Table 2. *Continued*

XU	1	2	3	4	5	6	7	8	9	10	11	12	13	14	15	16	17	18	19	20	21	22	23	24	25	26	27	28	29	30	31	32	33	34	35	36	37	38	39
Conus flavidus															1																								
Conus litteratus						2																																	
Conus lividus								2			1																												
Conus striatus						1																																	
Conus textile						1	1		2	2	1	1						1																					
Conus cf. coronatus												1																											
Conus sp.					3	3			1			2				1															1								
Cymatium sp.				1	1	1	1	2	1			1		1																									
Cypraea annulus				1	1	1		1				1			1					1																			
Cypraea arabica						1	1	1		2		3																											
Cypraea sp.							1	1	2																														
Cypraea spp.									1			1																											
Diodora sp.																												1											
Dolabella auricularia					8	2			11	4	7	1		1									1	1															
Dolabella sp.							3	10																															
Ellobium aurisjudae								1										1		1																			
Ellobium sp.								1				1						1																					
Hemitoma sp.					3	4		5	4		1	1				1			1	1																			
Lambis lambis				2	2	11	2		10	1					1	1				7	2	1			2			1											
Lambis millepeda											3											1		1															
Lambis sp.	3		1	2	4	1	8	12	4	2		2		1	1		1	1	7		2	1	1		1	1			1										
Littoraria scabra					11	1	1	1	1		3					1	1													1									
Littoraria sp.								1																					1										
Mitra sp.										1																													
Mitridae							2																																
Monodonta labio					1										1																								
Morula uva						1																																	
Muricidae					1			1																															
Nassarius crematus					1	2		1		1																													
Nassarius olivaceus						2																																	
Nassarius pullus							1																																

Table 2. *Continued*

XU	1	2	3	4	5	6	7	8	9	10	11	12	13	14	15	16	17	18	19	20	21	22	23	24	25	26	27	28	29	30	31	32	33	34	35	36	37	38	39
Nassarius sp.													1		2	2	4	2	2	1																			
Naticidae														1	1			1	1																				
Nerita albicilla				1	7	3	2	1		1																													
Nerita chamaeleon										1																													
Nerita planospira				3	3	9	6	8	9	6	3	4	2	4	2		2	1	1	1	1					1													
Nerita polita					3			5																															
Nerita cf. *polita*							1	1		2			1																										
Nerita undata	1				5	5	1	1																1															
Nerita sp.	1		1	6	6	4	2	1					3	2	2		1		2	1																			
Neverita peselephanti																			1																				
Oliva sp.											1																												
Patelloidea																																							
Planaxis sulcatus	2	2	2	15	4	7	2	9	8	1			2	3	3		2																						
Polinices mammilla															1																								
Polinices sp.							3					1																											
Pyrene sp.								1																															
Strombus aurisdianae					1	10	4	3	1		1											1									1								
Strombus bulla	1																																						
Strombus gibberulus					2	5			2		2	1	1				2	2				1																	
Strombus labiatus					1	5	2	4	2		5	1	2	1	1			1					1																
Strombus luhuanus	3	2	4	18	8	34	8	26	27	11	22	5	2	5	2		1	2	3		3	1	3	2															
Strombus urceus	1		1	1	4	6	2	8	8	3	3	2	4	1	1				1				1																
Strombus cf. *lentiginosus*																															1								
Strombus sp.	2	2		4	6	45	17	22	18	10	12	16	6	5	5	1		1	1			1			1		1	2	1			1							
Tectus fenestratus				1	2	5		3	4	3	3	2	1		2	2	1		1										3										
Telescopium telescopium		2		2	1			1	1																														
Terebralia sulcata	1			3	3	17	4	9	16	9	2	4	3	3	1		2	1	1	1																			
Terebridae													1																										
Thais cf. *alouina*						1																																	

Table 2. *Continued*

XU	1	2	3	4	5	6	7	8	9	10	11	12	13	14	15	16	17	18	19	20	21	22	23	24	25	26	27	28	29	30	31	32	33	34	35	36	37	38	39
Theodoxus (Clithon) oualaniensis											1																												
Trivia sp.																													1										
Trochidae																					1																		
Trochus maculatus									1		1																												
Turbo cinereus				1	12	37	10	9	4	4	1			2	2																1								
Turbo cf. *cinereus*					1																																		
Turbo sp.	2	2	4	5	1		3	1							2																								
Turritellidae														1					1																				
Vasum ceramicum								1																															
Vasum turbinellus																																							
Vexillum sp.								1										1																					
Gastropod																						2																	
Gastropod 1												1	2																										
Gastropod 3															1																								
Gastropod A	19																																						
Gastropod A2		32	2																																				
Gastropod A3			2					1														1	1				1	1											
Gastropod B	5			1	1						1	4		2								2	1	1	1				2										
Gastropod D		9																																					
Gastropod E					1				1								1																						
Gastropod J								1	1	1		2	1	3	1																								
Gastropod Q															1																								
Gastropod T													1				1																						
Gastropod V									1						1																								
Gastropod W				3				1												1																			
Gastropod X1														3																									
Gastropod Y1														1											1														
Gastropod Z																					1																		
Gastropod Z1																																							
TOTALS	40	62	21	157	436	1098	344	537	329	235	241	160	111	100	82	20	41	31	30	40	15	31	27	11	10	8	7	7	12	2	8	7							

Table 3. Vertebrate bone NISP data, Square A, Edubu 1.

XU:	1	2	3	4	5	6	7	8	9	10	11	12	13	14	15	16	17	18	19	20	21	22	23	24	25	26	27	28	29	30	31	32	33	34	35	36	37	38	39
Mammals																																							
Isoodon sp.								1						1				1																					
Peramelidae indet.						2	2	1			1			1				2																					
Macropus agilis				1																																1			
Macropodidae indet.							1				1																										1		
Phalanger sp.													1																										
Pteropididae								1		1																													
Canis				1		1	1											1	1																				
Sus scrofa		1		3	1	2			2		1						7	2	1																				
Melomys cf. *lutillus*						1	1		1	1	1																												
Rattus gestroi						1	1	2	3		2	2		1			1	1							1														
Rattus sp.	1				1	8	3	3	8	5	11		2	3			1	3	1		1	3	2	2		1													
Muridae indet.			1		5	5	7	3	17	15	24	1	17	19	12		4	5	2		2	5	7	6	3	1	1												
Reptiles																																							
Chelidae			3				1	1	23	1	13		2	1		1			1																				
Marine turtle				1																																			
Boidae						10	3	12	6										2	1																			
Colubroidea									1	2	1																												
Serpentes indet.				2	8	1	2	6	18	2	7		3		2						2	3																	
Varanidae														2						2																			
Scincidae											1																												
Agamidae						3															1																		
Lacertilian indet.									1																														
Fish																																							
Ariidae	1	1	1	1			1		2	1	1		1	2			1	1			2	1	2	1															
Labridae					1				1	1				2				1			1																		
Lutjanidae								1										1																					
Scaridae					1		1		2	1	1		1	2	1		1				2	1	2	1											2	1	1		
Serranidae									1	1													2												1				
Sparidae									1			1											1													1			

bivalves (4081.8 g), gastropods (12,162.3 g), chitons (31.2 g) and 7814.8 g of shell material that could not be identified to family, genus, or species level due to fragmentation and/or weathering. A total of 4260 individuals (MNI) was calculated for Square A (including unidentified bivalves and gastropods <1.0 cm long), represented mostly by gastropods (MNI=3606) and to a lesser extent bivalves (MNI=654). Land snails (family Camaenidae) formed an insignificant part of the shell assemblage (1.8 g, MNI=3).

More than 140 different mollusc taxa were identified, represented by at least 45 bivalve taxa and 99 gastropod taxa (Table 2). Of these, species-level identifications were possible for 43% (by weight) of the total shell assemblage. The top five ranked bivalve species in terms of MNI abundance are: *Anadara antiquata* (8.1%, MNI=53), *Polymesoda erosa* (6.0%, MNI=39), *Tellina remies* (2.9%, MNI=19), *Anodontia edentula* (2.8%, MNI=18) and *Anadara granosa* (2.0%, MNI=13). These species represent only 22% (by MNI) of the bivalve assemblage, with 27 bivalve taxa represented by a MNI of ≤5. The top five ranked gastropod species in terms of MNI abundance are: *Cerithidea largillierti* (59.6%, MNI=2159), *Strombus luhuanus* (5.3%, MNI=193), *Turbo cinereus* (2.3%, MNI=84), *Terebralia sulcata* (2.2%, MNI=78) and *Nerita planospira* (1.7%, MNI=62). These taxa represent 71% (by MNI) of the gastropod assemblage, with 62 gastropod taxa represented by a MNI of ≤5. A small proportion of the assemblage (N=214, 5%) consists of intact small specimens (maximum length <1 cm), suggesting that only a relatively small proportion of the assemblage was of non-economic utility.

Environmental data are available for 76 species, with 52 of these restricted to a single environment type (MNI=2644). Preliminary analysis indicates that shells were obtained from a variety of different environments: mud (MNI=2233), rocks and rocky areas (MNI=209), seagrass (MNI=121), coral reef (MNI=25), sand (MNI=37), muddy sand (MNI=18) and mangroves (MNI=2). Muddy environments were the main targeted shellfishing location, with only minor differences in habitat preference through time.

Fish. Fish bone represents 18.3% of the bone assemblage (by weight) and formed a consistent component of the diet of peoples camping at Edubu 1 (Table 3). The most common taxon of marine fish in terms of NISP values is Scaridae (parrotfish), followed by very low representations of Labridae (wrasse), Sparidae (sea bream), Serranidae (groupers) and Lutjanidae (snapper). These fish could have been obtained from local open-water reef habitats fringing Caution Bay. The lack of fishhooks recovered from Edubu 1 points towards use of spears or nets as fishing technology. While most fish remains were restricted chronologically to <2350-2500 cal BP in Square A, the deeper Scaridae bones in XUs 28-30 most likely date to 2600-2650 cal BP.

Marine turtle. The only marine turtle bone recovered from Square A was a fragment of phalanx in XU4 dating to younger than ca. 2350 cal BP.

Crustacean and sea urchin. Crustaceans (mostly a number of different crab taxa) and sea urchins represent consistent subsistence items throughout the history of site occupation. While crabs were a minor dietary item, the considerable amounts of sea urchin body and spine fragments point to sustained exploitation. Crabs are available from a range of marine habitats (e.g. rocks, coral reefs, mangroves), with sea urchins restricted more to coral reef contexts. Both crabs and sea urchins were probably collected from the intertidal zone at low tide.

Subsistence (terrestrial and freshwater)

Most of the terrestrial bone recovered from Square A reflects the results of human hunting and on-site discard practices. That is, the assemblage comprises bones of known prey species, and there is a significantly high proportion of burnt bone. The fact that rodent bone tends to be less burnt than other mammal bones may indicate contributions of natural on-site rodent deaths. The highly fragmented nature of the bone assemblage suggests that mechanical attrition rather than chemical solution has been the major agent of degradation, pointing to substantial pre-depositional loss, perhaps caused by trampling and foraging by pigs and dogs. Furthermore, the unburnt bone is moderately degraded, suggesting a significant attrition of the original assemblage, with an unknown quantity of bone probably lost to post-depositional degradation.

Endemic mammals. The mammal remains include a range of medium- and small-sized species characteristic of local savannah and gallery rainforest habitats, both ground-dwelling taxa – *Macropus agilis* (Agile Wallaby), *Isoodon* sp. (possibly Northern Short-Nosed Bandicoot, *Isoodon macrourus*), and native rodents (e.g. *Melomys* cf. *lutillus*, *Rattus gestroi*); and tree-dwelling taxa – *Phalanger* sp. (possibly Southern Lowland Cuscus, *Phalanger intercastellanus*) and Pteropididae (flying fox). While most of these mammals were recovered from upper levels in Square A, dated to <2350-2500 cal BP, some Agile Wallaby remains were found in the lower levels dated to 2600-2650 cal BP.

Introduced pig and dog. Pig remains were found in nine XUs between XU2 and XU19a dated to between <2350 and 2500 cal BP. Whether pig bones occur in the 2600-2650 cal BP levels of Squares B and C remains to be determined. Most pig bones in Square A are small fragments of teeth, but XU17a and XU19a each produced larger fragments of dentaries with teeth. Dog remains were also restricted to the upper levels of Square A (XUs 4, 6, 7, 18 and 19), dating to between <2350 and 2500 cal BP. There is no evidence of the introduced Pacific rat (*Rattus exulans*).

Reptiles. A range of reptile taxa was recovered from upper levels of Square A, dated to <2350-2500 cal BP. The most common are pythons (Boidae) and fanged snakes (Colubroidea), with minor representations of lizards such as goannas (Varanidae), dragons (Agamidae) and skinks (Scincidae). All of these reptiles are available from local savannah and grasslands.

Birds. While no bird bone was recovered from Square A, a few bird eggshell fragments (0.01 g) were found in XU23. These fragments are thin-walled and are consistent in character with megapode eggs. Clearly, hunting of birds for subsistence was of little interest to the occupants of Edubu 1.

Freshwater turtle and fish. Freshwater turtle (Chelidae) bone forms an inconsistent component of the faunal assemblage down to XU19a, dating to <2350-2500 cal BP. In contrast, bones of Ariidae (catfish) were identified only in a single level (XU5), indicating that catfish formed an insignificant component of the diet for the site inhabitants. Both the turtle and catfish remains indicate minor exploitation of nearby estuaries and/or freshwater creeks.

Ceramics

In Square A, ceramic sherds were recovered from XUs 1 to 32, dating from <2350 cal BP to at least 2600 cal BP (Figure 5). A total of 33 decorated sherds, excluding red slipping, was recovered. The ceramic sherds from XUs 1 to 15 are mainly plainwares, with decoration limited mostly to red-slipping. These upper-level ceramics date to younger than 2500 cal BP and fall outside the stylistic and chronological parameters of Lapita pottery found elsewhere at Caution Bay (McNiven et al. 2011). In contrast, decorated sherds found below XU16 exhibit features that are broadly characteristic of terminal Lapita ceramic assemblages elsewhere at Caution Bay, but also show stylistic signs that they were by then transforming out of Lapita into post-Lapita assemblages. Descriptions of all the ceramics containing body decorations other than red slip follow:

Vessel shapes. Sherds from three separate vessels are large enough to reveal information about vessel shapes:

1. A wide, shallow dish with a very slightly everted to vertical rim (orientation angle = 5°; see David et al. 2009:Figure 4 for definitions of pottery terms) and orifice diameter of 52

cm. It is made up of four conjoining sherds from XU16a and XU18a. The lip is flat and of a fairly uniform 10.6 mm width. The rim is typically 35.4 mm deep to its point of greatest curvature, where it then becomes the body of the dish. Plain sherds excavated from these same XUs are likely to also conjoin (Figure 6:top).

2. A small bowl with a slightly inverted, near-vertical rim (orientation angle = 355°). It consists of a set of three conjoining sherds from XU20a, two sets of two conjoining sherds from XU20a, and three individual sherds from XU19a and XU20a. The lip is rounded and of a uniform 6.7-6.8 mm width. The rim is typically 34.4 mm deep, after which it gently curves into the bowl body. Plain sherds excavated from these same XUs are likely to conjoin (Figure 6:bottom).

3. An indirect everted pot of unknown orientation angle (as there are no lip sherds). The aperture at the neck (the most constricted point at the base of the rim) is approximately 20 cm wide (internal measurement). At 5.9 mm thickness, the pot wall at the neck – the thickest part of the pot – is noticeably thin for what appears to be a moderately large jug or bowl (Figure 7e).

Body Decorations. The body decorations employed on the sherds described here can be divided into nine categories:

1. Comb dentate-stamped triple-triangles (note: throughout our work on the Caution Bay ceramics, we differentiate between tools used to make decorations – e.g. comb or shell, and the decorations themselves – e.g. indentations, which could be made by a number of tined or toothed tools such as combs or shells. Hence, we specify both tool [here comb] and decorative design [here dentate-stamped]). Two closely spaced comb dentate-stamped parallel lines curve around the pot a short distance below the top edge of the lip. These horizontal lines act as the upper margin of down-pointing dentate-stamped triangles, whose other two sides each have three parallel lines. Dentate-stamped triangles of this kind are repeated across the pot as a motif row. Twenty sherds, consisting of nine unconjoined sherds, plus 11 which conjoin into four sets, occur of this motif type. One conjoining set of four sherds, with internal surfaces red-slipped, comes from XU16a and XU18a and makes up part of one shallow dish (vessel shape 1 above); the other sherds, all coming from XU19a and XU20a, either conjoin or are very likely to have come from a small bowl with slightly inverted rim (vessel shape 2 above). The chrono-stratigraphic distribution, size, shape and surface texture of the sherds with this kind of decoration indicate that they all probably come from only two different vessels (Figure 6). Examples of similar (but not identical) decorative technique and design in confirmed late-Lapita sherds occur at Honiavasa on the edge of the Roviana Lagoon in New Georgia (Solomon Islands), illustrated in Felgate (2001:Figure 3 items HV.2.464, HV.4.175). However, the Edubu 1 sherds appear to represent a structurally simpler pattern of repeated triangles a short distance below the lip.

2. Irregular comb dentate-stamped horizontal lines. Two conjoining sherds, from XU12a and XU13a, contain a horizontal line below two other sub-horizontal lines meeting at a point along the rim. The lines consist of irregular, uneven comb dentate stamping.

3. Dentate-stamped rectilinear maze. A single non-rim sherd of this kind has been found from XU20a (Figure 7a). It is akin to established Lapita motifs such as the ones found

at site FAAH on the Willaumez Peninsula (New Britain) and illustrated in Specht and Torrence (2007:Figure 11b, 11c, 11d).

4. Dentate-stamped rectilinear lines on lip. A single thickened rim sherd is decorated with repeated sets of five parallel dentate-stamped lines separated by pairs of perpendicular lines. It was a surface find at the base of the slope along the northern edge of the site (Figure 7b).

5. Parallel wiggly lines created from shell valve end-impressions along the rim. Two such sherds come from XU6 to XU9b (Figure 7c).

6. Dentate-stamped triple-angled lines on lip. The outer edge of the lip is notched at regular intervals. The lip is noticeably thickened. A single sherd of this kind has been found from XU19a (Figure 7d).

7. Wiggly lines created from shell valve end-impressions to form triangles arced around a central design. Three conjoining sherds of this kind have been found, and are reminiscent of the banding triangles commonly found on Lapita ceramics (Figure 7e). Shell impressions are a feature of late-Lapita assemblages elsewhere in island Melanesia (e.g. Kirch 1997:155). These sherds were exposed during bulldozer construction works immediately following the archaeological excavations.

8. Row of lenticular, probably stick, side impressions on two body sherds from XU9b and XU18a.

9. Row of fingernail impressions on a body sherd from XU18b.

The Edubu 1 Square A decorated ceramic sherds include techniques and designs akin to terminal Lapita (in particular comb-impressed dentate stamping involving the use of more open, broad-tined combs; wiggly lines of shell end-impressions to create triangular designs separated by impressed rows; an interlocking rectilinear maze; and shallow stick-like impressions along lip edges). However, they also contain features more reminiscent of the immediate post-Lapita period in the Caution Bay region: there is a simplification of decorative forms particularly attested to by a predominance of simple geometric designs such as triple-lined dentate-stamped triangles below double rows a short distance below the lip, and a diminished area of surface decoration. The Edubu 1 Square A ceramic decorations imply a terminal-Lapita assemblage, as they are transforming out of the Lapita design system: they are still too similar to late Lapita to categorically classify them as something else, while at the same time indicating a repeated theme of reduction in design complexity and diminution of decorated area, indicative of the immediate post-Lapita period. This is entirely consistent with the radiocarbon determinations, and with the Lapita and post-Lapita ceramics from other sites at Caution Bay, which would place the ceramic assemblage from below XU13 as relating to that period when terminal Lapita (ca. 2500 cal BP) is transforming into post-Lapita, indicating the initial opening and breakdown of the Lapita decorative system.

Stone artefacts

The stone artefact assemblage from Square A comprises flaked artefacts (flakes, retouched flakes and cores), a flaked and ground adze, and an anvil. Nearly all (n=2546, 98%) artefacts were made from chert, which is locally available as small and large nodules across the Caution

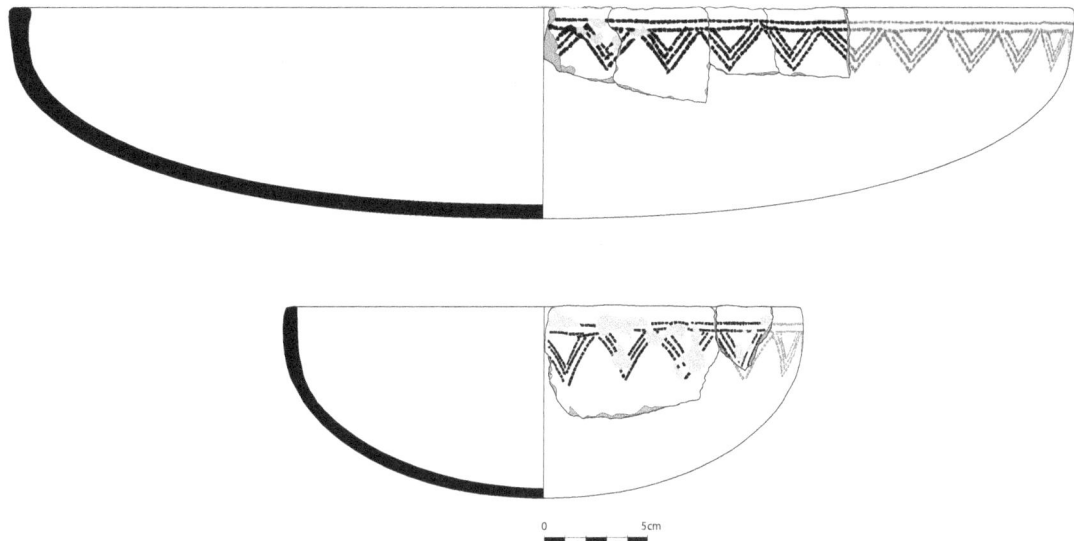

Figure 6. Ceramic bowls from Square A, Edubu 1, dating to 2350-2500 cal BP. Top: XUs 16-18. Bottom: XUs 19-20.

Bay landscape and the Port Moresby region more generally (Table 4) (Glaessner 1952:66-67; Mabbutt et al. 1965; Davies and Smith 1971:3303). In terms of fracture types for chert, most (83%) are flakes (complete and broken) (n=2115), followed by flaked pieces (n=374), cores (n=22) and other (n=34). The majority (57%) of chert artefacts reveal evidence of heat alteration (pot lid scars) from contact with fire – either hearths or natural landscape fires when the site was unoccupied. All of the 22 chert cores were reduced by freehand percussion, with no evidence of anvil-resting or bipolar reduction. That cores made of high-quality chert were reduced further than cores made from poor-quality chert is indicated by differences in mean core weight (15 g and 67 g respectively), mean core length (29 mm and 48 mm), mean number of platforms (3.5 and 2.5) and mean number of flake scars (nine and four). The small size of chert cores is reflected in the small size of complete chert flakes, which range in length from 2 mm to 26 mm (mean = 9.3 mm). Only 8% of complete chert flakes and 1.5% of chert flake platforms exhibit cortex, suggesting strongly that the early stages of core reduction took place elsewhere. The 12 obsidian artefacts are small flakes (mean weight, 0.3 g; mean length, 8 mm; length range, 4 mm-16 mm), showing no signs of bipolar reduction. The nearest source of obsidian is West Ferguson Island located 500 km to the east (Irwin and Holdaway 1996).

More formally manufactured implement types include retouched flakes and an adze. Retouched flakes comprise only 2.2% of the lithic assemblage and are represented by chert (n=55) and chalcedony (n=1). As with chert, chalcedony outcrops in the Port Moresby region (Glaessner 1952:66). The mean length of retouched chert flakes (14.8 mm) is larger than the mean length of unretouched chert flakes, indicating selection of larger chert flakes for manufacture into curated implements. A flaked and ground adze was recovered from XU20a (Figure 8). It is manufactured from volcanic stone (possibly basalt) and weighs 154.9 g. An anvil (also made from volcanic stone, possibly basalt) was recovered from XU2. The adze and anvil were likely imports, given that the nearest major outcrops of basalt are located 80-100 km to the northwest and southeast of Port Moresby (Glaessner 1952; Pieters 1978; Smith and Milsom 1984:165).

Whereas chert flakes were used throughout the history of site use, back to at least 2650 cal BP, obsidian artefacts have a more restricted occurrence in XUs 3 to 15, dating between <2350 and 2500 cal BP. While the adzes was recovered from the more recent levels, the lack of these implements in 2600-2650 cal BP levels may simply reflect issues of rarity and sampling.

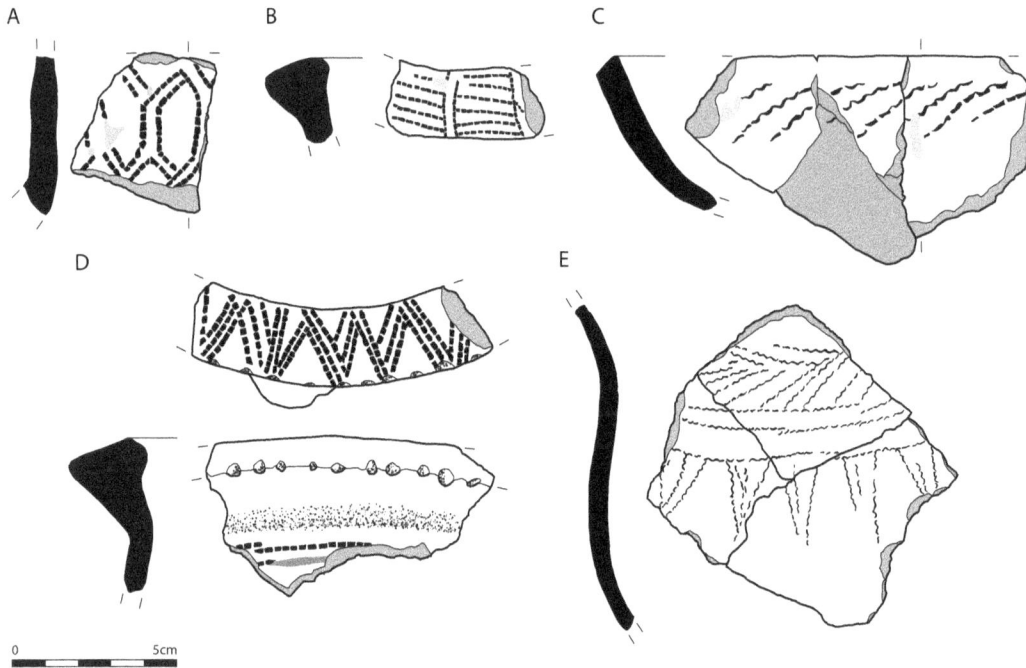

Figure 7. Decoration styles at Edubu 1 Square A. A: Dentate-stamped rectilinear maze dated to 2350-2500 cal BP. B: Dentate-stamped rectilinear lines on lip dated to 2350-2500 cal BP. C: Shell valve end-impressions on rim dated to <2350 cal BP. D: Dentate-stamped triple-angled lines on lip dated to 2350-2500 cal BP. E: Shell valve end-impressed triangles arced around central design.

Shell artefact

A fragment of a shell ring was recovered from XU13, dating to 2350 cal BP (Figure 9). The ring is approximately 30% complete and is finely made from what appears to be cone shell (*Conus* sp.). The surface of the ring exhibits smoothing and shaping by grinding. The ring is 10.5-10.7 mm high, 5.0-5.5 mm wide, and would have been 9 cm diameter (measured on the outside of the shell) when complete. Based on ethnographic information from the Port Moresby region, the shell ring may have been used as a body adornment and/or exchange valuable (Seligmann 1910).

Implications for Port Moresby regional land-use history

Extending the 2000 cal BP barrier

Before our Caution Bay study, the earliest evidence for people in the Port Moresby region was found at three sites – Nebira 4, Eriama 1 and Loloata Island. Nebira 4 is an open site located atop a low hill immediately north of Port Moresby, some 11 km inland from the coast. Excavations by Jim Allen in 1969 and 1970 revealed a 2.6 m cultural deposit of stone artefacts, shell artefacts and pottery (including red-slipped wares), along with bone and shell food remains (Allen 1972). Charcoal radiocarbon dates of 1760 ± 90 BP and 3340 ± 160 BP were obtained from Level 14, but well above basal cultural materials (Level 18). While Allen (1972:120) remarked that '[n]o archaeological reason can be given for this [dating] discrepancy', he rejected the earlier date, given the pottery assemblage similarities between Nebira 4 and the Oposisi site on Yule Island in the Gulf of Papua dated by Ron Vanderwal to 2000 years ago (cf. Vanderwal 1973). Eriama 1 is a small rockshelter located northeast of Port Moresby and 7.5 km from the sea. Excavations by Sue Bulmer in 1969 revealed a 0.9 m cultural deposit, with the lowest layer containing 'red slip pottery' and producing a charcoal radiocarbon date of 1930 ± 230 BP (ca. 1850 cal BP) (Bulmer 1975:56). Loloata Island is located 15 km southwest of Port Moresby and 1 km from the mainland. A 'shell sample from the lower midden layer' produced

Figure 8. Adze from XU20a, Square A, Edubu 1, dating to between 2350 and 2500 cal BP.

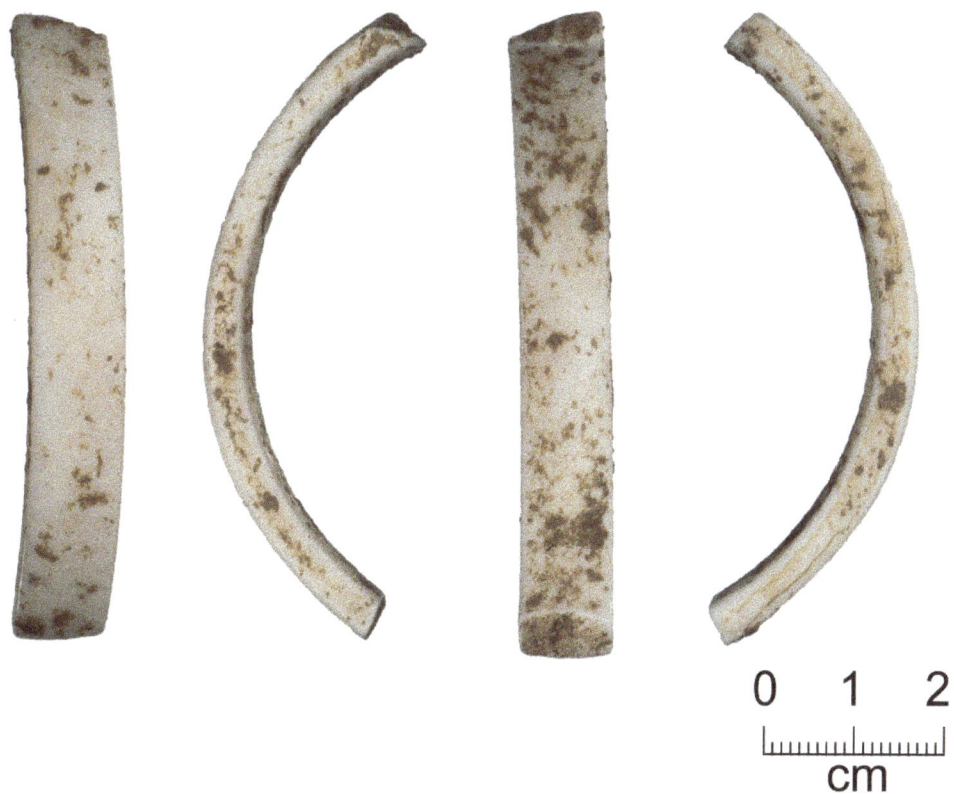

Figure 9. Shell ring from XU13, Square A, Edubu 1, dating to 2350 cal BP.

Table 4. Stone artefact raw materials, Square A, Edubu 1.

Raw material	#	%	Wt (g)	%
Chert	2546	98.2	1414.2	79.6
Chalcedony	19	0.7	4.5	0.3
Obsidian	12	0.5	3.1	0.2
Volcanic (basalt?)	9	0.3	345.1	19.4
Siliceous sandstone	2	0.1	0.1	<0.1
Quartz	2	0.1	0.1	<0.1
Quartzite?	1	<0.1	9.3	0.5
Fine-grained sedimentary	1	<0.1	0.2	<0.1
Fine-grained igneous	1	<0.1	0.1	<0.1
Totals:	**2593**	**100**	**1776.7**	**100**

a radiocarbon age of 2300 ± 100 BP (ca. 1900 cal BP) associated with 'red-slipped pottery' (Sullivan and Sassoon 1987:7). Based on site survey and excavation data, Allen (1977a:447) hypothesised that 'the existing archaeological evidence suggests the rapid occupation of both the coast and hinterland by pottery manufacturing people around 2,000 years ago'.

Recent excavation results from Caution Bay demonstrate that pottery-using (Lapita) peoples settled the Port Moresby region by 2900 cal BP, with pre-ceramic occupation extending back to at least 4200 cal BP (McNiven et al. 2011). As such, people settled the Port Moresby region well before the previously hypothesised 2000 cal BP for an initial migration by ceramicists to the region. Indeed, Edubu 1 was mostly, if not entirely, occupied before 2000 cal BP, between <2350 cal BP and 2650 cal BP. This expanded chronology for settlement of the region is not unexpected, given that Kukuba Cave, located 100 km along the coast northwest of Port Moresby, dates back to 4000 years ago (Vanderwal 1973). Indeed, it is probable that a human presence has been in the region since the Pleistocene, given archaeological evidence for well-established populations 49,000 years ago at Kosipe, located in the Owen Stanley Ranges 70 km north of Port Moresby (Fairbairn et al. 2006; Hope 2009; Summerhayes et al. 2010), and in the Papuan Gulf lowlands to the west by at least 13,000 years ago (David et al. 2007).

Terminal Lapita settlement expansions

Edubu 1, with terminal/transforming Lapita deposits, demonstrates that new Lapita sites were actively established at Caution Bay hundreds of years after the original arrival of Lapita peoples at 2900 cal BP (McNiven et al. 2011). Clearly, Caution Bay hosted a dynamic and developing community of Lapita peoples, with some sites such as Bogi 1 located 2.8 km west-northwest of Edubu 1 spanning the full temporal extent of Lapita occupation at Caution Bay for 400 years between 2900 and 2500 cal BP, and other sites such as Edubu 1 showing more temporally restricted Lapita occupation of less than 200 years (and possibly less than 100 years) between 2650 and 2500 cal BP. Significantly, Bogi 1 site, located downslope from Edubu 1 on the nearby shoreline of Caution Bay, reveals a major burst of activity at 2500-2650 cal BP (McNiven et al. 2011). The fact that Lapita sites located on both the shoreline and 1 km inland exhibit synchronous increases in site activity at 2500-2650 cal BP indicates broad-scale social linkages across what is clearly a complex and integrated community of Lapita peoples at Caution Bay, and that probably also involves local non-Lapita populations. While it is possible that both Bogi 1 and Edubu 1 were used by similar peoples as part of seasonal movements, the fact that both sites show synchronous increases in cultural discard indicates at the very least greater human activity and most likely more people living in the area.

Marine subsistence and terrestrial hunting

The lower sediment levels at Nebira 4 revealed exploitation of marine and terrestrial animals. Terrestrial animals include pig, dog and wallaby (*Macropus agilis*), while marine animals include turtle, dugong, fish (mostly estuarine catfish) and shellfish (mostly reef-dwelling *Chama* sp. and *Strombus* spp.). Interestingly, bandicoot and bird, which were present in upper levels, were not present in lower levels of Nebira 4. Allen (1972:123) concluded that early immigrants to the Port Moresby region possessed an 'economy based on mixed hunting, agriculture and fishing'. Yet it is clear that subsistence practices of early occupants at Nebira 4 focused more on aquatic animals than land animals, with fish accounting for 50% of bone assemblages (Allen 1972:118). While nearly all fish (92% of MNI) in Levels 16-19 were catfish, Allen (1972:118) argued that these high-saltwater-tolerant fish were probably 'caught in estuarine conditions' and not more characteristic riverine contexts located adjacent to the site (cf. White and O'Connell 1982:203). Bulmer (1971:57) concurs and commented that the 'earliest occupants of Nebira 4 had a predominantly maritime economy, with enormous quantities of fish, sea mammals, and shell fish consumed' (see also Bulmer 1979:18). Allen (1977b) acknowledged the early focus on marine resources, stating 'the subsistence patterns of these early migrants was oriented much towards the sea and the exploitation of sea resources, but that land hunting also contributed to the diet'. The nature of subsistence remains in the lower levels of Eriama 1 is difficult to determine given the disturbed stratigraphy and incomplete reporting of finds. However, remains from 'Crevice Layer 3' (from where the early date was obtained) revealed low amounts of shell (mangrove and freshwater species) and possibly small quantities of mammal (mostly wallaby) and fish bone (Bulmer 1978:Table 7.6, Appendix 7.2). The lower midden on Loloata Island contained 'abundant [marine] shell' and specialised marine subsistence is implied (Sullivan and Sassoon 1987:3).

Faunal remains from Nebira 4 and Eriama 1 reveal that the maritime peoples living in the Port Moresby region around 2000 years ago consumed pigs and wallabies. While wallabies were clearly hunted in the wild, Allen (1977b:37) hypothesises that pigs may have been 'husbanded rather than hunted in their feral state'. Yet the question remains as to what extent hunting of terrestrial animals reflected the location of both sites some distance from the coast (11 km and 7.5 km inland respectively). Interestingly, early ethnographic accounts from the late 19th century mention pigs and highlight the importance of wallaby hunting, particularly among the Koita, and the importance of fish, particularly among the Motu (Stone 1876:60; Lawes 1879:373, 375; Chalmers 1887:14-15; Turner 1878:482, 487, 495; see also Allen 1977a, 1991; Oram 1977; Vasey 1982). Turner (1878:481) observed that 'the food of the Motu consists principally of wallaby, fish, yams, bananas, cocoa-nuts, and sago' (see also Stone 1876:47). Today, local people of Caution Bay hunt Agile Wallaby (*M. agilis*) and Southern Lowland Cuscus (*P. intercastellanus*) (Woxvold 2008:69; Coffey Natural Systems 2009:12-47) (Figure 10). More specifically, in 'winter they live upon yams, bananas, and fish. In August the hunting season commences, and for two or three months they live almost entirely on the flesh of the wallaby' (Turner 1878:481). However, what is the history of this hunting and did the early ancestors of these coastal people similarly supplement diets with hunted animals or were they marine subsistence specialists? In this connection, Allen (1977a:450) posits that coastal locations with little access to terrestrial resources may have been too 'restrictive' for early maritime colonisers and their 'more general economy'. While the Loloata Island site is obviously located on the coast, unfortunately at this site subsistence evidence is too scant to shed detailed light on this question.

The faunal assemblage at Edubu 1 concurs in part with Allen's (1977a) hypothesis that early peoples of the Port Moresby region practised a mixed marine-terrestrial subsistence economy. While it is now clear that maritime (Lapita) peoples moved into the Port Moresby

Figure 10. Hunting party with three wallabies and hunting dogs returning to Boera village via north side of Moiapu creek, 17 October 2009. The wallabies were hunted a few kilometres inland of Edubu 1. Photo: Ian J. McNiven.

coast well before 2000 years ago, the view that these maritime colonists exploited a wide range of marine and terrestrial foods has been reinforced. In common with Nebira 4, the faunal assemblage at Edubu 1 included terrestrial hunting focusing on macropods with contributions by pig, snake and freshwater turtle, while the marine faunal suite included fish, sea urchins, turtle and an extensive array of shellfish. In this sense, terminal Lapita occupants of Edubu 1 practised a more mixed economy than the more marine specialised Lapita-period subsistence practices seen on nearby shoreline sites at Caution Bay, such as Bogi 1 (McNiven et al. 2011). While the more mixed terrestrial-marine subsistence regime at Edubu 1 can be attributed in large part to the site's location 1 km inland from the shoreline, the question of why these terminal-Lapita-period peoples chose to establish a site inland away from the coast remains to be determined. For example, was Edubu 1 part of the seasonal round of the same peoples who occupied shoreline sites such as Bogi 1, or was it used by more terrestrially oriented inland peoples who were in close interaction with coastal peoples, as was the case of inland-coastal relations during the early European contact period of the late 1800s-early 1900s? Whatever the case, it is clear that in the Caution Bay region a complex and diverse subsistence pattern existed for terminal-Lapita pottery-using peoples, whereby the ratio of terrestrial to marine foods in diets varied depending on context and site location. In this sense, Caution Bay Lapita sites have considerable potential to inform current debates on the degree of marine-protein specialisation by Lapita peoples of the Western Pacific and factors influencing the incorporation of wild and domesticated terrestrial animal foods (Field et al. 2009; Valentin et al. 2010). At this juncture, whether cultivated plant foods contributed significantly to Lapita diets at Caution Bay remains hypothetical, but it is plausible that plant food was cultivated given the likelihood of Lapita agriculture across the Western Pacific (Kirch 1997; Spriggs 1997; Kennett et al. 2006; Horrocks and Nunn 2007; Fall 2010; see also Fairbairn 2005:495) and major erosion and sedimentation at Edubu 1 (see below). The presence of pig in Edubu 1 suggests strongly that pigs were introduced to the region by at least 2500 years ago.

Firing and grassland formation

Directly related to the issue of the history of macropod hunting is the history of the development of savannah and grasslands – the major habitat of macropods such as Agile Wallabies. Early European visitors to the Port Moresby region in the late 19th century commented on savannah vegetation and burning of grasslands to aid hunting of wallabies and most likely pigs (Turner 1878:471, 487; Romilly 1889:164; Seligmann 1910:87) (Figure 11). In the Caution Bay region, such landscape burning to aid wallaby hunting continues today (Woxvold 2008:67-68; pers. obs. 2009-2010). Eden (1974) argues that the distribution of Port Moresby savannah and grassland vegetation cannot be accounted for simply by environmental factors and that these vegetation zones are also the product of human burning associated with shifting cultivation and hunting. In terms of the former, the potential for widespread impact is considerable, given that 'the life of individual fields' is usually only three to four years, and the principal local crops of banana (*Musa* sp.), taro (*Colocasia* sp.), sweet potato (*Ipomoea batatas*) and yam (*Dioscorea* sp.) involve considerable ground disturbance (Eden 1974:106). Oram (1977:83) suggested that the savannah and grasslands along the coast between Boera and Lea Lea (i.e. Caution Bay area) were 'probably as a result of human occupation'.

Historisicing the long-term process of local anthropogenic grassland formation, Bulmer (1975:66) hypothesised that the early maritime colonisers of the Port Moresby region 'may have helped to clear some of the lowland forest in the course of gardening and hunting on the river plains'. Swadling (1977:38) added that because of poor soils and low rainfall, the Port Moresby region was 'probably not considered a good place to live' prior to the introduction of horticulture and the associated clearance of forest, which subsequently created grasslands that supported rich stocks of wallabies. While both Bulmer and Swadling place the start of anthropogenic grassland formation back 2000 years, Bulmer (1975:21) hypothesises that these grasslands 'were widespread by 1,000 years ago', given archaeological evidence for local settlement patterns and extensive wallaby hunting within the past 800 years at sites such as Motupore (see Allen 1977a). Eden (1974:109) made the interesting point that while little evidence exists for the history and antiquity of anthropogenic deforestation and grassland formation in the Port Moresby region, 'it cannot be questioned that, even at rates of clearance significantly below those of today, the present savanna and grassland could readily have been created by cultivating peoples since about 2000 BP. The savanna and grassland may therefore represent the cumulative impact of cultivation and burning during such a period.'

The presence of macropod remains in Edubu 1 indicates that the necessary savannah and grassland habitats for wallabies were present in the Caution Bay area by at least 2600 years ago. Thus, it is possible that these savannah and grassland habitats were maintained, at least in part, by anthropogenic firing regimes similar to those recorded historically, and were initiated during the Lapita settlement of the coast sometime between 2900 and 2500 years ago when inland peoples became more attracted to this coastline for its new trade and other opportunities for social interaction generated by the newly established Lapita settlements at Bogi 1 and other nearby locations. Alternatively, landscape firing at Caution Bay may have a much longer antiquity: coring of Kosipe Swamp located 70 km inland indicates that people have been firing the landscape in the broader Port Moresby region for at least 40,000 years (Hope 2009). Furthermore, Hope et al. (1983:41) suggest that maintenance of grasslands and savannah in the Markham Valley on the northeast coast of Papua New Guinea by anthropogenic burning over the past 9000 years (Garrett-Jones 1979) opens the possibility of similar long-term firing in the Port Moresby region. Bulmer (1971:39) rightly points out that 'pollen samples need to be obtained in order to learn how long ago the forest was cleared and what other ecological

Figure 11. Late 19th century view of Port Moresby, showing grassy hills which reflect the long-term impact of human landscape burning (from Lawes 1886:663).

clues can be established concerning that period'. In this connection, pollen cores taken from mangroves fringing Caution Bay by Rowe and McNiven in 2010 will hopefully shed light on this question.

Erosion and coastal progradation

Bulmer (1971:41) hypothesised that 'clearance of the lowland forest would have accelerated the deposition of sediments on the river plains, and consequently the extension of the swamplands surrounding the river mouth'. In particular, she points to likely coastal progradation of 4.8 km at Galley Reach (including the mouth of the Laloki River) in the northern part of the study region (Bulmer 1978:13, 14, 73). While the rate of slope erosion and lowland deposition in the Port Moresby region is likely to be lower than many other regions of New Guinea due to low local rainfall (Mabbutt et al. 1965:110; Bleeker 1983:177), Bulmer rightly identifies an anthropogenic dimension to landscape evolution and shoreline progradation, with major potential consequences for long-term human settlement of the region.

Löffler (1977:8-9) notes that landscape burning 'makes the ground highly susceptible to both rain splash erosion and unconcentrated wash' and that 'rain splash erosion appears to be severe where frequent burning is practised and where it is responsible for the removal of large quantities of soil'. Indeed, 'slope wash has ... been responsible for a near complete removal of the soil cover on most of the slopes in the Port Moresby area, irrespective of rock type' (Löffler 1977:8; see also Spenceley and Alley 1986). In the Caution Bay study area, exposure of uplifted fossiliferous limestone deposits of probable Pleistocene age (Mabbutt et al. 1965:14, 30, 107) is also consistent with considerable erosion and stripping of overlying surface sediments (see also Löffler 1977:6).

Edubu 1 was formed by rapid accumulation of 90 cm of sediment mostly over a period of up to 300 years between approximately 2350 cal BP and 2650 cal BP. These sediments (SUs 1-3) were sitting on culturally sterile sediments (SU4) exhibiting a dramatically different layer orientation. While it is possible that SU4 sediments originally represented horizontal layers that have since been tilted through tectonic warping, it is more probable that the steep layer orientation reflects sediment accumulation on a steeper palaeo slope. This hypothesis is consistent with the nearby slope flanking the gully on the northern parts of the site. Whatever the case, the dramatic stratigraphic change between SU4 and SU3 represents a disconformity in soft sediments associated with an equally dramatic change in depositional context at Edubu 1. The surface of SU4 appears to be an erosion surface as its associated stratigraphy has not been homogenised and obliterated by soil formation and turbation processes. While detailed sediment analyses have yet to be undertaken, preliminary stratigraphic assessment from excavations across the Caution Bay coast to hinterland suggests that cultural deposition at Edubu 1 commenced soon after a period of erosion in the local landscape.

The most likely source for accumulating sediments forming SUs 1 to 3 is slightly more elevated land inland of Edubu 1. That is, eroded sediments from further inland moved downslope towards the coast and accumulated at Edubu 1. This sediment accumulation buried cultural materials left at the site by its occupants to create the archaeological deposit we see today. A direct correspondence between rapid sedimentation and cultural materials suggests strongly that upslope erosion and concomitant downslope sedimentation were associated with human landscape disturbance. Indeed, rapid sedimentation at Edubu 1 ceased once cultural materials stopped being discarded at the site perhaps 2000 years ago. We hypothesise that a key source of this landscape disturbance was firing of the local vegetation to maintain savannah and grassland habitats for macropods, particularly wallaby-hunting practices. It is also possible that ground disturbance was associated with gardening practices which accelerated downslope sedimentation (see Gosden and Pavlides 1994:169; Gosden and Webb 1994; Spriggs 1997:85-86, 88, 2010; Kennett et al. 2006:278). To what extent possible increased gardening activities reflected the introduction of agriculture or intensification of existing agricultural practices with the arrival of Lapita peoples is unknown. Previous hypotheses have linked the introduction of agriculture along the south coast with the arrival of pottery-using peoples 2000 years ago (e.g. Harris 1995:853) or ca. 2600 years ago (McNiven et al. 2006:74). Clearly, continued research into the history of agriculture in the region, including Torres Strait, will be critical to understanding associated landscape transformations, and requires further development of multi-proxy approaches integrating archaeological (e.g. Barham and Harris 1985; Parr and Carter 2003), geomorphological (e.g. Barham 1999) and palynological indicators (e.g. Rowe 2007; McNiven et al. 2010) (see also Fairbairn 2005).

Broader support for hypothesised increased landscape modification and sedimentation regimes linked to changing land-use practices comes from analysis of sediments at Waigani Lake east of Caution Bay and immediately north of Port Moresby (Figure 1). A dramatic increase in the deposition of fluvial clay sediments in the lake between 2540 ± 80 BP (ca. 2600 cal BP) and ca. 1200 years ago reflects increased input of flood waters and sediments from the nearby Laloki River, linked to increased effective precipitation (Osborne et al. 1993:607-608). However, Osborne et al. (1993:608) also suggest that increased sedimentation may be 'due to an increase in run-off generated by changing land-use or vegetation cover within the catchment area'. The fact that ca. 2600-2650 cal BP marks the establishment of Edubu 1 and a major increase in sedimentation provides increased support for the influence of humans on sedimentation regimes at Waigani Lake.

The consequences of landscape burning and slope erosion throughout the past 2500-3000 years in the Port Moresby region are likely to have had cumulative impacts on land-use practices, particularly in relation to responses to repositioning of shorelines associated with coastal progradation and sea level change. For example, Summerhayes and Allen (2007:103) hypothesise that Nebira 4 may have been much closer to the sea in the past. Further to the west in the eastern Gulf of Papua, Swadling et al. (1976:56) similarly discuss the potential impact of major coastal progradation over a distance of 20-25 km during the past 6000 years on settlement patterns and land-use strategies. In the Caution Bay region, Pain and Swadling (1980) document a series of palaeo shorelines in the form of 'sandspits' separated by mangrove forests across the lower reaches of the Vaihua River. While they associate shoreline development with lowering sea levels in the past 6000 years, it is also likely that deposition of terrestrial sediments in the intertidal zone from inland erosion associated with human land-use activities contributed to shoreline changes and mangrove development. While coastal progradation may have increased the distance of sites such as Edubu 1 to the coast by only a few hundred metres, the impact of mangrove development and changing marine resource availability on local subsistence practices may have been considerable. Analysis of sediment cores taken across mangrove forests at Caution Bay by two of us (CR and IM) will shed further light on the history of local shoreline development and how people contributed and responded to such changes.

Conclusion

The results of recent excavations at Caution Bay will transform understandings of Port Moresby region archaeology and understandings of the human history of southern coastal Papua New Guinea. Critically in terms of this chapter, the extended chronology for human settlement of the region provides considerable scope to understand changes in landscape engagements and transformations associated with the arrival of Lapita colonists 2900 years ago. Whereas elsewhere in Remote Oceania, Lapita colonists represent the first human presence in landscapes, in the Port Moresby region, like much of Near Oceania, Lapita colonists arrived and engaged with places already transformed by more than 1000 years of prior human settlement. In this sense, the establishment of long-term Lapita settlements in the Port Moresby region was a more complex, negotiated process compared with Remote

Oceania, as it involved interactions with existing social and environmental landscapes. While the nature of this process of negotiation remains to be more fully explored – a process towards which we make initial steps in this paper – it is clear that the result was a complex mosaic of local and immigrant peoples with different cultural values, processes of environmental and social interaction, and histories. While sharing the same geographical setting, the ways local and immigrant groups chose to engage, transform and define this setting will reflect these differences. As Gosden and Webb (1994:48) cogently note, 'Land use is culturally determined, arising from choices people make about how to provision the social system.' To what extent such differences were maintained, hybridised and transformed over succeeding generations and centuries is the focus of ongoing research. Clearly, understanding how such differences were manifested in different land and sea engagements and environmental transformations will require the combined expertise of archaeological and palaeoenvironmental investigators. The co-determining and mutually constitutive nature of human-environmental relationships will be evident in archaeological sites as much as it will be registered in pollen cores, a theoretical point we have come to appreciate better through the work of Peter Kershaw.

Acknowledgements

For assistance with shell analyses we thank Helene Tomkins, Janet Sypkens and Shoshana Grounds. Helpful comments on an earlier draft of this chapter were kindly provided by anonymous referees. Thanks to Cathy Carigiet for Figure 7 and drawing the sherds for Figure 6, and Kara Rasmanis and Toby Wood for kindly preparing the figures.

References

Allen, J. 1972. Nebira 4: an early Austronesian site in central Papua. *Archaeology and Physical Anthropology in Oceania* 7(2):92-124.

Allen, J. 1977a. Fishing for wallabies: trade as a mechanism for social interaction, integration and elaboration on the central Papuan coast. In: Friedman, J. and Rowlands, M.J. (eds), *The Evolution of Social Systems*, pp. 419-455. London: Duckworth.

Allen, J. 1977b. Management of resources in prehistoric coastal Papua. In: Winslow, J.H. (ed), *The Melanesian Environment,* pp. 35-44. Canberra: ANU Press.

Allen, J. 1991. Hunting for wallabies: the importance of Macropus agilis as a traditional food resource in the Port Moresby hinterland. In: Pawley, A. (ed), *Man and a Half: Essays in Pacific Anthropology and Ethnobiology in Honour of Ralf Bulmer*, pp. 457-451. The Polynesian Society Memoir 48. Auckland: The Polynesian Society.

Barham, A.J. 1999. The local environmental impact of prehistoric populations on Saibai Island, northern Torres Strait, Australia: enigmatic evidence from Holocene swamp lithostratigraphic records. *Quaternary International* 59:71-105.

Barham, A.J. and Harris, D.R. 1985. Relict field systems in the Torres Strait region. In: Farrington, I.S. (ed), *Prehistoric Intensive Agriculture in the Tropics*, pp. 247-283. BAR International Series 232.

Bleeker, P. 1983. *Soils of Papua New Guinea.* Canberra: ANU Press.

Bulmer, S. 1971. Prehistoric settlement patterns and pottery in the Port Moresby area. *Journal of the Papua New Guinea Society* 5:28-91.

Bulmer, S. 1975. Settlement and economy in prehistoric Papua New Guinea: a review of the archaeological evidence. *Journal de la Société des Océanistes* 31:7-75.

Bulmer, S. 1978. Prehistoric culture change in the Port Moresby region. Unpublished PhD thesis, University of Papua New Guinea.

Bulmer, S. 1979. Prehistoric ecology and economy in the Port Moresby region. *New Zealand Journal of Archaeology* 1:5-27.

Chalmers, J. 1887. *Pioneering in New Guinea*. London: Religious Tract Society.

Coffey Natural Systems 2009. PNG LNG Project: Environmental Impact Statement. Prepared by Coffey Natural Systems Pty Ltd, 126 Trenerry Crescent, Abbotsford Victoria 3067. Australia.

David, B., Fairbairn, A., Aplin, K., Murepe, L., Green, M., Stanisic, J., Weisler, M., Simala, D., Kokents, T., Dop, J. and Muke, J. 2007. OJP, a terminal Pleistocene archaeological site from the Gulf Province lowlands, Papua New Guinea. *Archaeology in Oceania* 42:31-33.

David, B., Araho, N., Barker, B., Kuaso, A. and Moffat I. 2009. Keveoki 1: exploring the *hiri* ceramics trade at a short-lived village site near the Vailala River, Papua New Guinea. *Australian Archaeology* 68:11-23.

David, B., McNiven, I.J., Richards, T., Connaughton, S., Leavesley, M., Barker, B. and Rowe, C. In press. Lapita sites in the Central Province of mainland Papua New Guinea. World Archaeology.

Davies, H.L. and Smith, I.E. 1971. Geology of eastern Papua. *Geological Society of America Bulletin* 82:3299-3312.

Eden, M.J. 1974. The origin and status of savannah and grassland in southern Papua. *Transactions of the Institute of British Geographers* 63:97-110.

Enright, N.J. and Gosden, C. 1992. Unstable archipelagos: south-west Pacific environment and prehistory since 30,000 B.P. In: Dodson, J. (ed), *The Naïve Lands: Prehistory and Environmental Change in Australia and the South-West Pacific*, pp. 160-198. Melbourne: Longman Cheshire.

Fairbairn, A. 2005. An archaeobotanical perspective on Holocene plant-use practices in lowland northern New Guinea. *World Archaeology* 37(4):487-502.

Fairbairn, A., Hope, G. and Summerhayes, G. 2006. Pleistocene occupation of New Guinea's highland and subalpine environments. *World Archaeology* 38:371-386.

Fall, P.L. 2010. Pollen evidence for plant introductions in a Polynesian tropical island ecosystem, Kingdom of Tonga. In: Haberle, S., Stevenson, J. and Prebble, M. (eds), *Altered Ecologies: Fire, Climate and Human Influence on Terrestrial Landscapes*, pp. 253-271. Terra Australis 32. Canberra: ANU E Press.

Felgate, M. 2001. A Rovania ceramic sequence and the prehistory of Near Oceania: work in progress. In: Clark, G.R., Anderson, A.J. and Vunidilo, T. (eds), *The Archaeology of Lapita Dispersal in Oceania: Papers from the Fourth Lapita Conference, June 2000, Canberra, Australia*, pp. 39-60. Terra Australis 17, Pandanus Books, Canberra.

Field, J.S., Cochrane, E.E. and Greene, D.M. 2009. Dietary change in Fijian prehistory: isotopic analyses of human and animal skeletal material. *Journal of Archaeological Science* 36:1547-1556.

Garrett-Jones, S.E. 1979. Evidence for changes in Holocene vegetation and lake sedimentation in the Markham Valley, Papua New Guinea. Unpublished PhD thesis, Australian National University, Canberra.

Glaessner, M.F. 1952. Geology of Port Moresby, Papua. In: Glaessner, M.F. and Rudd, E.A. (eds), *Sir Douglas Mawson Anniversary Volume: Contributions to Geology in Honour of Professor Sir Douglas Mawson's 70th Birthday Anniversary Presented by Colleagues, Friends and Pupils*, pp. 63-86. Adelaide: The University of Adelaide.

Gosden, C. and Pavlides, C. 1994. Are islands insular? Landscape vs. seascape in the case of the Arawe Islands, Papua New Guinea. *Archaeology in Oceania* 29:162-171.

Gosden, C. and Webb, J. 1994. The creation of a Papua New Guinean landscape: archaeological and geomorphological evidence. *Journal of Field Archaeology* 21:29-51.

Haberle, S.G. 1998. Late Quaternary vegetation change in the Tari Basin, Papua New Guinea. *Palaeogeography, Palaeoclimatology, Palaeoecology* 137:1-24.

Haberle, S., Hope, G.S. and van der Kaars, S. 2001. Biomass burning in Indonesia and Papua New Guinea: natural and human induced fire events in the fossil record. *Palaeogeography, Palaeoclimatology, Palaeoecology* 171:259-268.

Harris, D.R. 1995. Early agriculture in New Guinea and the Torres Strait divide. *Antiquity* 69:848-854.

Hiscock, P. 2008. *Archaeology of Ancient Australia.* London and New York: Routledge.

Hope, G.S. 1976. The vegetational history of Mt. Wilhelm, Papua New Guinea. *Journal of Ecology* 64(2):627-663.

Hope, G.S. 1983. The vegetational changes of the last 20,000 years at Telefomin, Papua New Guinea. *Journal of Tropical Geography* 4(1):25-33.

Hope, G.S. 2009. Environmental change and fire in the Owen Stanley Ranges, Papua New Guinea. *Quaternary Science Reviews* 28:2261-2276.

Hope, G.S. and Haberle, S.G. 2005. The history of the human landscapes of New Guinea. In: Pawley, A., Attenborough, R., Golson, J. and Hide, R. (eds), *Papuan Pasts: Cultural, Linguistic and Biological Histories of Papuan-Speaking Peoples,* pp. 541-554. Pacific Linguistics 572. Canberra: Research School of Pacific and Asian Studies, The Australian National University.

Hope, G.S., Golson, J. and Allen, J. 1983. Palaeoecology and prehistory in New Guinea. *Journal of Human Ecology* 12(1):37-60.

Horrocks, M. and Nunn, P.D. 2007. Evidence for introduced taro (Colocasia esculenta) and lesser yam (*Dioscorea esculenta*) in Lapita-era (c.3050-2500 cal. yr BP) deposits from Bourewa, southwest Viti Levu Island, Fiji. *Journal of Archaeological Science* 34(5):739-748.

Irwin, G. and Holdaway, S. 1996. Colonisation, trade and exchange: from Papua to Lapita. In: Davidson, J., Irwin, G., Leach, B., Pawley, A. and Brown, D. (eds), *Oceanic Culture History: Essays in Honour of Roger Green,* pp. 225-235. Dunedin: New Zealand Archaeological Society.

Jones, R. 1969. Fire-stick farming. *Australian Natural History* 16:224-228.

Kennett, D., Anderson, A. and Winterhalder, B. 2006. The ideal free distribution, food production, and the colonization of Oceania. In: Kennett, D. and Winterhalder, B. (eds), *Behavioral Ecology and the Transition to Agriculture,* pp. 265-288. Berkeley: University of California Press.

Kershaw, A.P. 1974. A long continuous pollen sequence from north-eastern Australia. *Nature* 251:222-223.

Kershaw, A.P. 1993. Palynology, biostratigraphy and human impact. *The Artefact* 16:12-18.

Kirch, P.V. 1997. *The Lapita Peoples: Ancestors of the Oceanic World.* Oxford: Blackwell.

Lawes, W.G. 1879. Ethnological notes on the Motu, Koitapu and Koiari tribes of New Guinea. *The Journal of the Anthropological Institute of Great Britain and Ireland* 8:369-377.

Lawes, W.G. 1886. Insular Australasia: New Guinea. In: Garran, A. (ed), *Picturesque Atlas of Australasia,* Vol. III, pp. 661-668. Sydney: Picturesque Atlas Pub. Co.

Lilley, I. 2008. Flights of fancy: fractal geometry, the Lapita dispersal and punctuated colonisation in the Pacific. In: Clark, G., Leach, F. and O'Connor, S. (eds), *Islands of Inquiry: Colonisation, Seafaring and the Archaeology of Maritime Landscapes,* pp. 75-86. Terra Australis 29. Canberra: ANU E Press.

Löffler, E. 1977. The impact of traditional man on landforms in Papua New Guinea. In: Winslow, J.H. (ed), *The Melanesian Environment,* pp. 3-10. Canberra: ANU Press.

Mabbutt, J.A., Heyligers, P.C., Scott, R.M., Speight, J.G., Fitzpatrick, E.A., McAlpine, J.R. and Pullen, R. 1965. *Lands of the Port Moresby-Kairuku Area, Territory of Papua and New Guinea.* Land Research Series No. 14. Melbourne: CSIRO.

McAlpine, J.R., Keig, G. and Falls, R. 1983. *Climate of Papua New Guinea*. Canberra: Australian National University Press.

McNiven, I.J., Dickinson, W.R., David, B., Weisler, M., von Gnielinski, F., Carter, M. and Zoppi, U. 2006. Mask Cave: red-slipped pottery and the Australian-Papuan settlement of Zenadh Kes (Torres Strait). *Archaeology in Oceania* 41(2):49-81.

McNiven, I.J., David, B., Aplin, K., Pivoru, M., Pivoru, W., Sexton, A., Brown, J., Clarkson, C., Connell, K., Stanisic, J., Weisler, M., Haberle, S., Fairbairn, A. and Kemp, N. 2010. Historicising the present: late Holocene emergence of a rainforest hunting camp, Gulf Province, Papua New Guinea. *Australian Archaeology* 71:41-56.

McNiven, I.J., David, B., Richards, T., Aplin, K., Asmussen, B., Mialanes, J., Leavesley, M., Faulkner, P. and Ulm, S. 2011. New direction in human colonisation of the Pacific: Lapita settlement of south coast New Guinea. *Australian Archaeology* 72:1-6.

Mooney, S.D., Harrison, S.P., Bartlein, P.J., Daniau, A.-L., Stevenson, J., Brownlie, K.C., Buckman, S., Cupper, M., Luly, J., Black, M., Colhoun, E., D'Costa, D., Dodson, J., Haberle, S., Hope, G.S., Kershaw, P, Kenyon, C., McKenzie, M. and Williams, N. 2011. Late Quaternary fire regimes of Australasia. *Quaternary Science Reviews* 30:28-46.

O'Connell, J. and Allen, J. 2004. Dating the colonization of Sahul (Pleistocene Australia-New Guinea): a review of recent research. *Journal of Archaeological Science* 31:835-853.

Oram, N. 1977. Environment, migration and site selection in the Port Moresby coastal area. In: Winslow, J.H. (ed), *The Melanesian Environment*, pp. 74-99. Canberra: ANU Press.

Osborne, P.L., Humphreys, G.S. and Polunin, N.V.C. 1993. Sediment deposition and late Holocene environmental change in a tropical lowland basin: Waigani Lake, Papua New Guinea. *Journal of Biogeography* 20:599-613.

Paijmans, K. 1975. *Explanatory Notes to the Vegetation Map of Papua New Guinea*. Land Research Series No. 35. Melbourne: CSIRO.

Pain, C. and Swadling, P. 1980. Sea level changes, coastal landforms and human occupation near Port Moresby. *Science in New Guinea* 7:57-68.

Parr, J.F. and Carter, M. 2003. Phytolith and starch analysis of sediment samples from two archaeological sites on Dauar Island, Torres Strait, northeastern Australia. *Vegetation History and Archaeobotany* 12:131-141.

Pieters, P.E. 1978. Port Moresby, Kalo, Aroa: Papua New Guinea. Sheets SC/55-6, -7 and -11. Explanatory Notes and 1:250,000 Geological Map. Dept. of National Development, Bureau of Mineral Resources, Geology and Geophysics. Dept. of Minerals and Energy, Papua New Guinea, Geological Survey of Papua New Guinea. Canberra: Australian Government Publishing Service.

Reimer, P.J., Baillie, M.G.L., Bard, E., Bayliss, A., Beck, J.W., Blackwell, P.G., Bronk Ramsey, C., Buck, C.E., Burr, G.S., Edwards, R.L., Friedrich, M., Grootes, P.M., Guilderson, T.P., Hajdas, I., Heaton, T.J., Hogg, A.G., Hughen, K.A., Kaiser, K.F., Kromer, B., McCormac, F.G., Manning, S.W., Reimer, R.W., Richards, D.A., Southon, J.R., Talamo, S., Turney, C.S.M., van der Plicht, J. and Weyhenmeyer, C.E. 2009. IntCal09 and Marine09 radiocarbon age calibration curves, 0-50,000 years cal BP. *Radiocarbon* 51(4):1111-1150.

Romilly, H. 1889. *From My Verandah in New Guinea: Sketches and Traditions*. London: David Nutt.

Rowe, C. 2007. A palynological investigation of Holocene vegetation change in Torres Strait, seasonal tropics of northern Australia. *Palaeogeography, Palaeoclimatology, Palaeoecology* 251:83-103.

Seligmann, C.G. 1910. *The Melanesians of British New Guinea*. Cambridge: Cambridge University Press.

Singh, G., Kershaw, A.P. and Clark, R. 1981. Quaternary vegetation and fire history in Australia.

In: Gill, A.M., Groves, R.A. and Noble, I.R. (eds), *Fire and the Australian Biota*, pp. 23-54. Canberra: Australian Academy of Science.

Smith, I.E. and Milsom, J.S. 1984. Late Cenozoic volcanism and extension in Eastern Papua. *Geological Society, London, Special Publications* 16:163-171.

Specht, J. and Torrence, R. 2007. Lapita all over: land use on the Willaumez Peninsula, Papua New Guinea. In: Bedford, S., Sand, C. and Connaughton, S. (eds), *Oceanic Explorations: Lapita and Western Pacific Settlement*, pp. 71-96. Terra Australis 26. Canberra: ANU E Press.

Spenceley, A.P. and Alley, N.F. 1986. Effect of human settlement on soil erosion in the Owen Stanley Range, Papua New Guinea. *Search* 17:213-216.

Spriggs, M. 1997. *The Island Melanesians*. Oxford: Blackwell.

Spriggs, M. 2010. Geomorphic and archaeological consequences of human arrival and agricultural expansion on Pacific islands: A reconsideration after 30 years of debate In: Haberle, S., Stevenson, J. and Prebble, M. (eds), *Altered Ecologies: Fire, Climate and Human Influence on Terrestrial Landscapes*, pp. 239-252. Terra Australis 32. Canberra: ANU E Press.

Stone, O.C. 1876. Description of the country and natives of Port Moresby and neighbourhood, New Guinea. *Journal of the Royal Geographical Society of London* 46:34-62.

Stuiver, M. and Reimer, P.J. 1993. Extended 14C database and revised CALIB radiocarbon calibration program. *Radiocarbon* 35:215-230.

Sullivan, M.E. and Sassoon, M. 1987. Prehistoric occupation of Loloata Island, Papua New Guinea. *Australian Archaeology* 24:1-9.

Summerhayes, G. and Allen, J. 2007. Lapita writ small? Revisiting the Austronesian colonisation of the Papuan south coast. In: Bedford, S., Sand, C. and Connaughton, S. (eds), *Oceanic Explorations: Lapita and Western Pacific Settlement*, pp. 97-122. Terra Australis 26. Canberra: ANU E Press.

Summerhayes, G.R., Leavesley, M. and Fairbairn, A. 2009. Impact of human colonisation on the landscape: a view from the Western Pacific. *Pacific Science* 63(4):725-745.

Summerhayes, G.R., Leavesley, M., Fairbairn, A., Mandui, H., Field, J., Ford, A. and Fullagar, R. 2010. Human adaptation and plant use in Highland New Guinea 49,000 to 44,000 years ago. *Science* 330:78-81.

Swadling, P. 1977. A review of the traditional and archaeological evidence for early Motu, Koita and Koiari settlement along the central south Papuan coast. *Oral History* 5(2):37-57.

Swadling, P. and Hope, G. 1992. Environmental change in New Guinea since human settlement. In: Dodson, J. (ed), *The Naïve Lands: Prehistory and Environmental Change in Australia and the South-West Pacific*, pp. 13-42. Melbourne: Longman Cheshire.

Swadling, P., Aitsi, L., Trompf, G. and Kari, M. 1976. Beyond the early oral traditions of the Austronesian speaking people of the Gulf and Western Central Provinces: a speculative appraisal of early settlement in the Kairuku District. *Oral History* V:50-80.

Turner, W.Y. 1878. The ethnology of the Motu. *The Journal of the Anthropological Institute of Great Britain and Ireland* 7:470-499.

Turney, C.S.M., Kershaw, A.P., Moss, P., Bird, M.I., Fifield, L.K., Cresswell, R.G., Santos, G.M., Di Tada, M.L., Hausladen, P.A. and Zhou, Y. 2001. Redating the onset of burning at Lynch's Crater (North Queensland): implications for human settlement in Australia. *Journal of Quaternary Science* 16(8):767-771.

Valentin, F., Buckley, H.R., Herrscher, E., Kinaston, R., Bedford, S., Spriggs, M., Hawkins, S. and Neal, K. 2010. Lapita subsistence strategies and food consumption patterns in the community of Teouma (Efate, Vanuatu). *Journal of Archaeological Science* 37(8):1820-1829.

Vanderwal, R.L. 1973. Prehistoric studies in central coastal Papua. Unpublished PhD thesis, Australian National University, Canberra.

Vasey, D.E. 1982. Subsistence potential of the pre-colonial Port Moresby area, with reference to the hiri trade. *Archaeology in Oceania* 17(3):132-142.

White, J.P. and O'Connell, J.F. 1982. *A Prehistory of Australia, New Guinea and Sahul.* Sydney: Academic Press.

Woxvold, I.A. 2008. Assessment and Impact Analysis of Terrestrial Biodiversity at the LNG Facilities Site, Central Province, Papua New Guinea. Report to Coffey Natural Systems, Australia.

6

Otoia, ancestral village of the Kerewo: Modelling the historical emergence of Kerewo regional polities on the island of Goaribari, south coast of mainland Papua New Guinea

Bryce Barker
School of Humanities and Communication, University of Southern Queensland, Toowoomba, Queensland
barker@usq.edu.au

Lara Lamb
University of Southern Queensland, Toowoomba, Queensland

Bruno David
Monash University, Clayton, Victoria

Kenneth Korokai
Kikori, Gulf Province, Papua New Guinea

Alois Kuaso
PNG National Museum and Art Gallery, Port Moresby, Papua New Guinea

Joanne Bowman
University of Queensland, St Lucia, Queensland

Introduction

This paper presents a model for the occupation of the Kikori River delta and the first archaeological results from excavations undertaken in Kerewo lands on the large river delta island of Goaribari, western Gulf of Papua (Gulf Province), Papua New Guinea (PNG) (Figure 1). The site of Otoia 1 is situated along the northwestern end of Goaribari. Ethnographically this region encompasses the lands of the Kerewo in the eastern Kiwai language area, who at the time of initial European contact in the 1870s, exerted socio-political control and/or competitive influence from the Turama River in the west, to Paia Inlet in the east, and upstream at least as far as Kopi in the north (e.g. Knauft 1993:27; Weiner 2006). The Kerewo are the largest tribal-linguistic group in the coastal Gulf of Papua region. Prior to the colonial period, Kerewo villages were organised around large men's longhouses (*dubu daima*), each of which was compartmentalised into clan sections, membership of which was determined through agnatic descent. Local oral traditions recall that at Otoia, for example, there were originally two longhouses, called Gewo and Ubo Gewo. Clan membership at Gewo consisted of Kibiri, Atenaramio, Karuramio, Hide'ere, Guei, Pinei and Neboru, while the Ubo Gewo clans were Neauri, Kurami, Gibi, Adia'amudae and Neboru. When clans became too large, they split into separate units and differentiated themselves by adopting names according to the position they occupied in the longhouse. For Kerewo, the three major sections of the longhouse were *tamu* (head of the house), *goho* (middle of the house) and *nupu* (back of the house), thus the original Karuramio clan became divided into Nupu Karuramio and so on (Kenneth Korokai [Neauri clan], pers. comm.; Weiner 2006:32).

Kerewo oral traditions state that Otoia was the origin village for all the Kerewo, being the settlement from which the different clans fissioned and from which all subsequent villages in the delta were established (see Figure 1). At the time of initial European contact in the late 19th century, Kerewo society was characterised by ritual headhunting and constant raiding of nearby regions, with little inter-marriage taking place across linguistic/tribal boundaries. Residential mobility was extremely limited, with residence restricted to a number of very large and easily defended villages on Goaribari Island and the Omati River (Weiner 2006:41).

The archaeological site of Otoia 1 is an abandoned village location, one of several abandoned ancestral villages of the Kerewo on the islands of the Kikori-Omati Rivers delta. The site was visited and historically documented by Alfred Haddon in 1914 and the Australian photographer and explorer Frank Hurley in 1921. People continued to live at the village until the early 1970s, when the last of the villagers relocated to other Kerewo villages and to colonial administrative centres such as Kikori.

Early European accounts go to some lengths to point out that the delta villages of the Kerewo and nearby groups were generally very large at the end of the 19th century, unlike those of peoples found further inland, with populations typically estimated at 1000-2000 people within individual villages in the delta villages (e.g. Ryan 1913), in contrast to settlements upstream, which typically numbered fewer than 100 inhabitants (e.g. MacGregor 1894a; Murray 1914:10; Ryan 1914:170, 172; Woodward 1920; Flint 1923; Cawley 1925; Liston-Blyth 1929). Thus Ryan (1913, cited in Goldman and Tauka 1998:59) reported that 'West of Vaimuru [Baimuru] are the villages of the Urama tribe. Their villages are situated on land between Era Bay and Pai'a inlet, consisting of seven villages, with a population of about 4,000 people'. In 1917, the Acting Assistant Resident Magistrate for the Delta Division, C.L. Herbert (1917:87), wrote that the two villages of Ebi-ka-o and Mai-aki, a few minutes apart by boat and located in the Kikori River delta some 20 km east of Goaribari, had a combined population of about 3000 people. In 1920, Woodward (1920:63) concluded that the Aird River delta supported around 6500 people; that the Kerewo numbered around 4000; and that the Urama villages supported 2000 people. William MacGregor (1893) likewise reported seeing numerous large villages in the Kikori and Omati Rivers delta in 1892:

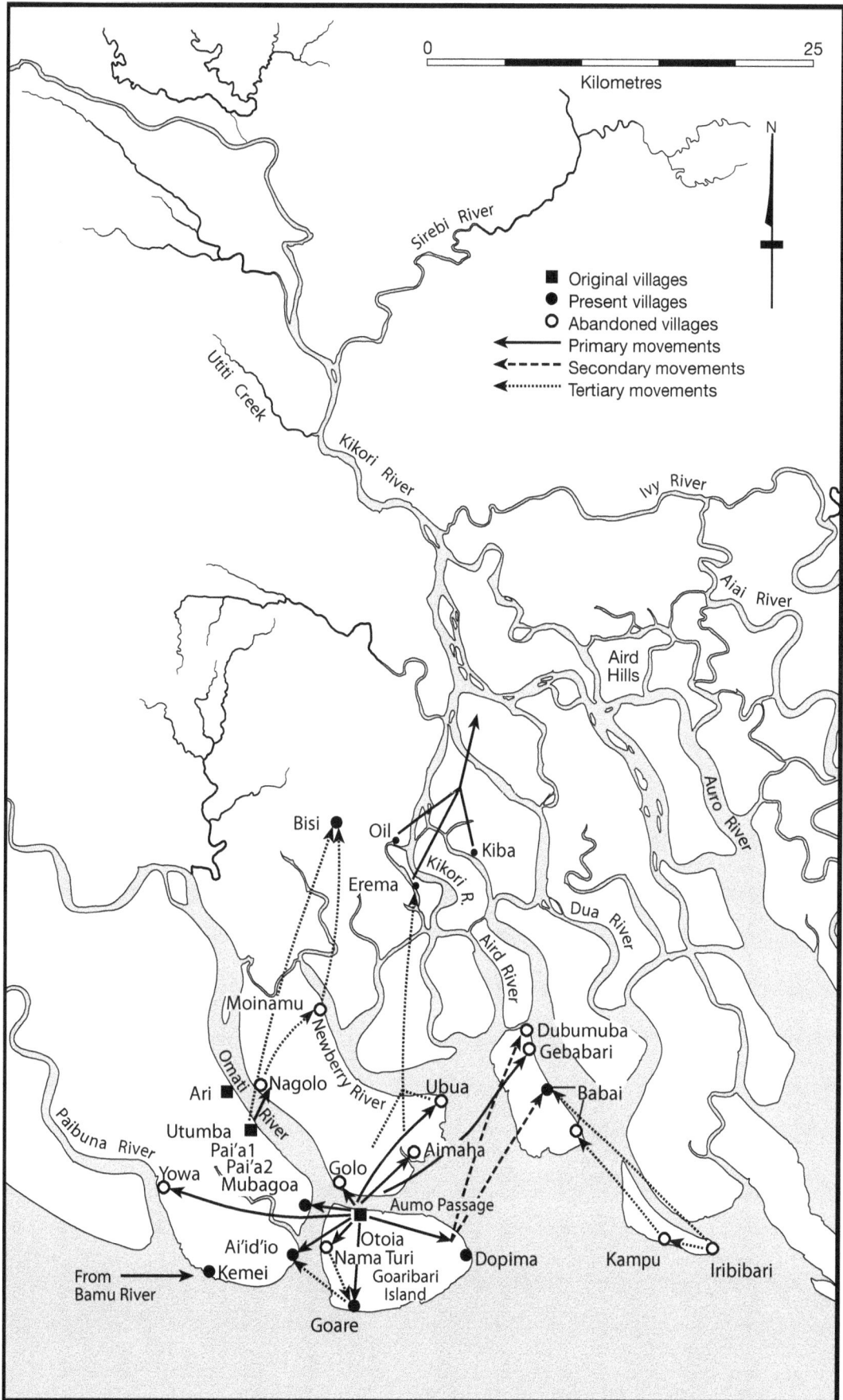

Figure 1. Map of the Kikori and Omati Rivers delta, showing Otoia as the original Kerewo village and the 'traditional Omati villages and migration history' according to oral traditions (after Goldman and Tauka 1998:63).

We proceeded to go through the Aumoturi channel, which cuts off Goaribari from the mainland. Entering it from the east end, we soon found ourselves in front of a large village Anawaida, [which seems to correspond with the village of Dopima in northeastern Goaribari] on its south bank. There are three or four very long houses, 300 to 400 feet each, and a number of smaller ones for the women and children. The village site was half swampy, but the coast there grows sago and some cocoanut trees. ... Near the western end of the strait there is another large village on the same bank Oteai [Otoia]. ... The Aumoturi joins a large river on the west side, which meets it at nearly a right angle, the two opening into the sea on the west side of Goaribari. We intended to examine this river, and it was decided that we should begin at its mouth. We accordingly steered for a large village on the right bank near the sea [probably Aiedio; see Figure 2]. It is a large and populous settlement, with a number of very long houses and many smaller ones. A large number of canoes came out to meet us ... About a mile and a-half further up the river there is, also on the right bank, a large village on a somewhat drier site [probably Mubagoa]. About four miles further up, on the same bank, is the great village of Baiaa [Pai-a], which seems to be the largest one I have seen in the Possession. Baiaa has over two scores of houses, many of them several hundred feet long. It is on land that would be about a yard above high water mark, and is firm enough to grow bread fruit and cocoanut trees. ... The village of Baiaa is about half a mile long, and presents quite an imposing appearance. ... On leaving our camp we found a large settlement on the right bank at a place called Naimesse. It was only a sago encampment. They had fairly good, small houses, not so large or substantial as those of the permanent villages. (MacGregor 1893:45-47)

In 1893, MacGregor (1894b, c) further reported that the villages of the Purari River delta a short distance to the east 'are numerous and very noisy', and that a 'large tribe was met with'

Figure 2. Longhouses in the village of Aiedio, opposite Goaribari (from *Papua Annual Report for the Year 1918-19*, Photo 4).

on the Gulf Province islands between the Era and Kikori Rivers (MacGregor 1894b:xix).

By contrast, in 1920 the Acting Resident Magistrate for the Delta Division, R.A. Woodward (1920), estimated that the total population of all the communities along the inland Kikori River, the 'Vero River' and the Tiviri Junction put together numbered a mere 500.

Although the delta region is a difficult environment in which to live, in the sense that almost everywhere the land is swampy and covered by saltwater mangroves rendering agriculture nigh impossible (see Haddon 1918; below), it clearly supported some of the largest and most densely populated ethnographic villages of the western Gulf Province lowlands (cf. Williams 1924). We suggest that a contributing factor was that the villages' position at the mouth of the Kikori and Omati Rivers placed them at the front end of redistributive networks inland to the north, and further to the west along the coast. This enabled the Kerewo to act as middlemen in aggrandising interaction networks with trade partners to the east, from which imported *hiri* ceramics were obtained. In their strategic positioning at the mouth of the major Kikori and Omati Rivers, the Kerewo of the delta region controlled the redistribution of pottery to the expansive populations residing along the Kikori and Omati Rivers. We argue that, initially at least, the Kerewo may themselves have depended to a significant extent on the pottery trade for access to inland products such as stone and supplementary plant produce to sustain themselves while living in a difficult environment from which they were supporting growing populations. Trade in supplementary plant produce is supported from oral accounts and the fact that many food plants will not grow in the delta because of its low-lying nature and periodic inundation by saltwater; even sago is not as prolific in the lower delta islands as elsewhere in the Gulf region.

That the *hiri* trade usually reached Goaribari via secondary means is supported by a range of historical and oral sources. Nigel Oram (1982) recorded that the Motu ceramic manufacturers divided the recipient *hiri* trade villages into four zones extending as far west as the Purari River (located 110 km east of Otoia). Present-day Kerewo oral traditions also state that pottery trade was indirect and came from their eastern neighbours, with no recollection of ever having been visited by Motu traders. Furthermore, Kerewo state that the Motu trade language was not spoken in the delta until introduced by missionaries at the turn of the 19th century (Kenneth Korokai pers. comm. 2008). This is supported by Dutton's (1982:82) linguistic study of the distribution of the Motu language in the Gulf, where he states:

> … although we know that some North-East Kiwai speakers, notably the Urama, traded with the Koriki of the Purari Delta, the traditional end point of the hiri, they do not seem to have traded with the Motu directly for it is reported that it was only after European contact that the Motu went to Urama [58 km to the east of Goaribari] and then they had to stop off at Maipua to find a man to translate Motu into Urama.

In July 1878, Henry M. Chester, the Police Magistrate at Thursday Island in Torres Strait, undertook an expedition to the south coast of New Guinea in the steamer *Ellengowan*. At the Motu village of Boera, he obtained the following information on customary *hiri* trade voyages from local clan leaders:

> We gleaned the following information from them. Their annual trading voyage commences in August, and extends about 100 miles to the westward. They call at all the villages to exchange their pottery ware for sago, and return to Boera with the first north-west wind in December. There are twelve villages to pass before arriving in the cannibal districts. Vaimuro is the last place of call, nine villages further on, and to them come people from three villages still further to the westward, Kerepo [Kerewa, aka

Otoia] being the last village with which they have intercourse. (Chester 1878:9)

We further suggest that *hiri* pottery redistribution networks may have contributed to the consolidation and possible expansion of headhunting cults for which the Kerewo were particularly feared during ethnographic times (see David et al. 2010). There is little doubt that the continuous raiding of neighbouring groups essential to Kerewo headhunting rituals kept neighbouring groups in constant fear of raiding parties (Knauft 1993; David et al. 2010). The sheer scale of such raiding can be seen from one account of a colonial punitive expedition sent to the Kerewo village of Dopima on Goaribari in the aftermath of the killing of the missionary James Chalmers and his associate the Rev. Oliver F. Tomkins on 8 April 1901:

> In a report on the massacre, the Rev. H.M. Dauncey says, 'in one of the dubus were over seven hundred skulls, and at another four hundred. Some of the other dubus were cleared before the party reached them, but I am within the mark in saying that there must have been ten thousand skulls in the twenty dubus burned'. (cited in Haddon 1918:180)

Indeed, David et al. (2010) suggest that persistent headhunting raids by the Kerewo in particular led to neighbouring delta and inland river-bordering tribal groups, themselves headhunters, relocating villages away from major waterways into more rainforest-hidden refugial locations, specifically for protection from headhunting raids.

In this context, through headhunting cults, in the Gulf Province river deltas actively configured regional polities, including settlement locations and regional alliances, reinforcing and defending hierarchical relationships between communities and individuals. Knauft (1993:196) thus states for headhunting practices in the broader contiguous Western-Gulf Province coastal region, including the Kerewo as the easternmost Kiwai-speaking group, that:

> Kiwai headhunting melded dimensions of warfare found variously among Marind, Purari and Asmat. ... coalitions of Kiwai in long-distance coastal raids could claim many victims, and some local groups were exterminated through headhunting. Kiwai were distinctive for their complex web of local and long-distance alliances. These provided opportunity for both substantial temporary coalition and large-scale death contracting and treachery. The detailed accounts obtained by Landtman (1917) illustrate how astute leaders could effectively ally with other settlements; sometimes they would even contract and aid third parties to travel long distances in fleets of canoes to carry out surprise attacks. Many heads and renown – as well as substantial payment – accrued to the attackers, while local political advantage went to the leader who contracted the killing ... The general sense one gets from the accounts of Landtman (1917), Riley (1925) and Beaver (1920) is that warfare among Kiwai themselves was driven by political disputes and revenge rather than ritual mandate.

We argue that because of the indirect nature of the trade, and the reliance on intermediary groups for access to imported pots and shell valuables, control over supply provided not only a powerful stimulus to largely command this interaction itself, but also a means to muster political and military support and to prevent competition from rival groups occupying the lower reaches of the Kikori River system. Over decades, generations, and some four to five centuries of permanent village settlement at the strategic mouths of the large river systems, the Kerewo and their neighbours in the Purari River delta where the westernmost regular direct *hiri* trade took place, grew into controlling coastal polities. These social entities manipulated and managed regional economies, territorial interactions and social relations through headhunting cosmologies that enabled powerful alliances to be formed and that spread fear among neighbouring populations. Kerewo domination of regional networks from a relatively resource-poor base involved control over incoming trade products, leading to further aggrandising of the Kerewo realm.

Previous archaeological research

In order to test this model, a series of excavations at ancestral village sites was planned in the Kikori delta. Our major aims were three-fold:

1. To determine whether the establishment of the large Kerewo villages coincided with or shortly followed the commencement of the ethnographic *hiri* trade, as known by the well-documented Motu genealogies for the origins of the *hiri*.

2. To determine whether the sequence of Kerewo villages known from local oral traditions concur with the sequence evidenced by the archaeological record.

3. To determine the extent to which the lower (coastal) Kikori River archaeological sequences (including the Goaribari villages) correspond with the mid Kikori River archaeological sequences. Ethnographically, imported trade ceramics entered the Kikori River via the coast, in exchange for mass-produced sago starch manufactured with tools made from stone pounders imported through inland trade systems. A key question asked of the upper, mid and lower Kikori River sequences thus concerns the antiquity of the articulating stone-sago-pottery production system and trade relations, and the effects of these trade relations on regional demography and village polities.

With these aims in mind, this paper represents our first results of excavations in the lower Kikori River coastal region. Here we present archaeological evidence for the age of the Kerewo origin village of Otoia.

Archaeological research in the Kikori River region was previously carried out by Sandra Bowdler (pers. comm., cited in David 2008:466) and James Rhoads (e.g. 1980, 1982) in the 1970s, and subsequently by Bruno David (e.g. 2008; David et al. 2007, 2008) from 2005 onwards. David Frankel and Ron Vanderwal (1982, 1985) undertook archaeological excavations at Kinomere village on the island of Urama shortly to the east in the 1980s. David et al. (2007) established a late-Pleistocene antiquity for settlement of the mid Kikori River with earliest radiocarbon dates of 13,000 cal BP, the only Pleistocene site yet found in the southern PNG lowlands.

David (2008) describes human occupation of the mid Kikori River system as a series of pulses separated by long periods of absence. These hiatuses in localised and possibly regional occupation occur between 8000 and 2750 years ago, 2750 and 2000 years ago, 2000 and 1450 years ago, and lastly between 950 and 500 years ago. The earliest archaeological evidence of village establishment in the mid Kikori River region coincides with the first appearance of ceramics from 1450 to 950 years ago, followed by a renewed period of village establishment and the presence of ceramics after 500 years ago. David (2008) suggests that this latest occupational pulse is likely linked to the onset of the ethnohistorically recorded Motu *hiri* trade in this part of PNG (David 2008; cf. Chalmers 1895; Seligman 1910; Dutton 1982).

Excavations at the lower Kikori River delta site of Emo, an ancient Porome village site in the nearby Aird Hills, also show a pattern of intermittent pulses of occupation beginning 1840 years ago and signalling the earliest appearance of pottery in that area. After 1530 years ago, pottery discard rates increase substantially, which corresponds closely with the first known appearance of pottery in the mid Kikori River region some 1450 years ago (David et al. 2010).

Site description

The island of Goaribari, like all Kikori River delta lands except for the highly localised Aird Hills, is very low-lying (maximum elevation = 1 m above sea level), consisting of accumulated sedimentary mud from the Kikori and Omati river systems. During those times of the year when king tides and storm events occur, large parts of Goaribari are inundated. This has precipitated abandonment and relocation of delta villages, at least in the recent historical past. A government Patrol report from 1924 thus states that 'Kerewa [Otoia] was at one time the parent village of the Goaribari District and the remains, broken sticks underwater were pointed out to me as part of the old DUBU [longhouse] which extended along the bank for at least 700 or 800 yards, by erosion of the river and big floods the DUBU broke away and many of the people left the village to form settlements in other parts of the district' (Woodward 1920, cited in Weiner 2006:22).

There are currently only two small villages (Goare and Dopima) remaining on Goaribari Island, both of which are periodically inundated. Due to encroaching sea levels, each of these villages has been moved progressively inland over the years, so that local people now refer to the current Goare village as Goare 5 and Dopima village as Dopima 3 (both sets of villages facing the sea along the southwestern and eastern sides of Goaribari respectively) (Figure 3). The dynamic nature of the river delta has also meant that the coastline of these islands is constantly changing over time, affecting settlement patterns and, with this, the archaeological record. Thus, large parts of the site of Otoia (which faces inland and thus is river- rather than sea-bordering) no longer exist, as much of the land encompassing the old village has eroded into the main channel of the Kikori River delta. The steep, deeply cut river bank at Otoia has exposed a line of house posts running for more than 100 m parallel to the river bank, approximately 1 m below the current ground surface (Figure 4). These were identified by Kerewo clan leaders as the posts of a men's longhouse (*dubu daima*) at Otoia. While visiting Otoia in the early 1900s, Haddon (1918:177) described one of these longhouse structures:

> The *dubu daima* is a very long pile-dwelling varying from about 100 to over 200 yards in length. The ridge is horizontal, or rises slightly at the front end and is supported by a central row of poles. There is a platform entrance usually at one end and several side entrances. A gangway extends along the whole length of the interior, on each side of which are a number of cubicles … I found the one at Dopima was nearly 201.3m (660 feet) long, 10m (33ft) wide and the floor was 1.98m (78in) above the ground.

Associated with these posts and eroding from the bank are high densities of artefactual material. These include stone artefacts such as adzes, axes and grindstones, but more prominently a great variety of organic artefacts, including remnants of canoe paddles, arrows, canoes, notched pieces from houses and a high density of cut wood (chips) identified by Kerewo as the debris from woodworking (archaeological organic artefacts are equally well-preserved in other nearby Kerewo river-bordering sites; e.g. Figures 5 and 6). No pottery was found eroding from this bank, although it is found elsewhere on Goaribari (e.g. at Goare), and at Aiedio across the Omati River channel.

The present ground surface at Otoia is muddy and heavily vegetated, dominated by tropical wet rainforest species including Nipa (*Nypa fruticans*) and coconut (*Cocos nucifera*) palms

Figure 3. Goare 2 village site, Goaribari, showing good preservation of longhouse posts in intertidal zone. This settlement was abandoned approximately 60 years ago. The current Goare 5 village can be seen in the background.

Figure 4. In situ longhouse posts at Otoia, exposed by tidal erosion. Note collapsed bank behind Hansen Iburi and the presence of posts behind the collapse.

interspersed with mounded crab (Ocypodidae) burrows. A single 1 m x 1 m test excavation square was positioned close to the river bank adjacent to eroding artefacts, as far away as possible from evidence of crab disturbance and on the most elevated location. The excavation square was thus located approximately 1 m from the river bank, close to a large coconut tree in

Figure 5. Well-preserved woven band made from plant matter from Goare 2.

A

0   5cm

B

0   5cm

Figure 6. A: Part of a house cross beam. **B:** Cut wooden piece, possibly floor slat.

a location that was once directly underneath a men's longhouse, as informed by oral traditions and the eroding cultural materials. We note that as most of the ancient village site of Otoia had already eroded into the adjacent river channel, only a very small undisturbed area remained available for excavation, and indeed the excavation square had almost completely eroded into the river channel on our return one year later (Figure 7).

Stratigraphy

The excavation square contains four distinct Stratigraphic Units (SU) (Figure 8). SU1 is a thin, 1 cm deep, mossy, surface vegetation layer. SU2 extends down to a maximum 32 cm below the ground surface, and consists entirely of densely matted coconut root with little or no soil deposit and no cultural material. SU3 is a culturally sterile red clay extending down to a maximum depth of 56 cm below ground. The upper levels of SU4 include the SU3-SU4 interface, which extends to a depth of 86 cm before giving way to the homogenous grey clayey mud of SU4 proper, which extends to the base of the excavation at a depth of 170 cm below ground.

Figure 7: The remaining portion of the Otoia 1 excavation square one year after excavation.

Chronology

Dating of the archaeological sequence was carried out on excavated wood and charcoal (Table 1). Radiocarbon dates on wood artefacts were conducted on in situ samples where the wood was clearly artefactual, identified as such through the presence of cut marks. The radiocarbon-dated charcoal samples were also retrieved in situ and plotted in three dimensions. The pattern of radiocarbon dates indicates that all the wooden artefacts date to approximately the past 200 years, with the oldest dates coming from XU36 at 165.3 cm depth, within a calibrated age range of 0-306 cal BP (highest probability within the 2 sigma range is 145-215 cal BP; see Table 1). Of the two charcoal radiocarbon determinations, one is in the 'modern' range near the top of the cultural layer in XU23, and the other is dated to 456-537 cal BP (highest 2 sigma probability) near the bottom of the cultural deposit in XU33 at 150.7 cm depth. In spite of the slight inversion of the oldest date on wood and the oldest charcoal sample, the absence of historically known fires in this wet tropical rainforest area indicates that all charcoal in this region is anthropogenic, and implies that the oldest radiocarbon date of 456-537 cal BP at Otoia is most likely to be indicative of village establishment.

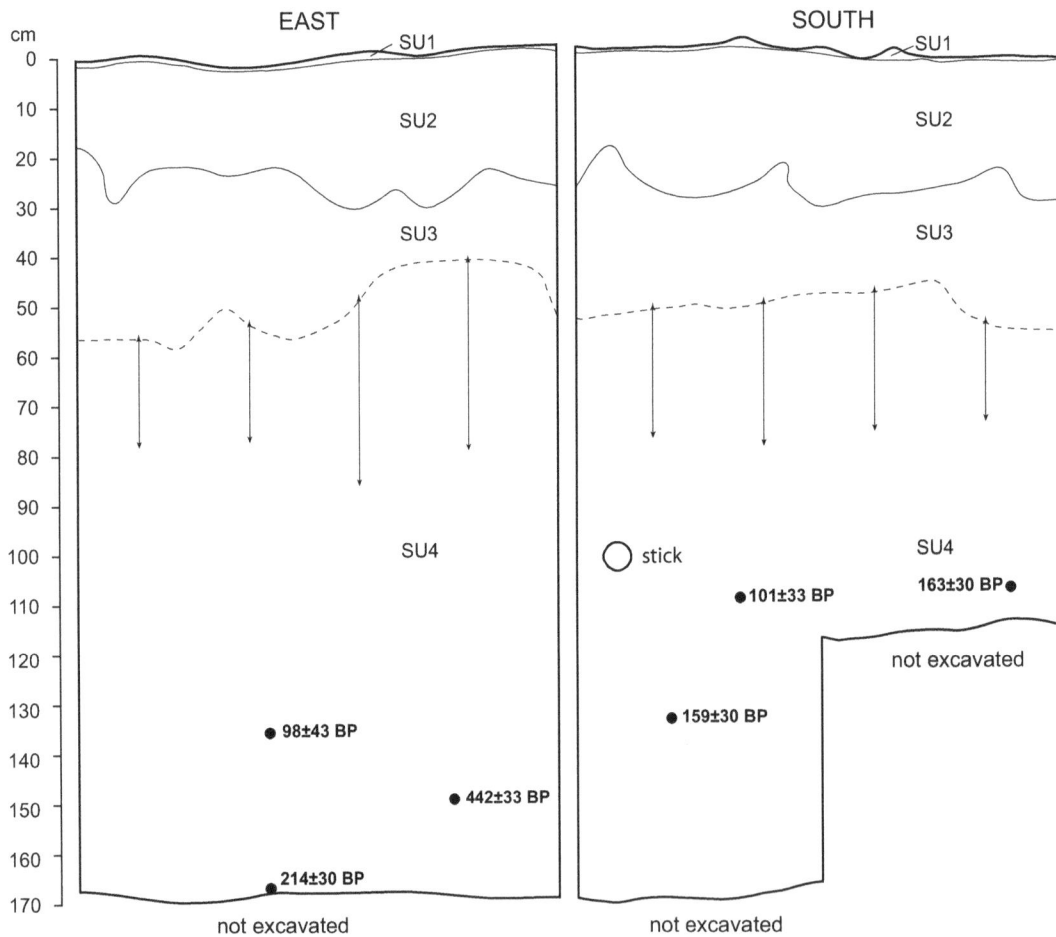

Figure 8. East and south section drawings, Otoia 1 excavated square.

All of the dated wooden artefacts appear to be contemporaneous at an archaeological time scale. That these artefacts all belong to a single village phase is supported by a radiocarbon date from an in situ sharpened wooden stake extending vertically from XU29 to XU36 (135 cm to 165 cm depth) and dated to sometime between 0-285 cal BP (164-231 cal BP at highest 2 sigma probability). A radiocarbon date obtained from an in situ house post eroding from the river bank 1 m from the excavation square indicates an age range of 0-304 cal BP (139-222 cal BP at highest 2 sigma probability). This is an almost identical age range to that of the lowest dated wood in the excavation square (145-215 cal BP at highest 2 sigma probability). The upper parts of this post had been exposed through water erosion caused by river activity, but still extended down into buried sediments for some 40 cm depth, the bottom extending to just below the maximum depth of the Otoia 1 excavation – that is, close to the same depth as XU36.

Table 1. Radiocarbon determinations, Otoia 1 Square 1. All radiocarbon determinations are AMS dates. Calibrations on Calib 6.0 (Stuiver and Reimer 1993), using INTCAL09 curve selection. *Top of post.

XU	Depth below surface (cm)	Laboratory #	Sample type	^{14}C age (years BP)	Calibrated age BP (95.4% probability)	
21	106.6	Wk-23998	wood	163 ± 30	0-36 68-118 **131-230** 244-287	(0.195) (0.116) **(0.512)** (0.177)
23	112.9	Wk-23058	charcoal	101 ± 33	0-0 **13-148** 188-196 212-269	(0.007) **(0.700)** (0.012) (0.281)
29	133.8	Wk-23059	coconut	98 ± 43	0-0 **11-150** 175-176 186-270	(0.013) **(0.666)** (0.002) (0.320)
29-36	135.0*	Wk-24000	wood	159 ± 30	0-37 65-118 124-158 **164-231** 243-285	(0.193) (0.139) (0.119) **(0.372)** (0.176)
33	150.7	Wk-23060	charcoal	442 ± 33	337-348 **456-537**	(0.023) **(0.977)**
36	165.2	Wk-23999	wood	214 ± 30	0-19 **145-215** 267-306	(0.143) **(0.511)** (0.346)
		Wk-25465	wood (longhouse post)	199 ± 32	0-30 **139-222** 259-304	(0.176) **(0.554)** (0.270)

We interpret the Otoia 1 chronometric chronology as indicating initial village establishment around 450-500 years ago, with construction of this particular longhouse at approximately 139-222 cal BP (see also Frankel and Vanderwal 1982, 1985, who report similar dates for the important nearby delta village of Kinomere). The deepest dated piece of wood in the excavation within XU36 revealed a comparable radiocarbon age for an in situ house post eroding from the river bank. We posit that the more recent dates reflect the period of occupation of the longhouse said by local clanspeople to have been located on the exact spot of the excavation.

Cultural materials

All of the excavated cultural material is found within SU4, with the greatest densities of stone artefacts, bone, cut wooden pieces, coconut shell, seeds and charcoal occurring between XU20 at 107.6 cm and XU32 at 149.6 cm depth (Table 2).

Table 2. Excavated material from Otoia 1.

XU	Stone artefacts #	Stone artefacts g	Bone g	Wood chips g	Wood chips NISP	Coconut shell	Other plant g	Seeds g	Charcoal g	Shell g	Crab g
1							1247				
2							1888				
3							2543				
4							2004				
5							1584				
6							4351				
7							3042				
8							1567				
9							1109				
10							2391				
11							3033				
12							1836				
13							1784				
14							1829				
15							1570				
16	2	4.58		0.54	133		478		0.19		
17				1.25	24		238	0.22	1.06		
18				3.48	55		252	0.61	1.98		
19	1	0.22		10.2	108		246				
20				31.68	207	x	196				
21				51.72	850	x	212	2.20	1.68		
22	1	1.28		144.71	2612		197	0.20			
23			0.01	234.63	3510	x	303	0.65	4.40		
24	1	0.19		217	5500	x	198	1.58	3.60		
25				294	3520	x	292	2.89	7.49		
26	1	0.02		141.5	1580	x	186	0.13	1.43		
27	1	3.99	1.84	363.38	6500	x	126	0.01	1.69		
28	2	2.25		250.69	2250	x	211	1.82	1.40		
29				148.73	1250	x	96	0.42	5.20	0.25	
30	1	0.39	0.01	72.23	400	x	17	0.71	3.36		
31			0.03	40.49	528	x	29	0.73	4.08	0.05	0.10
32				31.55	286	x	31	0.68	3.60		
33				25.17	233		8	0.25	1.73		0.15
34				19.58	204		4	0.22	2.17		
35				34.49	221		8	0.12	2.54		
36				36.92	245		5	0.21	2.50		
37	1	0.01		30.07	215		3	0.44	2.85		

Wooden chips

Cultural material in the site is overwhelmingly represented by plant remains consisting predominantly of cut wood chips. With no stone source occurring within the swampy Kerewo lands, during ethnographic times local material culture consisted predominately of wooden items. By analogy, the high density of cut wood chips found in the excavation and eroding from the river bank are thus likely to be the bi-products of wooden items, especially from house and/or dugout canoe manufacture. These wood chips are also precisely the size range and form of wooden manufacturing debris seen today at locations where houses or canoes have been recently constructed (Figures 9 and 10). The highest density of excavated wood chips came from XU27, with a NISP count of some 6500 chips (Table 2). Most of the wooden chips are angular with straight sides, usually forming a square or rectangular shape. The chips range from less than 1 cm to 12 cm in length, with most exhibiting clear evidence of having been cut.

Seeds

A large number of seeds were retrieved from the excavation. Taxonomic identifications were carried out by Joanne Bowman at the University of Queensland. In total, 23 types of seed and nut were recovered from Otoia 1. As a comprehensive floral reference collection is not yet available for the region, many of the excavated seeds could not be identified despite their distinctive morphologies. Identified taxa include *Cocos nucifera*, *Pandanus* sp. and specimens from the Cucurbitaceae and possibly the Malvaceae families. Coconut (*C. nucifera*) is the most abundant taxon recovered, with unidentified seed Types A, B, E and K also appearing

Figure 9. *Dubu daima* at Kaimare, Urama Island. Note the high density of house post supports (Hurley 1921).

Figure 10. Horizontal line of wooden artefacts eroding from the Otoia 1 cultural layer. Artefacts include wooden house pieces, pieces of dressed wood and large quantities of cut wooden chip.

in considerable quantities. Seed Type O is a single intact specimen from the Cucurbitaceae family recovered from XU25. The size of this seed suggests that it is from one of the smaller species of cucurbits, perhaps a variety of gourd. Types N and P are reinform seeds likely to be from the family Malvacae, possibly from the genus *Hibiscus*. Specimens labelled Type D are legume seed pods. Most other types had only a single occurrence in the assemblage. All of the identified plant species have either an ethnographically described subsistence or economic use and remain key resources today (Table 3).

Table 3. Summary of archaeobotanical analysis.

Taxon	NISP	Weight (g)
Cocos nucifera	52	91.2812
Pandanus sp.	1	22.3332
Cucurbitaceae (Type O)	1	0.0325
Malvaceae? (Type N)	1	0.0123
Malvaceae? (Type P)	1	0.0169
Fabaceae (Type D)	3	0.0295
Type A	11	1.1258
Type B	31	1.0532
Type C	1	1.5413
Type E	32	0.3673
Type F	1	0.0520
Type G	1	0.0371
Type H	1	0.0508
Type I	1	0.0082
Type J	1	0.0029
Type K	15	0.0544
Type L	1	0.0005
Type M	1	0.0203
Type Q	1	0.0287
Type R	2	0.1090
Type S	1	0.0508
Type T	2	0.0127
Type U	1	0.0133

Stone artefacts

The small sample of stone artefacts (N=11) retrieved from Otoia 1 was dominated by flaked pieces made on a coarse volcanic (basalt) material (Table 4). The two complete flakes represented the largest artefacts in the sample, and were also made on basalt. As there is no source of stone within Kerewo lands, the raw material had to have been imported from elsewhere, the nearest possible source being the Aird Hills some 37 km to the northeast. Although showing no signs of having been ground or polished, the coarse-grained nature of the stone and the fact that it is basalt suggests it would be most suited for adze or axe manufacture.

Table 4. Summary of stone artefact analysis.

	Flaked pieces	Flakes
Number	9	2
Mean weight (g)	0.58	3.85
Mean max. dimension (mm)	12.93	33.65
Mean length (mm)	-	17.5
Mean width (mm)	-	26.7
Mean thickness (mm)	-	4.45

Charcoal, bone and shell

Only small quantities of charcoal were present throughout the cultural layers. Similarly, bone, shell and crab remains have a very limited archaeological presence at the site. The bones are all of fish; as with the shell and crab fragments, they are too small to allow further taxonomic identification.

Discussion

The cultural material at Otoia 1 signals activity-specific discard and post-depositional factors mostly falling within the late pre-European contact to early contact periods of the 1800s to mid 1900s. All of the radiocarbon-dated material (except the charcoal sample from XU33) falls within this age range, with some minor chrono-stratigraphic inversions. A degree of stratigraphic integrity can also be surmised from the distribution of stone artefacts within the major cultural stratum spanning XU16 at 86.6 cm to XU37 at 169.8 cm depth. The presence of larger artefacts in the upper portion of the deposit suggests that downward movement of stone was limited in extent. The two oldest radiocarbon determinations also come from the lowest XUs. The in situ vertical-sharpened stake found between XUs 29 and 36 at depths spanning 135 cm to 165 cm below ground shows conclusively that major reworking of the deposit has not occurred.

The predominance of wooden artefacts and the low densities of stone artefacts probably reflect the lack of a local stone source in the immediate region. The unexpected absence of pottery in the excavation (given its presence on the surface at the nearby ancestral village sites of Aiedio and Goare) may be explained as a sampling issue relating to not only the size of the excavation but the spatial locality of the excavation square, placed as it was at the location of a men's longhouse. Ethnographic records among the Kerewo show that clay pots were used for cooking and the preparation of food took place in the women's houses (*upi daima*), which were spatially separated from the men's longhouses. Thus, the lack of pottery in the excavation and from the eroded river bank may relate to this gendered division of labour, space and artefacts. This explanation could also account for the small quantities of food remains (bone and shell) and charcoal retrieved from the excavation. The large quantities of cut wooden chips may be indicative of the construction of the longhouse itself (see Figure 9 for an indication of the large amount of woodworking that goes into the construction of a longhouse), or of subsequent woodworking by men in the longhouse, in particular of wooden sacred boards such as skull racks housed in the longhouse.

The radiocarbon age of 456-537 cal BP on charcoal closely mirrors the dates for the latest phase of village establishment and the late ceramic pulse in the mid Kikori River, arguably linked to the ethnohistorically recorded Motu *hiri* trade, as postulated by David (2008). Excavations undertaken by Frankel and Vanderwal (1982, 1985) at Kinomere on Urama Island, 58 km east of Otoia by sea, also similarly returned a basal radiocarbon determination for initial village establishment around 410 ± 80 BP (296-553 cal BP). The much earlier dates for occupation of the mid Kikori River and at the Emo site in the Aird Hills (David 2008; David et al. 2010) during earlier occupational pulses are from considerably more stable physical environments than the delta. While it is possible that earlier evidence of delta occupation has been destroyed by river flow and tidal erosion, cyclonic events and sea level rise, we argue that it is more likely that large villages in the delta region only emerged with the onset of regular large-scale trade partnerships in the form of the Motu *hiri* some 450-500 years ago (see David et al. 2010). That this environment was less than ideal for human occupation is supported by Haddon (1918:177), who states that 'Owing to the swampy nature of the country they [Kerewo] have poor gardens'. Although phases of pottery trade occurred well before 450-500 years ago along the mid Kikori River, such trade came with a different set of socio-historical contexts, and for much of the sequence appears to have been less intensive and/or regular than the ethnographically described *hiri* trade traceable genealogically to the past 450 years, and thus probably did not trigger the development of large, permanent delta villages at that time.

Conclusion

The contemporaneous onset of the large, domineering villages of Otoia (Kikori River mouth) and Kinomere (Purari River mouth) some 450-500 years ago, coincident in timing with the most recent ceramic pulse of the mid Kikori River, suggests significant causal relations between the establishment and growth of large, centralised village settlements at the mouths of large rivers, politico-economic control over redistribution networks, and the regular arrival of mass-produced (principally Motu ceramic and shell valuable) trade goods in the Gulf of Papua delta region. Although more work is now required to obtain a more robust chronology of village establishment for the past 450-500 years in the lower Kikori River and nearby river deltas, along with focused excavation of a series of pottery sequences in the West Papua-Western Province-Gulf Province region, our initial results clearly show how local and regional landscape history needs to consider not only physical environmental conditions, but just as importantly the network of social relations that enabled the cultural landscape to develop into its very particular configuration. In this sense, our initial results go some way to elucidating settlement temporal trends in this region, and to explaining the historical roots of the ethnographic situation through the workings of the Kikori River delta environment as a socialised, peopled landscape.

Acknowledgements

We gratefully thank Kerewo clan members for their invitation, welcome and support during the course of this research, in particular Kenneth Korokai, Hanson Iburi and Andrew Dairi; Jean-Michel Gereste, Jean-Jacques Delannoy, Patricia Marquet and Bernard Sanderre, Cathy Alex and the staff at Community Development Initiative (CDI) at Kikori for their support and wonderful assistance in the field; Jacinta John, Kongel Pombreol and the staff at CDI Port Moresby. We thank Laurence Goldman for permission to reproduce, and Kara Rasmanis for drafting, Figure 1. This project was undertaken with the generous assistance of ARC Discovery grant and QEII Fellowship DP0877782 and the Public Memory Centre, USQ.

References

Beaver, W.N. 1920. *Unexplored New Guinea: A record of the travels, adventures, and experiences of a resident magistrate amongst the head-hunting savages and cannibals of the unexplored interior of New Guinea*. London, Seeley, Service and Co.

Cawley, F.R. 1925. Description of country lying between the Purari River and the Era. In: *Papua Annual Report for the Year 1923-24*, pp. 18-19. Government Printer, Melbourne.

Chalmers, J. 1895. *Pioneer Life and Work in New Guinea 1877-1894*. Religious Tract Society, London.

Chester, H.M. 1878. *Narrative of Expedition to New Guinea*. Government Printer, Brisbane.

David, B. 2008. Rethinking cultural chronologies and past landscape engagement in the Kopi region, Gulf Province, Papua New Guinea. *The Holocene* 18(3):471-488.

David, B., Fairbairn, A., Aplin, K., Murepe, L., Green, M., Stanisic, J., Weisler, M., Simala, D., Kokents, T., Dop, J. and Muke, J. 2007. OJP, a terminal Pleistocene archaeological site from the Gulf Province lowlands, Papua New Guinea. *Archaeology in Oceania* 42, 31-33.

David, B., Pivoru, M., Pivoru, W., Barker, B., Weiner, J.F., Simala, D., Kokents, T., Araho, L. and Dop, J. 2008. Living landscapes of the dead: archaeology of the afterworld among the Rumu of Papua New Guinea. In: David, B. and Thomas, J. (eds), *Handbook of Landscape Archaeology*. Left Coast Press.

David, B., Geneste, J.M., Aplin, K., Delannoy, J.J., Araho, N., Clarkson, C., Connell, K., Haberle, S., Barker, B., Lamb, L., Stanisic, J., Fairbairn, A., Skelly, R., Rowe, C. 2010. The Emo Site (OAC), Gulf Province, Papua New Guinea: resolving long-standing questions of antiquity and implications for the history of the ancestral *hiri* maritime trade. *Australian Archaeology* 70:39-54.

Dutton, T. (ed), 1982. *The Hiri in history: further aspects of long distance Motu trade in central Papua*. Pacific Research Monograph 8. Australian National University.

Flint, L.A. 1923. Report on the patrol through the Samberigi valley, Mount Murray District, Delta Division. In: *Papua Annual Report for the Year 1921-22*, Appendix 2, pp. 141-152. Government Printer, Melbourne.

Frankel, D. and Vanderwal, R. 1982. Prehistoric research at Kinomere Village, Papua New Guinea, 1981: preliminary field report. *Australian Archaeology* 14:86-95.

Frankel, D. and Vanderwal, R. 1985. Prehistoric research in Papua New Guinea. *Antiquity* 59:113-115.

Goldman, L. and Tauka, R. 1998. Omati social mapping report. Unpublished report to Chevron Asiatic, Brisbane.

Haddon, A. 1918. The Agiba cult of the Kerewo culture. *Man* (OS) 18:177-183.

Herbert, C.L. 1917. Delta Division. In: *Papua Annual Report for the Year 1914-15*, pp. 86-89. Government Printer, Melbourne.

Hurley, F. 1921. Photograph album of Papua and the Torres Strait. National Library of Australia Digital Collections, Canberra. (PIC/8907/122 LOC Album 1067).

Knauft, B.M. 1993. *South Coast New Guinea Cultures*: History, comparison, dialectic. Cambridge University Press, Cambridge.

Landtman, G. 1917. *The Folk-Tales of the Kiwai Papuans*. Acta Societatis Scientiarum Fennicae, Vol. 47. Helsinki Finnish Society of Literature.

Landtman, G. 1927. *The Kiwai Papuans of British New Guinea: A Nature-born Instance of Rousseau's Ideal Community*. London Macmillian.

Liston-Blyth, A. 1929. Delta Division. In: *Papua Annual Report for the Year 1927-28*. Government Printer, Melbourne.

MacGregor, W.M. 1893. Despatch continuing the report of visit of inspection to the Western Division of the Possession. In: *Annual Report on British New Guinea from 1st July 1891, to 30th June 1892; with Appendices*, pp. 37-48. Government Printer, Brisbane.

MacGregor, W.M. 1894a. Despatch reporting visit of inspection to the Purari River district. In: *Annual Report on British New Guinea from 1st July 1893, to 30th June 1894; with Appendices*, pp. 22-29. Government Printer, Melbourne.

MacGregor, W.M. 1894b. British New Guinea. In: *Annual Report on British New Guinea from 1st July 1892, to 30th June 1893; with Appendices*, pp. xv-xxxiv. Government Printer, Melbourne.

MacGregor, W.M. 1894c. Despatch reporting inspection of the Gulf of Papua from Hall Sound to Port Bevan. In: *Annual Report on British New Guinea from 1st July 1892, to 30th June 1893; with Appendices*, pp. 24-36. Government Printer, Melbourne.

Murray, J.H.P. 1914. Pacification of the Territory and native affairs generally. In: *Papua Annual Report for the Year 1913-14*, pp. 8-14. Government Printer, Melbourne.

Oram, N. 1982. Pots for sago: the *hiri* trading network. In: Dutton, T.E. (ed), *The Hiri in history: further aspects of long distance Motu trade in central Papua*. Pacific Research Monograph 8, Research School of Pacific Studies, Australian National University, 1-33.

Rhoads, J.W. 1980. Through a glass darkly: present and past land use systems of Papuan sagopalm users. Unpublished PhD thesis, Australian National University, Canberra.

Rhoads, J.W. 1982. Prehistoric Papuan exchange systems: the hiri and its antecedents. In:

Dutton, T.E. (ed), *The Hiri in history: further aspects of long distance motu trade in central Papua*. Pacific Research Monograph 8, Research School of Pacific Studies, Australian National University, 131-51.

Riley E.B. 1925. Among the Papuan Headhunters: An account of the manners and customs of the old Fly River headhunters, with a description of the secrets of the initiation ceremonies divulged by those who have passed through all the different orders of the craft, by one who has spent many years in their midst. Philadelphia: J.B. Lippincott [Reprinted, 1982 New York: AMS Press].

Ryan, H.J. 1913. Magisterial report, Delta Division. Cited in Goldman, L. and Tauka, R. 1998. Omati social mapping report, Annexure 3. Unpublished report to Chevron Asiatic, Brisbane. Annexure 3.

Ryan, H.J. 1914. Patrol west of the Kikori and across the head waters of the Omati, Turama, Gama, and Awarra Rivers. In: *Papua Annual Report for the Year 1913-14*, p. 170. Government Printer, Melbourne.

Seligmann, C.G. 1910. The Melanesians of British New Guinea. Cambridge, Cambridge University Press.

Stuiver, M. and Reimer, P.J. 1993. Extended 14C database and revised CALIB radiocarbon calibration program. Radiocarbon 35:215-230.

Weiner, J.F. 2006. PNG-Queensland Gas Pipeline Social Mapping: Kaiam-Kopi-Omati-Goaribari. Unpublished report.

Williams, F.E. 1924. *The Natives of the Purari Delta. Anthropology Report* 5. Government Printer, Port Moresby.

Woodward, R.A. 1920. Delta Division, 1920-1921. In: *Papua Annual Report for the Year 1918-19*, Appendix 2, pp. 61-63. Government Printer, Melbourne.

7

Cranial metric, age and isotope analysis of human remains from Huoshiliang, western Gansu, China

John Dodson
Institute for Environmental Research, Australian Nuclear Science and Technology Organisation, Lucas Heights, NSW
jdd@ansto.gov.au

Fiona Bertuch
Australian Nuclear Science and Technology Organisation, Lucas Heights, NSW

Liang Chen
Northwest University, Xi'an, China

Xiaoqiang Li
Chinese Academy of Sciences, Xi'an, China

Introduction

The Chinese provinces of Gansu and Xinjiang are key places for understanding prehistoric exchange between West and East Eurasia. For the past 2000 years, this has been encapsulated in the term 'Silk Road' (which was, in fact, many roads), but goods and ideas have been exchanged across the region for much longer (e.g. Li et al. 2007). In this regard, the well-known existence of the Ürümqi mummies of Xinjiang, which are of Caucasian origin (e.g. Barber 1999), show that they had opportunity to interact with Mongoloid people much earlier.

Detailed examination of the hundreds of archaeological sites in the region has hardly begun and many of these are just places on a map with no further detail described. There are no regional archaeological surveys of Gansu or Xinjiang, as there are for Henan and Shandong (e.g. Liu 2004; Underhill et al. 2008) and those site surveys that have been published focus on grave sites and their inclusions, and are not usually independently dated. There are few detailed material studies from non-grave sites and few independent chronologies to put the region into a wider East Asian prehistory context. Much of what has been described, such as the presence of microliths, pottery and bronze artefacts, has been dated on the basis of comparative analyses from elsewhere, most particularly from eastern China. There is no guarantee that such transported timelines necessarily apply beyond where they were established.

Western China has numerous burial sites, and hundreds of human remains and tomb contents have been analysed. For example, more than 500 human skulls have been examined from Yumen (Gansu) and Datong (Qinghai). These are between 3000 and 3600 BP in age. These are consistently Mongoloid and probably indicate there was little if any western Caucasoid incursion this far east at that time (Tan et al. 2005).

Bronze objects in Xinjiang have been found in a number of burial sites that span 3150 BP to 1900 BP. Those that have human remains in the Tarim Basin and northern Xinjiang sites have remains of Caucasoid people. Undecorated pottery from this region spread eastward, and painted pottery probably entered from the east since these are in Xinjiang earlier than elsewhere (Tao 2001). The Xinjiang mummies date back to about 4000 BP and a westward migration of Mongoloid people from Gansu into eastern Xinjiang is thought to have occurred from about 2000 BC (Tao 2001). The Caucasoid people are generally credited with bringing wheat agriculture, new kinds of artefacts, horsemanship, mud-brick building and new religious practices into eastern Asia (Tao 2001; references in Chengwen and Yoshinori 2002).

Here we describe, date and analyse two skulls and an associated sheep bone to provide new data on the possible co-existence of Caucasoid and Mongoloid peoples in western Gansu, and to comment on health and diet.

Huoshiliang study site

The Huoshiliang archaeological site is at 40°15.6'N and 99°18.3'E in the Black River valley of western Gansu Province (Figure 1). It occurs as a largely surface scatter of artefacts across several thousand square metres among sand dunes. The cultural sediments are about 1.6 m in depth in some places. The site has not been systematically excavated.

Evidence of material culture includes macro-fragments of plain and painted pottery, microliths, bone of *Bos* (cattle) and *Ovis* (sheep), copper ore, bronze slag and charcoal (Figure 2). Some of this material may be lag deposit from mobile dune sands and the time-depth sequence is not clear.

Fine particles at the site include small particles of charcoal, pottery, bone and cereal seeds. We have radiocarbon dates on five charcoal samples collected over a 1.6 m depth profile, and these are consistently aged between 3500 BP and 3600 BP (Dodson et al. 2009). In addition, we have a date from a wheat seed, of 3635 ± 45 BP. We interpret the site as a population centre where cropping, animal husbandry and bronze smelting was carried out. We believe that the majority of copper ore for smelting came from the Baishantang mine site, which is located about 100 km north of Huoshiliang (Dodson et al. 2009). The area is now devoid of trees, but the abundant charcoal indicates the area had significant woodland at the time of occupation.

In the course of examining the site and collecting charcoal for dating in late 2007, two complete skeletons were seen on the sand surface near the Huoshiliang site. The skeletons were eroding out of the shifting sands, and would likely be destroyed as erosion continued. The skulls (Figure 3) were examined in the Department of Archaeology at Northwest University in Xi'an. A premaxilla sheep bone was also analysed.

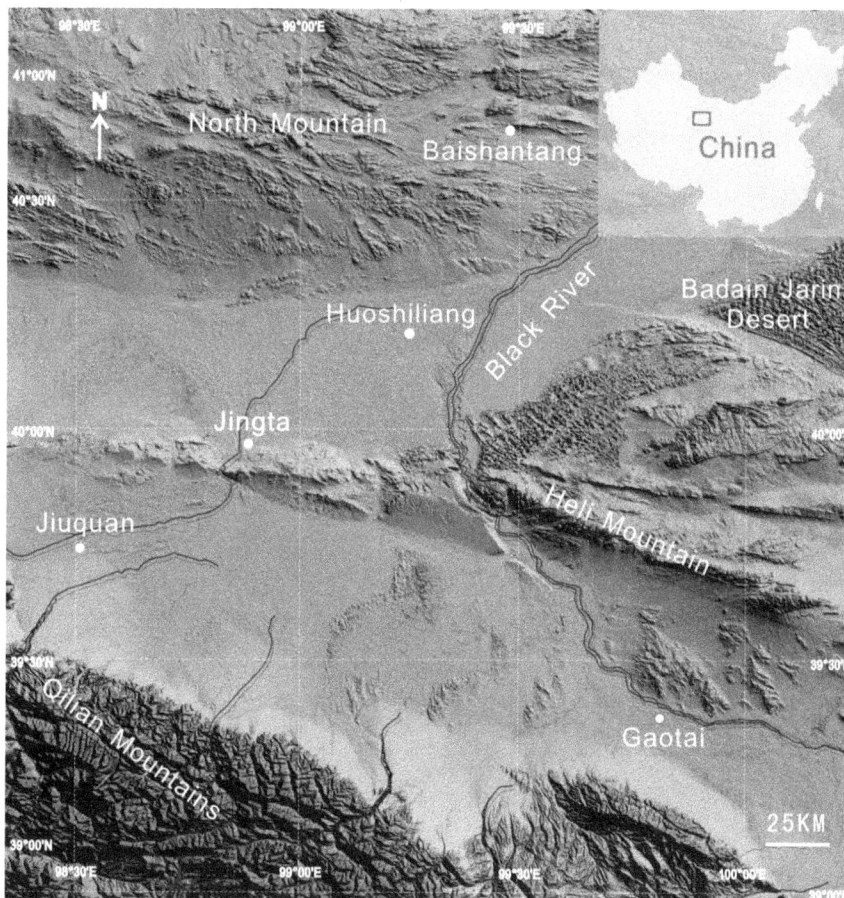

Figure 1. Map of northwestern China showing location of Huoshiliang.

Figure 2. View of site showing scatter of archaeological remains.

Methods

Using the standard techniques for determining Martin numbers, skull morphological features were measured and described (Martin 1928). Martin numbers are one of several systems for measuring anatomical features and this one was chosen because of its comprehensiveness. Comparative studies were made using the skull collection housed in the Department of Archaeology at North West Normal University in Xi'an and cluster analysis was used to

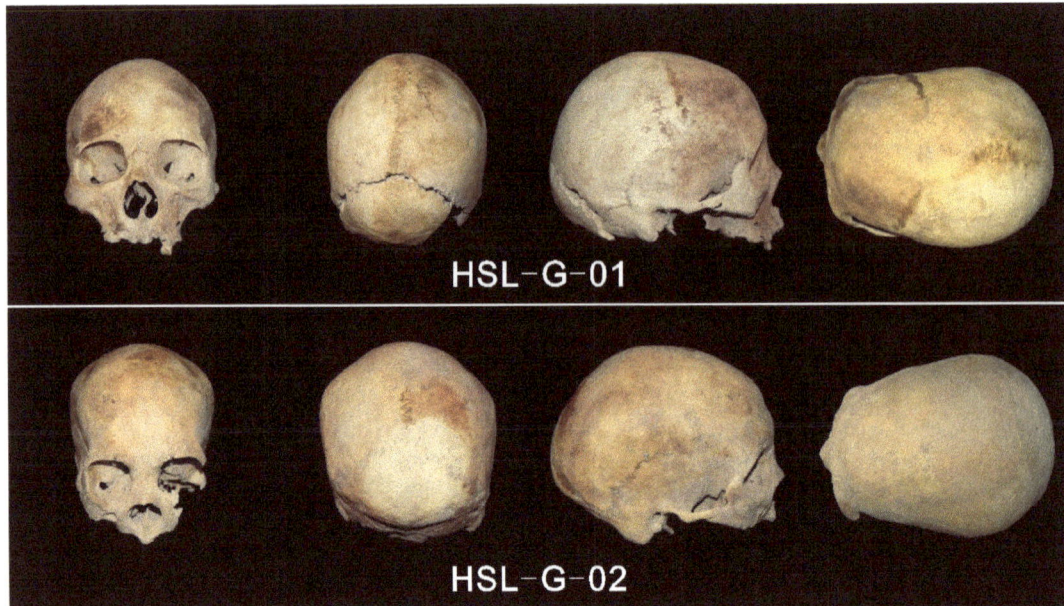

Figure 3. Photograph of skulls from the Huoshiliang archaeological site.

compare these with the Huoshiliang skulls.

About 10 g of bone from each skull and the sheep bone sample were transported to the Australian Nuclear Science and Technology Organisation (ANSTO) for analysis. The samples were pre-treated for radiocarbon dating in the AMS chemistry laboratories at ANSTO. Collagen was extracted to test whether they showed sufficient preservation and to identify effective removal of contamination to achieve reliable radiocarbon results.

The nitrogen percent of bone, the collagen percent of the sample, and the C:N atomic ratio of the extracted collagen were measured. ANSTO uses the ultrafiltration protocol (Brown et al. 1988; Bronk Ramsey et al. 2004; Higham et al. 2006) to pre-treat bone samples for radiocarbon dating. The ultrafiltration method has been shown to remove contamination more effectively than other methods (Bronk Ramsey et al. 2004). Ultrafiltration acts to remove material with a molecular weight below 30kD, which removes contaminants such as salts, fulvic acids and degraded collagen. The main steps in the ultrafiltration protocol used are:

- The bone sample is cleaned with a drill, washed with deionised water, dried and then crushed.
- The crushed bone samples are demineralised with 0.5M HCl.
- Humics are removed with 0.1M NaOH.
- Dissolved CO_2 is removed with 0.5M HCl.
- Samples are gelatinised with pH3 water (heated to 75°C for 20 hours).
- Samples are filtered through 100µm polyethylene eezi-filtersTM to remove insoluble residues.
- Eezi-filtered gelatine is then transferred to pre-cleaned Millipore 30kD ultrafilters and centrifuged until sufficiently filtered.
- The >30kD collagen solution is then freeze dried.
- Once the collagen had been extracted, the samples were processed to graphite as described by Hua et al. (2001).

Some problems have been encountered with the ultrafiltration method in the past and have been fully explored in Bronk Ramsey et al. (2004). This contamination originates from the glycerol which is added to the ultrafilter membrane during production. The ultrafiltration step in this method only adds carbonaceous contamination to the samples if the filters are not sufficiently cleaned. Quality assurance measures are routine in the ANSTO AMS chemistry laboratories to ensure that the use of ultrafilters does not pose a contamination risk for samples. Tests to ensure all carbonaceous contamination is removed include:

Measuring the quantity of carbon remaining on the ultrafilters after they have been cleaned. To achieve this, an ultrafilter is selected (randomly) for carbon content analysis. No measurable carbon was found, indicating that the ultrafilters had been cleaned satisfactorily.

A bone standard with a known age was run alongside the samples with unknown ages. This sample was selected was one that was a part of the VIRI international laboratory comparison study. The measured age from this bone standard matched the agreed age from the VIRI inter-comparison study, and from past measurements that the ANSTO AMS chemistry labs had attained. This further confirms that potential contamination from the use of ultrafilters was negligible.

The glycerol that coats the ultrafilters was also extracted and dated. The glycerol used on the batch of ultrafilters that was used in processing the samples resulted in a date of 1.06 pMC (with an error of ± 0.0034). This suggests no older age offsets, as might be expected if the samples were affected by contamination.

Samples OZL292-OZL294 were measured for $\delta^{13}C$ and $\delta^{15}N$ on a Elemental Analyser (EuroVector EA3000) and an Isotope Ratio Mass Spectrometer (GV Instruments IsoPrime). The reference materials used for the samples were as follows:

- $\delta^{13}C$ – IAEA C8 oxalic acid with an agreed value of −18.31‰ VPDB (used for graphite, bone and collagen) (Gonfiantini et al. 1995; Le Clercq et al. 2006).
- $\delta^{15}N$ – IAEA NO-3 with a consensus value $d^{15}N$ AIR = +4.7 ‰ (Bohlke and Coplen 1995) and IAEA N-2 with a consensus value of $d^{15}N$ AIR = +20.3 ‰ (bone and collagen) (Bohlke and Coplen 1995).
- 3:1 atomic ratio standard – Internal standard of 2-isopropylimidazole (bone and collagen).
- Collagen standard employed: Internal material check standard – un-denatured bovine achilles tendon collagen.

The ^{14}C content was measured on the STAR Accelerator at ANSTO and AMS ages were calculated after estimating fractionation effects from the $\delta^{13}C$ values determined on the same samples used for dating.

In addition, cereal seeds were collected from the site by sieving surface sands, and we counted a number of these to obtain a snapshot of the types and relative amounts of cereals grown at Huoshiliang.

Results

Table 1 gives measurements on the two skulls found at Huoshiliang.

Morphological characteristics
Skull No. 1. The skull is from a middle-aged female, probably between 40 and 45 years of age. Its morphology is oval and the cranial index is 75.64, meaning it is mesocrany in size. The length-height index of the skull is 74.79, putting it in the taller metriocrany range. The skull is acrocrany type as the breadth-height index is 98.86. The middle superscalar arch has a range below 1/2. The forehead is even and straight with a frontal index of 69.7. This places it as eurymetor type, with no suture in the middle forehead. The bregmatic and vertex part in the

coronal suture are microwave type, and both top and back of the coronal suture are of indented type. The mastoid process is small and the external occipital protuberance is slightly prominent. The orbital cavity is oval (index 73.56), belonging to the lower range in orbital cavity size. The upper part of the Apertura piriformis is heart-shaped, and below this it has a nest type form. The nasal base is superficially hollow, no nasal spine is apparent and the nasal index is 52.38, showing it to be of a broad nasal type. The nasal bridge is hollow type and the nasal bone is Type II. The Canine fossa show moderate growth, the Zygomatic bone is slender (i.e. there is no zygomatic-jaw node) and the jaw angle is clearly evident. Both sides of the skull have Parietal apertures, the sigittal crest is evident and the Palate form is 'V' type. The Upper facial index is 51.92, showing it to be meseny type, while the gnathic index is 92.31, showing it to be orthognath type.

Skull No. 2 The skull is from a female, probably with an age of about 35-40 years. Its morphology is oval and the cranial index is 74.73, meaning it is a shorter dolichocrany skull type. The middle superscalar arch has a range below 1/2. The forehead is moderate and the frontal index is 66.33, indicating it is metriometor type. No suture was apparent in the middle forehead. The bregmatic part of the coronal suture is deep wave type. Both the top and back part of the coronal suture are of indented type, while the vertex aperture part is microwave type. The Mastoid process is large and the external occipital protuberance is moderate. The orbital cavity is oval, with an index of 78.65, suggesting it is medium type. The supraorbital foramen is heart-shaped and its lower edge is obtuse. The nasal base is superficially hollow and the nasal spine is degree II in type. The nasal index is 54.36 and her nose is chamaerrhiny type. The nasal bridge is hollow type and the nasal bone is I type. The Canine fossa is middle range in degree. The zygomatic bone is middle range in height and breadth and no zygomatic-jaw node was apparent. The jaw angle is apparent. Both sides of the skull have Parietal apertures, a sigittal crest is evident, and the Palate form is 'V' type.

Interpretation

The morphology characters of the two skulls are very similar, in being both mesocrany and dolichocrany type. The skulls have simple coronal suture, an oval orbital cavity, medium-narrow

Table 1. Comparative measurements from the Huoshiliang skulls and nine ancient skull groups (female) (mm, degree, %). The numbers in brackets refer to the sample size for comparative measurements.

Martin #	Items	HSL	XC	QC	PP	SLY	HD	LW	SSJZ (H)	SSJ (K)	LJS
1	Cranial length (g-op)	174.49 (2)	174.74 (5)	173.97 (16)	177.5 (3)	180.50 (2)	183.9 (5)	178.58 (12)	174.1 (21)	175.1 (102)	177.30 (8)
8	Cranial breadth (eu-eu)	134.8 (2)	133.26 (5)	134.20 (16)	140.6 (3)	137.25 (2)	136.1 (5)	132.27 (10)	135.8 (21)	135.1 (103)	136.40 (8)
17	Cranial height (ba-b)	130.5 (1)	137.98 (5)	135.95 (15)	130.6 (3)	138.50 (1)	149.1 (1)	131.63 (16)	129.1 (18)	131.3 (99)	130.60 (8)
9	Minimum frontal breadth	91.6 (2)	86.26 (5)	89.89 (17)	90.7 (3)	90.47 (3)	91.1 (5)	87.42 (12)	89.0 (22)	88.9 (102)	89.20 (8)
45	Zygomatic breadth (zy-zy)	125.0 (1)	127.55 (2)	127.50 (12)	134.0 (3)	127.75 (2)	134.1 (3)	129.04 (16)	129.6 (17)	126.3 (93)	125.30 (7)
48	Upper facial height (sd)	67.8 (2)	69.88 (2)	67.70 (17)	67.5 (2)	70.70 (2)	70.6 (4)	70.73 (16)	71.0 (19)	71.7 (84)	72.10 (8)

Table 1. *Continued*

Martin #	Items	HSL	XC	QC	PP	SLY	HD	LW	SSJZ (H)	SSJ (K)	LJS
52	Orbital height right	33.5 (2)	32.38 (4)	32.71 (17)	33.1 (3)	31.90 (2)	33.4 (4)	33.13 (16)	34.6 (22)	34.1 (102)	34.60 (8)
51	Orbital breadth right	44.0 (2)	39.45 (4)	42.31 (17)	43.2 (3)	40.20 (2)	42.2 (4)	41.80 (16)	40.5 (21)	41.0 (103)	40.90 (8)
54	Nasal breadth	28.1 (2)	25.23 (4)	26.03 (17)	26.5 (3)	27.90 (2)	27.8 (4)	25.76 (17)	26.2 (20)	25.9 (99)	26.80 (8)
55	Nasal height (n-ns)	52.7 (2)	50.32 (5)	48.75 (17)	50.0 (3)	52.50 (2)	50.6 (4)	51.03 (16)	52.1 (21)	52.6 (98)	52.40 (8)
72	Facial angle	90.5 (2)	81.40 (4)	83.27 (15)	85.7 (3)	82.00 (1)	85.7 (3)	87.00 (12)	85.4 (18)	84.9 (76)	86.90 (8)

orbital type, an obtuse low edge of the supraorbital foramen, a shallow hollow of the nasal bases and obvious chamaerrhiny. The angle of jaws is down in gradient. They are of moderate range in their Canine fossa, and have hollow-type nasal bridges, evident sigittal crests, meseny and relatively flat degree faces. The two skulls therefore belong to the Mongoloid race.

To further examine their possible origin, an analysis of mean categories for the two cases against mean values for nine ancient groups was carried out. Ten measured items and eight indices and angles from nine ancient groups were selected for the comparison and these are shown in Table 1. In the analysis below, all measurements are based on female skulls, and the nine ancient people groups chosen for comparison are as follows:

(a) Zhou Dynasty people at Xicun in the south of Fengxiang, Shaanxi Province. These skulls are close to South and East Asia types of the Mongolian race.

(b) West Zhou skulls from Wotianma-Qucun in Shanxi Province. The skulls of the group are similar to the East Asia type, but include some factors close to North and South Asia Mongoloid types.

(c) Pengpu Bronze Age people in Guyuan County, Ningxia Province. This skull group belongs to the North Asia Mongoloid type.

(d) A group from the Zhou Dynasty in Shaolingyuan in Xi'an, Shaanxi Province. The skulls of the group are close to South Asia and East Asia types of the Mongolian race, and hence similar to the Xicun group.

(e) The skull group of Neolithic Hedang people in Foushan City, Guangdong Province. The skulls of the group are of South Asia type.

(f) The skull groups of Liuwan, Machang and Qijia cultures in Ledu, Qinghai Province. The skulls of these people are East Asia type.

(g) The skull group from the Han Dynasty in Shangsunjiazhai in Datong, Qinghai Province.

(h) The skull group from the Kayue culture group in Shangsunjiazhai (Datong), Qinghai Province.

(i) Skull group sample from the of Kayue Culture in Lijiashan, Xunhua, Qinghai Province.

The latter three skull groups are close to the East Asia type of the Mongoloid race and similar to modern Tibetan people.

The comparative data used for analysis are shown in Table 2. The Dij value between the two skulls from Huoshiliang and nine ancient groups was calculated using the following formula:

Where *i, j* indicating skull groups, *k* indicating measured items, *m* indicating the sample number for measured items. The smaller of *Dij* value, the closer the morphological set of

$$D_{ij} = \sqrt{\frac{\sum_{k=1}^{m} \left(xik - xjk\right)^2}{m}}$$

relationships of the skull groups. The results are shown in Table 3. The *Dij* values show that the skulls of Huoshiliang are most similar to the group of skulls from Liuwan, but differ from the Hedang group and are quite different from the Xicun group of the Zhou Dynasty.

A cluster analysis based on squared Euclidian distances was used to show the distribution of D*ij* values between the Huoshiliang group and other groups. The results are shown in Figure 4. This suggests that the Hedang group (6) has greater differences from all other groups, and Pengpu (4) is also relatively different from the other groups. The other eight groups are divided into two categories; with groups 8, 9, 10, 7, 1 as a category, and group 2, 3, 5 as the other. Morphological character is relatively similar in the group containing 8, 9 and 10. Group 1 belongs to this category and indicates that the Huoshiliang people group are members of the East Asia race. They thus have a close relationship with modern residents in North China. Group 2, 3, 5 is a mixture of South and East Asia Mongoloid people. So the values of *Dij* and CA analysis all show that the morphology of female skulls in Huoshiliang group is close to the East Asia type of the Mongoloid race.

Table 4 shows the bone protein yield, $\delta^{13}C$, $\delta^{15}N$, C:N ratio and radiocarbon results. The chemical indicators from the pre-treatments confirmed that the collagen that was extracted

Table 2. The key measurement data of skulls No. 1 and No. 2 from Huoshiliang site.

Martin No.	Measured Items	Skull 1(♀)	Skull 2(♀)	Average	Examples	Standard error
1	Cranial length (g-op)	174.5	184.0	179.25	2	6.72
8	Cranial breadth (eu-eu)	132.0	137.5	134.75	2	3.89
17	Cranial height (b-ba)	130.5	—	130.5	1	—
21	uricular height (po-po)	110.8	—	110.8	1	—
9	Frontal breadth (ft-ft)	92.0	91.2	91.6	2	0.57
7	Foramen magnum length (ba-o)	31.0	—	31.0	1	—
16	Foramen meanum breadth (FOR. MA. B)	24.5	—	24.5	1	—
25	Cranial sagittal arc (n-o)	367.0	365.0	366.0	2	1.41
26	Nasion-bregma chord (n-b)	122.0	107.0	114.5	2	10.61
27	Bregma-lambda chord (b-l)	132.0	142.0	137.0	2	7.07
28	Lambda -opisthion chord (l-o)	113.0	116.0	114.5	2	2.12
29	Nasion-bregma chord (n-b)	105.5	97.5	101.5	2	5.66
30	Bregma-lambda chord (b-l)	118.0	129.8	123.9	2	8.34

Table 2. *Continued*

Martin No.	Measured Items	Skull 1(♀)	Skull 2(♀)	Average	Examples	Standard error
31	Lambda -opisthion chord (l-o)	91.0	99.6	95.3	2	6.08
23	Cranial horizontal circumference (g,op)	500.0	—	500.0	1	—
24	Cranial transverse arc (po-b-po)	307.0	—	307.0	1	—
5	Basion-nasion length (n-enba)	97.4	—	97.4	1	—
40	prosthion to Endobasion length (pr-enba)	90.0	—	90.0	1	—
48	Upper facial height (n-pr)	65.7	65.0	65.4	2	0.49
48	Upper facial height (n-sd)	67.8	67.7	67.8	2	0.07
45	Bijugal breadth (Facial breadth) (zy-zy)	125.0	—	125.0	1	—
46	Bimaxillary breadth (zm-zm)	98.0	—	98.0	1	—
	sub zm-ss-zm	25.0	—	25.0	1	—
	Bimaxillary breadth (zm-zm)	97.2	—	97.2	1	—
	sub zm_1-ss-zm_1	23.0	—	23.0	1	—
43(1)	Bifrontal breadth (fmt-fmt)	103.2	105.0	104.1	2	1.27
50	Vordere Interorbital breite (mf-mf)	17.6	19.0	18.3	2	0.99
MH L	Zygomatic height left (fmo-zm)	42.0	—	42.0	1	—
MH R	Zygomatic height right	41.8	42.3	42.1	2	0.35
MB L	Zygomatic breadth left (zm-rim)	20.8	—	20.8	1	—
MB R	Zygomatic breadth right	22.6	25.0	23.8	2	1.70
54	Nasal breadth	27.5	28.7	28.1	2	0.85
55	Nasal height (n-ns)	52.5	52.8	52.7	2	0.21
SC	Simotic chord	11.0	8.0	9.5	2	2.12
SS	Simotic subtense	2.2	2.6	2.4	2	0.28
51 L	Orbital breadth left (mf-ek)	43.5	—	43.5	1	—
51 R	Orbital breadth right	43.5	44.5	44.0	2	0.71
51a L	Orbital breadth left (d-ek)	40.6	—	40.6	1	—
51a R	Orbital breadth right	40.5	42.4	41.5	2	1.34
52 L	Orbital breadth height left	34.2	—	34.2	1	—
52 R	Orbital breadth height right	32.0	35.0	33.5	2	2.12
03	Interorbital breadth	55.0	—	55.0	1	—
SR	Rhinion height	12.2	—	12.2	1	—
60	Maxillo-alvaolar length (pr-alv)	50.0	43.0	46.5	2	4.95
61	Maxillo-alveolar breadth (ekm-ekm)	61.8	54.0	57.9	2	5.52
62	Palatal length (ol-sta)	44.9	42.6	43.8	2	1.63
63	Palatal breadth (enm-enm)	36.3	38.6	37.5	2	1.63
12	Maximum Biasterionic breadth (ast-ast)	103.8	115.2	109.5	2	8.06
	(po-po)	112.6	—	112.6	1	—
11	Biauricular breadth (au-au)	116.0	—	116.0	1	—
44	Biorbital breadth (ek-ek)	100.0	—	100.0	1	—
FC	Innere Biorbitalbreite (fmo-fmo)	95.7	100.0	97.9	2	3.04
FS	Nasal orbital internal breadth and height	11.2	17.5	14.4	2	4.45
DC	Interorbital breadth (d-d)	20.8	21.5	21.2	2	0.49
DN	Dacryon-nasion salient	3.5	5.8	4.7	2	1.63
DS	Dacryal subtense	6.5	7.3	6.9	2	0.57
NLÐ	Nasal bone length (n-rhi)	25.8	29.0	27.4	2	2.26
RP	Rhinion- Alveolar length (rhi-pr)	40.4	37.0	38.7	2	2.40

Table 2. *Continued*

Martin No.	Measured Items	Skull 1(♀)	Skull 2(♀)	Average	Examples	Standard error
32	Profile angle of the frontal bone from nasion (∠n-m FH)	86.0	82.0	84.0	2	2.83
	Profile angle of the frontal bone from glabella (∠g-m FH)	81.0	77.0	79.0	2	2.83
	Bregmatic angle (∠g-b FH)	46.0	47.0	46.5	2	0.71
72	Total facial angle (∠n-pr FH)	89.0	92.0	90.5	2	2.12
73	Nasal prognathism (∠n-ns FH)	91.0	94.0	92.5	2	2.12
74	Alveolar Prognathism (∠ns-pr FH)	76.0	80.0	78.0	2	2.83
75	profilwinkel des Nasendaches (∠n-rhi FH)	70.0	75.0	72.5	2	3.54
77	Naso-malar angle (∠fmo-n-fmo)	154.0	143.0	148.5	2	7.78
SSA	Zyyo-maxillary angle (∠zm-ss-zm)	125.0	—	125.0	1	—
SSA	∠zm1-ss-zm1	129.0	—	129.0	1	—
	Winkel des gesichsdreiecks I (∠n-pr-ba)	76.0	—	76.0	1	—
	Winkel des gesichsdreiecks II (∠pr-n-ba)	64.0	—	64.0	1	—
	Winkel des gesichsdreiecks III (∠n-ba-pr)	40.0	—	40.0	1	—
	Nasal bridge angle	19.0	17.0	18.0	2	1.41
8:1	Cranial index	75.64	74.73	75.19	2	0.64
17:1	Cranial length –height index	74.79	—	74.79	1	—
17:8	Cranial height –breadth index	98.86	—	98.86	1	—
9:8	Forehead breadth index	69.7	66.33	68.02	2	2.38
16:7	Index of occipital foramen	79.03	—	79.03	1	—
40:5	Gnathic index	92.31	—	92.31	1	—
48:17 pr	Vertical cranial index	50.34	—	50.34	1	—
48:17 sd		51.95	—	51.95	1	—
48:45 pr	Upper facial index (K)	50.34	—	50.34	1	—
48:45 sd		51.92	—	51.92	1	—
48:46 pr	Middle facial index (V)	67.04	—	67.04	1	—
48:46 sd		69.18	—	69.18	1	—
54:55	Nasal index	52.38	54.36	53.37	2	1.40
52:51 L	Orbital index	78.62	—	78.62	1	78.62
52:51 R		73.56	78.65	76.11	2	3.60
52:51a L	Orbital index	84.24	—	84.24	1	—
52:51a R		79.01	82.55	80.78	2	2.50
54:51 L	Nasal orbital index	63.22	—	63.22	1	—
54:51 R		63.22	64.49	63.86	2	0.90
54:51a L	Nasal orbital index	67.9	—	67.9	1	—
54:51a R		67.73	67.69	67.71	2	0.03
SS:SC	Nasal base index	20.0	—	20.0	1	—
61:60	Alveolar index	123.6	125.58	124.59	2	1.40
63:62	palatal index	80.85	90.61	85.73	2	6.90
45: (1+8)/2	Transverse cranial index	81.57	—	81.57	1	—

Table 2. *Continued*

Martin No.	Measured Items	Skull 1(♀)	Skull 2(♀)	Average	Examples	Standard error
17: (1+8)/2	High level index	85.15	—	85.15	1	—

Table 3. The Dij values used for comparison of the Huoshiliang group to nine ancient groups (female).

	Xicun	Qucun	Pengpu	Shaolingyuan	Hedang	Liuwan unite	Shangsunjiazhai (Han)	Shangsunjiazhai (Kayue)	Lijiashan
HSL	18.23	13.06	14.89	15.15	27.43	10.31	15.31	13.78	14.06

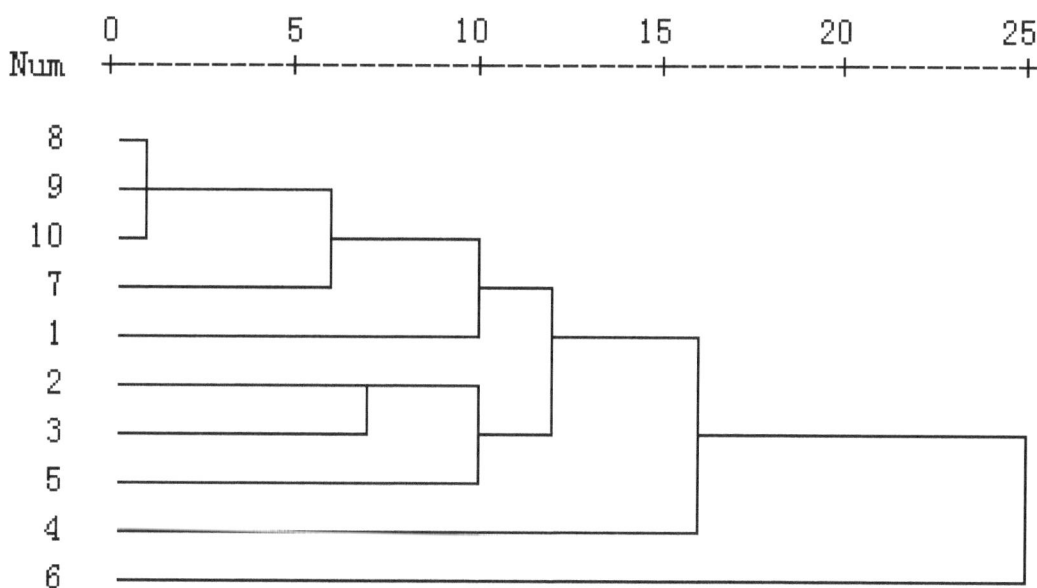

Figure 4. Hierarchical analysis map between the Huoshiliang skulls and other groups. Where 1 = Huoshiliang; 2 = Xicun; 3 = Qucun; 4 = Pengpu; 5 = Shaolingyuan; 6 = Hedang; 7 = United Liuwan group; 8 = Shangsunjiazhai (Han); 9 = Shangsunjiazhai (Kayue); and 10 = Lijiashan.

from the bone samples was sufficiently preserved for reliable dating. The samples also exhibited an acceptable level of nitrogen in the whole bone sample. Measuring the percentage of nitrogen of the whole bone allows us to estimate the quantity of collagen present before chemical treatment commences. For nitrogen, this usually ranges from approximately 4% in a fresh bone to below 0.2% in a poorly preserved bone (Tisnérat-Laborde et al. 2003). The nitrogen levels of the bones (ca. 4%) was acceptable, although the yield for OZL292 was surprisingly low. The small amount of collagen extracted from that bone appears not to be the result of poor preservation, as the percentage of nitrogen of the whole bone was of an acceptable value (4.1%) and the C:N ratio (3.23) of the collagen was satisfactory. This unusual result is most likely explained by a loss of collagen during the pre-treatment steps. In general, the C:N ratios of the collagen were within the acceptable range (ca. 3%), indicating that the collagen preservation was acceptable for dating (see Deniro 1985) and contaminants had been effectively removed.

The skull ages and the sheep bone overlap in age in the ± 2σ range (Table 4 shows the ± 1σ values and ± 2σ calibrated range values), which probably puts them all within an age range of 1700-2017 cal BC. The calibrated ages are based on Reimer et al. (2004). Other radiocarbon dates from the site have yet to be published and these are based on charcoal (five samples) and a charred wheat seed (one sample). These all have calibrated ages within the range 1733 to 2135 cal BC and the dates are therefore consistent within the site and the bone ages are consistent with these. The sheep bone sample has an indistinguishable radiocarbon age from the human skulls, suggesting they were contemporaneous.

The δ¹³C values of the human skulls are about the same, but they are low negative values (ca. -8 to -9‰). The sheep bone had a value of ca. -18‰.

The seed analysis was based on 13,472 identified seeds (Table 5). The relative ease of finding cereal seeds suggests that crops were a staple part of the diet and people at the site were farmers. The seeds were dominated by broomcorn and foxtail millet types, with the latter declining in proportion above about 60 cm depth, when wheat and oat make an increasing contribution. Broomcorn millet is the dominant seed type in the upper layers, suggesting this was by far the dominant local crop, but an increase in crop diversity in the upper layers is suggested by the seed numbers.

Discussion

Table 4. Bone protein, δ13C, δ15N and radiocarbon results. Calibrated ages are from Reimer et al. (2004).

Sample name	Bone protein yield (%)	Whole bone N (%)	δ¹³C (⁰/₀₀) graphite collagen	δ¹⁵N (⁰/₀₀)	Collagen C:N ratio	AMS date number	Age BP	Calibrated age (BC) ± 2σ
HSG-01	0.5	4.22	-8.2 ± 0.1 -8.71	6.1	3.19	OZL 292	3515 ± 45	1955-1737
HSG-02	4.1	4.1	-9.0 ± 0.1 -8.86	9.8	3.23	OZL 293	3590 ± 45	2124-1995
HSG-03	6.5	3.93	-17.5 ± 0.1 -18.51	11.8	3.20	OZL 294	3515 ± 40	1946-1776

Table 5. Distribution of seed types with depth at Huoshiliang.

Sample depth (cm)	*Panicum miliaceum* Broomcorn millet	*Setaria italica* Foxtail millet	*Triticum* Wheat	*Avena sativa* Oat	*Hordeum vulgare* Barley	Other	Total seeds
0-20	95.74	2.13	1.49	0.11	0	0.54	900
20-40	90.03	9.47	0.41	0.01	0	0.09	10973
40-60	92.08	5.87	1.76	0.07	0	1.72	1256
60-80	76.21	18.97	2.41	0	0.69	0	221
80-100	73.24	26.76	0	0	0	0	52
100-120	77.61	22.39	0	0	0	0	52
120-140	72.22	27.08	0	0	0	0	13
140-160	100.00	0	0	0	0	0	5

The correspondence of the radiocarbon ages of the skulls, sheep bone and other materials from the site indicate that Huoshiliang was occupied by East Asian people between about 1860 BC and 2020 BC. The skull dates are the first definitive early ages of Mongoloid people in this part of Central Asia. Earlier people from the region were possibly Caucasoid (e.g. Barber 1999), but since the remains analysed in this study are from a sample of two, it remains to be demonstrated whether the site was wholly occupied by Mongoloid people, whether there was mixed occupation, or whether there was merely the opportunity for exchange.

Elsewhere, we have published evidence of bronze technology (Dodson et al. 2009) and we have also noted that wheat, barley and oats at Huoshiliang are among the oldest in China

(Li et al. 2007). It is probable that this mix arose from western Asia and is thus indicative of at least strong east-west interaction by around 2000 BC. The seed evidence shows that millets, the quintessentially north China and Yellow River valley crops, dominated agriculture at Huoshiliang.

Stable isotope analyses on bone can reveal much about diet and living conditions of individuals (Larsen 1998). In this case, the human bones had low negative δ^{13}C values, which indicates that C4 plants, and possibly protein from animals which fed on C4 plants, were the mainstay of the diet. There has been no systematic study of the abundant animal bone at the site, although sheep bone fragments are relatively common. However, the human ^{15}N values (6-10‰) are low and suggest that very little animal protein was consumed. The sheep value may be inflated due to fertilising pastures or manuring. Millet agriculture originated in the Yellow River valley of China from about 8000 BP; these are C4 plants, while wheat, oats and barley are C3 plants. We surmise that millet was the underpinning mainstay of the human food chain at Huoshiliang around 2000 BC. The sheep bone had a more negative δ^{13}C value, suggesting part of its diet included C3 plants.

Several other studies in the Yellow and Wei river systems of northern and northeastern China, where millet agriculture originated, have measured similar δ^{13}C values in human bone samples and concluded that millets must have been the mainstay of the food chain to humans (Pechenkina et al. 2005; Hu et al. 2008). Barton et al. (2009) recently argued that the domestication of dogs and pigs in northern China was accompanied by a shift to less negative δ^{13}C values, which were associated with broomcorn and foxtail millet forming part of their diet.

The sheep bone from Huoshiliang has a more negative δ^{13}C value, but one that is less negative than C3 plants in general, suggesting its diet was a mix of C3 and C4 plant foods. Perhaps sheep protein was not a large part of the Huoshiliang women's diet.

Skull No. 1 showed periodontitis. More than 50% of the tooth roots were exposed. For the left top jaw, P1, P2, M1 show abscess, and some corroded holes in the cheek side had diameters of 3.8 mm, 4.5 mm and 4.0 mm respectively. In Skull No. 2, all teeth had fallen out in the upper jaw. The alveolar had atrophied and closed completely. Both skulls thus show severe mouth disease and poor teeth health. This may be evidence of a narrow diet based on millet. Larsen argued some time ago (Larsen 1998) that the agricultural transition from hunter-gather society was probably associated with greater sedentism, narrower diet and a proneness to poor dental health and even earlier death.

Conclusion

The morphological characteristics of the two skulls from Huoshiliang show they are female and are representatives of the East Asia type of the Mongoloid race. The cluster analysis shows that the skull morphologies are similar to those of groups that were in China from the Neolithic to Han Dynasty times. The individuals and the archaeological site in which they were found have an age between 1860 BC and 2020 BC, and evidence of pottery, bronze and agriculture suggests they were part of a complex society with strong technological links to both western and eastern Asia. While the range of crops became more complex with the introduction of wheat, oats and barley, millets remain the dominant food plant, and this is reflected in the stable-isotope data obtained from the bones. The two individuals may have lived before a diversification of additional crops appeared at the site. The lack of animal protein and reliance on millet consumption may have contributed to poor oral health.

Acknowledgements

We thank the Chinese Academy of Sciences and the Australian Nuclear Science and Technology Organisation for support of the project. Ms Hu Songmei (Archaeological Institute of Shaanxi Province, Xian) kindly identified the sheep-bone sample.

References

Barber, E.W. 1999. *The Mummies of Ürümchi.* London: W.W. Norton Press.

Barton, L., Newsome, S.D., Chen, F.H., Wang, H., Guilderson, T.P. and Bettinger, R.L. 2009. Agricultural origins and the isotopic identity of domestication in northern China. *Proceedings of the National Academy of Sciences* 106:523-5528.

Bohlke, J.K. and Coplen, T.B. 1995. Interlaboratory comparison of reference materials for nitrogen isotope ratio measurements, taken from an IAEA Technical report, Reference and intercomparison materials for stable isotopes of light elements. *IAEA-TECHDOC-825.* September 1995.

Bronk Ramsey, C., Higham, T., Bowles, A. and Hedges, R. 2004. Improvements to the Pretreatment of Bone at Oxford. *Radiocarbon* 46:155-163.

Brown T.A., Nelson, D.E., Vogel, J.S. and Southon, J.R. 1988. Improved Collagen Extraction by Modified Longin Method. *Radiocarbon* 30:171-177.

Chengwen, G., and Yoshinori, Y. 2002. *Study on the Bronze Culture of the Yangtze River Valley.* Beijing: Science Press.

Deniro, M.J. 1985. Postmortem preservation and alteration of *in vivo* bone collagen isotope ratios in relation to palaeodietary reconstruction. *Nature* 317:806-809.

Dodson, J., Li, X.Q., Zhou, X.Y. and Levchenko, V. 2009. Bronze in two Holocene archaeological sites in Gansu, NW China. *Quaternary Research* 72:309-314.

Gonfiantini, R., Stichler, W. and Rozanski, K. 1995. Standards and intercomparison materials distributed by the International Atomic Energy Agency for stable isotope measurements, *IAEA-TECDOC-825*, pp.13-29.

Han, K.X., Tang, J.Z. and Zhang, F. 2005. *The racio-anthropological study on ancient west-north area, China.* Shanghai: Fudan University Press.

Higham, T.F.G., Jacobi, R.M. and Bronk Ramsey, C. 2006. AMS Radiocarbon Dating of Ancient Bone Using Ultrafiltration. *Radiocarbon* 48(2):179-195.

Hu, Y., Wang, S., Luan, F., Wang, C. and Richards, M.P. 2008. Stable isotope analysis of humans from Xiaojingshan site: implications for understanding the origin of millet agriculture in China. *Journal of Archaeological Science* 35:2960-2965.

Hua, Q., Jacobsen, G.E., Zoppi, U., Lawson, E., Williams, A.A., Smith, A.M. and McMann, M.J. 2001. Progress in Radiocarbon Target Preparation at the ANTARES AMS Centre. *Radiocarbon* 43:275-282.

Larsen, C.S. 1998. *Post-Pleistocene human evolution: bioarchaeology of the agricultural transition.* 14th International Congress of Anthropological and Ethnological Sciences, Williamsburg, Virginia, July 26-August 1, 1998.

Le Clercq, M., van der Plicht, J. and Gröning, M. 2006. New 14C reference materials with activities of 15 and 50 pMC. *Radiocarbon* 40(1):295-297.

Li, X., Dodson, J., Zhou, X., Zhang, H. and Masutomoto, R. 2007. Early cultivated wheat and broadening of agriculture in Neolithic China. *The Holocene* 17:555-560.

Liu, L. 2004. *The Chinese Neolithic: trajectories to early states.* Cambridge: Cambridge University Press.

Martin, R. 1928. *Lehrbuch der Anthropologie in systematischer Darstellung mit besonderer Berücksichtigung der anthropologischen Methoden für Studierende, Ärzte und Forschungsreisende.* Jena: Fischer.

Pechenkina, E.A., Ambrose, S.H., Xiaolin, M. and Benfer Jr, R.A. 2005. Reconstructing northern Chinese Neolithic subsistence practices by isotopic analysis. *Journal of Archaeological Science* 32:1176-1189.

Reimer, P.J. and 27 others. 2004. IntCal04 terrestrial radiocarbon age calibration, 0-26 cal kyr BP. *Radiocarbon* 46:1029-1058.

Tan J., Han, Z. and Kang, F. 2005. *The ancient residents of Northwest China ethnic studies.* Shanghai: Fudan University Press.

Tao, S. 2001. *Papers on the Bronze Age archaeology of northwest China.* Beijing: Science Press.

Tisnèrat-Laborde, N., Valladas, H., Kaltnecker, E. and Arnold, M. 2003. AMS Radiocarbon Dating of Bones at LSCE. *Radiocarbon* 45:409-419.

Underhill, A.P., Feinman, G.M., Nicholas, L.M., Fang, H., Luan, F., Yu, H.G. and Cai, F. 2008. Changes in regional settlement patterns and the development of complex societies in southeastern Shandong, China. *Journal of Anthropological Archaeology* 27:1-29.

8

Not for the squeamish: A new microfossil indicator for the presence of humans

Mike Macphail

Department of Archaeology and Natural History, Research School of Pacific and Asian Studies,
The Australian National University, Canberra, ACT
mike.macphail@anu.edu.au

Mary Casey

Casey and Lowe Pty Ltd, Marrickville, NSW

Matthew Kelly

Archaeological and Heritage Management Solutions, Annandale, NSW

Introduction

Considerable efforts have been made to find proxy indicators for humans at sites lacking direct archaeological evidence such as pottery, tools and the remains of built structures. Indirect evidence of human activity, such as charcoal and pollen records showing deforestation, continues to be equivocal (cf. Ellison 1994), although pollen of introduced crop and ornamental plants are an important exception (Macphail 1999; Macphail and Casey 2008; Prebble and Wilmshurst 2009). Buried seeds with gnaw marks of the introduced Pacific rat (*Rattus exulans*) provide equally reliable evidence for detecting initial human colonisation of islands in Remote Oceania (Wilmshurst and Higham 2004).

In this note, we describe and illustrate additional specimens of a microfossil species, *Cloacasporites sydneyensis*, which is strongly associated with that most basic of human activities, defecation. At present, the microfossil has only been recorded from historical archaeological sites in Sydney. Our twofold aim is to (i) alert the wider archaeological community to the existence of another proxy for humans, and (ii) thereby test whether the microfossil also occurs in archaeological sites elsewhere, in particular in non-European contexts.

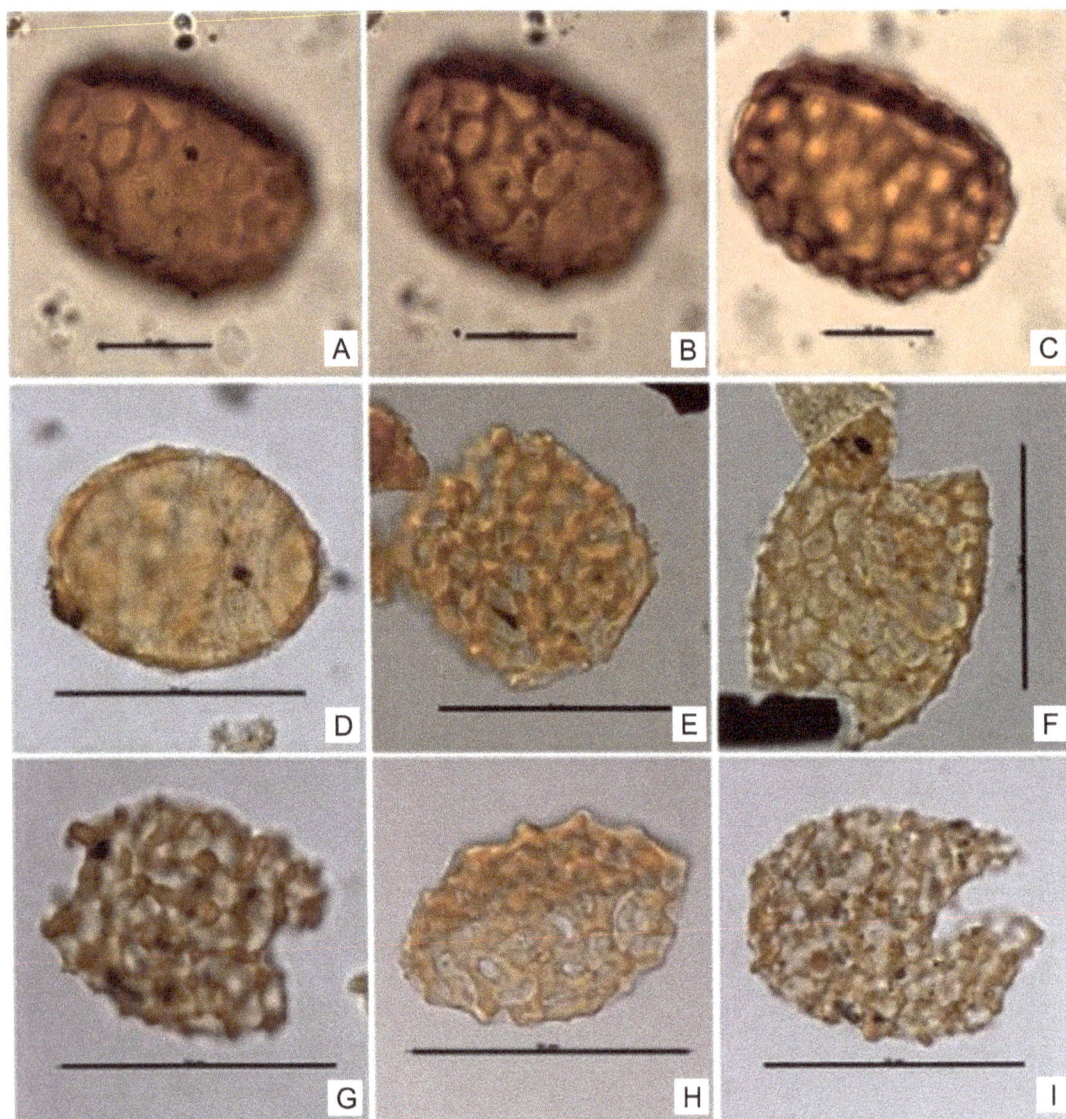

Figure 1. A-C. Upper fill in brick-lined cesspit (CTX F-1186), 66 Howard Street, Quadrant Development Site, Mountain Street and Broadway, Sydney (scale bar = 20 µm). **D.** Lower fill in wood-lined cesspit (CTX 8037), Darling Walk Development Site, Darling Harbour (scale bar = 50 µm). **E-F.** Fill in sandstone cesspit fill (CTX 8737), House 9, Darling Walk Development Site, Darling Harbour (scale bar = 50 µm). **G-H.** Mud mortar (CTX 8.020), Development Site at 185-193 Gloucester Street, the Rocks (scale bar = 50 µm). **I.** Mud mortar (CTX 8.012), Development Site at 185-193 Gloucester Street, the Rocks (scale bar = 50 µm).

Cloacasporites sydneyensis (Macphail and Casey 2008)

Description:

Monad, quasi-isopolar, subspherical to ellipsoidal; aperture if any obscured; laterally biconvex to concavo-convex, amb elliptical to subcircular; wall not stratified c. 0.8-4 µm thick; reticulate-rugulate, sculptural elements varying in width and thickness, c. 1-8 µm thick, weakly aligned longitudinally, coalescing to enclose subcircular to subangular lumina up to 16 µm in maximum diameter, or breaking down into irregular rugulae and verrucae; 52 (71) 84 µm x 30 (44) 56 µm (20 specimens measured).

Type specimens:

Geoscience Australia CPC 3987 (Holotype) and 39788-39789 (Paratypes).

Derivation of Name:

From Sydney, the capital city of New South Wales, where the microfossil was first recorded.

Affinity:

Unknown; presumed to be the egg case of an unidentified animal parasite.

Discussion

As well as fossil spores, pollen and algal cysts, most sedimentary deposits preserve numerous microfossils whose source(s) remain unknown (see Van Geel 2001:206). Nevertheless, some of these are useful for interpreting the past because of their strong empirical link with a particular depositional environment or archaeological context. *Cloacasporites sydneyensis* is an example of the latter.

This microfossil has been recorded in low to trace numbers at eight historical archaeological sites in Sydney (references in Macphail and Casey 2008). In almost all instances, the archaeological contexts are cesspits (Figure 1 A-C), drains used for the disposal of human sewage or other structures that are likely to have been contaminated with sewage. The latter include sediment infilling a well in the yard of a demolished terrace house near Wentworth Park – an area which is recorded as being flooded by sewage backing up along Blackwattle Creek during storms and high-tide surges into Blackwattle Bay during the late 19th century.

The specimens illustrated in Figure 1, D-F reinforce the association of *Cloacasporites sydneyensis* with Europeans, since these specimens were preserved in cesspits attached to demolished 19th century terrace houses on the Darling Walk Development Site, Darling Harbour (Casey & Lowe 2009; M.K. Macphail unpubl. data). The specimens illustrated in Figure 1, G-I extend the record of the microfossils even closer to the site of the first European settlement in Australia, since these specimens were preserved in mud mortar used in the foundations of a ca. 1840 building at 185-193 Gloucester Street, the Rocks (Archaeological & Heritage Management Solutions 2007; M.K. Macphail unpubl. data). Pollen of aquatic herbs and native shrubs occurs in the same samples and the combined data hint at sewage-pollution of the source of the water used to make the mud mortar, presumably the Tank Stream which had become one of Sydney's 'main sewers' by 1835 (de Vries-Evans 1987).

As noted by Macphail and Casey (2008), *Cloacasporites sydneyensis* is highly unlikely to be of plant origin, but the source remains unknown. The microfossil is not present in every cess-fill, including some that preserve pollen of cereal and other edible plants eaten by the colonial inhabitants, e.g. the cesspit attached to the demolished 1830s Wool Pack Inn at Old Marulan on the Southern Tablelands south of Sydney (Macphail 2008). This indicates the microfossil is less likely to be the egg case of flies, e.g. the blowfly (Calliphoridae species), or other insects attracted to human faeces. Accordingly, we have proposed it is the egg case of a parasite living in the human gut or that the host was present in/on one of the many animals eaten by Europeans during the colonial period. Time will tell whether *Cloacasporites sydneyensis*, like *Rattus exulans*, was one of many human commensals inadvertently introduced into the southwest Pacific region.

References

Archaeological & Heritage Management Solutions Pty Ltd 2007. *185-193 Gloucester Street, The Rocks, NSW. A research design and excavation methodology for proposed historical archaeological excavation at the site.* Report to Stamford Windsor Ltd.

Casey & Lowe Pty Ltd. 2009. *Non-Indigenous Archaeological Assessment, Darling Walk, Darling Harbour.* Report to Lend Lease Development.

De Vries-Evans, S. 1987. *Historic Sydney as Seen by its Early Artists.* Sydney: Angus and Robertson.

Ellison, J. 1994. Paleo-lake and swamp stratigraphic records of Holocene vegetation and sea-level changes. Mangaia, Cook Islands. *Pacific Science* 48:1-15.

Macphail, M.K. 1999. A hidden cultural landscape: Colonial Sydney's plant microfossil record. *Australasian Historical Archaeology* 17:79-115.

Macphail, M.K. 2008. Pollen analysis of soil samples from Old Marulan, Southern Tablelands, NSW. Report to Banksia Heritage & Archaeology Pty Ltd.

Macphail, M.K. and Casey, M. 2008. News from the Interior: what can we tell from plant microfossils preserved on historical archaeological sites in colonial Parramatta. *Australasian Historical Archaeology* 26:45-69.

Prebble, M. and Wilmhurst, J. 2009. Detecting the initial impact of humans and introduced species on island environments in Remote Oceania using palaeoecology. *Biological Invasions* 11:1529-1556.

Van Geel, B. 2001. Non-pollen palynomorphs. In: Smol, J.P., Birks, H.J.B. and Last, W.M. (eds), *Tracking Environmental Change Using Lake Sediments*; Volume 3: Terrestrial, algal and silicaceous indicators, pp. 99-119. Dordrecht, Kluwer.

Wilmshurst, J.M. and Higham, T.F.G. 2004. Using rat-gnawed seeds to independently date the arrival of Pacific rats and humans in New Zealand. *The Holocene* 14:801-806.

9

Science, sentiment and territorial chauvinism in the acacia name change debate

Christian A. Kull
School of Geography and Environmental Science, Monash University, Clayton, Victoria
christian.kull@monash.edu

Haripriya Rangan
Monash University, Clayton, Victoria

Introduction

The genus *Acacia*, as Peter Kershaw has often told us, may be widely present in the landscape, but its pollen is seldom found in any abundance. The pollen grains are heavy and probably not capable of long-distance transport, and even where they dominate the vegetation, their pollen is greatly under-represented. Compounding the problem, *Acacia* pollen tends to break up into individual units that are difficult to identify. However, as we hope to show in our contribution celebrating Peter's work, the poor representation of acacias in palaeoenvironmental records is more than compensated by its dominating presence in what has been described as one of the longest running, most acrimonious debates in the history of botanical nomenclature (Brummitt 2011).

Few would imagine botanical nomenclature to be a hotbed of passion and intrigue, but the vociferous arguments and machinations of botanists regarding the rightful ownership of the Latin genus name *Acacia* give an extraordinary insight into the tensions that arise when factors such as aesthetic judgement, political clout and nationalist sentiments dominate the process of scientific classification. After much lobbying and procedural wrangling, on July 16, the last day of the 2005 International Botanical Congress in Vienna, botanists approved a decision to allow an exception to the nomenclatural 'principle of priority' for the acacia genus. With increasing demand by botanists to split apart the massive cosmopolitan and paraphyletic genus into several monophyletic genera, the Vienna decision conserved the name acacia for the members of the new genus from Australia. Normal application of the rules of priority would instead have kept the name acacia for a subset of the trees native to the Americas, Africa and Asia. The Vienna decision was unprecedented in the number of species affected and in the amount of public indignation generated across the world. Many professional and amateur botanists, horticulturalists and naturalists, particularly those working in Africa, Asia and Central America (Luckow et al. 2005), were incensed by the decision. In eastern and southern Africa, where the iconic acacias dominate the savannah landscape, popular newspapers such as Nairobi's *Sunday Nation* announced in a headline "Did you know it is illegal to call this tree acacia? Australia claims exclusive rights to the name" (Githahu 2006).

This essay argues that the ongoing debate and controversy over the acacia genus name is a reflection of a deeper crisis in botanical taxonomy and nomenclature arising from the use of molecular systematics in classification. The splitting of genera and the shifting of species from one genus to another have not only revived older debates in botany regarding classification systems, but also put a great deal of pressure on genus names themselves. We show how the acacia name debate reveals these tensions and contradictions arising from molecular systematics and how rhetoric centred on a variety of non-scientific and non-rational factors, such as aesthetic judgment, sentiments of belonging, territorial chauvinism and politics (lobbying, vote-rigging, etc), came to dominate the procedures of botanical nomenclature.

In the following sections, we offer a brief review of the history of the science and practice of botanical nomenclature, and show how there have been longstanding tensions between folk- or place-based classification systems and universal, scientific approaches to plant classification. After explaining the relevant conventions and rules set out by the International Code of Botanical Nomenclature, we describe how the controversy over the ownership of the genus name acacia has developed over the past two decades. We draw on arguments published in scientific journals and the popular media, on interviews with botanists and participant observation of the nomenclature sessions during the 2011 Melbourne IBC to show how sentiments, chauvinisms and egos have dominated the debate and prevented any 'scientific' resolution or compromise emerging from within the conventions of international botanical nomenclature. The essay concludes by arguing that the acacia name controversy and other potential naming crises emerging from molecular systematics can only be resolved by recognising and incorporating the social histories of attachment in plant names in processes of botanical nomenclature.

What's in a name? Taxonomic debates over systems of classification

> Juliet: *What's in a name? That which we call a rose*
> *By any other name would smell as sweet,*
> *So Romeo would, were he not Romeo called,*
> *Retain that dear perfection which he owes,*
> *Without that title.*
> *Romeo and Juliet, Act 2, Scene 2*

Notwithstanding Juliet's impetuous claim, the tragic ending of this Shakespearian drama underscores the importance of names and lineages in ordering society and social interactions. Bowker and Star (1999:326) note that 'seemingly purely technical issues like how to name things and how to store data in fact constitute much of human interaction and much of what we come to know as natural'. By naming and classifying things, humans construct a nature that is not just based on objective or observable characteristics, but which also reflects a variety of aesthetic sentiments, cultural traditions and place-based associations and attachments.

Names and categories matter a lot in botanical nomenclature and classifications. People in every part of the world have developed different systems for classifying the plants around them. These 'folk' classifications play a central role in providing a material and emotional sense of particular places and regions (Dear 2006). Hence, plant names and their classifications can vary from one place to another, change over time, and vary from one perspective to another. In some cases, a particular plant species may have multiple names within a region depending on how it is used, and in other cases, many different plant species may be called by the same name. But as Gledhill (2008) points out, despite the cultural richness of common names for plants, their immense diversity can make it difficult for those who seek to identify plants according to some kind of larger or 'universal' order so as to compare their characteristics or catalogue their uses.

Going back over two millennia in European history, natural philosophers and botanists sought various principles and criteria that would reveal the hidden order of the rich diversity of plant life in nature. Pavord (2005) traces these efforts back to Theophrastus, a disciple of Aristotle, who attempted to define plants both in philosophical terms of their essential being and in terms of the physical characteristics that could be used to classify them. She notes that his collected works were repackaged in different forms by subsequent Roman scholars, taken up and expanded on by Arab scholars well into the 14th and 15th centuries, and further elaborated on by Italian, Swiss, German and English natural philosophers in the 16th and 17th centuries. Every period of enquiry raised the question of which physical characteristics or behaviours of plants could be used for classification. Scholars of medicine were among the first to draw on folk methods of classifying plants according to their uses as food, dyes, medicines, or poisons, and compiled them in volumes that were known as 'herbals'. Such methods continue to be found today in various herbal reference and guide books for lay people (see, for example, Foster and Johnson 2006), or in classification systems based on the phytochemical properties of plants.

The discipline of taxonomy (derived from Greek, meaning 'arrangement') grew rapidly from the 16th century onwards, alongside the expansion of European maritime exploration, trade and colonialism in the Old and New World. Much of the biota of these regions was unknown to European naturalists. Sixteenth and seventeenth compilations of plants by physicians such as Garcia da Orta for India, Nicolás Monardes for the West Indies, and Cristobal Acosta and Jacobus Bontius for the East Indies described the plants they encountered in terms of their morphological features, such as leaf characteristics, fruit types or flower structure, along with details of their local environments and uses (Cook 2005; Pavord 2005). In doing so, they attempted to combine and adapt local systems of plant classification to similar systems followed in Europe.

As mercantile colonialism gathered pace during the 17th and 18th centuries, European countries were often in direct competition with one another to capture profits from trade in exotic or useful tropical plants that could be cultivated in their colonies. These pressures added to the motivations of naturalists to seek new methods for identifying and classifying plants that could be universally comprehended and applied in different places (Browne 1996). By the end of the 18th century, various naturalists in France, Germany and England had developed some common conventions of naming and ordering plants. Pavord notes that John Ray's *Methodus plantarum emendata*, published in 1703, provided the six basic rules for classifying plants that ever since have underpinned the discipline of taxonomy: 'Plant names should be changed as little as possible to avoid confusion and mistakes; the characteristics of a group must be clearly defined and not rely on comparisons; characteristics must be obvious and easy to grasp; groups approved by most plantsmen should be preserved; related plants should not be separated; the characteristics used to define should not be unnecessarily increased.' (2005:392)

By the mid 18th century, botany and taxonomy underwent another radical change with the introduction of Carl Linnaeus's classification system. Linnaeus's botany sought, in effect, to develop systems of standardised information exchange that would serve as both a knowledge framework and an instrument for identifying plants of potential economic value to

European colonisers (Müller-Wille 2005).[1] In this sense, his approach reflected the new levels of abstraction required by Europe's emerging modern states with their imperial ambitions for control, communication and legibility across their territories (Scott 1998). Linnaeus developed what he called 'artificial' taxonomies of the natural world that reflected the social order and religious ideas of his times (Williams 1980; Dear 2006). God's empire of nature was divided into three kingdoms – vegetable, animal and mineral. Life forms coming under the vegetable and animal kingdoms were hierarchically grouped into classes, orders, families, genus, species and varieties. Linnaeus proposed a binomial system of identification in Latin, with the genus name preceding the special descriptive name for the species (Koerner 1996). He went on to propose a new method, the '*systema sexuale*', for grouping plants based on the number, size, arrangement and shape of reproductive organs (stamen and carpels) within in their flowers, and their sexual behaviour (Schiebinger 1996).

The binomial nomenclature system outlined in Linnaeus's *Species Plantarum* was rapidly adopted during the 18th and 19th centuries as European states competed with each other to launch numerous scientific expeditions for collecting and documenting plants in new lands or in remote parts of their colonies and imperial territories. This was also the period when colonial territories around the world were used for large-scale commercial cultivation of plants as raw materials for burgeoning industrial production in Europe (Brockway 1979; Bonneuil 2002; Parry 2004; Schiebinger 2004; Schiebinger and Swan 2005). However, not all naturalists agreed with Linnaeus's hierarchical ordering or sexual system of classification. Comte de Buffon claimed that Linnaeus's hierarchically organised categories were not properly grounded in understandings of plants and animals in their particular environments. He argued that 'species' was the only category that could be given a clear philosophical definition, i.e. species could be defined on the basis of membership in a common breeding community. In contrast to Linnaean taxonomy, Buffon's approach echoed the methods used in early compilations of plants by Portuguese, Spanish and Dutch physicians in the East and West Indies by focusing on the morphological description of species, their characteristic behaviours and habitats, and their uses to human beings (Cook 2005; Dear 2006). Spanish Creole naturalists in the Americas also opposed the Linnaean system imposed on them by metropolitan botanists, claiming that its abstract mode of classification disregarded important local conditions such as the plant's location, flowering season, climatic requirements and soil characteristics. They asserted that plants needed to be identified and understood biogeographically in terms of their distinctive physical and moral climates, and criticised their imperial overlords for using Linnaean taxonomy to impoverish their colonial subjects by transferring plants to other parts of their territorial empire for economic exploitation (Lafuente and Valverde 2005).

Even though Linnaeus's binomial system of naming plants is now accepted as the starting point for present nomenclature systems, his method of hierarchical ordering and grouping has been routinely criticised for its artificiality, focus on a few selected subsets of characteristics, and lack of contextual references. Pavord observes that '[f]rom Bentham and Hooker in 1862-63, to Cronquist in 1988, eight major systems of plant classification have been proposed in the last hundred years alone' (2005:400). Following the acceptance of Darwin and Wallace's ideas of evolution in nature, evolutionary taxonomy attempted to provide a historical reinterpretation of the Linnaean taxonomic system as a relatively stable and effective format for explaining and predicting genealogical similarity and variations among species. But this was challenged in

---

1   Müller-Wille (2005) notes that Linnaeus developed his nomenclatural reform and classification system at a time when Sweden had unsuccessfully attempted and later given up its ambitions to gain colonial territories in Africa, Asia, or the New World. However, Linnaeus was inspired by the political-economic ideology that Sweden's prosperity depended on substituting imports with domestic equivalents, or by importing foreign plants and products and subsequently acclimatising them within Swedish territory.

the 20th century by several botanists who, like many critics of Linnaeus before them, argued against classifications based on similarities of some intuitively determined subset of characters and proposed a phenetic system that determined overall patterns of similarity and dissimilarity between species based on all characteristics. Phenetics, in turn, was accused of being too cumbersome and reliant on subjective choices of statistics for producing measures that give different classifications (Maclaurin and Sterelny 2008). Every proposal for a new system of classification offers different philosophical reasons for grouping species into new genera and for 'lumping' or 'splitting' them into large or small categories. And each of these classification systems then requires new names and combinations as species are moved from one genus to another or regrouped in new genera (Bonneuil 2002; Pavord 2005).

Over the past three decades, however, the emergence and widespread use of DNA analysis has enabled plant scientists to look beyond morphology and work out evolutionary relationships not visible through outward characteristics. Given the long history of debate over 'subjective' criteria for plant classification, DNA-based phylogenetics has been heralded by botanists as offering a more rigorous scientific basis for taxonomy. Molecular systematics or cladistics has become the dominant system used in plant classification (Winston 1999; Soltis et al. 2007; Maclaurin and Sterelny 2008). The system seeks to group plants based on *monophyly* or evolutionary descent from a common ancestry. Like other reformulations of classification systems, the use of genetic analysis to generate monophyletic trees gives rise to the rearrangement of species in different genera, tribes, sub-families and families.

Molecular systematics represents a monumental shift in both the philosophy and methods of plant classification because it fundamentally challenges the physical observations and experiences that most common or folk systems, as well as traditional botanical systems, have relied on and used. As Yoon (2009) observes, what may appear to most people as a naturally coherent grouping, such as fish, ceases to exist under this new system of classification because not all species show a clear evolutionary relationship. Classification systems based on physical observation may consider lotuses and waterlilies as closely related, but a molecular systematist would argue that they have little in common based on evolutionary relationships, and that lotuses are more related to proteas and plane trees than to waterlilies.

Reclassifications based on monophyly have given rise to substantial upheavals in the ordering and naming of species. Plant genus names are like human surnames, and provide a sense of familiarity and historical continuity of relationships with other species that carry the same genus name. Most people – and this cuts a wide swathe, including amateur naturalists and gardeners, plant breeders, foresters, ecologists and botanists who are used to associating particular Latin binomial names with species that they study, cultivate, use or sell - may find the changes to genus names unnecessary or objectionable and resist reclassification by insisting on retaining the older name.[2]

The current debate over the name acacia illustrates many of the tensions and contradictions that have arisen from taxonomy's move towards abstract genetic-based science. While cladistics relies on the latest scientific advancements and technologies in molecular genetic analysis to justify the splitting of the acacia genus into new genera, the controversy over the resulting name changes reflects a range of concerns that some taxonomists regard as subjective and unscientific. These include aesthetics tastes, sensibility of place, territorial chauvinism and personal and institutional politics. In the following sections, we outline the rules of botanical nomenclature, followed by a description of the genesis and evolution of the acacia name war, and analysis of

---

2   As one botanist pointed out, 'There is a whole other debate and gnashing of teeth whenever botanists change names. We are the villains to many horticulturalists for example. Just listen to any gardening program or TV garden show when there is a difficult to pronounce botanical name or if a taxonomic change has been made. We encounter this resistance all the time' (anon. interview 2009).

the arguments marshalled on either side of the debate.

Nomenclatural rules and conventions

Taxonomists usually follow a set of standardised rules in naming plants (Bailey 1933; Winston 1999; Spencer et al. 2007). These rules are recorded in a central register called the International Code of Botanical Nomenclature, which is maintained by the International Association for Plant Taxonomy. Each edition, which is published after the meetings of the International Botanical Congress in a series called *Regnum Vegetabile*, provides an update of these rules and decisions of changes to nomenclature that were proposed and endorsed by the Congress. For instance, the 'Vienna code', which was published in the series *Regnum Vegetabile* (Vol. 146, 2006), includes nomenclature decisions from the 2005 Congress.

The fundamental rule of nomenclature is that the first person to scientifically describe a species has the privilege of naming it. For plants, this involves placing a specimen in a herbarium (the 'type' specimen) and publishing a technical description of it (Gledhill 2008). The epithet given to a species may reflect distinctive physical characteristics – like *Acacia grandifolia* (for large leaves) and *A. microsperma* (for small seeds), and may also commemorate people, places and cultures. For example, *Acacia baileyana* memorialises the Australian botanist F.M. Bailey, *Acacia farnesiana* is named after the famous garden estate of Cardinal Farnese in Rome, and *Acacia koa* honours the indigenous Hawaiian name of the tree.

However, if a species has been officially described more than once with different names, or if two species described separately are later determined to be one and the same, then the rule specifies that the oldest name should be used. This is known as the *rule of priority*. For example, the black wattle from southeastern Australia was known for much of the 19th and early 20th centuries by names such as *Acacia decurrens* var. *mollis* or *A. mollissima*, based on specimens described in European herbaria (Brenan and Melville 1960). Now it is named after an American naturalist, Edgar A. Mearns, who collected naturalised specimens in Kenya between Thika and Nairobi while on a hunting and scientific safari with Theodore Roosevelt in 1909. Mearns, best known as an ornithologist, died in 1916, unable to process his findings (Richmond 1918), but his African botanical collection made its way to the National Botanical Gardens in Brussels. Here, the Director, Émile de Wildeman, published a description of the specimen in 1925 in *Plantae Bequaertianae*, honouring Mearns in the name of the plant he believed to be a new African species, *Acacia mearnsii* (de Wildeman 1925). De Wildeman's species description, and hence Mearns's name, later achieved taxonomic priority because the older scientific names for the black wattle were found to be invalid.[3] Hence *A. mearnsii* was the oldest legitimate name available, and thus it took priority. Brenan and Melville (1960:38) lamented that this acacia from Australia 'must bear the misconceived and not especially relevant name *mearnsii*'.

Following on from this rule, the next is that the name of a genus should be taken from the name given to one of its member species that has been designated as the 'type' representing the genus. The type species is designated by a botanist when publishing his or her conception and description of a genus; there are no rules as to which must be chosen, but in practice the type species is usually one that is well-known, is widespread, or holds an old, established name. For example, the type for the genus acacia is the widely known Afro-Asiatic thorn

---

3   Specifically, the name *A. mollissima* was rendered inappropriate as it was found that the herbarium specimen it described (its 'type') was actually *A. pubescens*. As far as *A. decurrens* var. *mollis*, this name is still technically legitimate but most botanists consider the black wattle a distinct species from *A. decurrens* (green wattle), not a variety (Brenan and Melville 1960).

tree, *A. nilotica*, described in 1753 by Linnæus.[4] As botanists examined samples of Australian wattles, they included them in the already existing acacia genus because of their morphological similarities to known African and American acacias. In contrast, when l'Héritier described gum tree specimens, he created a new genus *Eucalyptus* based on its uniqueness (Brooker and Kleinig 2007).

The above two rules would be sufficient were it not for the fact that the science of classifying living things is a difficult and continually evolving endeavour. On the one hand, obscure early publications on particular plants are sometimes rediscovered after alternative names have become commonly used. On the other hand, developments in taxonomy can lead researchers to propose different ways of grouping plants into species and genera. Both factors can contribute to a cascading effect of name changes. This then leads to a third rule of nomenclature called 'conservation'.

Conservation is a special clause in the rules of nomenclature – an infrequently used exception to the rules of priority – that is used to protect certain botanical names (typically well-known ones) from being changed due to new developments in taxonomic science or due to the technicalities of botanical nomenclature. For example, in 1980 it was discovered that the widely used Latin name for wheat, *Triticum aestivum*, would have to change to *Triticum hybernum*. These two species were described separately by Linnaeus in 1753, but later became seen as just two varieties of a single species, widely called *T. aestivum*. However, as was rediscovered in 1980, the earliest person to merge the species had given the name *T. hybernum* to the combined species, and by strict application of the rule of botanical nomenclature, this name would achieve priority. In order to avoid changing the widely used botanical name of wheat, the name *Triticum aestivum* was 'conserved' at the 1987 International Botanical Congress in Berlin (Hanelt et al. 1983).

Technically speaking, 'conservation' applies to the name given to a particular herbarium specimen. That is, a particular name is permanently attached to a particular specimen, normally the 'type' specimen that defines a species and perhaps its genus. In cases where one conserves the name of a specimen that was previously not the type specimen, it is referred to as 'retypification'.

According to the rules, the decision to conserve must be justified by establishing that it would serve the interests of maintaining nomenclatural stability and avoid disadvantageous name changes. Cases for conservation and rejection are made in the journal *Taxon*, and are considered by specialist committees of the International Botanical Congress's Nomenclature Section before being approved by the congress as a whole.

A final convention in botanical nomenclature is that new species names resulting from the above procedures do not officially exist until they are published as 'combinations', that is, until a scientific publication appears with the combined new genus and species name. According to one plant taxonomist (anon. interview 2007), some botanists are hesitant to be the first to publish an unpopular new name, partly out of a sense of not wanting to step on another botanist's turf, while others might see it as an opportunity to 'get their names on combinations'.[5] The peer-review process for scientific literature may serve different roles in these situations by either

---

4   The first specimen described was named *Mimosa scorpioides* by Linnaeus but is now universally accepted as a synonym of *Acacia nilotica*. Botanist Philip Miller formally adopted the genus name *Acacia* (already in wide use) in 1754, hence its notation as '*Acacia* Mill'. However, it was only in the 1800s that *Acacia* became widely accepted, following George Bentham's broad definition of *Acacia* (for details, see Maslin et al. 2003a; Orchard and Maslin 2003).

5   This statement refers to the convention among botanists where the first published source of a new combination must be cited when establishing the 'authority' of a plant's name in scientific publications.

retarding the appearance of unpopular new combinations or enforcing their use.[6] It should also be noted that one may continue to use old names if one justifies the taxonomic reasons for doing so. For instance, if one insists that the acacia genus does not need to be split, one can continue using the name acacia in its broad sense (see Pedley 2004; Smith et al. 2006; Robin 2007).

The acacia genus, the Battle of Vienna and beyond

The modern acacia genus is a broad classification that emerged in the 1840s through the efforts of botanist George Bentham (Ross 1980; Maslin et al. 2003b). Thanks to his position at the Royal Botanic Gardens in Kew, Bentham was exposed to plants from around the world (Bentham 1842; Bentham and Mueller 1864). The broad genus he defined now includes more than 1300 species worldwide, of which approximately 1000 are found in Australia. It is classified in the *Mimosoideae* subfamily of the *Fabaceae* family, better known as the legume family for its seed pods.

Over the past few decades, botanists have argued that the acacia genus was too massive and not monophyletic, and hence needed to be split along the lines of sub-genera identified by Vassal (1972). Molecular genetic analysis added weight to these claims, showing, for example, how Australian wattles are more closely related to the tribe Ingeae (which includes the genera *Albizia*, *Calliandra* and *Paraserianthes*) than to other acacia sub-groups typified by *Acacia nilotica* or *Acacia senegal* (Figure 1; Clarke et al. 2000; Murphy et al. 2003; Jobson and Luckow 2007; Brown et al. 2008; Murphy 2008). Les Pedley (1986), of the Queensland Herbarium, first proposed a three-way split, which was later modified to include two minor new genera from the Americas (Figure 2; Figure 3). Because the type species for the old, broad genus was *Acacia nilotica*, the name *Acacia* was to be given to the subset of the old genus that contained *A. nilotica* and some 160 other pan-tropical acacias. This left the two new genera in need of names. Based on the rules of priority, Pedley recovered two genus names from the dustbin of botanical history for the remaining species: *Senegalia*, which applied to 200 tree species mainly from Africa, and *Racosperma*, for the roughly 1000 species mostly found in Australia and nearby islands. In his proposal, Pedley published 33 combinations for *Racosperma* and two for *Senegalia*, including *Racosperma auriculiforme* (for *Acacia auriculiformis*), *Racosperma mearnsii* (for *Acacia mearnsii*), and *Racosperma koa* (for *Acacia koa*).

The genus names suggested in Pedley's original proposal were adopted in very few publications, an exception being the *Flora of New Zealand* (Webb et al. 1988), which used these to describe a number of introduced Australian wattles as *Racosperma*. Most botanists resisted using the new name, arguing that further evidence was needed from molecular research, but also out of some discomfort about the implications for name changes (Maslin et al. 2003a, b; Murphy 2008).

Even as molecular evidence accumulated to back Pedley's proposed division of the genus, many botanists remained reluctant to accept the split. The main reason, it so happened, was the ungainly name *Racosperma*. Some called it an 'abomination' in comparison to the more elegant and euphonic name acacia (Woodford 2002; Pedley 2004:4). Some South African botanists gloated over the appropriateness of a harsh-sounding name for species that had been declared 'alien invasive weeds'.[7] Montgomery (2006) wrote in *South African Gardening*:

---

6   This was our personal experience with a previous publication in an ecological journal. A reviewer who had taken the Vienna decision on board wrote, 'The authors have to take into account that the genus *Acacia* now only refers to Australian species. Other species have now been assigned to other genera.' See also Boy (2005:27).

7   In a similar vein, an article in the South African magazine *Veld and Flora* plays on trans-Indian Ocean rivalries when titling an article about efforts to control invasive *A. pycnantha* 'Golden wattle loses its lustre'. The first line mentions the tree's status as Australia's floral emblem; much of the rest describes efforts to control it by introducing a gall wasp from its home range (Hoffmann 2001:58).

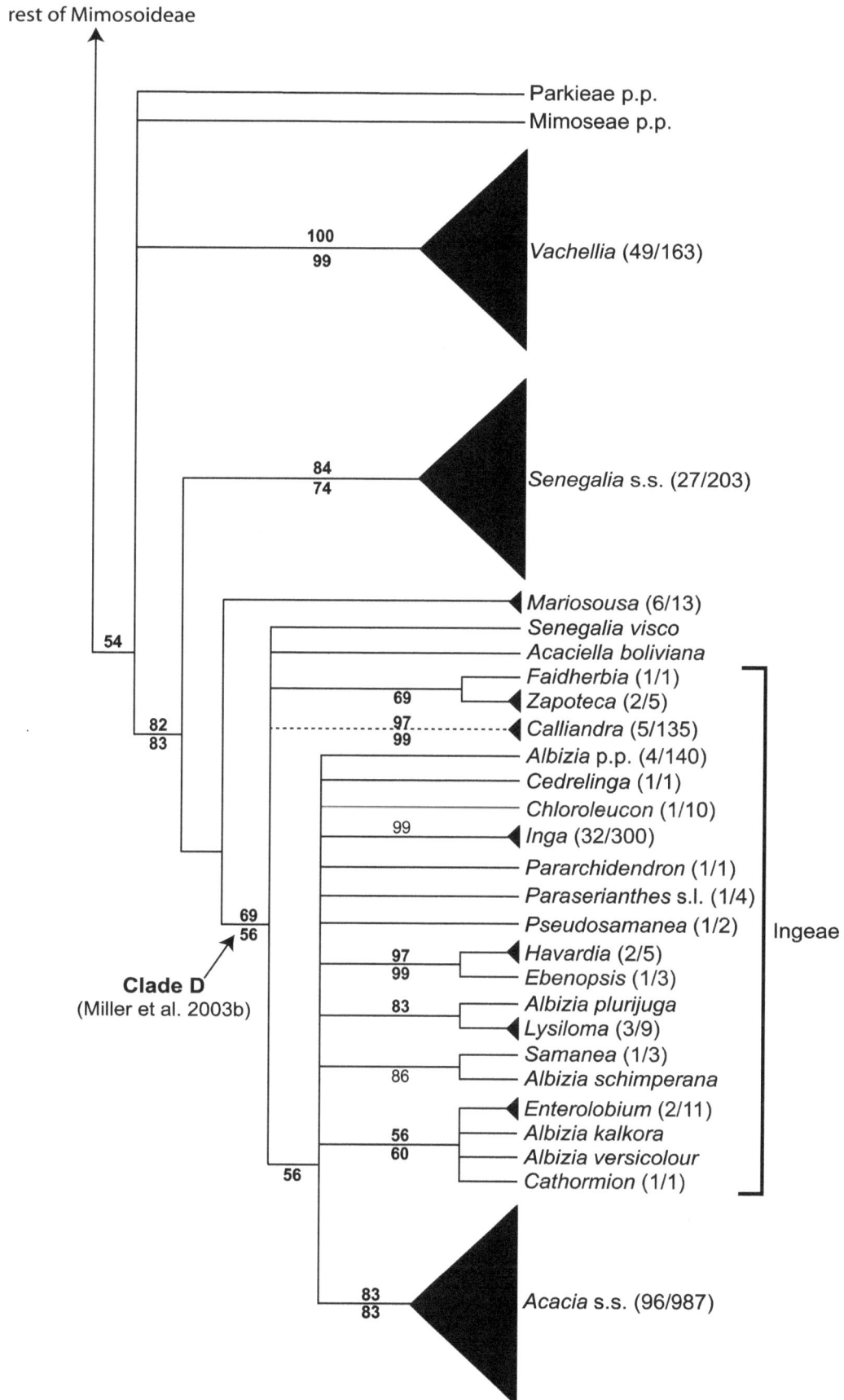

Figure 1. A 'genetic family tree' created from DNA analysis of different acacia species and nearby genera. This tree demonstrates how different sections of the old acacia genus are closer to other species in the tribe Ingeae than to each other.
Source: first published in Brown et al. (2008:741), reproduced with permission of authors.

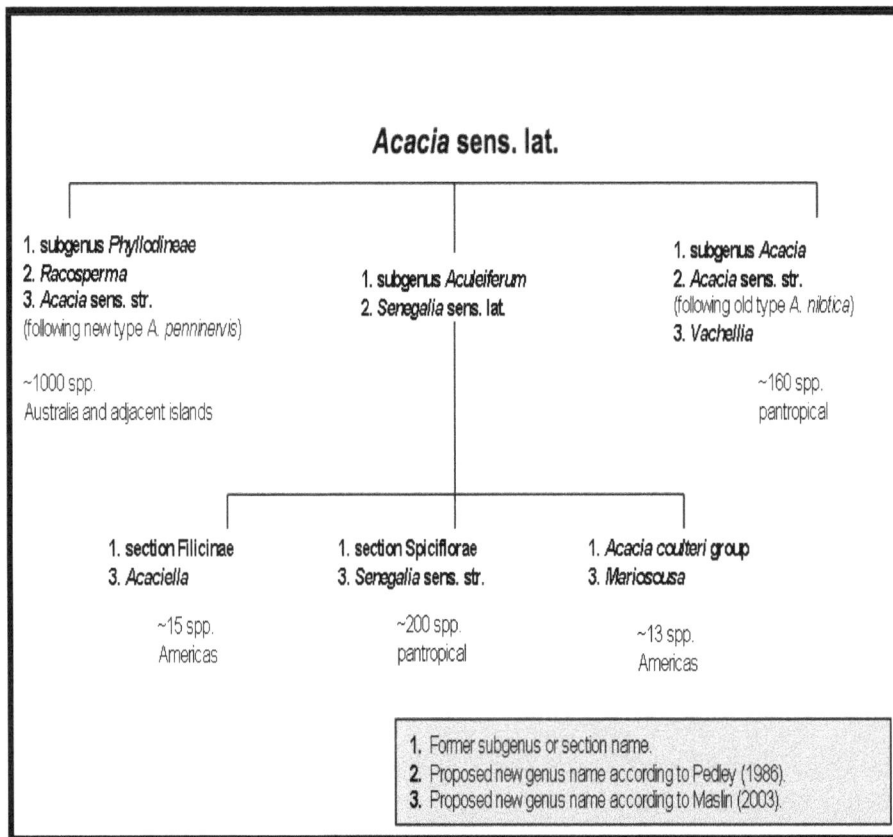

Figure 2. The different names of the proposed divisions of the acacia genus.
Source: based on Maslin et al. 2003a.

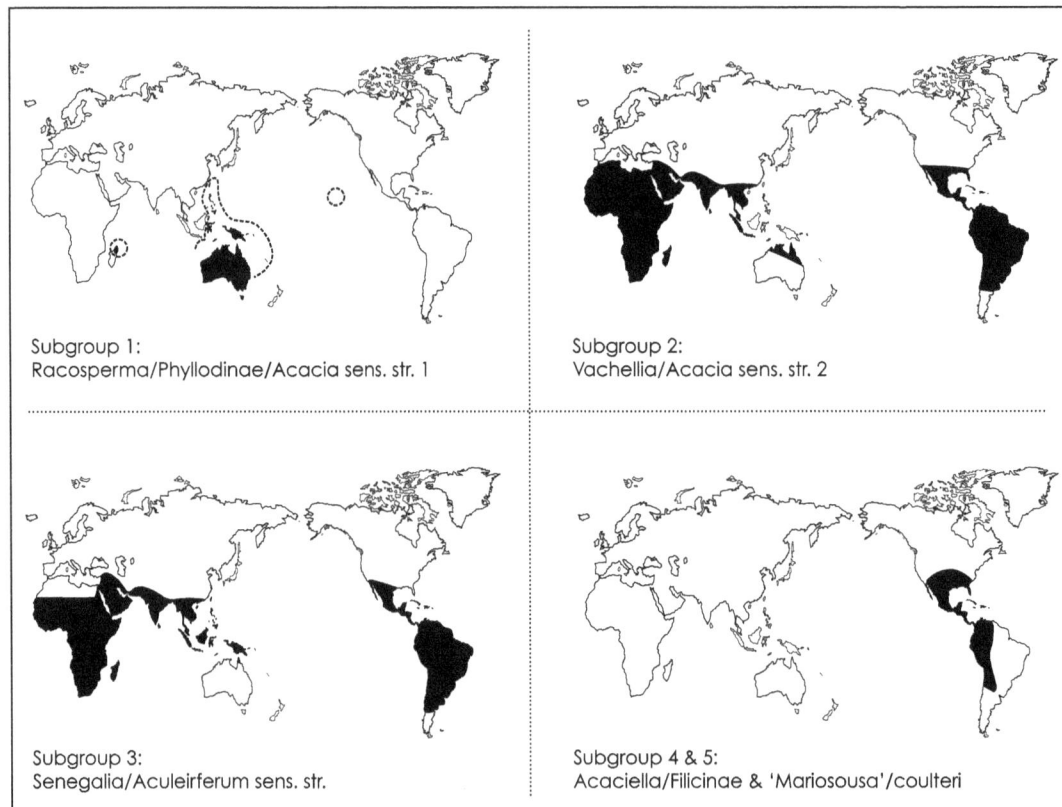

Figure 3. The geography of acacia's taxonomic revision: native distribution of different acacia subgroups.
Source: based on Maslin et al. 2003a.

It was further decided that Australian acacias should fall under the genus *Racosperma*. The local botanical community gave a chuckle. Twelve Australian acacias (wattles) including the black wattle (*Acacia mearnsii*) are designated Category 2 invasive alien weeds in this country. To give our unwanted cousins a new name (*Racosperma mearnsii*) was viewed locally as positive. (Montgomery 2006)

In contrast, Pedley's proposal to rename a different set of acacias *Senegalia* did not meet with the same resistance because the name sounded reasonably mellifluous and referred to a geographical area where the genus was widespread (interview anon. 2009).

Resistance to the use of the genus name *Racosperma* inspired Bruce Maslin, of the Western Australian Herbarium, together with Tony Orchard, of the government's Australian Biological Resources Study (ABRS), to propose an alternative solution that would maintain the familiar moniker acacia for the Australian species. Orchard and Maslin's (2003) proposal called for a retypification of the acacia genus. Specifically, the proposal called for the conservation of the name *Acacia penninervis* (commonly known as the hickory wattle), implying that it would become the type specimen for the genus that includes it. This species was chosen because it already served as the type for the *Acacia* sub-genus *Phyllodineae*. As a result, the consequences of splitting the acacia genus would be different: the name *Acacia* would apply to the mainly Australian part of the genus, while the group including the former type species, *A. nilotica*, would take on the oldest valid alternative name, *Vachellia*. *Senegalia* was to remain unaffected (see Figure 2).

While the Committee for Spermatophyta (a specialist committee of the Nomenclature Section of the IBC) was deliberating on Orchard and Maslin's proposal, Pedley (2003) published some 800 combinations for *Racosperma*, allegedly in reaction to the claim that very few *Racosperma* combinations had been made. While the secretary of the committee commented on the 'surprising' nature of Pedley's action (Brummitt 2004:828), some botanists regarded it as a gauntlet thrown down to challenge Orchard and Maslin's proposal. By this point, the issue was already entwined with egos and the lobbying of key individuals (interviews anon. 2011).

The case for retypification was debated at the 17th International Botanical Congress, held in Vienna in 2005. On 16th July, the last day of the Congress, the Nomenclature Section voted on the recommendation of specialist committees to change the type species of acacia from *A. nilotica* to *A. penninervis*. Following the complex procedural rules of nomenclature revision (Smith et al. 2006; van Rijckevorsel 2006; Moore 2007; Glazewski and Rumble 2009; McNeil and Turland 2010), a 60% supermajority vote was required to *overturn* this recommendation. The headings 'Africa' and 'Australia' were penned on a whiteboard as the vote was prepared, accentuating the geographic symbolism (Moore 2007:114). When votes were tallied, a majority (54.9%) of the Nomenclature Section members had voted against the recommendation, but this was short of the 60% supermajority requirement. Hence, the recommendation for retypification was allowed to pass and be ratified by the plenary session.

The result of the vote meant that the Australian plants would retain the name *Acacia* following a split of the genus, and many of the non-Australian species would be renamed *Vachellia*. Shortly after the decision, new combinations were published for several American species under the name *Vachellia* (Seigler and Ebinger 2006) as well as *Senegalia* (e.g. Seigler et al. 2006), and the names have begun to appear in Latin American floras (Maslin 2011).

The botanic community was sharply split, with one faction, described as 'the tropical botanist community', saying it was 'immensely disappointed' by the decision (Smith et al. 2006:225), a sentiment also held by *Racosperma* promoter, Les Pedley (2004).[8] The Vienna decision touched a sensitive nerve of botanists in eastern and southern Africa, who were outraged by the outcome. 'So we won but we lost,' said Dr Siro Masinde, Director of the East African Herbarium in Nairobi, referring to the supermajority needed to overturn the recommendation.[9] Numerous members of the biological, forestry and environmental communities in South Africa expressed their indignation over the naming coup (interviews 2007). Opposition was also voiced by the Chairperson of the Botanical Society of Namibia (Hoffman 2006). *Veld and Flora*, the Journal of the South African Botanical Society, was overwhelmed by letters of protest.[10] Likewise, *Swara*, the journal of the Kenya-based East African Wild Life Society, carried a detailed analysis, with the opening proclamation:

> It is official: Africa no longer has any indigenous Acacias. The continent's magnificent, archetypal thorn trees – which, as Acacias, have long been universally synonymous with Africa's great savannahs and bush lands – have been formally stripped of their names (Boy 2005:25).

Media reports further sensationalised the news. In addition to the *Sunday Nation* article (mentioned in our introduction), the *Kruger Park Times* ran an article, 'Africa to lose all its acacias', and the *South African Gardening* magazine opined 'Brand Acacia goes to Australia' (Montgomery 2006). In Australia, in contrast, the mood was jubilant. The story appeared on the national broadcaster ABC's popular radio program Australia All Over. The *Sydney Morning Herald* declared:

> September 1 has many names. Some welcome it as spring's dawn, a time to celebrate nature's renewal. For others it's Wattle Day. But it will never be Racosperma Day. (Macey 2005)

In the aftermath of the Vienna decision, the opponents of retypification complained about 'the way the Australians ... conned the Vienna conference' (Cameron 2006:51). Moore (2007:109) called it 'an attempt at minority rule' and blamed inconsistent and confusing implementation of procedural rules (see also Boy 2005; Smith et al. 2006; van Rijckevorsel 2006; Moore et al. 2010, 2011). Moore (2007: 112) noted that African and South American taxonomists – who formed the bulk of the opposition – were under-represented in the debate because of their relatively small presence at the Vienna meeting due to limited budgets for international travel. In contrast, he claimed, the pro-retypification lobby, which had 'substantial backing from Australian botanists', was prominent in meetings and in notice-board postings (see also Smith et al. 2006:225). African representatives accused Australian botanists of stacking the debate by mobilising various resources, including funding delegates to attend the meeting and vote in their favour.[11] The editor-in-chief of *South African Gardening* commented on the

8   According to one botanist, 'there were actually many Australian botanists against the proposal (see Luckow et al. 2005 for names, including Lyn Craven and Mike Crisp). On the other side, many international botanists were supporting the Australian Acacia proposal... It was not really divided along geographical or political lines in my opinion, more on whether people agreed or disagreed with the arguments for application of the conservation of names measures in the ICBN. Some institutions had their own internal votes before Vienna and others left the decision up to a single delegate to cast their votes. I think the media and subsequent argument has skewed that sense of sentiment and politics' (interview 2009).

9   Interview 2006. Masinde, along with most African botanists, voted by proxy.

10   A page of letters was printed in the March 2006 issue (p. 51). According to the subsequent issue, these letters were 'the tip of the iceberg in responses' (*Veld and Flora*, June 2006, p. 72); this issue printed four more letters (p. 73).

11   Interview, Dr. Siro Masinde, Nairobi, September 28, 2006 (see also Boy 2005). Such perceptions were also described by Dr. Najma Dharani, author of *Field Guide to Acacias of East Africa* (interview Nairobi 2006), and in Moore (2007) and Smith et al. (2006).

Vienna decision saying: 'It is an extraordinary tale of botanical intrigue... a team of Australian botanists pulled off the world's greatest branding coup... with ferocious public support, massive documentation, and a superb public relations machinery.' Referring to the sporting rivalry between the two countries (in cricket and rugby), she added: 'the South Africans underestimated the Australians (again!)' (Montgomery 2006).

O'Neill (2007), an Australian botanist, hailed the Vienna decision with relief for having escaped the ugly genus name *Racosperma*: 'Had the proposal failed, Australia's national floral emblem, the golden wattle, might now be *Racospermum pycnantha*' [sic]. The unstated assumption for the supporters of retypification was that by celebrating the wattle as a national symbol, Australia had gained the right to retain the euphonic name Acacia for its species.

In the run-up to the following International Botanical Congress (held in July 2011 in Melbourne, Australia) these arguments and tensions resurfaced, many of them focused on perceived injustices in the procedures behind the Vienna decision.[12] Gerry Moore, of New York's Brooklyn Botanic Garden, led the charge in seeking to annul the decision made in Vienna (Moore et al. 2010, 2011). A website was dedicated to documenting support (www.acaciavote.com, last accessed 28 July 2011). Others defended the Vienna decision (Thiele et al. 2011). In what might be interpreted as a rearguard action, Bruce Maslin encouraged members of the Australian Systematic Botany Society and of the 50-year-old Acacia Study Group (member of the Association of Societies for Growing Australian Plants) to attend the Melbourne IBC and vote to maintain the Vienna decision. Some botanists outlined new, compromise, solutions to rename acacia (Brummitt 2010, 2011; Turland 2011; Table 1). Dick Brummitt, of Kew Herbarium, the ex-Chair of the Committee for Spermatophyta, said that while the Vienna decision was sound, the unprecedented uproar (and numbers of species affected) needed an unorthodox solution. In the event, the highly charged sessions of the Nomenclature Section of the Melbourne IBC did not overturn the Vienna decision nor accept any of the compromise proposals (Smith 2011).[13] As Gideon Smith, of South Africa's National Biodiversity Institute, remarked during the deliberations, the issue of which genus has the right to retain the name acacia will remain divisive and the controversy is unlikely to go away.

Analysing the arguments: From rational criteria to rhetoric and politics

The formal rules of botanical nomenclature are, in principle, set up to alleviate the tensions over names caused by the evolving science of taxonomy. According to L.H. Bailey, author of *How Plants Get Their Names* (1933), scientific decisions over taxonomy come first, and any consequences for names follow:

The naming of plants under rules of nomenclature is an effort to tell the truth. Its purpose is not to serve the convenience of those who sell plants or write labels or edit books; it is not commercial. Serving the truth it thereby serves everybody. (p. 39)

---

12   B. Maslin and D. Brummitt, pers. comm. 2011, and email from Bill Aitcheson, leader of the Acacia Study Group, 10 May 2011. See also Maslin (2011) and acacia name change blog postings at christiankull.net (last accessed 28 July 2011).

13   The meeting started with a contentious discussion of the 60% supermajority rule. It then ratified the Vienna Code as printed using a card vote (373 yes, 172 no). The large 'no' vote reflected opposition to acacia retypification. Brummitt's compromise proposal was considered both unconventional and unacceptable; Turland's compromise proposal to create *Protoacacia* and *Austroacacia* (Table 1) was seen as a possible way out, but the name *Protoacacia* was questioned because it seemed to imply some evolutionary meaning. An alternative proposal by Paul van Rijckevorsel (Utrecht) accepted *Acacia* for the Australian species but sought a different name to replace *Vachellia*, tentatively *Africacia*, but this was critiqued, as the distribution of *Vachellia* includes Asia and Latin America (never mind that in this proposal 'victory' would remain with the retypification proponents). Brian Schrire (Kew) suggested, in the spirit of getting a compromise, a different alteration of Turland's proposal, replacing *Protoacacia* with *Acanthacacia* (representing 'thorns'). Turland's proposal (with Schrire's modifications) was, however, rejected by a 70% majority.

Table 1. The naming consequences of different proposals to the IBC.

Sub-genera of old inclusive genus *Acacia*	Nomenclatureal consequence of Pedley (1986) and later research	Proposal by Orchard and Maslin (2003) approved at Vienna IBC in 2005; contested by Moore et al. (2010, 2011)	Proposed compromise by Brummitt (2010)	Proposed compromise by Turland (2011) (includes keeping use of *Acacia* for old, inclusive genus)	Result of Melbourne IBC in 2011
subgenus *Phyllodinae* (>1000 species of wattles in and around Australia – typified by *A. penninervis*)	*Racosperma*	*Acacia*	*Acacia* for general use *Acacia* (*Racosperma*) for specialist use	*Australacacia*	*Acacia*
subgenus *Acacia* (ca. 160 species of thorn trees in Africa, Asia, Latin America – typified by *A. nilotica*)	*Acacia*	*Vachellia*	*Acacia* for general use *Acacia* (*Vachellia*) for specialist use	*Protoacacia* (amended at Melbourne IBC 2011 to *Africacacia* then *Acanthacacia*)	*Vachellia*

and, unaffected by the conflict, but not to be forgotten....

subgenus *Aculeirferum* (>200 species in Africa, Latin America, Asia – typified by *A. senegal*)	*Senegalia* (as well as *Mariosousa* and *Acaciella*)	*Senegalia* (as well as *Mariosousa* and *Acaciella*)	*Acacia* for general use (except *Mariosousa* and *Acaciella*) *Acacia* (*Senegalia*) for specialist use	*Senegalia* (as well as *Mariosousa* and *Acaciella*)	*Senegalia* (as well as *Mariosousa* and *Acaciella*)

Bailey's views continue to be espoused by many professional botanists. As one remarked, 'Botanists like me, in the end, typically just want the rules applied strictly. It is the science that matters, and the rules, and names, follow' (interview anon. 2007). Yet the existence of procedures for the 'conservation' of names is an acknowledgement that inconvenience, even tension, may arise from strictly following the rules.

The official case for conservation of a name is meant to rest on objective evidence that such an action would minimise disruption and stabilise nomenclature for well-known, widely used names. In the acacia debate, both sides formally argued their case in the journal *Taxon* using arguments about numbers of name changes, the magnitude of inconvenience and the economic value of these well-known species. These rational arguments, however, were supplemented by particular kinds of rhetoric – centred on aesthetic judgements, place-based sentiments and feelings over process and politics – to bolster the claims for which genus should rightfully bear the name acacia. Below, we review both the more objective arguments, as well as the rhetoric and politics in the preparatory technical documents prepared before the Vienna meeting and in the flurry of commentaries and editorials that emerged afterwards, up through the re-hashing of the issue at Melbourne.

The formal case *for* retypification was laid out by Orchard and Maslin (2003, 2005). First, they argued that a change to *Racosperma* would create a high level of inconvenience. They pointed out that acacias were so dominant in the flora of Australia that fully 6% of all plant species would need to change to *Racosperma* unless retypification took place. They pointed out that 1274 species would be subject to name changes were the acacia genus split into five new genera without retypification (Figure 2). By changing the type species from *Acacia nilotica* to *A. penninervis*, the number of name changes would be reduced to 392 species (231 names, for the new genera *Senegalia*, *Acaciella* and *Mariosousa*, would have to change either way). By retaining the genus *Acacia* for the more numerous Australian species, they argued that potential confusion and disruption would be minimised in terms of relabelling work for herbaria, nature walks, legislation, textbooks and databases around the world. They also made an issue of the grammatical impacts of using *Racosperma*: due to this word's neuter gender, its use would necessitate spelling changes to numerous species names, replacing the suffix –*a* with –*um*, such as *Acacia pycnantha* to *Racosperma pycnanthum*, or *Acacia aneura* to *Racosperma aneurum* (Maslin and Orchard 2004a). The retypification lobby less convincingly added that it would be easier to have Africa, Asia and the Americas rename (nearly) all their acacias than only renaming some of them.

The opponents of retypification made their official case in an article co-authored by 37 botanists from around the world, with the central argument that 'in a case as contentious and hotly debated as this one… simple priority should prevail' (Luckow et al. 2005:515; see also Smith et al. 2006). They claimed that the case for retypification based on the magnitude of inconvenience was made on spurious grounds, and that instead of counting numbers of species, it was necessary to count the number of countries or the total human population in the zones affected by retypification of the genus. They pointed out that adopting the name *Racosperma* would only affect Australia's 20-odd million people and people living in a few small Pacific islands. In contrast, retypification of the acacia genus to represent the Australian species would force name changes across several continents with a combined population of more than two billion people, and, as a result (among other things), necessitate updates to many more national floras and databases.

A second set of arguments focused on economic disruption. Retypification proponents pointed out that the commercial importance of Australian species far outweighs that of the rest. There are more than two million hectares of Australian acacia plantations outside Australia, with many species widely known in industrial forestry, in agroforestry and as ornamentals (Midgley and Turnbull 2003). They argued that 'large scale name changes would not only burden the above industries and activities with large overhead costs, but would also take considerable time and effort' (Maslin et al. 2003a:13).

In response, the critics of retypification accused the proponents of trying to monopolise the name '*Acacia* for a developed country at the expense of widespread changes across numerous developing countries' that could least afford the cost of such name changes (Luckow et al. 2005:516; see also Boy 2005; Moore 2006b). They noted that non-Australian acacias had substantial economic importance both in terms of numbers of people using the trees for various subsistence, medicinal and commercial uses, and in terms of their symbolic value for the tourism sector in eastern and southern Africa (Luckow et al. 2005). Moore (2006b:72) pointed out that:

> the Acacias are charismatic trees of the African savanna…, their silhouettes against the setting sun a savanna icon. It is a brand name, as valuable as champagne, and one that serves as the logo for one of Africa's largest banks. It is a key marketing attraction for the entire African tourist industry.

References to symbolic value and iconic status indicate that the arguments surpassed mere discussions of nomenclatural stability and minimisation of disruption. Other rhetorical arguments were made to bolster the claims. For example, the opponents of retypification asserted that since the Greek root word for acacia is *akis*, meaning thorn, it was ludicrous to retain this genus name for the largely thornless acacias of Australia (Moore 2006b). They also evoked the usage of common names, pointing out that Africa's thorny acacias are widely recognised and known as 'acacia' the world over, whereas the Australian acacias are commonly known as 'mimosas' in Europe and as 'wattles' in Australia.[14] Why, they asked, should one conserve the genus name *Acacia* for the Australian species when few refer to them by that name? Instead, they pointed out, it would be simpler to use the genus name *Racosperma* for nearly all the native species in Australia and neighbouring countries (Pedley 2004; Luckow et al. 2005; Boy 2005; Moore 2006b).

Sentiments of place, merged at times with national or continental chauvinism, were also widely called on to justify both positions. Orchard and Maslin (2005) noted that acacia flowers were a prominent element of the winter and springtime landscapes of the populated parts of Australia, when their brilliant yellow blossoms provided spectacular splashes of colour against the rich green hues in pastures and eucalyptus forests. They pointed to the iconic status of wattles for the nation of Australia. As Libby Robin and Jane Carruthers have documented (Robin 2002, 2007, 2008; Carruthers and Robin 2009; Carruthers et al. 2011), the wattle was crowned as the floral symbol for the young federation of Australia that was seeking to establish its distinctive identity in relation to Britain. According to Robin (2008:3), wattles were seen to symbolise 'a fair and equal Australia' because of their widespread presence and distribution across the whole continent. The golden wattle, *Acacia pycnantha*, was finally formally named the national floral emblem in 1988 during bicentennial celebrations of European settlement. It appears on the national coat of arms (Figure 4), in the design of the Order of Australia medals, and in the green and gold colours of the national sports teams. The first day of spring (1st September) in Australia is called Wattle Day, when politicians pin sprigs of wattle bloom on their lapels (as they also did in October 2002 to commemorate the Australian victims of the terrorist bombings in Bali). All of these factors, proponents of retypification asserted, were additional reasons to justify keeping the Latin name *Acacia* for Australian species.

Their opponents made similar arguments. Botswana's Andy Moore (2006a:51) noted that acacias are 'the icons of the savannah' and rhetorically stated: 'Put out a picture of an acacia silhouette against a setting sun, and ask anyone to say whether this makes them think of Africa, Australia, or Antarctica.' Similar assertions claiming the African-ness of acacia were made in *Swara* (Boy 2005) and in letters to *Veld and Flora* (Cameron 2006:51). As the exchanges between the two sides grew increasingly vitriolic, Pedley (2004) complained:

> the rest of the world will not only have to abandon the name *Acacia*, but will have to accept transfers of species to at least two genera, both with unfamiliar names. The lovely flat-topped trees of the African veldt will be *Acacia* no more, but *Vachellia*. About an equal number of African species will go to *Senegalia*. The situation in Asia is similar... the situation in the Americas is worse. (Pedley 2004:4)

By the time the Vienna meeting approached, the so-called rational arguments and application of rules for retypification of acacia had disintegrated into a slanging match over its iconic status in Africa and Australia and its association with a sense of national or continental identity. The secretary of the Committee for Spermatophyta accused both sides of chauvinism (Brummitt 2004). Pedley (2004:4) stated it bluntly, claiming that:

---

14   Continental Europeans use the common name 'acacia' to refer to an unrelated American tree (*Robinia pseudoacacia*). Acacias introduced to Australia tend not to be given the common name 'wattle': *Acacia nilotica* goes by the common name 'prickly acacia', while *A. farnesiana* is 'mimosa bush'.

Figure 4. Australia's current '1912' coat of arms, with stylised golden wattle Acacia pycnantha under the shield, emu and kangaroo.
Sources: www.itsanhonour.gov.au/coat-arms/index.cfm and www.anbg.gov.au/emblems/commonwealth-coat-of-arms.html, last accessed 13 March 2009.

> Preserving the name *Acacia* exclusively for Australian species smacks somewhat of jingoism, inverse colonialism, or a sort. Australia is in a great position. It is a rich country with a well educated botanical public that can absorb name changes with a minimum of fuss...

Bruce Maslin denied that his push for retypification was 'just about jingoism' (Woodford 2002:1). In their response to Pedley, Maslin and Orchard said the 'emotive argument' about the loss of flat-topped trees of the African veldt was 'misleading', and dismissed his accusations, saying 'Such comments are inappropriate and serve no useful purpose and as such do not really warrant a response' (2004b:10-11).

Despite accusing each other of being jingoistic or emotive about the retypification issue, neither party wanted to acknowledge that no amount of justification regarding species numbers, grammatical inconveniences, or economic value could truly make a case for one side or the other. As Dick Brummitt (2010:1925) commented, 'there are strong practical (if not nomenclatural) arguments on both sides; the magnitude of changes required either way is unprecedented'. Hence, both sides invoked sentiments about euphonic, well-deserved names and place-based chauvinistic rhetoric to stand their ground and to gain the sympathy of non-scientific audiences. When the decision to retypify was passed in Vienna, the winners returned to their scientific high ground, saying that their case had been validated on rational criteria, while the opponents highlighted non-scientific biases in procedure, irrationality and unequal resources and power. As the altercations over acacia clearly reveal, not only is there tension in the science of classification, but sentiment – otherwise known as 'subjectivity' or 'irrationality' – played an important role in framing the arguments over the naming, labelling and reclassification of the genus. While Brummitt (2010) called them 'mere emotional outpourings', they have had no mere role in initiating and fuelling the conflict.

Conclusion

| First Bruce: | *This here's the wattle – the emblem of our land. You can stick it in a bottle or you can hold it in yer hand.* |
| All: | *Amen!*[15] |

The battles of Vienna and Melbourne reveal the welter of human sentiment that is invested in the name 'acacia'. More generally, they highlight the disgruntlements that can surface with each new scientific development in botanical taxonomy. They demonstrate how these feelings took on particular rhetorical forms in the acacia battles – about the euphony of words, senses of place, territorial chauvinisms and procedural and personal politics. Botanical nomenclature is much more than a set of rules and conventions; it reflects a particular institutional and ideological history about the classification and naming of plants, and it is permeated by sentiment. Although taxonomists say that the botanical nomenclature process accords names on the basis of objective rules, the reality, at least in the case of acacia, was far from being so. In the end, despite not being quite so simple or clear cut, the acacia name war *was* seen by many as an ambit by 'Australia' to claim the euphonic name acacia for its native wattles, with an outraged opposition, particularly 'African', defending the rules of priority to maintain acacia for its thorn trees.

Despite the decision in Vienna, several authors writing about African acacias have pointedly stuck with the broad genus name *Acacia* (e.g. Cameron 2006; Dharani 2006). Rupert Watson (2007:193) expresses the long-drawn resistance to altering the genus name *Acacia* in the acknowledgements to his book *The African Baobab*:

> I end with a botanical footnote. I am well aware that Africa has lost the battle with Australia for the right to use the genus name of *Acacia*; however, I continue to use the old names for members of this genus, not through any sense of scientific stubbornness, but simply because these will continue in use in Africa for many years to come.

Watson's matter-of-fact assertion points to what Bowker and Star (1999:67-68) call the 'fault line' between scientific and folk classification. It is not a Great Divide, they say, but 'a fracture that is constantly being redefined and changing its nature as the plate of lived experience is subducted under the crust of scientific knowledge. This fault line is the ways in which temporal experiences – history, events, development, memory, evolution – are registered and expressed', by formal systems of classification. Taxonomic rules and procedures for naming plants may claim to be based on scientific objectivity and operate above the subjectivity of local and regional traditions, but they cannot avoid the sentiments of their own practising members. Taxonomists, it seems, are just as emotional as non-scientists when it comes to naming plants.

Debates surrounding the renaming of plants are bound to be fractious, messy and contradictory because of the experiences, memories and place-based sentiments that scientists, as people, and people as non-scientists (as 'folk' or 'community'), bring to the taxonomical

---

15   This proclamation, made by a stereotyped khaki-clad, knee-socked white bloke holding a wattle branch in bloom, features in an iconic sketch of Australia in the British comedy series, *Monty Python*, Episode 22 'How to recognise different parts of the body', filmed on 25 September 1970 and first aired 24 November 1970 (en.wikipedia.org/wiki/List_of_Monty_ Python's_Flying_Circus_episodes).

exercise. Bowker and Star (1999:326) argue for flexible classifications 'whose users are aware of their political and organizational dimensions and which explicitly retain traces of their construction... The only good classification is a living classification'. We would echo their view by saying that the only good name is a living name. As Helmreich (2005:119) points out in the case of naming the algae species found around Hawaii, 'it is ironic that, through oral traditions, the Hawaiian names have been perpetuated and usually *accurately applied* to the individual species, whereas three-fourths of the scientific names have been changed in the past 90 years'. The living scientific name for *Acacia* is *Acacia* in its broad sense, regardless of whether it is properly classified according to the rules of cladistics or any other system.[16] This, in part, is why the acacia name change has generated such furore and emotion among members of the international botanical community and generated outpourings of sentiment among non-botanists in Australia, Africa and many other parts of the world.

Brummitt (2010, 2011) noted that the unprecedented uproar over the acacia name required unconventional approaches to taxonomic rules so as to accommodate cases where particular plant names are associated with strong sentiments. While he suggests that bending the rules this one time would be 'unlikely to impact on other names in the future' (2010:1925), we suggest that aesthetic sentiments, territorial chauvinisms and personal agendas will always play a role in debates over classification and nomenclature. Molecular systematics has put many plant classifications, and hence names, under pressure, as have other scientific advances in the past and as others will no doubt do in the future. The crises that result, for acacia and other names, may only be truly resolved by finding ways to recognise and incorporate people's feelings for sounds, places and traditions in plant names – even in Latin names. Denying their importance by invoking the pretext of 'scientific objectivity' will only undermine the ability of the International Code of Botanical Nomenclature to serve as a universal system into the future.

Acknowledgements

We are grateful to various botanists interviewed for this piece for sharing their insights regarding botanical nomenclature and molecular systematics, and for engaging with us in enlightening discussions and correspondence. We thank Michelle Aitken for contributing interview material from South Africa, and the anonymous referees who provided useful comments and suggestions for revising the essay. As always, the errors and the opinions expressed remain our own.

References

Bailey, L.H. 1933 [1963]. *How Plants Get Their Names*. New York: Dover.

Bentham, G. 1842. Notes on Mimoseae, with a synopsis of species. *London Journal of Botany* 1:318-528.

Bentham, G. and Mueller, F. 1864. *Flora Australiensis*. London: Lovell Reeve and Co.

Bonneuil, C. 2002. The manufacture of species: Kew Gardens, the Empire, and the standardisation of taxonomic practices in late nineteenth-century botany. In: Bourguet, M.-N., Licoppe, C. and Sibum, H.O. (eds), *Instruments, Travel and Science*, pp. 189-215. London: Routledge.

Bowker, G. and Star, S.L. 1999. *Sorting Things Out: Classification and its consequences*. Cambridge: MIT Press.

---

16   For instance, proponents of a 'phylocode' advocate a completely new way to organise botanical nomenclature (Foer 2005), while proponents of a new, information-technology based biology predict a revolution in research based on a new 'Global Names Architecture' (Patterson et al. 2010).

Boy, G. 2005. The name of the acacia. *Swara (East African Wildlife Society)* 28(3):24-27.

Brenan, J.P.M. and Melville, R. 1960. The Latin name of the Black Wattle. *Kew Bulletin* 14:37-39.

Brockway, L.H. 1979. *Science and Colonial Expansion*. New York: Academic Press.

Brooker, M.I.H. and Kleinig, D.A. 2007. *Field Guide to Eucalypts*, 3rd ed. Melbourne: Bloomings Books.

Brown, G.K., Murphy, D.J., Miller, J.T. and Ladiges, P.Y. 2008. *Acacia s.s.* and its relationship amongst tropical legumes, tribe Ingeae (Leguminosae:Mimosoideae). *Systematic Botany* 33:739-751.

Browne, J. 1996. Biogeography and Empire. In: Jardine, N., Secord, J. and Spary, E. (eds), *Cultures of Natural History*, pp. 305-321. Cambridge: Cambridge University Press.

Brummitt, R.K. 2004. Report of the Committee for Spermatophyta: 55. Proposal 1584 on *Acacia*. *Taxon* 53(2):826-829.

Brummitt, R.K. 2010. Acacia: a solution that should be acceptable to everybody. *Taxon* 59(6):1907-1933.

Brummitt, R.K. 2011. *Acacia*: Do we want stability or total change? *Taxon* 60(3):915.

Cameron, B. 2006. We were conned! *Veld and Flora* March 2006, p. 51.

Carruthers, J. and Robin, L. 2009. Taxonomic imperialism in the battles for Acacia: identity and science in South Africa and Australia. *Transactions of the Royal Society of South Africa* 65(1):48-64.

Carruthers, J., Robin, L., Hattingh, J., Kull, C.A., Rangan, H. and van Wilgen, B.W. 2011. A native at home and abroad: the history, politics, ethics and aesthetics of Acacia. *Diversity and Distributions*. DOI:10.1111/j.1472-4642.2011.00779.x.

Clarke, H.D., Downie, S.R. and Seigler, D.S. 2000. Implications of chloroplast DNA restriction site variation for systematics of *Acacia* (Fabaceae: Mimosoideae). *Systematic Botany* 25(4):618-32.

Cook, H. 2005. Global economies and local knowledge in the East Indias: Jacobus Bontius learns the facts of Nature. In: Schiebinger, L. and Swan, C. (eds), *Colonial Botany*, pp. 100-118. Philadelphia: University of Pennsylvania Press.

de Wildeman, É. 1925. Acacia mearnsii. *Plantae Bequaertianae* 3(1):61-62.

Dear, P. 2006. *The intelligibility of nature*. Chicago: University of Chicago Press.

Dharani, N. 2006. *Field Guide to Acacias of East Africa*. Cape Town: Struik Publishers.

Foer, J. 2005. What if we decide to rename every living thing on Earth? Pushing phylocode. *Discover* (discovermagazine.com/2005/apr/pushing-phylocode/).

Foster, S. and Johnson, R. 2006. *Desk Reference to Nature's Medicine*. Washington: National Geographic Society.

Githahu, M. 2006. Did you know it was illegal to call this tree acacia? *Sunday Nation* (Nairobi) June 25.

Glazewski, J. and Rumble, O. 2009. A rose is a rose but is an Acacia an Acacia? Global administrative law in action. *Acta Juridica* 9:374-394.

Gledhill, D. 2008. *The Names of Plants*. 4th ed. Cambridge: Cambridge University Press.

Hanelt, P., Schultze-Motel, J. and Jarvis, C.E. 1983. Proposal to conserve Triticum aestivum L. (1753) against Triticum hybernum L. (1753) (Gramineae). *Taxon* 32(3):492-498.

Helmreich, S. 2005. How scientists think; about 'natives', for example. A problem of taxonomy among biologists of alien species in Hawaii. *Journal of the Royal Anthropological Institute* NS 11:107-28.

Hoffman, L. 2006. Letter. *Veld and Flora* March 2006, p. 51.

Hoffmann, J. 2001. Golden wattle loses its lustre: biological control of Australian acacias in South Africa. *Veld and Flora* 87(2):58.

Jobson, R.W. and Luckow, M. 2007. Phylogenetic study of the genus Piptadenia (Mimosoideae: Leguminosae) using plastid trnL-F and trnK/matK sequence data. *Systematic Botany* 32(3):569-75.

Koerner, L. 1996. Carl Linnaeus in his time and place. In: Jardine, N., Secord, J.A. and Spary, E.C. (eds), *Cultures of Natural History*, pp. 145-162. Cambridge: Cambridge University Press.

Lafuente, A. and Valverde, N. 2005. Linnaean botany and Spanish imperial biopolitics. In: Schiebinger, L. and Swan, C. (eds), *Colonial Botany*, pp. 134-147. Philadelphia: University of Pennsylvania Press.

Luckow, M., Hughes, C., Schrire, B., Winter, P., Fagg, C., Fortunato, R., Hurter, J., Rico, L., Breteler, F.J., Bruneau, A., Caccavari, M., Craven, L., Crisp, M., Delgado, A., Demissew, S., Doyle, J.J., Grether, R., Harris, S., Herendeen, P.S., Hernández, H.M., Hirsch, A.M., Jobson, R., Klitgaard, B.B., Labat, J.N., Lock, M., MacKinder, B., Pfeil, B., Simpson, B.B., Smith, G.F., Sousa, M., Timberlake, J., van der Maesen, J.G., van Wyk, A.E., Vorster, P., Willis, C.K., Wieringa, J.J. and Wojciechowski, M.F. 2005. Acacia: the case against moving the type to Australia. *Taxon* 54(2):513-9.

Macey, R. 2005. Oh baby, look what the change in weather blew in. *Sydney Morning Herald*, September 2, 2005.

Maclaurin, J. and Sterelny, K. 2008. *What is biodiversity?* Chicago: University of Chicago Press.

Maslin, B. 2003. Classification and phylogeny of *Acacia*: a synopsis http://www.worldwidewattle.com/infogallery/taxonomy/classification.php, last accessed 2008 December 18.

Maslin, B. 2011. Acacia and the IBC. *Australian Systematic Botany Newsletter* 146: 2-6.

Maslin, B. and Orchard, T. 2004a. Most Australian wattles likely to remain Acacia. *Acacia Study Group Newsletter* 92:5-7.

Maslin, B. and Orchard, T. 2004b. Response to Pedley's paper titled 'Another view of Racosperma'. *Acacia Study Group Newsletter* 93:7-12.

Maslin, B., Miller, J.T. and Seigler, D.S. 2003a. Overview of the generic status of Acacia (Leguminosae: Mimosoideae). *Australian Systematic Botany* 16:1-18.

Maslin, B., Orchard, A.E. and West, J.G. 2003b. Nomenclatural and classification history of Acacia (Leguminosae: Mimosoideae), and the implications of generic subdivision. www.worldwidewattle.com/infogallery/taxonomy/nomen-class.pdf, last accessed 7 June 2011.

McNeill, J. and Turland, N.J. 2010. The conservation of *Acacia* with *A. penninervis* as conserved type. *Taxon* 59(2):1-4.

Midgley, S.J. and Turnbull, J.W. 2003. Domestication and use of Australian acacias: case studies of five important species. *Australian Systematic Botany* 16:89-102.

Miller, J.T., Grimes, J.W., Murphy, D.J., Bayer, R.J., & Ladiges, P.Y. 2003. A Phylogenetic Analysis of the Acacieae and Ingeae (Mimosoideae: Fabaceae) based on trnK, matK, psbA-trnH, and trnL/trnF Sequence Data. *Systematic Botany*, 28(3):558-566.

Montgomery, K. 2006. Brand acacia goes to Australia. *South African Gardening*, July 2006, www.sagardening.co.za/forum/forums/thread-view.asp?tid=25@posts=3andstart=1, accessed 22 Nov. 2007.

Moore, A. 2006a. Economic and cultural pillage. *Veld and Flora*, March 2006, 51.

Moore, A. 2006b. Re-typing *Acacia*. *Veld and Flora*, June 2006:72-3.

Moore, G. 2007. The handling of the proposal to conserve the name Acacia at the 17th International Botanical Congress – an attempt at minority rule. *Bothalia* 37(1):109-118.

Moore, G., Smith, G.F., Figueiredo, E., Demissew, S., Lewis, G., Schrire, B., Rico, L. and Van Wyk, A.E. 2010. Acacia, the 2011 Nomenclature Section in Melbourne, and beyond. *Taxon* 59(4):1188-1195.

Moore, G., Smith, G.F., Figueiredo, E., Demissew, S., Lewis, G., Schrire, B., Rico, L., van Wyk, A.E., Luckow, M., Kiesling, R. and Sousa, M.S. 2011. The *Acacia* controversy resulting from minority rule at the Vienna Nomenclature Section: much more than arcane arguments and complex technicalities. *Taxon* 60(3):852-857.

Müller-Wille, S. 2005. Walnuts at Hudson Bay, Coral reefs in Gotland: The colonialism of Linnaean botany. In: Schiebinger, L. and Swan, C. (eds), *Colonial Botany*, pp. 34-48. Philadelphia: University of Pennsylvania Press.

Murphy, D.J. 2008. A review of the classification of *Acacia* (Leguminosae, Mimosoideae. *Muelleria* 26:10-26.

Murphy, D.J., Miller, J.T., Bayer, R.J. and Ladiges, P.Y. 2003. Molecular phylogeny of Acacia subgenus Phyllodineae (Mimosoideae: Leguminosae) based on DNA sequences of the internal transcribed spacer region. *Australian Systematic Botany* 16:19-26.

O'Neill, G. 2007. Flowering flow of ideas leads to solving thorny questions. *UniNews (University of Melbourne)*, 5-19 February 2007.

Orchard, A.E. and Maslin, B.R. 2003. Proposal to conserve the name Acacia (Leguminosae: Mimosoideae) with a conserved type. *Taxon* 52:362-3.

Orchard, A.E. and Maslin, B.R. 2005. The case for conserving Acacia with a new type. *Taxon* 54(2):509-12.

Parry, B.C. 2004. *Trading the Genome: Investigating the Commodification of Bio-Information*. New York: Columbia University Press.

Patterson, D.J., Cooper, J., Kirk, P.M., Pyle, R.L. and Remsen, D.P. 2010. Names are the key to the big new biology. *Trends in Ecology & Evolution* 25: 686-691.

Pavord, A. 2005. *The Naming of Names: The search for order in the world of plants*. London: Bloomsbury.

Pedley, L. 1986. Derivation and dispersal of Acacia (Leguminosae), with particular reference to Australia, and the recognition of Senegalia and Racosperma. *Botanical Journal of the Linnean Society* 92:219-54.

Pedley, L. 2003. A synopsis of Racosperma C.Mart. (Leguminosae: Mimosoideae). *Austrobaileya* 6(3):445-496.

Pedley, L. 2004. Another view of *Racosperma*. *Acacia Study Group Newsletter* 90:3-5.

Richmond, C.W. 1918. In memoriam: Edgar Alexander Mearns. *The Auk: A Quarterly Journal of Ornithology* 35(1):1-18.

Robin, L. 2002. Nationalising nature: wattle days in Australia. *Journal of Australian Studies* 73:13-26 and notes 219-23.

Robin, L. 2007. *How a Continent Created a Nation*. Sydney: UNSW Press.

Robin, L. 2008. Wattle nationalism. *National Library of Australia News*, January 2008:3-6.

Ross, J.H. 1980. A survey of some of the pre-Linnean history of the genus Acacia. *Bothalia* 13(1 and 2):95-110.

Schiebinger, L. 1996. Gender and natural history. In: Jardine, N., Secord, J.A. and Spary, E.C. (eds), *Cultures of Natural History*, pp. 163-177. Cambridge: Cambridge University Press.

Schiebinger, L. 2004. *Plants and Empire*. Cambridge: Harvard University Press.

Schiebinger, L. and Swan, C. (eds), 2005. *Colonial Botany*. Philadelphia: University of Pennsylvania Press.

Scott, J. 1998. *Seeing Like a State*. New Haven: Yale University Press.

Seigler, D.S. and Ebinger, J.E. 2006. New combinations in the genus *Vachellia (Fabaceae: Mimosoideae)* from the New World. *Phytologia* 87:139-178.

Seigler, D.S., Ebinger, J.E and Miller, J.T. 2006: The genus *Senegalia (Fabaceae: Mimosoideae)* from the New World. *Phytologia* 88:38-93.

Smith, G.F., van Wyk, A.E., Luckow, M. and Schrire, B. 2006. Conserving *Acacia* Mill. with a conserved type. What happened in Vienna? *Taxon* 55(1):223-5.

Smith, B. 2011. Wattle it be? Name claim for Africa or Australia. *The Age* (Melbourne), July 25.

Soltis, D.E., Chanderbali, A.S., Kim, S., Buzgo, M. and Soltis, P.S. 2007. The ABC Model and its Applicability to Basal Angiosperms. *Annals of Botany* 100(2):155-163.

Spencer, R., Cross, R. and Lumley, P. 2007. *Plant Names: A Guide to Botanical Nomenclature*, 3rd ed. Melbourne: Royal Botanic Gardens and CSIRO Publishing.

Thiele, K.R., Funk, V.A., Iwatsuki, K., Morat, P., Peng, C.-I , Raven, P.H., Sarukhán, J. and Seberg, O. 2011. The controversy over the retypification of *Acacia* Mill. with an Australian type: a pragmatic view. *Taxon* 60(1):194-198.

Turland, N.J. 2011. A suggested compromise on the nomenclature of *Acacia*. *Taxon* 60(3) (online early).

van Rijckevorsel, P. 2006. Acacia: what did happen at Vienna? *Anales del Jardín Botánico de Madrid* 63(1):107-10.

Vassal, J. 1972. Apport des recherches ontogéniques et séminologiques à l'étude morphologique, taxonomique et phylogénique du genre *Acacia*. *Bulletin de la Société d'Histoire Naturelle de Toulouse* 108:105-247.

Watson, R. 2007. *The African Baobab*. Cape Town: Struik.

Webb, C.J., Sykes, W.R. and Garnock-Jones, P.J. 1988. *Flora of New Zealand. Volume IV: Naturalized Pteridophytes, Gymnosperms, Dicotyledons*. Christchurch: DSIR.

Williams, R. 1980. Ideas of nature. In: Williams, R. (ed), *Problems in Materialism and Culture*, pp. 67-85. London: Verso.

Winston, J.E. 1999. *Describing Species*. New York: Columbia University Press.

Woodford, J. 2002. Wattle they call it? Icon with a name that came out of Africa. *Sydney Morning Herald*, November 9:9-10.

Yoon, C.K. 2009. *Naming Nature: The Clash Between Instinct and Science*. New York: WW Norton & Company.

10

Nature, culture and time: Contested landscapes among environmental managers in Skåne, southern Sweden

Lesley Head
Australian Centre for Cultural Environmental Research (AUSCCER), University of Wollongong, NSW, Australia
lhead@uow.edu.au

Joachim Regnéll
Kristianstad University, Kristianstad, Sweden

Introduction

Our increased understanding of 'Man's Role in Changing the Face of the Earth' (Thomas 1956) is one of the key scientific achievements of the second half of the 20th century. Human activities now appropriate more than one third of the Earth's terrestrial ecosystem production, and between a third and a half of the land surface of the planet has been transformed by human development (Vitousek et al. 1997). Humans are inextricably embedded in all earth surface processes, and often dominate them. These findings are increasingly being recognised in political and policy spheres, most notably in contemporary debates about climate change (IPCC 2007). Peter Kershaw's work has been an influential component of this achievement, particularly in alerting us to a much longer potential timeframe of human entanglement through hunter-gatherer use of fire. He has forced us to think differently about cultural landscapes, and his research findings have persistently challenged the ideal of pristine wilderness.

Although research now clearly shows the variety of ways in which culture and nature are closely embedded, to the point of challenging their constitution as separate entities, most Western jurisdictions still attempt to manage them separately. For example, environmental management frequently distinguishes between natural heritage, managed by biological scientists, and cultural heritage, managed by archaeologists and related professionals. There are many spatial implications of this dualistic approach: 'sites' or 'reserves' tend to be conceptualised and managed separately from their broader landscape context; and cultural and natural heritage values tend to be opposed rather than complementary. Where humans are constituted as not belonging to nature, they and their activities are physically excluded from protected areas. This has had particular implications for indigenous peoples, who have often been excluded from their own country. The management challenges of the 21st century require us to take integrative approaches to landscape management, cognisant of both human and nonhuman activities and processes, and scales from local to global. An important but relatively undeveloped area of research is to understand how environmental managers in government bureaucracies are experiencing and negotiating these challenges.

In this paper, we focus on managers of the landscapes of the county of Skåne, southern Sweden. This is the most densely occupied part of Sweden and preserves a visible signature of human activity dating back thousands of years. It is the sort of place for which the term 'cultural landscape' was coined (Birks et al. 1988; Sporrong 1995; Emanuelsson et al. 2002). Although it is somewhat contested, there is strong evidence in the environmental history of southern Sweden that these forests have developed in the past few thousand years, with agricultural activities, particularly grazing, as an integral part of them (e.g. Berglund 1969, 1991; Cooper 2000; Bradshaw 2004, 2005; Bradshaw and Lindbladh 2005). Environmental histories in this region are built up from a combination of palaeoecological work and historical geography using old maps. If dualistic approaches to the management of culture and nature persist in such a demonstrably hybrid context as the cultural landscapes of Skåne, it demonstrates their resilience, and it is extremely significant to wider international debates. On the other hand, if alternative approaches are being developed here, there may be lessons for other parts of the world that are still using a separationist paradigm.

The main methodological window we use in the paper is interviews with professional environmental managers employed by the County Administrative Board (*Länsstyrelsen*), bearing responsibility for the management of nature reserves and cultural heritage management. Although environmental managers are powerful agents in decision-making and policy about such landscapes, there has been little systematic analysis of their understandings and practices.

We first contextualise our approach using recent culture/nature debates in geography and related disciplines. We then provide more specific background to the Swedish environmental context. An outline of the legislative and bureaucratic context of environmental management in Sweden shows how the situation in which these environmental professionals work already imposes certain divisions on them.

We then explain our methodology and focus on the results of interviews in which we analysed how these professionals negotiate questions of nature and culture in their work. These negotiations occur differently in relation to different habitats/landscapes. We compare forests and traditional agricultural landscapes, showing that the attraction of the primeval remains strong in relation to deciduous forests.

Dualistic and hybrid approaches to human-nature relations

In parallel to the accumulating scientific evidence of pervasive human influences in earth system processes, there has been substantially increased interest within the humanities and social sciences in non-human worlds. Ideas of hybridity and networks are being utilised to more effectively understand interactions between culture and nature, or more specifically to dissolve the distinction between them. The most well-known recent elaboration of ideas of hybridity has been in the work of Haraway (1991), Latour (1993) and Whatmore (2002) and their attempts to break free of the binary categories of society and nature. A further influence on these debates in the past few decades has been the increasing political voice of indigenous peoples. Indigenous people's struggle for representation has become an important influence on thinking about environment, nature and landscape (Howitt et al. 2006).

The clearest example of the critique of nature converging on a practical environmental issue is in the postcolonial reassessment of the wilderness ideal and associated environmental imaginaries (Cronon 1996; Head 2000; Braun 2002; Baldwin 2004). Definitions of wilderness have a long history of change, and a shift from negative to positive connotations. The 19th century romantic wilderness ideal – of timeless, unchanging and remote landscapes – underpinned conservation and national parks policy in frontier societies such as the United States and Australia over the past century. The challenge came from diverse lines of evidence, including palaeoecological and archaeological demonstrations of long histories of human occupation in changing environments, and indigenous voices for whom so-called wilderness areas are home (McNiven and Russell 1995; Langton 1998). In Australia, the settler encounter with indigenous understandings of land and country has profoundly challenged management frameworks (Howitt 2001; Adams 2004, 2008; Howitt and Suchet-Pearson 2006). One management response has been the development of 'cultural landscapes' as a land management category, as seen, for example, in the World Heritage listing process (Head 2010).

The power of the wild has been exerted not only in New World contexts where indigenous inhabitants could be conceptually and physically erased. Mels (1999, 2002) has shown how it informed the establishment of national parks in Sweden, with problematic consequences for both the Saami lands of the Arctic north (see also Beach 2001, 2004), and the forest parks of Skåne, such as *Söderåsen*. Mels argues that the concept of nature promoted in national parks and through the Swedish EPA[1] is one heavily informed by biological science views that exclude humans. The park principle 'remains committed to an image of parks as spaces of natural science rather than social convention' (Mels 1999:174). Parks such as Söderåsen, Stenshuvud and Dalby Söderskog had long histories of cultural engagement and transformation, indeed were 'to a substantial degree the product of human practices' (Mels 1999:170). They responded in unexpected ways to management plans that fenced them and left them to take care of themselves.

> How can the national parks of Skåne be defined as natural landscapes requiring careful protection from human intervention, while simultaneously needing active management to maintain a pastoral condition, to remove "unnatural" species and to provide unobstructed scenic viewpoints? Or should the parks and their nature be seen as cultural products? This alternative would bring about insurmountable problems for the park ideal, because its nature by definition is of a "natural", not "cultural" kind. (Mels 1999:173)

Participants in our study work mostly with nature reserves rather than national parks, but there are a number of areas of comparison with Mels' work. The national context, which he identified as extremely important, has also changed somewhat in recent years, with much habitat and species management now subject to European Union commitments such as Natura 2000.

It is beyond the scope of this paper to resolve the long-standing dualisms in Western thought. Rather, we explore how both the dualisms and challenges to them are being experienced and negotiated in the working lives of these environmental managers. As Castree argues:

---

1   *Naturvårdsverket.*

The baroque jargon of academia may confidently declare that there never was a Maginot line dividing natural things from social things. But in several walks of life people continue to speak and act as though such a divide were self-evident... there is a continuing need for close analysis of nature-talk in any and all realms of society. (Castree 2004:191)

By focusing here on a particular set of 'nature-talk', we illustrate the practical challenges that lie ahead of all of us. We can also identify the organisational and disciplinary spheres of influence where separationist ideas remain spatially powerful.

The Swedish context

Legislation

There are two major sets of laws covering Swedish landscape management, *Miljöbalken*, the Environmental Code, and *Kulturminneslagen*, the Heritage Conservation Act. The conservation of areas, species of plants and animals are regulated by the Environmental Code, whereas the conservation of ancient remains, churches, other buildings, and place names are within the Cultural Heritage Law (for overviews in English see Ministry of the Environment, Sweden (2000); National Heritage Board, Sweden (2000)).

In the Environmental law section 2, chapter 7, Nature conservancy, there are three major instruments of spatial conservation.

- National parks (§2-3) are areas owned by the state for the purpose of preserving larger areas of a certain landscape type in their natural or in all essentials unchanged conditions.
- Nature reserves (§4-8) have the purpose of preserving biodiversity, conserving valuable nature areas or meeting the demands of outdoor recreation.
- Cultural reserves (§9) have the purpose of preserving valuable cultural landscapes.

Swedish legislation concerning nature conservation was established in 1909 with the formation of nine national parks, ranging from large remote mountain areas in northern Sweden to small deciduous woodlands in southern Sweden. All were supposed to be 'natural' areas, with no or only slight interference by humans, and were therefore not to be managed in any way. A new nature conservation law in 1965, *Naturvårdslagen*, introduced the concept of nature reserves, and in 1967 the Swedish Environmental Protection Agency (*Naturvårdsverket*[2]) was formed.

In 1999, the parliament established 15 environmental objectives to emphasise the change from defensive politics against environmental threats such as pollution, to offensive politics, with the aim of handing over to the next generation a society where environmental problems had been solved. To reach these objectives, a large number of environmental laws were revised and brought together in the Environmental Code. Various national authorities are in charge of the objectives and the broader issues related to the objectives. The Swedish EPA is, for example, responsible for the 'natural environment', while the National Heritage Board is in charge of the 'cultural environment'. And the Swedish Board of Agriculture rules the objective 'A varied Agricultural landscape', whereas the Swedish Forest Agency is responsible for 'Sustainable

---

2   *Naturvård* and *naturskydd* are two Swedish terms which both may be translated nature conservation. However, the word *skydda*, means 'protect, shelter, defend', while *vårda* means 'take good care of' (with a slightly more positive meaning than the word *sköta*, which means 'manage', 'take care of'). *Naturskydd* is the older term from 1904, while *naturvård* did not come into use until 1958 (the Swedish National Encyclopedia).

Forests'. As most Swedish landscapes combine elements of several of these interests, for example where the forests have been part of the agricultural economy for thousands of years, there is considerable overlap between the interests of the conservation authorities. This overlap in responsibilities provides potential for both fruitful cooperation between managers of different backgrounds, and also conflict. A 16th objective, on biodiversity, was adopted in 2005 (SEOC 2006).

Organisation

Most of the policies of the national authorities are implemented at the regional level by 21 county councils. In county Skåne, this means that the issues of 'natural' and 'cultural' environment are handled within the Nature Conservation Section (*Naturskyddssektionen*) and the Culture Conservation Section (*Kulturmiljösektionen*) of the Environmental Department (*Miljöavdelningen*), while agricultural issues are managed by the Agricultural Department (*Jordbrukssektionen*). Forestry issues, including forestry measures within protected areas, are handled by the regional offices of the Swedish Forest Agency (*Skogsstyrelsen*), which are not located at the county councils.

When it joined the European Union in 1994, Sweden also signed EU nature legislation, including the Habitats Directive. Article 3 of this directive states that 'a coherent European network of special areas of conservation shall be set up under the title Natura 2000'. There are now more than 60,000 km² of Natura 2000 sites in Sweden, from the alpine in the north, through the boreal zone, to the nemoral zone of southernmost Sweden. The Natura 2000 concept is presented to the public as follows:

> Not all kinds of habitats and species of Natura 2000 can be preserved in the same way. Management will vary depending on which kind of values are to be preserved. Sometimes active management or restoration is needed, sometimes no alterations should be made. The basic principle is that a meadow should remain a meadow through mowing, while the forest will continue to develop to natural forest by being excluded from forestry. (Naturvårdsverket 2003) (translated to English by JR)

This provides an interesting example of a theme that reappears in our interview material – how the open cultural landscape (grasslands) is contrasted to the supposedly natural forest landscape.

Methods

The now well-established tradition of qualitative research into environmental values and behaviours in cultural geography, anthropology and related disciplines has produced analyses of different, often conflicting understandings of nature (Harrison and Burgess 1994; Trigger 1997; Trigger and Mulcock 2005; Gill 2006) between groups, as well as wider societal discourses (Dekker Linnros and Hallin 2001). Important recent studies have examined environmentalists (McGregor 2004). Among the wide range of groups that can be considered stakeholders in environmental issues, managers and bureaucrats within government agencies are among the least studied. An important Nordic exception is Emmelin's (2000) questionnaire study of professional cultures within Nordic environmental administrations.

Our interviewees came from the Nature Conservation Section (*Naturskyddssektionen*) and the Culture Conservation Section (*Kulturmiljösektionen*) of the Environmental Department (*Miljöavdelningen*). The initial research strategy was to compare the experiences and thinking of 'nature' and 'culture' professionals working in broader landscape management, where a clear nature/culture delineation was likely to be problematic. We undertook semi-structured interviews with 13 people from these sections, identified by their section directors as having responsibility for landscape issues and reserve selection and management policy. Although small in number, this represents an almost complete sample of the designated group, the exceptions being several people who were on leave or otherwise unavailable during the interview period, the winter of 2005-06. Interviews were done in the offices of *Länsstyrelsen* in Malmö and Kristianstad.

Ten were seen by themselves and their colleagues as 'nature people', three as 'culture people', the proportion reflecting the dominance of nature people in issues to do with landscape management. (We excluded cultural heritage managers with exclusive responsibility for building conservation.) Nature people tended to have an educational background in biology, ecology and/or physical geography, while culture people had studied archaeology, history and/or cultural geography. A number of individuals had generalist backgrounds. There was a mixture of ages, from recent graduates to an imminent retiree, and nine men and four women. Such direct and full access was only possible with the support of the section directors, although they declined to be interviewed themselves. This unique research opportunity also created some dilemmas. For example, participants were aware that although they would be formally anonymous, their colleagues would likely be able to identify their opinions. All were happy to proceed on this basis, a number stating adamantly that they would stand by their statements.

Interviews were conducted in English by the two authors, who bring both outsider and insider perspectives to the Swedish situation in general, and to Skåne environments in particular. Initial questions covered participants' training and background, their current job responsibilities, landscape management strategies, the ways the culture/nature distinction is important (or not) in their work environment, departmental organisation, and issues of scale (county/nation/EU). Questions about specific examples they had worked on were used to lead into more conceptual discussions. Interviewees were all well educated in English, but were encouraged to switch to Swedish when necessary. Interviews were transcribed in full, sections translated into English as needed, and analysed for the dominant themes, which are discussed below. The terms nature (*natur*) and culture (*kultur*) have more or less parallel meanings in Swedish and English, but a number of other terms needed conceptual as well as linguistic clarification. These are explained in the following sections as necessary.

Nature and culture

All interviewees expressed great passion for their work and commitment to the broader endeavour in which they were engaged. Most expressed frustrations of one sort or another at the bureaucratic and legislative barriers to effective work. A number had suggestions about how the organisation could work better to manage landscapes in a more integrated way. The widespread feeling of being understaffed, overworked and frequently reorganised would be shared in many similar organisations today. As one participant laughed when asked how his thinking had changed over time, 'As I said, I haven't had time to think.'

Many could trace their involvement in environmental work to childhood experiences of country life, and/or involvement in community environmental organisations. All recognised the complexities of nature/culture entanglements, with a variety of different positions on whether and where a line should be drawn between them. We outline the diversity of views here, and then show how they are expressed in particular environments.

Nature nature and culture nature

There was a strong sense among participants of the pervasive and long-term influences of humans on the Skåne landscape. This has led to a vernacular expression that 'there is no nature in Skåne'.[3] While participants did not usually go to those lengths, most saw Skåne as having a different sort of nature. The relevant comparison was either spatial – the north of Sweden, where 'real' nature is considered to still exist; or temporal – a past time before significant human impacts. These distinctions were summarised by one participant as 'nature nature' (untouched nature, in the north), 'nature culture' (*natur påverkad* = impacted nature, nature formed by people) (this includes avenues, fields, grazing areas, meadows, stone walls, open ditches and earth walls, i.e. a variety of agricultural situations that have become important contexts for biodiversity preservation), and 'culture culture' (cultural heritage sites such as houses, buildings, churches, archaeological sites).

The middle ground idea of 'nature culture', or 'culture nature', had a variety of expressions in the interviews, coming from both nature and culture people:

> all nature is human made, at least in this province… people have used the landscape during the… last 10,000 years.

> a natural landscape doesn't exist in this part of Sweden.

> I don't think there is so much nature in Skåne… I think you have to go to the north of Sweden to see nature… that doesn't have as much impact from human beings… because all the woods are in some way cultivated.

> In Skåne, where you have high cultural values, you also have high nature values, many times, they're connected… also in the forest, not only in the grass land.

All made reference to the history of the landscape and its utilisation by people, first as hunter gatherers but more particularly as agriculturalists. One identified the enclosure period as the relevant temporal boundary, with traditional land uses before that time falling under the umbrella of 'nature'. The relationship with humans can also change the status of non-humans, as in the example 'horses are nature when grazing, but culture when someone is riding them'. Despite the acknowledged entanglements between nature and culture, only two participants mentioned the possibility of nature being in the city.

There was a widespread feeling that the difference between a *naturreservat* (nature reserve) and *kulturreservat* (cultural reserve) is a purely administrative one, the only difference being that nature reserves rarely include buildings. Many argued that there should just be one type of reserve in which both natural and cultural features could be protected. For example, 'it's not necessary to point out whether it's a nature reserve or a culture reserve, it's just a reserve, because all nature is humanmade, at least in this province'. Differences arose, however, in the specifics of management, whether inside or outside reserves. The dilemmas of managing various aspects of the continuing human presence leads to a set of contested landscapes, the most contested of which are the deciduous forests:

> beech forest, they're cultural in Skåne… they are no more natural than *rapsfält* [a field of oil seed rape].

Although we did not pursue it in the interviews, and do not explore it further in this paper, it is worth noting that the ideal of a pristine, untouched nature in the north of Sweden is itself

---

3   *Det finns ingen natur i Skåne.*

a highly contested notion.

> It's a little bit different, between northern and southern Sweden, because in the north, I think... most land is nature in that case, but in the southern part, we have just a small piece of nature and the rest is the humanmade landscape.

For example, such views erase the presence and aspirations of Saami people (or include them in nature) (Beach 2001, 2004) and also ignore the agro-industrial nature of forestry in the boreal forests. This wilderness ideal can be seen to represent 'a flight from history' in much the same way as Cronon has argued for colonial societies such as the USA (Cronon 1996; see also McNiven and Russell 1995 for Australia).

The deciduous forests

The most contested of the landscapes under discussion are the deciduous beech and oak forests, confined in Sweden to the more temperate areas of the far south. For this and other reasons, including its time as part of Denmark, Skåne is often referred to as the most 'continental' part of Sweden. Preservation and restoration of these deciduous forests is high on the EU agenda given their decimation in most areas of continental Europe.

Despite a long history of human occupation, interaction and transformation, forests are more likely than open landscapes to be thought of, or managed, as 'nature nature'. The main exception is the cultural heritage people, and some older ecologists with long practical management experience. This distinction was argued by one participant to be reinforced by the (northern) Stockholm perspective of *Naturvårdsverket*, under the influence of traditional understandings of biology:

> *Naturvårdsverket* don't see humans as a natural part..., especially not when it comes to forest habitats. The grassland, they understand that they do need to have humans that have cattle and so on, but they don't see the human as a natural part of the dynamics of a forest.

Southern forests are seen as different to northern taiga ones because the human activities have led to the characteristics that are now valued:

> [the northern ones have] been impacted but the values you have there are not dependent on the impact of humans, on the contrary.

The differences between the northern boreal and southern deciduous forests are further seen in the way people used the following terms. *Urskog* – old wood, untouched forest, primeval forest – was seen not to exist any more. *Naturskog* is natural forest, or a nature wood. 'Nature wood is a wood that hasn't been used by man for a very, very, very long time, but we don't actually have those forests in Skåne, in Sweden maybe up in the mountains.' Some of the study participants see themselves as trying to re-establish *naturskog*, forest with a minimum of human impact, in order to still have biologically interesting forest. *Betesskog* (wood pasture, or grazed forest) is seen as the most cultural type of forest, or as one person described it, a 'fruit garden' for pigs:

> the main purpose with the beech trees was to produce acorns/mast [for] the pigs in the

forest…They were similar to a fruit garden, after the same principles the acorns were central in that type of management… timber, [was] not so important.

The idea of some forests as agricultural landscapes, as described above, is not widely favoured. The same participant showed us a cartoon he uses regularly in seminars that summarises the critique that he feels he has to defend against. The caption translates as '(Enclosed) wood pasture is a bad mixture of forest and pasture not good for either trees or cows'. Thus the conceptual purity of the forest is maintained by excluding both humans and their grazing animals.

Fri utveckling

The concept of *fri utveckling* (= free development = hands off management) is an important key to understanding the differences in attitude and practice. The concept is frequently used in the guidelines for the county councils regarding management plans for Natura 2000 areas. All deciduous-forest habitats of Sweden, with the exception of '9070 Fennoscandian wood pastures', are supposed to be managed by the hands-off regime (*Naturvårdsverket* 2007), sometimes, however, with the somewhat contradictory advice to clear areas around light-demanding large oaks.

In contrast to detailed discussion of the forest history, where both the long continuity of human influence in the forest and late 19th and 20th century changes in the forested landscape (such as the introduction of forestry and the decline in forest grazing) are described, free development is more promoted than discussed. A specific example of this lack of discussion, when linking the description of former land use and biological values to the proposed management style, is the management plan of the national park Söderåsen. After saying that 45% of the area contains cultural elements such as ancient fields, clearance cairns and stone walls (p. 10), the first stated aim of the national park is to protect the 'natural vegetation' for 'free, natural development' (p. 11) (*Naturvårdsverket* 2003).

Invoking the currency of biodiversity, plant ecologists among our sample were particularly keen to allow mixed deciduous forests to develop without human interference. The pressure to do this came not only from *Naturvårdsverket* but from the requirements to fulfil Natura 2000.

We have to leave a lot of forest for internal dynamics or free development, sort of. To come closer to a natural state of forest… no large extraction of wood, increasing the dead wood in the area, and having a forest with several layers… so going from one farm-like forest to… what could be emerging to be a natural forest.

The opposite view, held mostly by the cultural heritage people, is that beech forest left to its own devices will change to something else. If you want it to stay as a beech forest, it is necessary to actively intervene, through grazing, for example. They see the biologists as denying the long cultural history of the forests. The cultural heritage people want to preserve the history of the forests, but increasingly feel they have to use the language of biodiversity to support their arguments. Thus they talk about a different suite of species being protected under a managed forest regime. The National Heritage Board has also promoted the concept of 'biological culture-heritage' (*biologiskt kulturarv*) during recent years to include the biological remains of former land-use meadows, pollards and grazed forest (e.g. Emanuelsson 2003).

In fact, even those who were in favour of re-establishing multi-aged forest stands recognised that *fri utveckling* was not, in fact, free or hands off, but another type of management:

In the main part of Sweden you get most of the values if you leave the [spruce and pine] forest alone, but down here it's not so, it's not that easy to decide what will be the best in the long run, because you have the hands of man all over the landscape, and it's been so for a very long time.

A number of participants felt that, whatever the rhetoric, *fri utveckling* was in fact favoured because it is cheaper.

Preserving the past vs hurrying up nature

Although several of the 'nature tribe' refer in an almost romantic sense to forests before people, there is, as discussed in the introduction, little empirical support for this view in the environmental history of southern Sweden. In fact, the cultural people were more likely to invoke the past in forest management discussions, with the plant ecologists more often referring to the present and the future. In a context where environmental issues are understood as urgent, the ecologists see a need to reestablish this biodiversity more quickly than the several hundred years it might otherwise take. Thus they talk about the 'fast development of the natural' and 'hurrying up nature'. In another instance, an ecologist said that she did not think often of the people who had lived there in the distant past. 'I'm more interested in animals and the present, I see more that than I see … the people before.'

There is also debate over whether this means a new sort of nature is being created.

The grazing disturbance is very important I think, because if you are leaving it to free development, you are creating [a] type of landscape of which we know very little, and [which] is completely new. The grazing disturbance has a very, very long continuity, and I think it's very important from many aspects.

For another nature person, the establishment of such new communities was important because Skåne is the only part of Sweden where climatic conditions allow the possibility of deciduous forest:

To keep some areas, we would like to suggest that the areas should develop freely, with[out] any management plan, and that would be… natural forest, not a virgin forest, because we don't have virgin forests in this part of the country…

(Interviewer) Is it creating something new then…?

Yes, no it's quite new, well it existed during the Bronze Age, perhaps, but not later, because it has been used by people during all the time…We have to start from the beginning.

Restoration work was seen as urgent, and also as requiring investment in areas of currently low biological value but with great potential.

We have to start with forests that today are used for forestry, and… with very low biodiversity, to be able to create a high diversity, because it takes about one hundred years to reach that stage of forest development, but we can't wait until, let's say, until 2015 to start with that work, because in that case those areas that we today are interested in are gone, or have developed in another direction.

They acknowledged a need to educate the public about what would become a multi-aged forest, since more open, managed and tidy forests are very popular for recreational purposes, particularly in spring.

> Sometimes we get in conflict with *friluftslivet*[4] because we want to have a lot of dead wood for the insects and the bugs and we need to have big old trees lying everywhere but people that are going out with the dog for a walk… they like a beech forest with high nice trees and so on, and the sun.

Time in the job and lengthy field experience influenced nature people towards favouring more hands-on management.

> The people that work with the management of nature reserves … outside… see what happens… they think that it should be managed much more than those that are working with the plan… but I think it's changing.

The traditional agricultural landscape

The more obviously agricultural landscape is the area referred to as culture nature or nature culture. We distinguish here between two parts of this landscape: first, grazing areas and meadows, in which there is widespread acceptance of the need to maintain traditional practices, or some proxy thereof, in order to maintain biodiversity; and second, human constructions such as stone walls, which have become sites of biodiversity maintenance and are contested in terms of whether they should be managed for this or for cultural heritage.

Grazing lands and hay meadows
There is widespread recognition among ecologists that traditional management, or some replica thereof, is important to biodiversity conservation in the so-called semi-natural grasslands.

> The main reason for restoring these man-made grasslands is their exceptionally high species richness at small spatial scales… A prerequisite for keeping high species richness is to continue grazing, as the number of species drastically decline on grasslands when abandoned. (Lindborg 2006:957)

These open environments were the least contested among our participants, who all acknowledged the integral role of cultural activities. For example, one passionate animal ecologist with current responsibility for developing the management plan for a nature reserve in coastal grasslands talked of the importance of maintaining grazing in order to prevent reed encroachment on important bird habitat.

> … apart from all the plants that you wouldn't have there because the reed is so tall… the waders are dependent on… the short grass for foraging and… breeding… This marsh land here is so special, so like the highest values is for… when it comes to animals, the waders, to keep them.

---

4   Outdoor recreation.

Cultural remains as biodiversity habitat

More contested are avenues, stone walls, open ditches, earth walls and pollarded trees, all of which are obviously the outcome of past human activity. Because of their longevity in the landscape they have become, or protect, habitats supporting biodiversity. The cultural people consider the nature people to be appropriating these remains to the domain of nature, with insufficient recognition of their cultural heritage value. One of our discussions concerned the example of avenues of trees (*allés*).

> You have the trees, they are planted by man… with one purpose… it's a very big part of an open landscape, you see them [from a] long [way]. But these trees are not part of culture any more, it's nature now… When we want to preserve this avenue or… take down some trees, we can say OK, we think you should have that kind of tree and not that one, we are all the time on… nature's area, we have to play with the nature people, but they are making the final decision.

For this participant, responsibility for failing to recognise these more vernacular parts of the cultural landscape lay with the government who, in the laws of '*biotopskydd*' (biodiversity protection), had designated such remains important for biodiversity. By the same token, their cultural significance is not necessarily recognised in the cultural heritage law, which focuses on buildings and archaeological sites more than 100 years old.

An ecologist recognised the cultural remains (earth walls, pollarded trees, mounds) in the landscape she was responsible for, but 'those things don't really have to be managed in any different way, they are just kind of there'. These remains were contentious not only in what should be done about them, but in an organisational sense, with both nature and culture people accusing the others of neglecting or not understanding them.

The different attitudes to forested and open agricultural landscapes have some parallels in two societal discourses of resistance to the Öresund Link between Sweden and Denmark (Dekker Linnros and Hallin 2001), suggesting they are indicative of more widely entrenched understandings. The first, referred to as 'Fertile Earth[5]', emphasises 'conservation of soil as an important resource for future agriculture' (Dekker Linnros and Hallin 2001:394). It is grounded in a vision of Skåne 'where small-scale, dispersed patterns of settlement and a 'balanced' relationship with nature are the prerequisites for achieving a 'Good Society' (p. 396)'. The second, 'Protect Nature', calls for the protection of original, untouched nature. Although this discourse does not include a narrative about society, it is reinforced by the (human) spatial practice of 'being out in nature'. For example, some of the strongest advocates of this discourse are members of the Field Biologists. Although Dekker Linnros and Hallin were discussing these discourses in relation to Skåne as a whole, rather than particular environments within it, it is clear that the forest continues to be a site, perhaps the last possible site, where the Protect Nature discourse has purchase.

Currencies

Knowledge and understanding of environmental history was not in itself broadly accepted as a basis for future management strategies, so several people commented that they had to make their arguments for continuation or reestablishment of historical practices using the currency of species protection. For example, one advocate of grazing practices explained how his argument for putting cattle in a nature reserve was strengthened by the discovery that the beetle

---

5   In Swedish, the same word *jord* is used for both 'soil' and 'earth'.

läderbagge (*Osmoderma eremita*) was favoured in other areas by more open forest conditions. In these situations, it could often become a contest between different suites of species. Plant communities were considered to be treated as more important in the management plan process than a holistic view of landscape, and more important than ancient remains.

Cultural people felt they had no equivalent of IUCN red-listed (threatened) species, nor of quantitative measures of significance. While a large number of buildings are protected by the law of '*byggadsminne*' (built heritage), there was argued to be a lack of strong laws 'in the middle', for the cultural landscape.

> When you're dealing with a cultural landscape, you're not dealing with objects, one there and one there, you have communication between all parts of the landscape... I think it's easier to just pick one tree or one forest or something like that but we have to deal with the structures and the... process in the landscape.

This was also connected to what was seen as the greater political power of nature and the nature people, which went along with greater funding.

On the other hand, nature people saw themselves as having considerably broadened their perspective in the past decade or so. Whereas in the past they had focused on flora and perhaps birds, today's understanding of biodiversity was seen as much broader, considering 'sites that are important to beetles, butterflies, birds, mushrooms, lichens, mosses' as well.

Species are also the currency of labour allocation, since each endangered species needs an action plan. One participant said with some weariness, 'in Skåne, we are privileged with approximately 100 of these action plans'. While this is seen as an advance over times in the past when biodiversity values were not recognised at all, it would sometimes be preferable 'to look at a higher level at the landscape':

> If you look at meadows, you have a vast diversity of species that are threatened, each of these gets their own action plan instead of the habitat getting an action plan. So if you are trying to save one species, you are threatening another. For example, we have *Crex crex*, '*kornknarr*' which is a bird that likes meadows in a degenerating phase, but in the same place you have a lot of plants, that need yearly management, so they have a problem!

Conclusions

Dualistic conceptions of nature and culture remain firmly entrenched in the management of Skåne landscapes, although there are also sources of challenge.

Sweden's legislation and administrative organisation provides the setting for significant divisions between nature and culture. Both 'nature people' and 'culture people' demonstrated an understanding of Skåne as a hybrid landscape that has experienced human entanglement and influence over many thousands of years. In contrasting this with the north of Sweden, both groups demonstrated the power of an ideal, untouched (or less touched) nature existing somewhere else. Culture people and nature people differed most obviously in how they used the past and invoked time. Culture people tended to value the past for its own sake and as a guide to future activity. Nature people's temporal reference was more often the future, including when this involved creating new landscapes and forests. In this respect, they recognised and acknowledged the role of humans in contemporary management practice. None of our participants was a passionate advocate of *fri utveckling*, if this is understood as completely 'hands off' management. Where they were in favour of a version of it, they recognised it as a managed process involving considerable human investment. Those who opposed it included nature people whose many years of field experience had led them to see human activities in forests as essential to both biodiversity and cultural protection. Each group was inclined to conceptually appropriate contested areas such as stone walls for their respective 'sides', but in practical management terms, the barriers to working together on these sorts of sites do not seem insurmountable. Their perspectives have been shown to be influenced by educational background, childhood experiences and grappling with practical issues on the ground in the course of their work.

However, there are differences in the way different landscapes are understood. There is greater resistance to recognising the reality of the human role in forests and wooded landscapes than open landscapes such as grasslands and meadows. The ideal of the primeval virgin forest so powerful in popular culture holds sway among science-oriented environmental managers beyond what can be argued on the historical evidence. We speculate that it is connected to the power of trees in both scientific and popular imaginations (Jones and Cloke 2002), and to the more visible connection between open landscapes and human activity in the form of agriculture.

The historical power of 'nature' as traditionally understood within the biological and ecological sciences, relative to culture history, has been exacerbated by EU agreements such as Natura 2000, which focus on the (non-human) species as the relevant currency. This is a somewhat paradoxical outcome given that traditional understandings of nature are often argued to set humans apart from and above the rest of the natural world. The intention of such agreements has been the protection of vulnerable species in rapidly changing landscapes across national and other boundaries. It would be counterproductive, however, if increasingly separationist approaches to particular species' protection occurred at the expense of dynamic and resilient total landscapes.

To return to the theme of the volume, Peter Kershaw's work has been influential in international thinking about peopled landscapes and the challenges of the Anthropocene. Given the importance of disciplinary background as an influence on our research participants, one clear implication is on how we train future managers. It is increasingly recognised within ecology that past frameworks and conceptual understandings need revision to meet the challenges of the future (e.g. Hobbs et al. 2006, 2009). A variety of disciplinary perspectives will continue to be important, enhancing the capacity of students to approach wider cross-cultural issues. There are no simple solutions to these challenges, but understanding how they play out in different organisational settings will continue to be an important complement to palaeoecological research.

Acknowledgements

This research was undertaken when LH was King Carl XVI Gustaf Visiting Professor of Environmental Sciences at Högskolan Kristianstad, Sweden. We thank Högskolan Kristianstad for financial and logistic support. We thank all our interviewees for their enthusiastic participation.

References

Adams, M. 2004. Negotiating nature: collaboration and conflict between Aboriginal and conservation interests in NSW. *Australian Journal of Environmental Education* 20:3-11.
Adams, M. 2008. Foundational Myths: Country and Conservation in Australia. *Transforming Cultures eJournal* Vol. 3 No. 1.

Baldwin, A. 2004. An ethics of connection: social-nature in Canada's boreal forest. *Ethics, Place and Environment* 7:185-194.

Beach, H. 2001. World Heritage and Indigenous Peoples – the Example of Laponia. In: Sundin, B. (ed), *Upholders of Culture Past and Present* (Royal Swedish Academy of Engineering Sciences (Kungl. Ingenjörsvetenskapsakademien-IVA) pp. 90-98. Stockholm: Elanders Gotab.

Beach, H. 2004. Political ecology in Swedish Saamiland. In: Anderson, D. and Nuttall, M. (eds), *Cultivating Arctic Landscapes. Knowing and Managing Animals in the Circumpolar North*, pp. 110-123. New York: Berghahn Books.

Berglund, B.E., 1969. Vegetation and human influence in South Scandinavia during Prehistoric time. In: Berglund, B.E. (ed), Impact of Man on the Scandinavian landscape during the Late Post-Glacial. *Oikos Suppl* 12:9-28.

Berglund, B.E. (ed), 1991. The cultural landscape during 6000 years in southern Sweden – The Ystad project. *Ecological Bulletins 41*.

Birks, H.H., Birks, H.J.B., Kaland, P.E. and Moe, D.E. (eds), 1988. *The cultural landscape: past, present and future*. Cambridge: Cambridge University Press.

Bradshaw, R.H.W. 2004. Past anthropogenic influence on European forests and some possible genetic consequences. *Forest Ecology and Management* 197:203-212.

Bradshaw, R.H.W. 2005. What is a natural forest? In: Stanturf, J.A. and Madsen, P. (eds) *Restoration of Boreal and Temperate Forests*, pp. 15-30. Boca Raton: CRC Press.

Bradshaw, R.H.W., Hannon, G.E. and Lister, A.M. 2003. A long-term perspective on ungulate-vegetation interactions. *Forest Ecology and Management* 181:267-280.

Bradshaw, R.H.W. and Lindbladh, M. 2005. Regional spread and stand-scale establishment of *Fagus sylvatica* and *Picea abies* in Scandinavia. *Ecology* 86:1679-1686.

Braun, B. 2002. *The Intemperate Rainforest. Nature, Culture, and Power on Canada's West Coast.* Minneapolis: University of Minnesota Press.

Castree, N. 2004. Nature is dead! Long live nature! *Environment and Planning A* 36:191-194.

Cooper, N.S. 2000. How natural is a nature reserve?: an ideological study of British nature conservation landscapes. *Biodiversity and Conservation* 9:1131-1152.

Cronon, W. 1996. The trouble with wilderness; or, getting back to the wrong nature. In: Cronon, W. (ed), *Uncommon Ground*, pp. 69-90. New York: W.W. Norton.

Dekker Linnros, H. and Hallin, P.O. 2001. The discursive nature of environmental conflicts: the case of the Öresund link. *Area* 33:391-403.

Emanuelsson, M. 2003. *Skogens Biologiska Kulturarv: Att Tillvarata Föränderliga Kulturvärden.* Stockholm: Riksantikvarieämbetet.

Emanuelsson, U., Bergendorff, C., Billqvist, M., Carlsson, B. and Lewan, N. 2002. *Det Skånska Kulturlandskapet.* Lund: Naturskyddsföreningen i Skåne.

Emmelin, L. 2000. Nordisk miljöförvaltnings professionskultur och några aktuella frågeställningar i miljöpolitken. (Professional Culture in the Nordic Environmental Administrations and Some Current Issues in Environmental Policy.) *Tidskrift for samfunnsforskning* 41:3.

Gill, N. 2006. What is the Problem? Usefulness, the cultural turn, and social research for natural resource management. *Australian Geographer* 37:5-17.

Haraway, D. 1991. *Simians, Cyborgs, and Women.* London: Free Association Books.

Harrison, C. and Burgess, J. 1994. Social constructions of nature: a case study of conflicts over the development of Rainham Marshes. *Transactions, Institute of British Geographers NS* 19:291-310.

Head, L. 2000. *Second Nature. The history and implications of Australia as Aboriginal landscape.* Syracuse: Syracuse University Press.

Head, L. 2010. Cultural Landscapes. In: Hicks, D. and Beaudry, M. (eds), *The Oxford Handbook of Material Culture Studies.* Oxford: Oxford University Press, pp. 427-439.

Hobbs, R.J., Arico, S., Aronson, J., Baron, J.S., Bridgewater, P., Cramer, A.A., Epstein, P.R., Ewel, J.J., Klink, C.A., Lugo, A.E. and USDA F.S. 2006. Novel Ecosystems: theoretical and management aspects of the new ecological world order *Global Ecology and Biogeography* 15:1-7.

Hobbs, R.J., Higgs, E. and Harris, J.A. 2009. Novel ecosystems: implications for conservation and restoration. *Trends in Ecology and Evolution* 24:599-605.

Howitt, R. 2001. Frontiers, Borders, Edges. Liminal Challenges to the Hegemony of Exclusion. *Australian Geographical Studies* 39:233-245.

Howitt, R., Connell, J. and Hirsch, P. (eds), 2006. *Resources, Nations and Indigenous Peoples.* Oxford: Oxford University Press.

Howitt, R. and Suchet-Pearson, S. 2006. Rethinking the building blocks: ontological pluralism and the idea of 'management'. *Geogr. Ann.* 88B:323-335.

IPCC 2007. *Climate Change 2007: Climate Change Impacts, Adaptation and Variability. Summary for Policymakers.* Working Group II Contribution to the Intergovernmental panel on Climate Change Fourth Assessment Report. http://www.ipc.ch/(Accessed 7.4.07).

Jones, O. and Cloke, P. 2002. *Tree Cultures. The place of trees and trees in their place.* Oxford: Berg.

Langton, M. 1998. *Burning Questions: Emerging Environmental Issues for Indigenous Peoples in Northern Australia.* Centre for Indigenous Natural and Cultural Resource Management, Northern Territory University, Darwin.

Latour, B. 1993. *We Have Never Been Modern.* New York: Harvester Wheatsheaf.

Lindborg, R. 2006. Recreating grasslands in Swedish rural landscapes – effects of seed sowing and management history. *Biodiversity and Conservation* 15:957-969.

Mels, T. 1999. *Wild landscapes: the cultural nature of Swedish national parks.* Doctoral Thesis. Lund University Press.

Mels, T. 2002. Nature, home and scenery: the official spatialities of Swedish national parks. *Environment and Planning D: Society and Space* 20:135-154.

McGregor, A. 2004. Sustainable development and 'warm fuzzy feelings': discourse and nature within Australian environmental imaginaries. *Geoforum* 35:593-606.

McNiven, I. and Russell, L. 1995. Place with a past: reconciling wilderness and the Aboriginal past in World Heritage areas. *Royal Historical Society of Queensland Journal* 15:505-19.

Ministry of the Environment (Sweden) 2000. *The Swedish Environmental Code*, Ministry Publication Series, Ds 2000:61.

National Heritage Board (Sweden) 2000. Heritage Conservation Act (SFS1998: 950). pdf file at http://hilebrand.raa.se/laws/hcact.asp

Naturvårdsverket 2003. *Natura 2000 Värdefull natur i EU.* Leaflet 91-620-8131-4, 16pp.

Naturvårdsverket 2007. *Naturvårdsverket skog1rev.pdf and skog2rev.pdf* (http://www. naturvardsverket.se/sv/Arbete-med-naturvard/Skydd-och-skotsel-av-vardefull-natur/ Natura-2000/Vagledning/Art--och-naturtypsvisa-vagledningar-for-Natura-2000-/ (Accessed June 2007)).

Sporrong, U. 1995. *Swedish Landscapes.* Stockholm: Swedish Environmental Protection Agency.

Swedish Environmental Objectives Council (SEOC) 2006. *Sweden's Environmental Objectives – buying into a better future, De Facto 2006. A progress report from the Swedish Environmental Objectives Council.* Stockholm: SEOC.

Thomas, W.L. (ed), 1956. *Man's Role in Changing the Face of the Earth.* Chicago. University of Chicago Press.

Trigger, D. 1997. Mining, landscape and the culture of development ideology in Australia. *Ecumene* 4:161-180.

Trigger, D. and Mulcock, J. 2005. Forests as spiritually significant places: nature, culture and belonging in Australia. *The Australian Journal of Anthropology* 16:306-20.

Vitousek, P.M., Mooney, H.A., Lubcheco, J. and Melillo, J.M. 1997. Human domination of earth's ecosystems. *Science* 277:494-499.

Whatmore, S. 2002. *Hybrid Geographies*. London: Sage.

II. Biogeography and Palaeoecology

11

The rise and fall of the genus *Araucaria*: A Southern Hemisphere climatic connection

Marie-Pierre Ledru
Institut de Recherche pour le Développement, Institut des Sciences de l'Evolution de Montpellier, Université de Montpellier 2, France
marie-pierre.ledru@ird.fr

Janelle Stevenson
The Australian National University, Canberra, ACT

Introduction

Understanding tropical sensitivity and its link with higher latitudes is a major issue for both climatologists and climate modellers. Moreover, changes in the floristic composition of tropical forests through time are of interest to ecologists wanting to understand the evolution of tropical biodiversity. Araucariaceae is a very ancient family of conifers dating to the Triassic. Its maximum diversity was reached during the Jurassic and Cretaceous periods, becoming extinct in the Northern Hemisphere at the end of the Cretaceous. Today, the genus *Araucaria* includes 19 species, 13 of which are endemic to New Caledonia, with another six distributed across Norfolk Island, eastern Australia, New Guinea, Argentina, Chile and southern Brazil (Enright and Hill 1995; Kershaw and Wagstaff 2001).

Among these 19 Southern Hemisphere species of *Araucaria*, five are endangered and three have a vulnerable status (UICN). All species have restricted distributions, which are relicts of a past expansion. These large trees, with a massive central trunk that commonly reach heights of 30-60 m, are mainly restricted to moist forests in the wet cool tropics (Figure 1) with a mixed lower strata of angiosperms. When temperatures are high, as in New Guinea or near Rio de Janeiro in Brazil, *Araucaria* grows at higher elevation (Kershaw and McGlone 1995).

Araucaria forest expansion and contraction during the last glacial/interglacial cycle is recorded in three long pollen records, all at a similar latitude: southern Brazil, northeastern Australia and New Caledonia. In northeastern Australia, at Lynch's Crater on the Atherton Tableland, the pollen record extends back 230,000 years and shows that rainforest expansion took place during wetter interglacial periods, replaced by drier rainforest and sclerophyll vegetation during drier glacials. The expansion of *Araucaria* forest is seen twice in this record, from ca. 190,000-130,000 years ago and from 80,000-45,000 years ago (Kershaw et al. 2007).

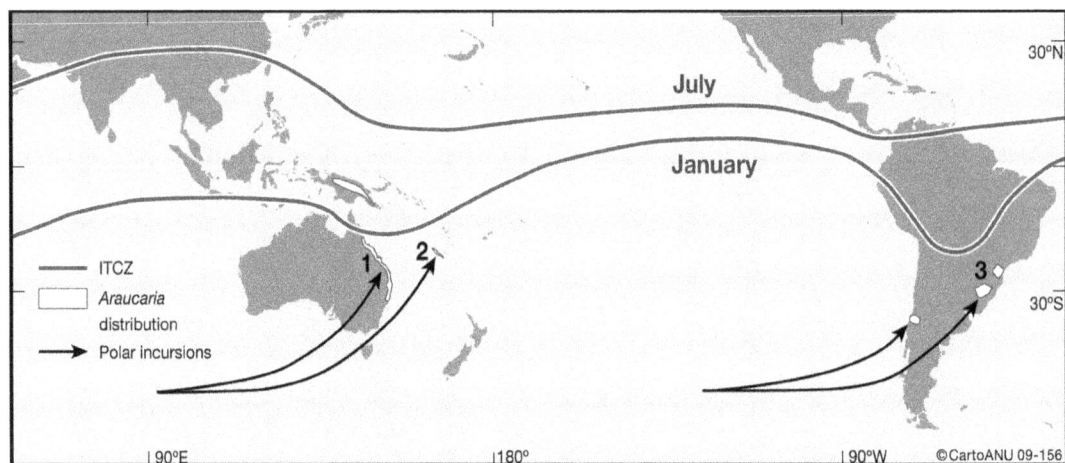

Figure 1. Modern distribution of Araucaria in the Southern Hemisphere (from Enright and Hill 1995), trajectories of the polar advections (black arrows) towards the continents, northern and southern extent of the Inter Tropical Convergence Zone (ITCZ) and localities of the palaeorecords: 1) Lynch's Crater, 2) Lake Xere Wapo, 3) Colônia Crater.

The disappearance of *Araucaria* at 45,000 years ago, synchronous with an increase in fire and a shift towards open sclerophyllous forest or savanna, was attributed to the burning activities of people in the absence of any significant change in global climate cyclicity (Kershaw 1978, 1986). The comparative marine record off the northeast Australian coast, ODP 820, illustrates that at a regional scale there was a trend of decreasing *Araucaria* pollen and increasing pollen from sclerophyll taxa over the past 250,000 years (Moss and Kershaw 2007), and attests to the continuous presence of fires in the broader landscape for the past 250,000 years. It is thought, therefore, that what are seen at 45,000 years ago in the Lynch's Crater record are changes that were already in motion before the human colonisation of Australia, but that these changes were accelerated by human impact (Kershaw et al 2007). Detected in both the Lynch's Crater record and the ODP 820 record is a non-Milankovitch 30,000-year frequency that is present in most major attributes and dominates both gymnosperm and charcoal records (Kershaw et al. 2003, 2007; Moss and Kershaw 2007), and which is thought to represent a modification of the precessional signal due to ENSO variability (Beaufort et al. 2003).

A long comparative record from New Caledonia contains a similar disappearance of *Araucaria* forest (Stevenson and Hope 2005). However, as this island was only colonised by people 3000 years ago, human disturbance is not a factor in explaining this vegetation change, nor is fire a component of *Araucaria* decline. As human impact is excluded until the late Holocene, the ENSO hypothesis was retained in spite of some chronological matching uncertainties.

In the pollen record from Colônia Crater in southern Brazil, neither human impact nor ENSO can be inferred to explain the disappearance of *Araucaria;* human arrival on the South American continent is estimated to be around 12,000 years ago and the site is located on the Atlantic side of the continent where the ENSO phenomenon is weak and dependent on other intrahemispheric linkages (Lau and Zhou 2003). Indeed, a spectral analysis performed on arboreal pollen frequencies from the Colônia Crater record shows only one signal, that of precession (Ledru et al. 2009). In addition, fire was not a driving factor, as no charcoal particles were recovered from the sediments. Consequently, a change in Southern Hemisphere climate is the remnant hypothesis to explain the decline of *Araucaria* at all three sites 45,000 years ago. In this paper, we explore this phenomenon in an attempt to establish the major contributing factor to *Araucaria* expansion and contraction during the late Quaternary.

Climate control on the modern distribution and composition of *Araucaria* forests

Australia

In Australasia, five species of *Araucaria* are observed today; one species is restricted to Norfolk Island, two species are restricted to New Guinea, and two species are in northeastern Australia (*A. bidwillii* and *A. cunninghamii*). Both species are found in the cooler subtropical forests of southeast Queensland, but apart from a few isolated occurrences of *A. bidwilli*, all of the araucarian forest patches further north, extending as far as New Guinea, are dominated by *A. cunninghamii* (Kershaw and McGlone 1995; Kershaw and Wagstaff 2001; Kershaw and Walker 2007). While found in the cooler tropical forests of the northeast Queensland's mountain tops, where the short dry season is attenuated by fog and cloud interceptions, *A. cunninghamii* is also an important emergent in the lowland dry vine forests of southeast Queensland and northern New South Wales, as well as growing on a number of islands off the northeast Australian coast (Enright 1995). The habitat of the species is therefore highly diverse, growing on a variety of substrates both in Australia and New Guinea, under average annual rainfall regimes varying from 850 mm to >4000 mm and in locations with mean annual temperatures from as low as 11°C to as high as 26°C (Enright 1995).

The Australian climate is dominated by the seasonal migration of the subtropical high system, the synoptic scale manifestation of the descending limb of the Hadley circulation (Hobbs 1999). During austral winter, the subtropical highs occupy their most northerly position over the Australian continent. This directs dry easterly trade winds over the north of the continent, while frontal lows embedded within the mid-latitude westerlies are directed over southern Australia, which experiences its wet season. During austral summer, the subtropical highs migrate towards the pole, occupying a position near Australia's southern margin. The subsiding air beneath the anticyclone brings primarily dry summer weather to southern Australia, as the westerlies are pushed south of the continent (Gentilli 1972). At the same time, the Inter Tropical Convergence Zone (ITCZ) moves south of the equator, bringing monsoon rains to Australia's north.

The Atherton region in northeastern Australia has a tropical climate, with an average annual rainfall of 1420 mm. The majority of Atherton's rainfall occurs during summer between December and March, a time when the monsoon trough is close to this region, with the intensity and frequency of rainfall attributed to the location of the ITCZ and the South Pacific Convergence Zone (SPCZ). Eighty percent of the region's precipitation falls during the summer and less than 10% in winter. However, frequent cloud and drizzle at higher elevations like Lynch's Crater mean that moist conditions are maintained throughout the year. The climate of the Atherton Tableland region is also affected by the ENSO phenomenon, with El Niño phases inducing a strong decrease in seasonal rainfall due to changes in the position of the convergence zones: the ITCZ and the SPCZ.

The prevailing winds at Atherton are east to southeasterly, with the strongest winds (excluding cyclones) occurring during April and August. During the winter months, the region can experience incursions of cooler air brought by westerly winds (Baines 1980) (Figure 1).

New Caledonia

Of the 13 species that grow in New Caledonia today, 11 grow in the southern massif (22°S), where topography and altitude ensure that annual rainfall is in excess of 1500 mm (Jaffré 1995). The southern massif is dominated by an ultramafic geology, resulting in a landscape with a characteristic vegetation type known locally as the maquis minier; a sclerophyllous, evergreen, light-demanding formation associated with ultramafic rocks. This formation can be composed of either shrubs or a woody-herbaceous combination with a dense layer of sedges that may be

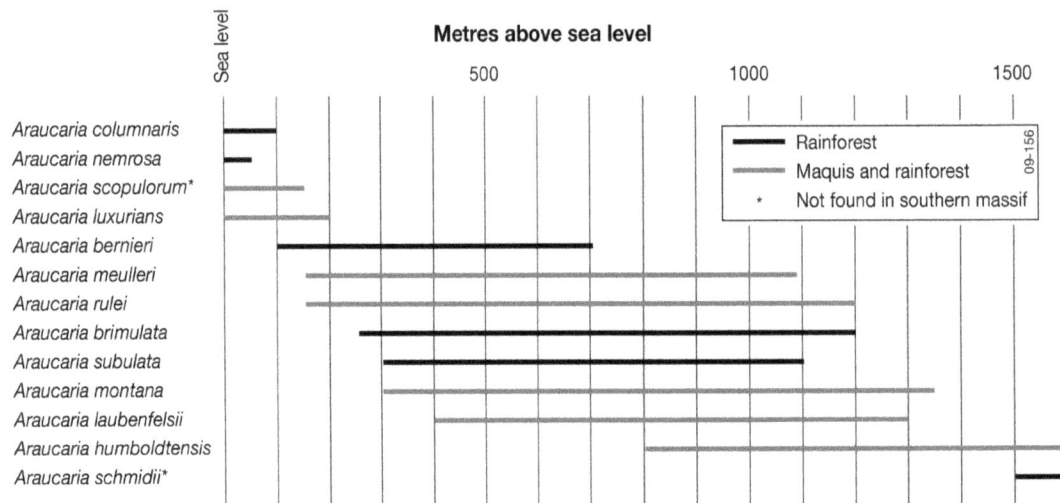

Figure 2. Distribution of the species of Araucaria in New Caledonia.

locally dominated by an arborescent stratum of *Araucaria* spp. or *Agathis ovata*. However, within this landscape and on this same substrate can be found complex evergreen rainforest, which also has a significant *Araucaria* component; Figure 2 illustrates the altitudinal and vegetation associations of the New Caledonian *Araucaria*. The bulk are found between 100 m and 1300 m altitude (seven species), with four restricted to below 200 m, and only one between 1500 m and 1600 m (Figure 2, Jaffré 1995).

The climate of New Caledonia is controlled primarily by the southeast trade winds, which in conjunction with the island's topography result in a wet east coast (annual rainfall >3000 mm) and a much drier west coast (annual rainfall <1500 mm) (Renson 1981). From November to April, the winds become more easterly and tropical depressions from the north dominate. The wettest months are during the warm season, January to March, which receive 57% of the annual precipitation. The remainder is distributed more or less equally over the other nine months. During the cool season (April to July), precipitation is irregular and occasional intrusions of cold polar air from the westerlies are observed (see Figure 1). These strong, cold winds come up through the Tasman Sea and lead to cool winter rainfall.

South America

Two species of Araucaria grow today in South America, as opposed to 17 in Australasia (Enright and Hill 1995). In South America, the two species of *Araucaria* are majestic trees that attain a height of 50 m and a diameter of 2 m. In southern Brazil, the *Araucaria angustifolia* forest is located in an area influenced by polar advection, providing moist and cool conditions throughout the year (Ledru 1993). In Chile, *Araucaria araucana* almost disappeared due to the intensive logging of past decades, and now shows a patchy distribution (Veblen et al. 1995). However, it grew in the cool and moist temperate regions of central Chile, at high elevations near the coast or deeper inland. No pollen records have ever been studied within the Chilean araucaria forest. This zone is under the influence of the westerlies from the southwestern Pacific, bringing regular precipitation throughout the year (Figure 1) (Garreaud 2000).

Cold surges are among the most energetic influences on the tropical circulation by the extratropics, with surface topography playing a central role, as shown by both observation and simulation (Markgraf 2001). In the absence of a significant mountain range, as in southern South America, cold surges are not deflected equatorward, but rather continue eastward. Key elements in the South American scenario are the cold anticyclones that move from the southeastern Pacific into southern Argentina and a centre of low pressure that deepens in the southwestern Atlantic. The advance of the cold-air incursion along the subtropical Andes is set up by the topographic blocking of the synoptic-scale flow. The strong temperature gradient produces the acceleration of the low-level winds parallel to the Andes range. This is a well-known extreme and abrupt event in Brazil that can damage the coffee plantations (Hamilton and Tarifa 1978).

In Brazil, *Araucaria angustifolia* grows at high elevations between 1500 m and 1800 m near Rio de Janeiro (20°S) or between 500 m and 900 m in the subtropics, latitude 25°S to 30°S. In the subtropical area, an expansion of *Araucaria angustifolia* occurred during the past 1000 years from refugia located either in the gallery forest or in the vicinity of a water source (Behling 1997a, b; Bitencourt and Krauspenhar 2006). The existence of these forests in southern Brazil is determined by the modern climate and the mean position of the polar fronts during winter, providing permanent drizzle (Ledru et al. 1994). The climate of this area is characterised by the absence of a dry season due to frequent and intense shifts of the polar fronts, providing regular fog and cloud cover in winter, along with low temperatures and occasional frost (Ledru 1993; Marengo et al. 1997; Garreaud 2000, 2001). While in the mountain range near Rio de Janeiro, the same phenomenon is induced by the uplift of warm Atlantic Ocean moisture, leading to almost continuous cloud cover and cooler temperatures, with the *Araucaria* forest assimilated into a cloud forest formation at higher altitude.

Araucaria forest expansion during the last glacial

Lynch's Crater, northeastern Australia

Lynch's Crater (17°37'S, 145°70'E) lies at an altitude of about 760 m above sea level (asl). Human activities over recent decades have destroyed much of the swamp surface and the main vegetation in the crater today is composed of introduced pasture species. A complex rainforest dominates the regional landscape with different facies according to location, drainage and rainfall distribution – from mesophyll vine forest, mesophyll palm forest, *Melaleuca* open woodland, to notophyll and microphyll vine forests. More open woodland with species of *Eucalyptus* is related to edaphic and topographic conditions. The crater is approximately 500 m in diameter, with recovered deposits extending to 64 m in depth and consisting mainly of lake muds, except for the top 14.5 m, where they are replaced by fresh to oxidised swamp sediments.

The pollen record contains two glacial-interglacial cycles, with 11 major changes in the composition of the rainforest (Kershaw 1978, 1986; Moss and Kershaw 2000; Kershaw et al. 2003, 2007). Although not illustrated in Figure 3, *Araucaria* frequencies are high between ca. 170,000 and 130,000 years ago. From 130,000 to 80,000 years ago, they are low, with a median value of less than 2%. During this time, rainforest angiosperms dominate the pollen spectrum, with the pollen of *Araucaria* virtually disappearing just before 80,000 years ago. High values are once again observed between ca. 80,000 and 45,000 years ago, with a median of 9.5% over 42 samples. The pollen frequency of *Araucaria* then suddenly declines at 45,000 years ago, coincident with an increase in charcoal particles. A short and weak return of *Araucaria* pollen to the record is observed at around 17,000 years ago.

Lake Xere Wapo, New Caledonia

Lake Xere Wapo (22°17.5′S, 166°58.5′E), at an altitude of 220 m, is located on the Plaine des Lacs, within the southern massif, where the average annual rainfall is 3000 mm. Here, the local climate is controlled largely by topography, with rain or fog occurring above 200 m on most afternoons, even during the dry season.

This shallow lake is approximately 0.85 km² in area and surrounded by gentle slopes that are mostly covered in a ferritic soil mantle, in places characterised by an iron pan crust known as 'sols cuirasse'. The surrounding vegetation is a bushy maquis dominated by either *Gymnostoma deplancheanum* or *Dacrydium araucarioides*. Five species of *Araucaria* are found at this altitude within the Plaines des Lacs, but not in the immediate vicinity of the lake.

Establishing a robust chronology for Lake Xere Wapo has been difficult, but based on the best model produced so far, three main phases are observed in the *Araucaria* pollen frequencies (Stevenson and Hope 2005). Between ca. 130,000 and ca. 80,000 years ago, the landscape moved between forest and maquis fairly frequently, with fire important in both landscapes, although more abundant during times of maquis dominance. The median value of *Araucaria* during this period is 2.2%, with two outliers of 12% and 8%.

From ca. 80,000 years ago to possibly 45,000 years ago, the pollen frequencies suggest *Araucaria* was more dominant in the landscape surrounding Lake Xere Wapo, however fire remained an integral part of the ecology. The median value of *Araucaria* pollen during this time is 7.5%, with four of the 13 samples being above 10%.

The most profound changes in the record occur after 45,000 years ago to the present, when *Araucaria* pollen frequencies fall to their lowest levels, with a median value of 0.9% and with only three of the 19 samples having values greater than 2%. Of note is that this change occurred in the absence of fire, suggesting that other forms of disturbance, such as an increase in climate variability, may be responsible (Stevenson and Hope 2005).

Colônia Crater, Brazil

Colônia is a meteor crater 3.5 km in width, filled with 392 m of organic sediments, located within the city of São Paulo (23°52'S, 46°42'20"W, 900 m asl). Its location near the Atlantic Ocean and in the Sierra do Mar mountain range means the vegetation is highly sensitive to changes in sea level and temperature. Today, the climatic features of the Colônia region are characterised by a mean annual precipitation of 1700 mm, a mean winter temperature of ca. 15°C and a three-month winter dry season.

There are no *Araucaria* forests in the vicinity of Colônia today. Only single trees of *Araucaria angustifolia* are observed in a landscape dominated by semi-deciduous Atlantic rainforest. Compared with the *Araucaria* pollen frequencies of Australia and New Caledonia, here the frequencies that indicate *Araucaria* within the landscape are an order of magnitude lower. This is in part related to *Araucaria* being, in general, a low pollen producer with large grains that are poorly dispersed. Consequently, pollen remains mostly in the vicinity of the source, leading to relatively low frequencies in pollen records, with pollen rain studies revealing ratios of up to 80% under the tree cover, to a single grain in the middle of a mire surrounded by *Araucaria* forest (Ledru 2002).

The Colônia record, as illustrated in Figure 3, has four main pollen zones in *Araucaria* pollen frequency. All eight samples from the tail end of MIS 6, 135,000-130,000 years ago, contain *Araucaria* pollen; these are also the highest frequencies for the entire record. Only two out of the next 10 samples that cover the last interglacial, 130,000-120,000 years ago, have *Araucaria* pollen, and from ca. 120,000 to 80,000 years ago, only 15 samples, out of a total of 56 samples, contain *Araucaria* pollen grains. However, there is a shift between ca. 80,000 and 60,000 years ago, with 22 samples out of a total of 35 containing *Araucaria* pollen grains. From 60,000 to 40,000 years ago, the presence of Araucaria declines, with only seven samples out of 35 containing *Araucaria* pollen grains. The past 40,000 years record the lowest frequencies of arboreal pollen in general, along with the disappearance of *Araucaria*. Only two samples out of 73 have Araucaria pollen – between ca. 18,000 and 16,000 years ago, at the end of the glaciation (Ledru et al. 2009).

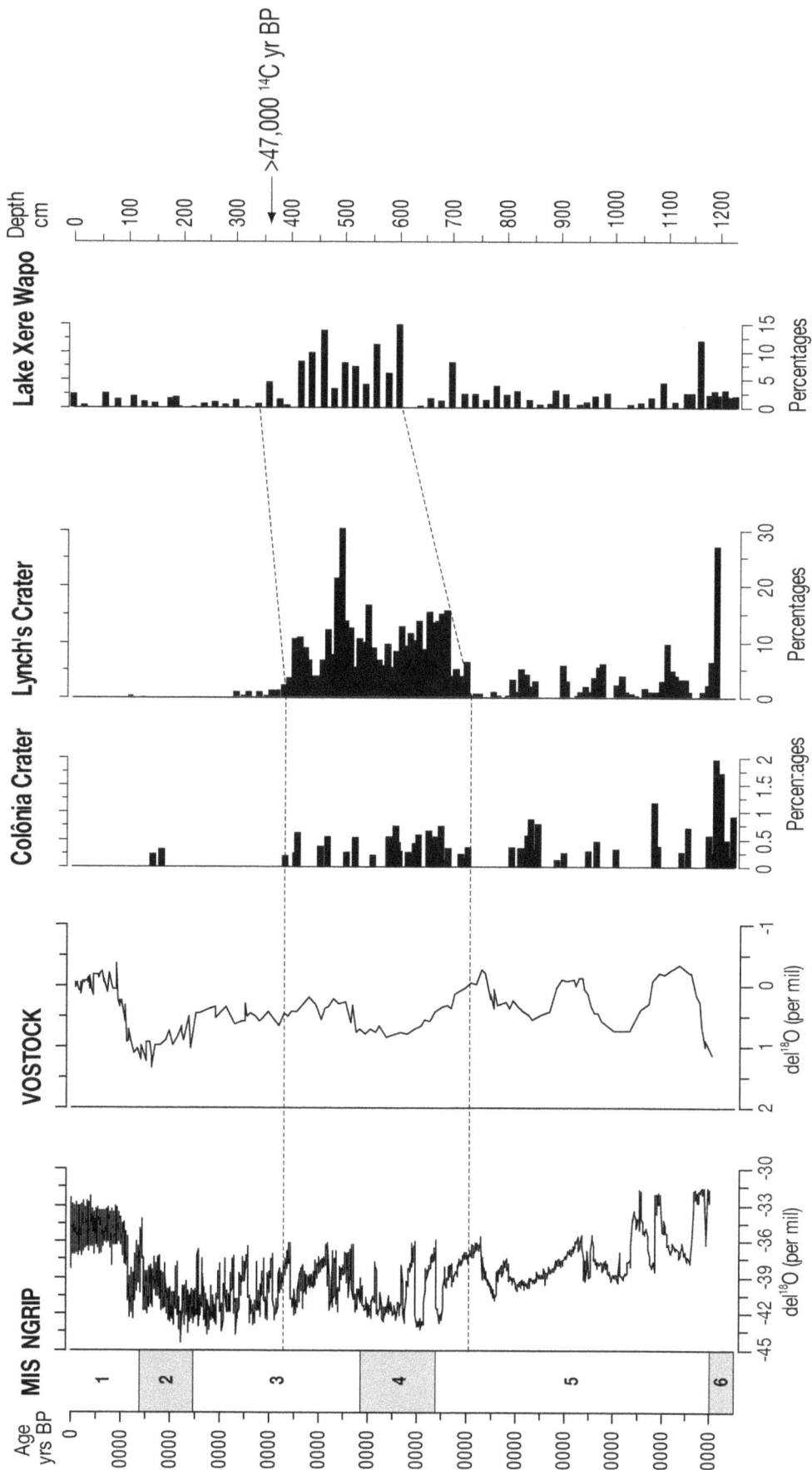

Figure 3. Changes in Araucaria frequency and charcoal accumulation for the past 130,000 years at Colônia Crater, Lynch's Crater, Lake Xere Wapo against the records of temperature change recorded at Vostok (Antarctica) and NGRIP (Greenland) (Dansgaard et al. 1993; Petit et al. 1999).

Discussion

Between 85,000 and 60,000 years BP, a drastic drop in temperature is observed in Antarctic ice cores (Figure 3), characterised by ice expansion and lowering of sea level by 70 m (Shackleton 1987; Petit et al. 1999). Climatic reconstructions for this period suggest that moisture in the tropics was provided by a steep temperature gradient between the southern latitudes and the equator (Vimeux et al. 1999; Delaygue et al. 2000), with the overall result being, relative to present, a change in the frequency and intensity of northward moving polar advections, which induced a shift of the winter rainfall zone.

The modern climate in the three studied areas (northeastern Australia, southwestern Pacific and southeastern Brazil) is primarily directed by the seasonal interplay between the subtropical high-pressure cells and the migration of the easterlies associated with the ITCZ, resulting in summer rainfall and a dry winter. However, the dry winter season can be attenuated by a northward shift of the westerlies, with the three study regions today subjected to abrupt, short and extreme climatic events brought about by this equatorward migration of polar air masses (Garreaud 2000, 2001). The signature of the cold surges over the tropics is much clearer in South America than in Australia, mainly as a consequence of favourable topography, specifically the narrow and tall Andean Cordillera, which extends continuously from south to north, providing an ideal barrier to channel these surges northward. In Brazil and Chile, the location of *Araucaria* forest today is defined by the intensity and frequency of these polar advections, and where winter rainfall is reliable.

The chronology of the three pollen records presented here is limited by the range of the radiocarbon method (ca. 40,000 years). Where possible, empirical methods of chronological reconstruction were used, such as precession-cycle-based tree maxima at Colônia Crater, or matching with nearby marine cores, such as at Lynch's Crater. Therefore, the time intervals that are discussed are not definitive and are subject to refinement with improvements in dating methods and/or the increase of comparable palaeoclimatic records. While taking this qualification into account, the changes in *Araucaria* at this stage appear synchronous (Figure 3). These three sites in three different regions of the Southern Hemisphere show the same directional change. Before 130,000 years ago, *Araucaria* frequencies are high, falling as the record enters the last interglacial (130,000-120,000 years ago). Between 120,000 and 80,000 years ago, *Araucaria* pollen frequencies decreased or were not represented at all in samples. Between 80,000 and 40,000 years ago, a full expansion of *Araucaria* forest is observed around Lynch's Crater and Lake Xere Wapo, with a more consistent presence recorded in the Colônia Crater record between 80,000 and 60,000 years ago. The pollen record of Colônia shows a progressive contraction of *Araucaria* forest after 60,000 years ago until its disappearance shortly after 40,000 years ago, while in Australia and New Caledonia, the frequencies of *Araucaria* pollen started to decrease shortly before 40,000 years ago. A last short increase in *Araucaria* pollen frequencies is observed during the late glacial at Colônia and at Lynch's Crater, while they remain low and constant up to the present at Lake Xere Wapo. The higher pollen frequencies between 80,000 and 40,000 years ago at Lynch's Crater and Lake Xere Wapo suggest that *Araucaria* was dominant in these landscapes, and that the summers were cooler and winters moister than in the preceding time interval.

Keeping in mind the uncertainty of the chronology, the major expansion phase of *Araucaria* between ca. 80,000 and 60,000 years ago may be linked to the corresponding cold stage 4 (76,000-62,000 years ago). While the role of human activity in the subsequent decline of *Araucaria* at 45,000 years ago at Lynch's Crater may still be under debate, we know that with the late arrival of people in South America (ca. 12,000 years ago) and New Caledonia (ca. 3000 years ago), human impact is not a component in these landscapes. Consequently, global climatic change is the most likely driver of this common feature in the three records.

Indeed, studies from Brazil suggest that the westerly storm track migrated northward several times during the late glacial, each time favouring the development of *Araucaria* at a different latitude. These records lack the time depth of Colônia Crater, but suggest that *Araucaria* forests were modified several times over the past 20,000 years. For instance, *Araucaria* forest is observed at 19°S at 15,000 years ago (Ledru 1993), at 23°S at 17,000 years ago (Ledru et al. 2009), and at 25°S today (Ledru et al. 1994). These shifts have been interpreted as a response to increased winter precipitation and a strongly reduced winter dry season (Ledru et al. 1994). The mechanism for this has been put forward as the weakening and poleward shift of the westerlies, giving more space to the tropical easterlies and the ITCZ seasonal shifts, which results in a summer rainfall regime (Ledru et al. 1994).

The phases of expansion and retraction of the *Araucaria* forests of the Southern Hemisphere provide a good bio-indicator for detecting change in the tropical hydrological system. In other southern regions where no *Araucaria* forests are observed today, palaeorecords that cover late MIS 5 to MIS 3 with a similar length and resolution are scarce. However, in South Africa, hyrax middens and a marine record, MD962094, attest to a cooler and moister climate between late MIS 5 and MIS 4. The increase in precipitation observed in these records has been related to Antarctic ice expansion and a northward shift of the westerlies band track between ca. 86,000 and 59,000 years ago (Stuut et al. 2002; Chase and Meadows 2007; Chase 2010). The termination of this moist and cool phase has been interpreted as a poleward shift of the westerlies, with an associated decrease in winter precipitation (Chase 2010). Further evidence in support of this scenario comes from the interior of Australia. Research into the palaeohydrology of Australia's mega-lakes, in combination with other palaeoclimatic proxies, has concluded that while there were multiple sources of precipitation leading to the formation of these lakes over the last glacial cycle, Southern Ocean sources of precipitation were an integral component until 47,000 years ago (Cohen et al. 2011). After this time, there is evidence from shoreline data and other palaeoclimatic data for increasing aridification and a decreasing contribution of moisture from Southern Ocean sources (Cohen et al. 2011), the time period that sees the contraction of *Araucaria* in the Lynch's Crater record.

In the equatorial Pacific, changes in sea surface temperatures of between 2°C and 3.5°C at 70,000 years ago precede change in Northern Hemisphere ice volume by 3000 years (Lea et al. 2006). In addition, deuterium analyses of an Antarctic ice core (Vostok) showed that the obliquity cycle, which controls insolation at 60°S, induced strong changes on both atmospheric and ocean circulations during cold stages 2, 4 and 6 (Vimeux et al. 1999).

Therefore, we suggest that during MIS 4, the rainfall environment that favoured an expansion of *Araucaria* in the southern tropics might have contributed to the observed increase in snow accumulation in Antarctica. This hypothesis is further supported by models (NASA/GIS AGCM) which show that present-day moisture sources for Antarctica originate in the subtropics and mid-latitudes of the Southern Hemisphere and from the intertropical zone during the Last Glacial Maximum (Delaygue et al. 2000). Our results suggest that this moisture was mostly distributed from the latitudinal band comprising (at least) 16°S to 23°S during stage 4 and the early part of stage 3. This was made possible because of a northward shift of the westerlies band track over the southern tropics, with the progressive decline in *Araucaria* after 45,000 years ago most likely the consequence of a reorganisation of the ocean-atmosphere forcing in the Southern Hemisphere.

Conclusion

In spite of the difficulties in establishing reliable chronologies, the simultaneous presence and expansion of *Araucaria* within the three study areas, located between 19°S and 23°S, and composed of two fragments of Gondwana, attest to a similar response to Southern Hemisphere climatic changes and an equal reaction of the species to the installation of a permanent cool and wet climate between at least 80,000 and 40,000 years ago. Consequently, we suggest a Southern Hemisphere climatic cause, rather than a human or ENSO-based scenario, to explain the decline of the *Araucaria* at these three sites.

It is hypothesised that the strong cooling in Antarctica that induced the shift of the westerly band tracks towards the equator between 86,000 and 60,000 years ago strengthened the winter rainfall system and thus lead to an expansion of *Araucaria* across the Southern Hemisphere subtropics. However, after 40,000 years ago, the poleward shift of the westerlies band, in combination with the intensification of the northern circulation and the ITCZ seasonal shifts on the distribution of precipitation in the tropics, resulted in weaker and less frequent northward polar advections, which had profound consequences for *Araucaria* forests in northern Australia, New Caledonia and Brazil. The consequent dominance of an easterly circulation system induced a drier winter season at low latitudes, causing the decline of *Araucaria* in the tropics, with forests only surviving in regions where cloud cover or winter rains meet their needs. In some southern tropical regions, a last tropical expansion of the *Araucaria* was observed during the late glacial, approximating the Antarctic Cold Reversal, when abrupt and short climatic changes were induced by strong differences between Northern and Southern Hemisphere temperature gradients (Broecker 1998; Stocker 2003).

While this synthesis illustrates how vulnerable this ancient tree is without suitable protection of its current remnant distributions, it also provides an interesting hypothesis for the past importance of the westerlies in the Southern Hemisphere hydrologic cycle, a topic of interest to fully understand the potential impacts of climate change on tropical environments and of the contribution of southern low-latitude moisture on Antarctic ice-expansion phases.

Acknowledgements

Thanks to Peter Kershaw for having created and stimulated the Southern Connections meetings and post meetings within the palynological community. We also thank two anonymous referees for their constructive discussion and comments.

References

Baines, P.G. 1980. The dynamics of the southerly buster. *Australian Meteorological Magazine* 28:175-200.

Beaufort, L., de Garidel-Thoron, T., Linsley, B., Oppo, D. and Buchet, N. 2003. Biomass burning and oceanic primary production estimates in the Sulu Sea area over the last 380 kyr and the East Asian monsoon dynamics. *Marine Geology* 201:53-65.

Behling, H. 1997a. Late Quaternary vegetation, climate and fire history from the tropical mountain region of morro de Itapeva, SE Brazil. *Palaeogeography Palaeoclimatology Palaeoecology* 129:407-422.

Behling, H. 1997b. Late Quaternary vegetation, climate and fire history of the *Araucaria* forest and campos region from Serra Campos Gerais, Paraná state (South Brazil). *Review of Palaeobotany and Palynology* 97:109-121.

Bitencourt, A.L.V. and Krauspenhar, P.M. 2006. Possible prehistoric anthropogenic effect on *Araucaria angustifolia* (Bert.) O. Kuntze expansion during the Late Holocene. *Revista Brasileira de Paleontologia* 9:109-116.

Broecker, W.S. 1998. Paleocean circulation during the last deglaciation: A bipolar seesaw? *Paleoceanography* 13:119-121.

Chase, B.M. and Meadows, M.E. 2007. Late Quaternary dynamics of Southern Africa's winter rainfall zone. *Earth-Science Reviews* 84:103-138.

Chase, B.M. 2010. South African palaeoenvironments during Marine Oxygen Isotope stage 4: a context for the Howiesons Poort and Still Bay industries. *Journal of Archaeological Science* 37:1359-1366.

Cohen, T.J., Nanson, G.C., Jansen, J.D., Jones, B.G., Jacobs, Z., Treble, P., Price, D.M., May, J.-H., Smith, A.M., Ayliffe, L.K. and Hellstrom, J.C. 2011. Continental aridification and the vanishing of Australia's megalakes. *Geology* 39:167-170.

Delaygue, G., Masson, V., Jouzel, J. and Koster, R.D. 2000. The origin of Antarctic precipitation: a modelling approach. *Tellus* 52B:19-36.

Dansgaard, W., Johnsen, S.J., Clausen, H.B., Dahljensen, D.S., Gundestrup, N., Hammer, C.U., Hviberg, C.S., Steffensen, J.R., Sveinbjörnsdottir, A.E., Jouzel, J. and Bond, G. 1993. Evidence for general instability of past climate from a 250-kyr ice core record. *Nature* 364:218-220.

Enright, N.J. and Hill, R.S. 1995. *Ecology of the Southern conifers*. Melbourne University Press, 342 pp.

Enright, N.J. 1995. Conifers of Tropical Australia. In: Enright, N.J. and Hill, R.S. (eds), *Ecology of the Southern conifers*, pp. 197-222. Melbourne University Press.

Garreaud, R.D. 2000. Cold air incursions over subtropical South America: mean structure and dynamics. *Monthly Weather Review* 128:2544-2559.

Garreaud, R.D. 2001. Subtropical cold surges: regional aspects and global distribution. *International Journal of Climatology* 21:1181-1197.

Gentilli, J. 1972. *Australian climate patterns*. Nelson, Melbourne, 285 pp.

Hamilton, M.G. and Tarifa, J.R. 1978. Synoptic aspects of a polar outbreak leading to frost in tropical Brazil, July 1972. *American Meteorological Society* 106:1545-1556.

Hobbs, J.E. 1999. Present climates of Australia and New Zealand. In: Hobbs, J.E., Lindesay, J.A. and Bridgman, H.A. (eds), *Climate of the Southern Continents: Present, Past and Future*, pp. 63-105. Wiley and Sons, Chichester.

Jaffré, T. 1995. Distribution and Ecology of the Conifers of New Caledonia. In: Enright, N.J. and Hill, R.S. (eds) *Ecology of the Southern conifers*, pp. 171-196. Melbourne University Press, Melbourne.

Kershaw, A.P. 1978. Record of last interglacial-glacial cycle from northeastern Queensland. *Nature* 272:159-61.

Kershaw, A.P. 1986. Climatic change and aboriginal burning in North-East Australia during the last two glacial/interglacial cycles. *Nature* 322:47-9.

Kershaw, A.P. and McGlone, M.S. 1995. The Quaternary of the southern conifers. In: Enright, N.J. and Hill, R.S. (eds), *Ecology of the Southern conifers*, pp. 30-63. Melbourne University Press, Melbourne, Australia.

Kershaw, A.P. and Wagstaff, B. 2001. The southern conifer family Araucariaceae: history, status and value for paleoenvironmental reconstruction. *Annual Reviews of Ecology and Systematics* 32:397-414.

Kershaw, A.P., van Der Kaars, S. and Moss, P.T. 2003. Late Quaternary Milankovitch-scale climatic change and variability and its impact on monsoonal Australasia. *Marine Geology* 201:81-95.

Kershaw, A.P. and Walker, D. 2007. Quaternary vegetation and environments of the North-

East Queensland volcanic provinces. *XVII INQUA Cairns 2007 Post Excursion B2 guide.* Monash University, 128 pp.

Kershaw, A.P., Bretherton, S.C. and van der Kaars, S. 2007. A complete pollen record of the last 230 ka from Lynch's Crater, north-east Australia. *Palaeogeography, Palaeoclimatology, Palaeoecology.* 25:23-45.

Lau, K.-M. and Zhou, J. 2003. Anomalies of the South American summer monsoon associated with the 1997-99 El Niño-southern oscillation. *International Journal of Climatology,* 23:529-539.

Lea, D.W., Pak, D.K., Belanger, C.L., Spero, H.J., Hall, M.A. and Shackleton, N.J. 2006. Paleoclimate history of the Galapagos surface waters over the last 135,000 yr. *Quaternary Science Reviews* 25:1152-1167.

Ledru, M.-P. 1993. Late Quaternary environmental and climatic changes in central Brazil. *Quaternary Research* 39:90-98.

Ledru, M.-P. 2002. Late Quaternary history and evolution of the Cerrados as revealed by palynological records. In: Oliveira, P.S. and Marquis, R.J. (eds), *The Cerrados of Brazil: Ecology and natural history of a neotropical savanna*, pp. 33-52. Columbia University Press, New York.

Ledru, M.-P., Behling, H., Fournier, M., Martin, L. and Servant, M. 1994. Localisation de la forêt d'araucaria du Brésil au cours de l'Holocène. Implications paléoclimatiques. *Comptes Rendus de l'Académie des Sciences de Paris* 317:517-521.

Ledru, M.-P., Mourguiart, P. and Riccomini, C. 2009. Related changes in biodiversity, insolation and climate in the Atlantic rainforest since the last interglacial. *Palaeogeography, Palaeoclimatology, Palaeoecology* 271:140-152.

Marengo, J.A., Cornejo, A., Satyamurty, P., Nobre, C. and Sea, W. 1997. Cold surges in tropical and extratropical South America: The strong event in June 1994. *Monthly Weather Review* 125:2759-2786.

Markgraf, V. (ed) 2001. *Interhemispheric Climate Linkages.* Academic Press, 488 pp.

Moss, P.T. and Kershaw, A.P. 2000. The last glacial cycle from the humid tropics of northeastern Australia: comparison of a terrestrial and a marine record. *Paleogeography, Palaeoclimatology, Palaeoecology* 155:155-176.

Moss, P.T. and Kershaw, A.P. 2007. A late Quaternary marine palynological record (oxygen isotope stages 1 to 7) for the humid tropics of northeastern Australia based on ODP Site 820. *Paleogeography, Palaeoclimatology, Palaeoecology* 251:4-22.

Petit, J.-R., Jouzel, J., Raynaud, D., Barkov, N.I., Barnola, J.-M., Basile, I., Bender, M.L., Chappellaz, J., Davis, M.E., Delaygue, G., Delmotte, M., Kotlyakov, V.M., Legrand, M., Lipenkov, V.Y., Lorius, C., Pépin, L., Ritz, C., Saltzman, E. and Stievenard, M. 1999. Climate and atmospheric history of the past 420,000 years from the Vostock ice core, Antarctica. *Nature* 399:429-436.

Renson, S. 1981. Le Climat de la Grand Terre et des Iles. In: Mathieu-Daudé, J. (ed), *Atlas de Nouvelle Calédonie. Editions du Cagou, Hatchette Calédonie*, New Caledonia, 92 pp.

Shackleton, N.J. 1987. Oxygen isotopes, ice volume and sea level. *Quaternary Science Reviews* 6:183-190.

Stevenson, J. and Hope, G. 2005. A comparison of Late Quaternary forest changes in New Caledonia and Northeastern Australia. *Quaternary Research* 64:372-383.

Stocker, T.F. 2003. South dials North. *Nature* 424:496-499.

Stuut, J.-B.W., Prins, M.A., Schneider, G.J., Weltje, J.H.F., Jansen, G. and Postma, A. 2002. A 300 kyr record of aridity and wind strength in Southwestern Africa: Inferences from grain-size distributions of sediments on walvis ridge, SE Atlantic. *Marine Geology* 180:221-233.

Veblen, T.T., Burns, B.R., Kitzberger, T., Lara, A. and Villalba, R. 1995. The ecology of the conifers of Southern South America. In: Enright, N.J. and Hill, R.S. (eds), *Ecology of the Southern conifers*, pp. 120-155. Melbourne University Press, Melbourne, Australia.

Vimeux, F., Masson, V., Jouzel, J., Stievenard, M. and Petit, J.R. 1999. Glacial-interglacial changes in ocean surface conditions in the Southern Hemisphere. *Nature* 398:410-412.

12

When did the mistletoe family Loranthaceae become extinct in Tasmania? Review and conjecture

Mike Macphail
Department of Archaeology and Natural History, College of Asia and the Pacific, The Australian National University, Canberra, ACT
mike.macphail@anu.edu.au

Greg Jordan
University of Tasmania, Hobart, Tasmania

Feli Hopf
The Australian National University, Canberra, ACT

Eric Colhoun
University of Newcastle, Newcastle, NSW

Introduction

'Mistletoe' is the common name for a diverse group of hemi-parasitic shrublets that grow attached to and within the branches of trees and shrubs. Vidal-Russell and Nickrent (2008a) and Nickrent et al. (2010) infer that the mistletoe habit has evolved five times in the sandalwood order Santalales. The first of these clades is the family Misodendraceae, which is endemic to southern South America and whose species grow mainly on *Nothofagus*. The habit evolved three times within the Santalaceae – in the cosmopolitan tribe Visceae, which includes the 'archetypal' European Mistletoe *Viscum album*, in tropical American species of Santaleae that were formerly placed into a separate family, the Eremolepidaceae, and in the tropical tribe Amphorogyneae. The third clade comprises all members of the Loranthaceae except the early diverging genera *Nuytsia* and *Atkinsonia*. The family is restricted to the Southern Hemisphere except for a few genera growing north of the equator in the tropics and around the Mediterranean.

Two mistletoe clades are extant in Australia. These are (1) Visceae (three genera including *Viscum*), which is restricted to rainforests, monsoon forests and woodlands along the northern and eastern margins, and (2) Loranthaceae (12 genera), which is widely distributed across mainland Australia, with hosts ranging from coastal mangrove forests to mulga (*Acacia aneura*) woodlands in the arid zone (www.anbg.gov.au/abrs/online-resources/flora/redirect.jsp). Two Australian species are root parasites and for this reason give the appearance of being stand-alone shrubs or small trees – *Nuytsia floribunda*, which is endemic to southwest Western Australia, and *Atkinsonia ligustrina*, which is confined to exposed habitats in the Blue Mountains of NSW (Barlow 1984).

No mistletoes now occur in Tasmania. The reason(s) for this remain obscure given the wide ecological tolerance of many genera within the Loranthaceae and their observed dispersal over long distances by birds (www.anbg.gov.au/mistletoe/remote-islands.html). In this review, we illustrate and discuss the implications of Loranthaceae-type pollen recovered from a range of offshore and onshore sites around Tasmania (Figure 1). Unlike many fossil angiosperm pollen types, the morphology of these specimens is sufficiently distinctive to allow them to be assigned to this family and one of the six fossil species possibly to an extant genus.

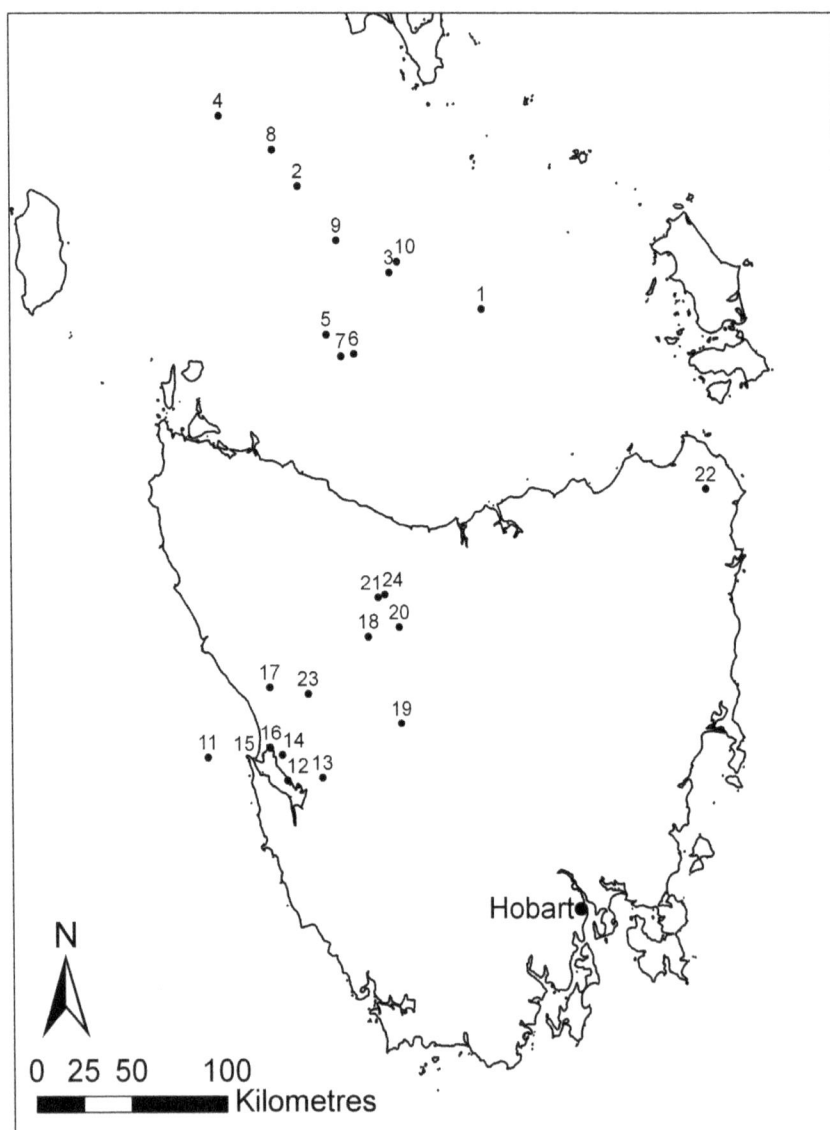

Figure 1. Location of sites mentioned in the text.

Loranthaceae pollen

Modern Loranthaceae pollen comprise two morphologically distinct types, one of which is a ±convex triangular, oblate grain with simple to complex tricolpate (rarely syncolpate) apertures, e.g. *Alepis*, *Amyema*, *Amylotheca*, *Decaisnia*, *Peraxilla* and *Nuytsia*, and the other of which is a sphaeroidal, triporate to tricolporoidate grain ornamented with stout spines (baculae, echini), e.g. *Lepidoceras* (Chile) and *Tupeia* (NZ) (see Erdtman 1966; Heusser 1971; Moar 1993; Beug 2004).

Pollen produced by Loranthaceae genera (Figures 2a-o) are easily distinguished from other Australian pollen morphotypes. Key characteristics are: (a) the oblate, triangular to lobate shape, (b) thin exine except across the poles where distinctively sculptured or thickened exine may form a bluntly triangular 'bridge' (Pocknall and Mildenhall 1984) or 'apical cushion' (Mildenhall and Pocknall, 1989), respectively, and (c) apertures (colpi) located at the tips (angles) of rounded to truncated apices. The usually gaping, colpate to demicolpate apertures vary in length from relatively short, in, for example, *Amylotheca*, to extremely long, in, for example, *Amyema* and *Peraxilla*. Ornamentation in the mesocolpial regions varies from psilate-scabrate to verrucate-baculate. In some genera, the tectum breaks down to form a pseudo-reticulum, e.g. *Nuytsia*; more rarely, the colpi are bordered by very short parallel rods (striae) that may extend close to the poles, e.g. *Alepis* and *Peraxilla*.

Australian fossil pollen types displaying the same morphological characteristics have been assigned to, or compared with, six formally described species: '*Amylotheca*' *pliocenica* Cookson 1957, *Gothanipollis bassensis* Stover and Partridge 1973, *Gothanipollis gothani* Krutzsch 1959, *Gothanipollis perplexus* Mildenhall and Pocknall 1984, *Tricolpites simatus* Stover and Partridge 1973, and *T. thomasii* Cookson and Pike 1954. Examples of these morphospecies and their time distribution in southeast Australian sedimentary basins are shown in Figure 3 and Table 1, respectively. Additional specimens are illustrated in Stover and Partridge 1973, Martin 1978, Pocknall and Mildenhall 1984, Hill and Macphail 1985, Mildenhall and Pocknall 1989, Macphail et al. 1994 and Macphail 1999. *Cranwellia striata* (Couper) Srivastava 1966 and a related but more coarsely striate morphospecies, *C. costata* Mildenhall 1978 are assumed to be fossil Loranthaceae, e.g. by Askin (1990). The relationship has yet to be confirmed. However, in terms of amb shape and ornamentation, we note that both morphospecies more closely resemble pollen of Krameriaceae, a monogeneric family that is endemic to the southwest United States, Argentina and Chile (see Plate 31: Figure 378 in Heusser 1971).

Table 1. First (FA) and Last (LA) appearance of Loranthaceae-type morphospecies in southeast Australian sedimentary basins (north to south). [1] Data from Macphail (1999), [2] Data from Partridge (1999), [3] Data from Partridge (2002).

Fossil species	Murray [1] FA	Murray [1] LA	Gippsland [2] FA	Gippsland [2] LA	Bass [3] FA	Bass [3] LA
Tricolpites simatus	middle Eocene	late Miocene	middle Eocene	late Eocene	middle Eocene	late Eocene
Tricolpites thomasii	late Eocene	early Oligocene	late Mid. Eocene	late Eocene	late mid. Eocene	late Eocene
Cranwellia striata	middle Eocene	late Miocene	late Eocene	late Eocene	late Eocene	late Eocene
Gothanipollis bassensis	middle Eocene	late Miocene	late early Eocene	early Miocene	middle Eocene	early Miocene
Gothanipollis cf. *gothani*	late Eocene	late Miocene	not recorded	not recorded	not recorded	not recorded
Gothanipollis perplexus	late Miocene	early Pliocene	not recorded	not recorded	not recorded	not recorded
'*Amylotheca*' *pliocenica*	late Miocene	early Pliocene	late Pliocene	Quaternary?	not recorded	not recorded

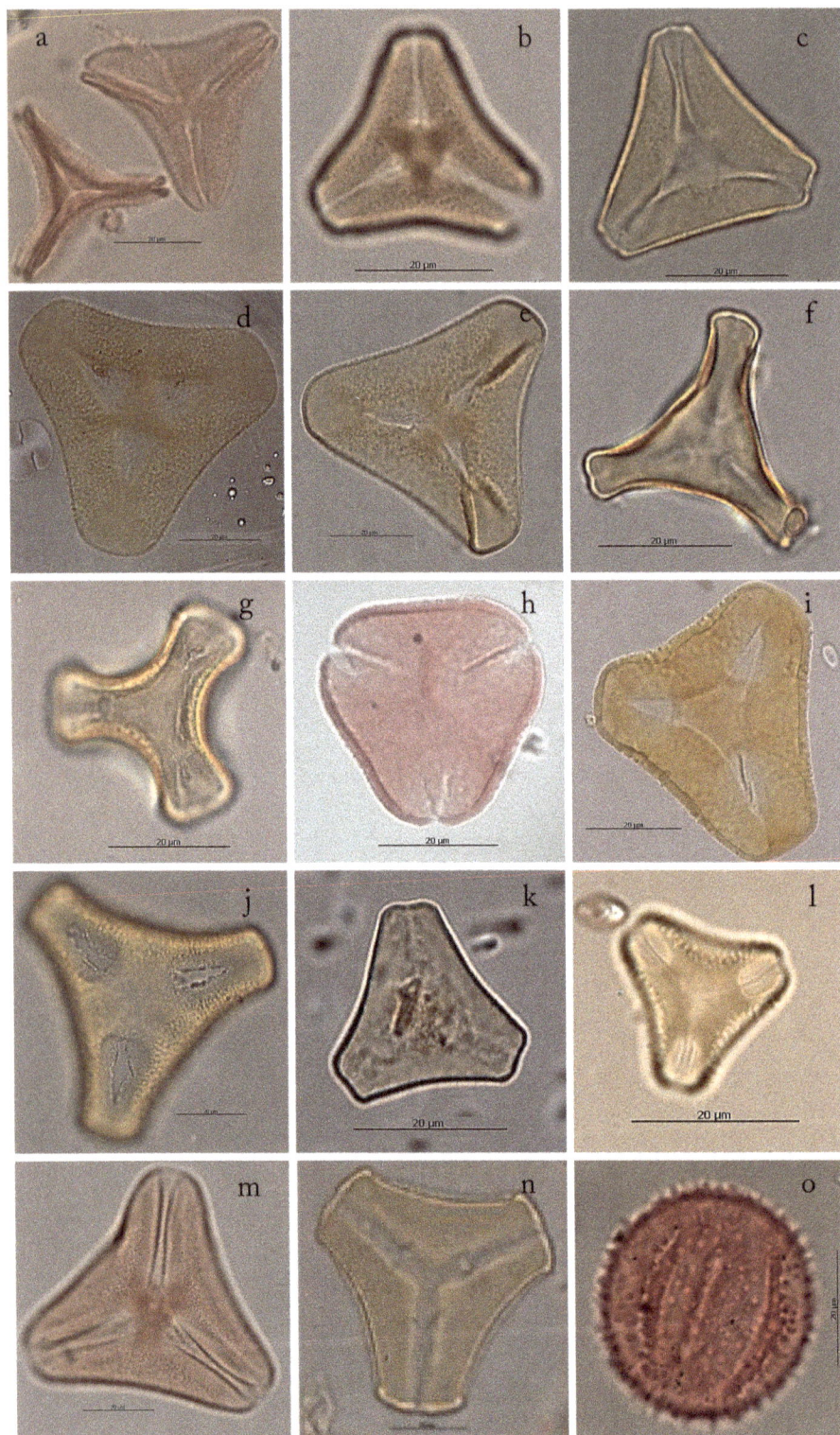

Figure 2. Photomicrographs of modern Loranthaceae pollen. **a.** *Amyema pendula* (Sieb. Ex Spreng.) Tiegh., NSW. **b.** *Amyema congener* (Sieb. Ex Schult. and Schult.f.) Tiegh., Queensland.**c.** *Amyema miquelli* (Lehm. Ex Miq.) Tiegh., NSW. **d.** *Amylotheca dictyophleba* (F. Muell.) Tiegh. Queensland. **e.** *Decaisnina signata* (F. Muell. ex benth.) Tiegh., Western Australia. **f.** *Dactyliophora novae-guineae* (F.M. Bail.) Dans, New Guinea. **g.** *Helixanthera sessiliflora* Danser, Philippines. **h.** *Ileostylus micranthus* (Hook.f.) Tiegh., New Zealand **i.** *Lysiana* sp. cf. *L. spathula* (Blakey) Barlow, Western Australia. **j.** *Lysiana murrayi* (Tate) Tiegh., Western Australia. **k.** *Muellerina eucalyptoides* (DC) Barlow, NSW. **l.** *Nuytsia floribunda* (Labill.) R.Br., Western Australia. **m.** *Peraxilla colensoi* (Hook.f.) Tiegh., New Zealand. **n.** *Plicosepalus curviflorus* Tiegh., South Africa. **o.** *Tupeia antarctica* (Forst.f) Cham,. and Schlect., New Zealand.

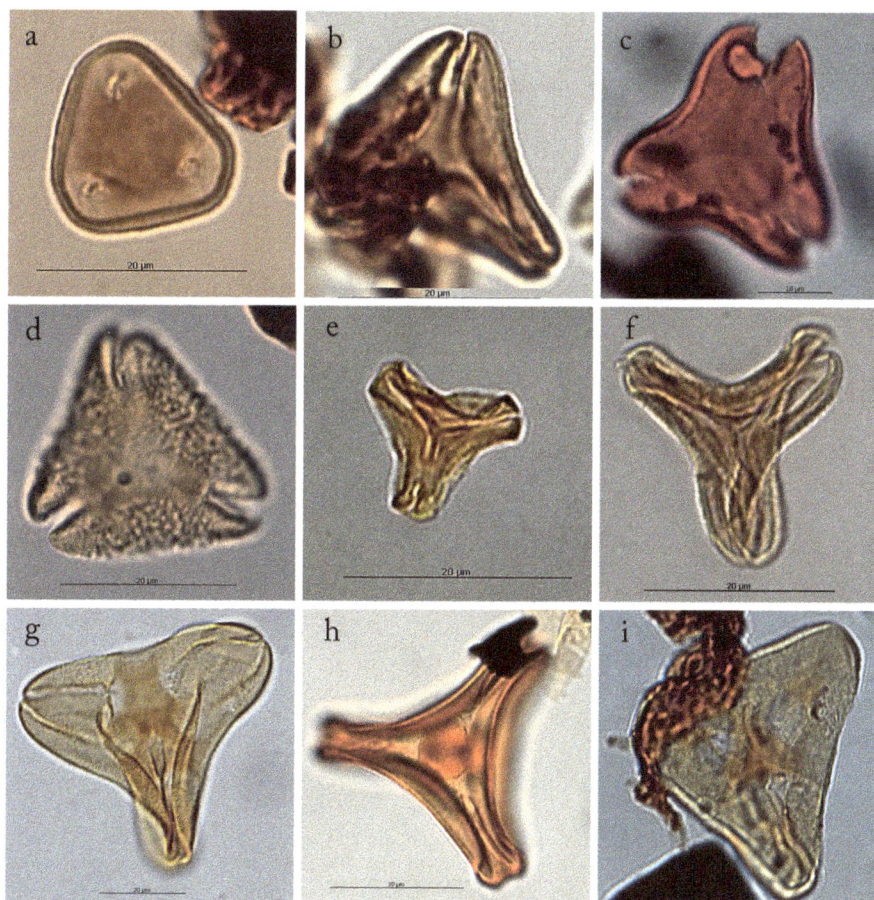

Figure 3. Fossil Loranthaceae pollen (SE Australia). **a.** *Anacolosidites sectus.* Late Eocene, BMR DDH 36-839 121-122 m, Darling Basin. **b.** *Tricolpites simatus.* Oligocene-Early Miocene, Marma-1 268-269 m, Murray Basin. **c.** *Tricolpites simatus.* Late Eocene, Hatfield-1 384 m, Murray Basin. **d.** *Tricolpites thomasii.* Late Eocene, Booligal-1 362-363 m, Murray Basin. **e.** *Gothanipollis bassensis.* Early Oligocene, HEC DDH 5825 131 ft., Tasmania. **f.** *Gothanipollis* cf. *gothani.* Oligocene-Early Miocene, Leaghur-1 177-178 m, Murray Basin. **g.** *Gothanipollis perplexus.* Early to late Miocene, Scotia-1 9.60 m, Murray Basin. **h.** *Gothanipollis* cf. *perplexus.* Early-Late Miocene, Tyntynder West-2 61-64 m, Murray Basin. **i.** *'Amylotheca' pliocenica.* Early Pliocene, Cal Lal-1 80-81 m, Murray Basin.

A comparison with pollen produced by extant Loranthaceae genera indicates that *Gothanipollis bassensis* has no close living analogue in Australia due to its very small size (<20 µm) and strongly lobate shape. Nevertheless, this species and its larger relative *Gothanipollis* cf. *gothani* bear a general resemblance to pollen produced by extant species such as *Amyema congener* and *Dactyliophora novae-guineae* (Figures 2b, f): The same qualification applies to *Tricolpites simatus* and *T. thomasii.* Ornamentation on the latter species includes coarse striae bordering the colpi (cf. *Decaisnina signata* and *Peraxilla colensoi*: (Figures 2e, m) but, as far as we are aware, no living Loranthaceae produce pollen with medium to coarse reticulate sculpturing of exine in the mesocolpial regions. The largest morphotypes found in southeast Australia, *'Amylotheca' pliocenica* and *Gothanipollis perplexus* closely resemble pollen of *Amylotheca*, and *Peraxilla*, respectively (cf. Martin, 1978; Pocknall & Mildenhall, 1984). The latter genus is endemic to New Zealand.

Chronostratigraphic distribution of fossil Loranthaceae pollen in Tasmania

At present, there are no records of Loranthaceae at Tertiary or Quaternary sites in southeast Tasmania, and the selection of fossil Loranthaceae pollen types illustrated in Figures 4a-o comes from sites in the north and west of the state: The stratigraphic distribution of these species in

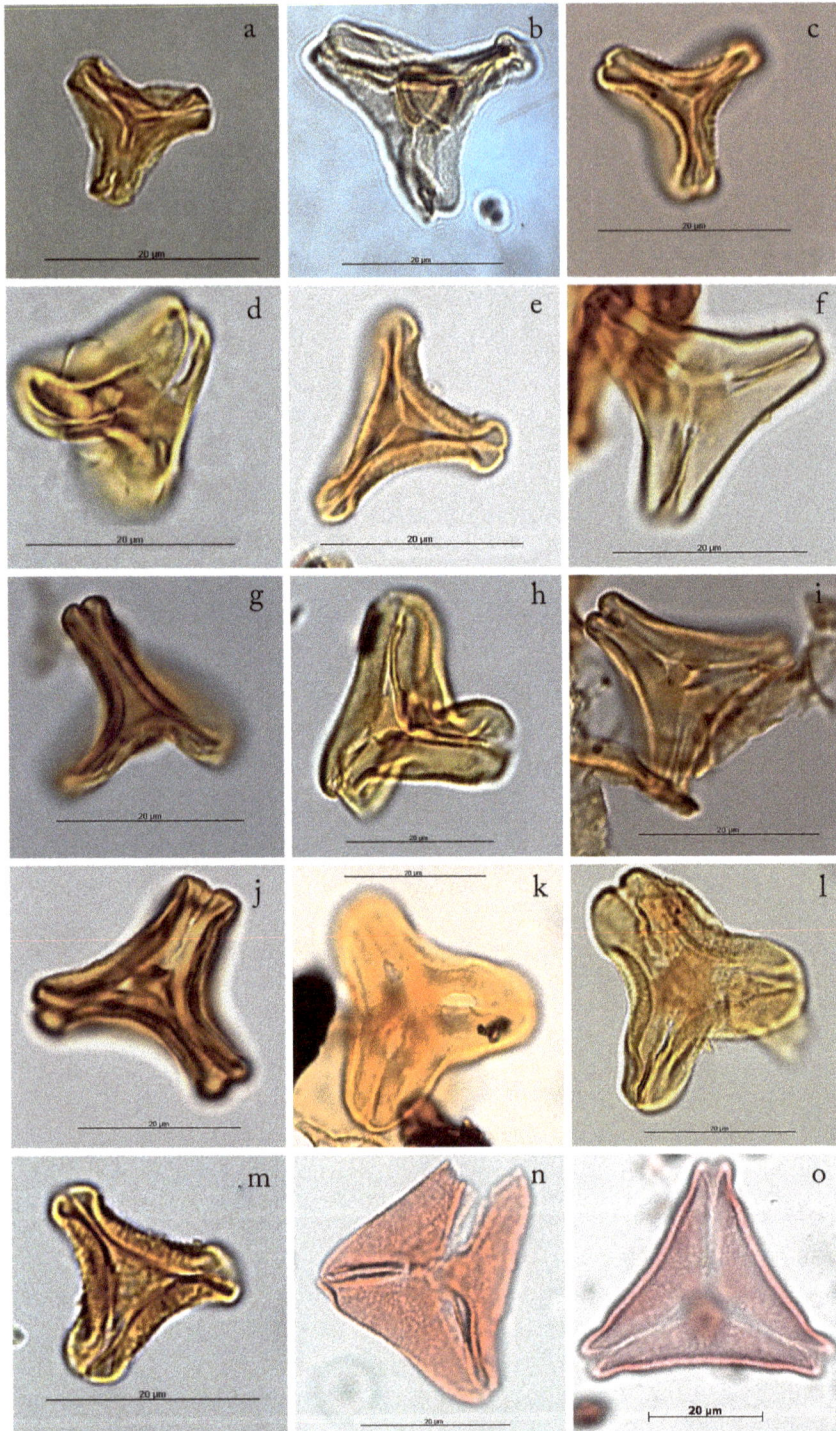

Figure 4. Fossil Loranthaceae pollen (Tasmania). **a.** *Gothanipollis bassensis.* Early Oligocene, Lemonthyme Creek DDH5825 131 ft. **b.** *Gothanipollis* cf. *bassensis.* Early Oligocene, Lemonthyme Creek DDH5825 131 ft. **c.** *Gothanipollis bassensis.* Early Eocene, outcrop Macquarie Harbour Beds, Strahan. **d.** *Gothanipollis* cf. *gothani.* Middle Pleistocene, Darwin Crater core 61.5 m. **e.** *Gothanipollis* cf. *gothani.* Early-Middle Pleistocene, outcrop Regatta Point, Strahan **f.** *Gothanipollis* cf. *gothani.* Middle Pleistocene, Darwin Crater 61.5 m. **g.** *Gothanipollis* cf. *gothani.* Middle Pleistocene, Darwin Crater core 61.0 m. **h.** *Gothanipollis* cf. *gothani.* Early Oligocene, outcrop, Lea River. **i.** *Gothanipollis* cf. *gothani.* Early Eocene, outcrop Macquarie Harbour Beds, Strahan. **j.** *Gothanipollis* cf. *gothani.* Early Oligocene, Wilmot Dam DDH 4558 77.5 ft. **k.** *Gothanipollis* cf. *perplexus.* Early Oligocene, Lemonthyme Creek DDH5825 97 ft. **l.** *Gothanipollis* cf. *perplexus.* Early Oligocene, Lemonthyme Creek DDH5825 131 ft. **m.** *Gothanipollis* cf. *T. thomasii.* Early-Middle Pleistocene alluvium Regatta Point. **n.** *Gothanipollis* aff. *'A.' pliocenica.* Early-Middle Pleistocene alluvium, Regatta Point. **o.** Loranthaceae. Last Glacial, Core GC 320 cm, Lake St. Clair.

Table 2. Chronostratigraphic distribution of Loranthaceae pollen in the Bass and Sorell Basins and onshore sites in Tasmania.

Basin/Area	Well/locality	Age	Zone	Morphospecies	Reference	Comment
offshore Bass	Chat-1	late Eocene	middle *Nothofagidites asperus*	*Gothanipollis bassensis*	Partridge et al. 2002	data from Morgan 1986
	Cormorant-1	late Eocene	middle *Nothofagidites asperus*	*Tricolpites simatus, T. thomasii*	M.K. Macphail, unpubl.	see Partridge et al. 2002
	Dondu-1	late Eocene	middle *Nothofagidites asperus*	*Gothanipollis bassensis, Tricolpites simatus*	Partridge et al. 2002	data from Stover 1973
	Kon Kon-1	late Eocene	middle *Nothofagidites asperus*	*Tricolpites simatus, T. thomasii*	Partridge et al. 2002	data from Stover 1973
		middle Eocene	lower *Nothofagidites asperus*	*Gothanipollis bassensis*	Partridge et al. 2002	unconfirmed record
	Narimba-1	late Eocene	middle *Nothofagidites asperus*	*Tricolpites thomasii*	Partridge et al. 2002	data from Morgan 1986
		middle Eocene	lower *Nothofagidites asperus*	*Tricolpites simatus*	Partridge et al. 2002	
	Pelican-2	early Eocene?	*Proteacidites asperopolus?*	*Tricolpites thomasii*	Partridge et al. 2002	middle Eocene?
	Pelican-5	late Eocene	middle *Nothofagidites asperus*	*Tricolpites simatus*	Partridge et al. 2002	data from Morgan 1986
	Toolka-1	late Eocene	middle *Nothofagidites asperus*	*Tricolpites simatus*	Partridge et al. 2002	data from Stover 1974
	Yolla-1	late Eocene	middle *Nothofagidites asperus*	*Tricolpites thomasii*	Partridge et al. 2002	see Partridge et al. 2002
		early Oligocene	upper *Nothofagidites asperus*	*Gothanipollis bassensis*	Partridge et al. 2002	
	Yurongi-1	late Eocene	middle *Nothofagidites asperus*	*Tricolpites thomasii*	Partridge et al. 2002	data from Stover 1973
		middle Eocene	lower *Nothofagidites asperus*	*Gothanipollis bassensis, Tricolpites thomasii*	Partridge et al. 2002	
offshore Sorell	Cape Sorell-1	early Eocene?	middle *Malvacipollis diversus*	*Tricolpites simatus*	M.K. Macphail, unpubl.	caved specimen
	Coal Head	late Quaternary	-	*Gothanipollis* cf. *gothani*	M.K. Macphail, unpubl.	reworked specimen
West Coast	Darwin Crater	middle Pleistocene	core samples at 22.0-44.7 m, 61.0-61.5 m. Absent above 22 m.	*Gothanipollis* cf. *bassensis, G.* cf. *gothani and perplexus*	M.K. Macphail, unpubl. E.A. Colhoun, unpubl.	interval between 44.7 and 61.0 m not yet analysed
	King River	middle Pleistocene	Regency site at 0, 15, 40, 60 cm	*Gothanipollis perplexus*	E.A. Colhoun, unpubl.	Fitzsimmons et al. 1990
	Lowana Road	late Early Eocene	*Proteacidites asperopolus*	*Gothanipollis* cf. *bassensis*	M.K. Macphail, unpubl	collected by G. Jordan
	Regatta Pt.,	early? Quaternary	*Tubulifloridites pleistocenicus?*	*Gothanipollis bassensis, Tricolpites simatus*	Hill and Macphail 1985	reworked specimen
	Strahan	late early Eocene	*Proteacidites asperopolus*	*Gothanipollis bassensis, G.* cf. *gothani*	Macphail et al. 1993a	Macquarie Harbour Beds
	Zeehan	Oligo-Miocene	*Proteacidites tuberculatus*	*Gothanipollis* cf. *gothani*	M.K. Macphail, unpubl.	collected by M. Pole

Table 2. *Continued*

Basin/Area	Well/locality	Age	Zone	Morphospecies	Reference	Comment
Northeast	Loch Aber	mid late Eocene	*Nothofagidites asperus*	*Gothanipollis* cf. *bassensis*	M.K. Macphail, unpubl.	
	Cradle Mt.	recent	core LDGS at 7, 33, 38 cm	*Loranthaceae*	Dyson 1995	Dove Lake
	Lake St. Clair	late Pleistocene	core CG at 320 cm	*Loranthaceae* cf. *Amyema*	F. Hopf, unpubl.	Narcissus Bay
northwest Central Plateau	Lemonthyme Creek	early Oligocene	*Proteacidites tuberculatus*	*Gothanipollis bassensis, G.* cf. *gothani, G. perplexus, G.* cf. *perplexus.*	Macphail et al. 1993b M.K. Macphail, unpubl.	HEC DDH 5825
	Lea River	Oligo-Miocene	*Proteacidites tuberculatus*	*Gothanipollis* cf. *gothani*	M.K. Macphail, unpubl.	collected by R. Hill
	Tyndall Ranges	Holocene	-	*Gothanipollis* cf. *gothani*	Macphail and Colhoun 1982	peat in summit tarn
	Wilmot Dam	early Oligocene	*Proteacidites tuberculatus*	*Gothanipollis bassensis, G.* cf. *gothani*	Macphail and Hill 1994	HEC DDH 4558

the Bass Basin and onshore sites in Tasmania is given in Table 2. We emphasise that absence of evidence is not necessarily evidence of absence. Reasons are small morphospecies such as *Gothanipollis bassensis* are easily overlooked, especially if partly obscured by plant detritus, and also can be lost during processing if the organic extracts are filtered through 10 to 20 μm sieve cloth. Nevertheless, the records are adequate to confirm (numerical ages after Ogg et al. 2008):

1. The oldest reliable records of Loranthaceae species in Tasmania (and probably Australia as a whole) are specimens of *Gothanipollis bassensis* and *Tricolpites simatus* preserved in the late early Eocene (ca. 50.5-51.5 million years ago) Macquarie Harbour Formation at Regatta Point and Lowana Road near Strahan, western Tasmania. Specimens of *Tricolpites simatus* in the same sections appear to pre-date the first appearance of the species in the offshore Bass and Gippsland Basins in Bass Strait, although the converse applies to *Gothanipollis bassensis*. Both species grew in complex evergreen mesotherm to megatherm (subtropical-tropical) rainforest communities (Macphail 2007) that began developing at relatively high (60-65°S) palaeolatitudes in southern Australia following rapid greenhouse warming of the planet by 5°C to 10°C at the Paleocene/Eocene boundary ca. 55.8 million years ago (Paleocene-Eocene Thermal Maximum) and subsequent hyperthermal events that characterised the Early Eocene (see Dickens 2009). Presumed reworked specimens occur in Quaternary deposits at Coal Head in Macquarie Harbour and in cuttings along the road between Strahan and Lowana Road, e.g. behind the Cool Store at Regatta Point (Macphail et al., 1993a)

2. Except in the offshore Bass Basin and onshore extensions in northern Tasmania, middle to late Eocene deposits are rare (see Macphail 2007) and there is insufficient evidence to determine whether Loranthaceae species such as *Tricolpites thomasii* ever extended southwards into the mountainous interior or along the west or east coasts. Stratigraphic evidence from the Bass Basin indicates *Tricolpites thomasii* became extinct at about the major cooling event that marks the Eocene/Oligocene boundary at 33.9 million years ago (references in Wei 1991; Exon et al. 2004). *Tricolpites simatus* is recorded in early Oligocene sediments at Lemonthyme Creek (see Plate 5, Figure D in Macphail and Hill 1994): *Gothanipollis bassensis* survived into Oligo-Miocene time in western Tasmania and also in the Bass, Gippsland and Murray Basins in mainland southeast Australia.

3. *Gothanipollis perplexus* and a complex of morphotypes centred on this species and *G. gothani* first appear in the early Oligocene in Tasmania, with the majority of records coming from the disputed glacial sequences at Lemonthyme Creek and Wilmot Dam in the Mersey-Forth Valleys (cf. Paterson 1965; Paterson et al. 1967; Fitzsimons et al. 1993; Macphail et al. 1993b; Macphail and Hill 1994). These and other records from correlative sediments from the Lea River and near Zeehan provide compelling evidence that at least three species of Loranthaceae were growing in *Nothofagus* warm temperate rainforest in western Tasmania during the late Paleogene: It is possible that the host trees included *Nothofagus* spp. (cf. Vidal-Russell and Nickrent 2008b).

4. *Gothanipollis perplexus* may have reached Tasmania by trans-oceanic dispersal, since the morphospecies first occurs in late Eocene deposits in New Zealand (Pocknall and Mildenhall 1984) but is absent in late Eocene deposits at Prydz Bay, East Antarctica (see Macphail and Truswell, 2004; Truswell and Macphail 2008). The Lemonthyme Creek records of *Gothanipollis perplexus* and *G.* cf. *perplexus* are significantly older than the earliest records of *Gothanipollis perplexus* in the epicontinental Murray Basin in southeast Australia (Table 2). It is unlikely any of the currently known *Gothanipollis* species evolved in Tasmania, although the migration route(s) are unknown.

5. Fossil Loranthaceae pollen have not been recorded in the late Oligocene-early Miocene cool-climate flora at Monpeelyata (920 m elevation) on the eastern Central Plateau (Macphail et al. 1991), nor at the one known late Pliocene site in western Tasmania, the Linda Valley palaeosol (Macphail et al. 1995).

6. At least one fossil species, '*Amylotheca*' *pliocenica*, which first appears in the late Pliocene in eastern Australia and New Guinea, did not extend as far south as Tasmania or the Bass and Gippsland Basins (Macphail 1997, 2007). The reason for this may be cool to cold climates, since *Amylotheca* is restricted to coastal rainforest from central NSW northwards into southern New Guinea (Barlow 1966; Henty 1995). Fossil specimens are found in late Pliocene assemblages from the northwest Murray Basin (Scotia Province) to the Atherton Tableland in northeast Queensland Australia (Kershaw and Sluiter 1982; Macphail 1999).

7. Specimens belonging to the complex of morphotypes centred on *Gothanipollis bassensis*, *G. gothanii* and *G. perplexus* are rare but ubiquitous in (1) the middle Pleistocene Regency interglacial sequence in the King River Valley near Queenstown (Fitzsimons et al. 1990), and (2) in the lower ca. 40 m of a 62 m thick sequence of lacustrine clays at Darwin Crater in the Andrew River Valley ca. 20 km east of Macquarie Harbour (Howard and Haines, 2007). The lake sediments show normal magnetic polarity, the Brunhes-Matuyama Boundary is not recorded (Barton 1987) and, based on K/Ar fission-track dates of 0.82 million years ago on a siliceous impact glass (Darwin Glass) associated with the crater, are middle to late Pleistocene (Lo et al. 2002): The section has yet to be fully pollen-analysed, but Loranthaceae pollen commence at 22.0 m and are consistently present down to 44.7 m, and occur between 61.0 m and 61.5 m, immediately above the contact with poorly sorted, coarser crater fill deposits (Colhoun 1988; Colhoun and Van der Geer 1998; E.A. Colhoun and M.K. Macphail, unpubl. data). The section between 44.7 m and 61 m has not been analysed but we anticipate Loranthaceae pollen will be present.

8. The Regency sequence represents the replacement of montane scrub rainforest by *Nothofagus cunninghamii*-Podocarpaceae cool temperate rainforest. Specimens of *Gothanipollis perplexus*

and *G.* cf. *perplexus* occurring between c. 22 m and 45 m in Darwin Crater mostly occur in cool temperate rainforest pollen assemblages dominated by Huon Pine *Lagarostrobos franklinii*), associated with lesser amounts of Subantarctic Beech (*Nothofagus cunninghamii*) and Celery-top Pine (*Phyllocladus aspleniifolius*) and occasional tree-fern (*Cyathea* spp.) spores, but are absent in assemblages representing lower, more open vegetation types such as wet scrub/heath and herbfield (E.A. Colhoun, unpubl. data). A not unreasonable interpretation is that mistletoes were growing on trees at the margins of temperate rainforest during the middle Pleistocene and the host species may have included *Lagarostrobos* and *Nothofagus*.

9. The top 20 m of sediment in Darwin Crater is believed to have accumulated between Oxygen Isotope Stages (OIS) 1 to OIS 7, indicating that Loranthaceae became locally? extinct in the Andrew River Valley sometime before the OIS 7 Interglacial, i.e. before 0.24-0.25 million years ago.

10. Four specimens of Loranthaceae pollen have been found in last glacial and postglacial sediments in western Tasmania – at Lake St. Clair on the Central Plateau (F. Hopf, unpubl. data), at Dove Lake, Cradle Mountain (Dyson 1995), and at an unnamed tarn on the summit of the Tyndall Ranges (Macphail and Colhoun 1982). Since the two Dove Lake specimens occur in near surface muds, it is equally likely that the Lake St. Clair and Tyndall Ranges specimens have been long-distance transported, e.g. by migrating birds.

Discussion

All mistletoe lineages within the Santalales appear to have evolved from root parasite ancestors (Vidal-Russell and Nickrent, 2008a, b). Molecular dating of genera divergence times indicates that the 'aerial' parasitic habit evolved in the Misodendraceae sometime between ca. 80 and 25 million years ago (early Campanian to late Oligocene), in the Viscaceae between ca. 72 and 43 million years ago (late Campanian to middle Eocene), in the Santaleae between ca. 46 and 53 million years ago (middle to early Eocene), in Amphorogyneae between ca. 53 and 22 million years ago (early Eocene to early Miocene), and the Loranthaceae at ca, 27 to 28 million years ago (mid Oligocene). On this model, '*Amylotheca*' *pliocenica* and? *Gothanipollis perplexus* are likely to represent hemi-parasitic shrublets growing on and within the branches of trees and shrubs: Whether the Eocene parent plants of *Gothanipollis bassensis*, *G.* cf. *gothani*, *Tricolpites simatus*, *T. thomasii* and early Oligocene parent plant(s) of *G.* cf. *perplexus* were root parasites or represent wholly extinct clades of aerial parasites is unknown, and likely to remain so without macrofossil evidence.

Pollen evidence for the origins of the Loranthaceae per se is complicated by the uncertain evolutionary relationship of the family to late Cretaceous taxa, in particular *Cranwellia*, a morphogenus that first occurs in the Campanian in Antarctica (Askin 1990), and *Aquilapollenites*, one of many extinct angiosperm genera within the *Triprojectacites* group that dominated Northern Hemisphere Maastrichtian palynofloras (see Farabee 1993). However, there is compelling pollen evidence in the form of an extinct morphospecies of *Anacolosa* (*Anacolosidites acutullus* Cookson and Pike 1954) that one clade within the Santalales was present in northwest Australia during the late Cretaceous (see Macphail 2007). The *Anacolosidites* species that most closely resembles modern Loranthaceae pollen (*A. sectus* Stover and Partridge 1973: Figure 3a) does not appear before the middle Eocene in southeast Australia. Whether late Cretaceous morphotypes assigned to *Cranwellia* and *G. gothani* ssp. *plicus* in China (see Plate 150: Figures 1-3 in Song et al. 2004) are in fact fossil members of the Loranthaceae has yet to be confirmed.

Phylogenetic and fossil evidence (Vidal-Russell and Nickrent 2008a, b) suggests that the family evolved in East Gondwana (Australia-New Guinea, New Zealand, South America) during the late Cretaceous, with relatively late dispersals into Africa, India, Asia and Europe. For example, the genus *Loranthus* occurs in Europe and Asia, but the split between it and *Cecarria* (a monotypic genus spanning Australia to the Philippines) has been estimated at less than 30 million years ago. Several other clades of similar or younger age have members in both Australasia and Asia (Vidal-Russell and Nickrent 2008a, b). Nevertheless, morphospecies resembling modern Loranthaceae pollen have been recorded from early to late Eocene deposits in northern and central Europe, the United States and China (references in Muller 1981; Song et al. 2004) and it is possible that the family first appeared in both hemispheres about the same time but the ancient Northern Hemisphere lineages subsequently became extinct. Whether such speculations are an artefact of inadequate fossil data is unknown. However, it is tempting to hypothesise that the migration of birds and possibly bats (see Archer et al. 1991:137, 230) across the equator is a plausible reason for the near-contemporaneous first appearance of Loranthaceae in both hemispheres. This conjecture is reinforced by records of Loranthaceae pollen at sites where alternative modes of transport are improbable. Examples are the now-submerged Ninetyeast Ridge in the remote Indian Ocean where *Gothanipollis* cf. *gothani* occurs in Oligocene sediments (see Kemp and Harris 1975), and the summits of mountains in tropical South America, where Loranthaceae pollen has been recovered from modern snow samples (see Reese et al. 2003).

The data demonstrate that fossil species of Loranthaceae in Tasmania were tolerant of megatherm and then mesotherm climates during the Paleogene. For example, macro- and microfossils from late-early Eocene deposits at Regatta Point and Lowana Road confirm that the tropical mangrove palm *Nypa* lined tidal channels within Macquarie Harbour (Partridge 1976; Pole and Macphail 1997). Other mesotherm-megatherm rainforest taxa associated with *Tricolpites simatus* and *Gothanipollis bassensis* include *Anacolosa*, Arecaceae (*Arecipites*, *Dicolpopollis*), *Ascarina*, Cunoniaceae, *Dysoxlum*, *Freycinetia*, *Ilex*, Meliaceae and Strasburgeriaceae (*Bluffopollis scabratus*). In contrast, Loranthaceae pollen (*Gothanipollis bassensis*) found at the middle-late Eocene Loch Aber site in northeast Tasmania (M.K. Macphail, unpubl. data) grew in floristically diverse, warm temperate rainforest dominated by *Nothofagus* and gymnosperms that are now extinct in Tasmania. For example, trees included a species similar to *Nothofagus moorei*, now endemic to NSW, Podocarpaceae, including *Acmopyle*, now endemic to New Caledonia and Fiji, and an extinct *Araucaria* belonging to the South American section *Columba*: Average leaf lengths in the Loch Aber fossil flora are smaller than in early Eocene floras, consistent with a general decrease in mean air temperatures from the early to middle Eocene (Hill and Carpenter 1991; Carpenter et al. 1994).

Major global cooling that followed subsidence of the South Tasman Rise and Drake Passage to abyssal depths and formation of the Circumantarctic Current during the Eocene-Oligocene transition meant Loranthaceae species growing in cool temperate rainforest in the Mersey-Forth River Valleys had become resilient to even cooler (microtherm) conditions by about 33 million years ago (see Macphail et al. 1993b; Exon et al. 2004). Whether Loranthaceae were continuously present in Tasmania from the early Oligocene to Early Miocene and Quaternary is unclear due to lack of tightly dated palynosequences south of the offshore Bass Basin. Clusters of *Gothanipollis* spp. found in the basal 40 m of lake sediments in Darwin Crater and in the Regency Interglacial deposit indicate that Loranthaceae were growing in cool temperate rainforest in western Tasmania during the middle Pleistocene. We note that these rainforest communities differed from modern *Nothofagus cunninghamii*-Podocarpaceae closed forest in Tasmania in two minor aspects only – the tree-fern stratum was dominated by *Cyathea* spp., not *Dicksonia antarctica*, and in some areas the subcanopy stratum included several small tree and shrub genera that are now restricted to mainland Australia (*Haloragodendron*, *Quintinia*, *Symplocos*).

A number of taxa now endemic to South America became established in Tasmania during the Paleogene. Examples (fossil species in parentheses) are: *Nothofagus* subgenus *Nothofagus* (*Nothofagidites flemingii*) in the earliest Paleocene, *Embothrium* (*Granodiporites nebulosus* Stover and Partridge 1973) in the earliest Oligocene, and the tribe Mutisieae of the Asteraceae (*Mutisiapollis patersonii* Macphail and Hill 1994) and the ground fern *Lophosoria quadrapinnata* (*Cyatheacidites annulatus* Cookson 1947) by the early Oligocene. Given the observed close relationship of *Misodendron* with *Nothofagus* in southern South America, it seems reasonable to speculate that this mistletoe family also could have been present in Tasmania in the past.

Misodendron pollen are distinctive due to their sphaeroidal shape and apiculate sculpture. However, confirming the presence of this genus in Tasmania will be difficult since the small size (19-22 μm) of its pollen grains means specimens can be easily lost during processing, or overlooked, or the pollen type may have evolved after the split between *Misodendron* and other Santalaceae. Fossil specimens are likely to be assigned to the morphogenus *Compositoipollenites* Potonié ex Potonié 1960 (see Macphail and Cantrill 2006). At present, the only species of *Compositoipollenites* that has been formally described in Australia is *C. tarrogoensis* Truswell and Owen 1988 from a middle Eocene palynosequence at Bungonia on the Southern Tablelands of NSW (cf. Plate 38, Figure 462 in Heusser 1971, Figure 8 H-U in Truswell and Owen 1988). Similar morphotypes occur in late Maastrichtian sediments in the Bonaparte Basin in northern Australia, an early Eocene sample from the Cape Sorell-1 petroleum exploration well in the offshore Sorell Basin (Figure 1), and a probable early Eocene deposit in the Styx River Valley in southeast Tasmania (Macphail 2007). *Viscum* pollen are easily recognised by the presence of stout spines (see Tafel 35: 1-4 in Beug 2004), but, as yet, no fossil specimens have been recorded in Australia (G.S. Hope pers. comm.).

Reconstructing and interpreting the history of mistletoes in Tasmania will continue to depend on the slow accumulation of microfossil evidence, often from sites that have been uncovered by chance events (see Hill 1987). We consider it unlikely that palynologists working on Tertiary to late Quaternary deposits will have mis-identified or overlooked larger morphospecies such as *Gothanipollis perplexus* or '*Amylotheca' pliocenica*, and therefore conclude that, even when establishment was successful, Loranthaceae always were rare elements in the Tasmanian flora. Why this should be so is more likely to be a consequence of cool/wet climates than a lack of suitable host species. If correct, then any climatic barriers that have prevented the establishment of mistletoes during the Holocene may well be breached through warming/drying of Tasmania in the near future.

Acknowledgements

We express our thanks to Dr Mike Fletcher (Department of Archaeology and Natural History, Australian National University) for his useful comments on an earlier draft of the paper.

References

Archer, M., Hand, S.J. and Godthelp, H. 1991. *Riversleigh: The Story of Animals in Ancient Rainforests of Inland Australia*. Sydney: Reed Books.

Askin, R.A. 1990. Campanian to Paleocene spore and pollen assemblages of Seymour Island, Antarctica. *Review of Palaeobotany and Palynology* 65:105-113.

Barlow, B.A. 1984. Loranthaceae. In: George, A.D. (ed), *Flora of Australia* Vol. 22, pp. 68-131.

Canberra: Australian Government Publishing Service.

Barton, C. 1987. Palaeomagnetism. Age of the Darwin Crater. *Bureau of Mineral Resources Yearbook* (1987): 36-37.

Beug, H-J. 2004. *Leitfaden der Pollenbestimmung für Mitteleuropa und angrenzende Gebiete.* Munich: Verlag.

Carpenter, R.J., Hill, R.S. and Jordan, G.J. 1994. Cenozoic vegetation in Tasmania: macrofossil evidence. In: Hill, R.S. (ed), *History of the Australian Vegetation: Cretaceous to Recent,* pp. 276-298. Cambridge: Cambridge University Press.

Colhoun, E.A. 1988. Cainozoic Vegetation of Tasmania. *Department of Geography, University of Newcastle Special Paper* ISBN 0 7259 0621 9.

Colhoun, E.A. and van de Geer, G. 1998. Pollen analysis of 0-20 m of Darwin Crater, western Tasmania. In: Horle, S. (ed), *International Project on Palaeolimnology and Late Cenozoic Climate Newsletter* 11:68-89.

Dickens, G.R. 2009. Early Cenozoic hyperthermals: the sedimentary record of rapid global warming and massive carbon input. *CSEG Recorder* (February):28-32.

Dyson, W.D. 1995. A pollen and vegetation history from Lake Dove, western Tasmania. BSc. Honours thesis, University of Newcastle (unpubl.).

Erdtman, G. 1966. *Pollen Morphology and Plant Taxonomy: Angiosperms.* New York: Hafner Publishing.

Exon, N.F., Kennett, J.P. and Malone, M.J. 2004. 1. Leg 189 synthesis: Cretaceous-Holocene history of the Tasmanian Gateway. In: Exon, N.F., Kennett, J.P. and Malone, M.J. (eds), *Proceedings ODP Scientific Results* 189:1-37.

Farabee, M.J. 1993. Morphology of triprojectate fossil pollen: form and distribution in space and time. *Botanical Review* 59:211-249.

Fitzsimons, S.J., Colhoun, E.A., van de Geer, G. and Hill, R.S. 1990. Definition and character of the Regency Interglacial and Early-Middle Pleistocene stratigraphy in the King Valley, western Tasmania. *Boreas* 19:1-15.

Fitzsimons, S.J., Macphail, M.K., Colhoun, E.A., Kiernan, K. and Hannan, D. 1993. Glacial climates in the Antarctic region during the late Paleogene: evidence from northwest Tasmania: Comment and reply. *Geology* 21:958-959.

Heusser, C.J. 1971. *Pollen and Spores of Chile.* Tucson: University of Arizona Press.

Hill, R.S. 1987. Discovery of *Nothofagus* fruits corresponding to an important Tertiary pollen type. *Nature* 327:56-58.

Hill, R.S. and Macphail, M.K. 1985. Reconstruction of the Oligocene vegetation at Pioneer, northeast Tasmania. *Alcheringa* 7:281-299.

Hill, R.S. and Carpenter, R.J. 1991. Evolution of *Acmopyle* and *Dacrycarpus* (Podocarpaceae) foliage as inferred from macrofossils in south-eastern Australia. *Australian Systematic Botany* 4:449-479.

Howard, K.T. and Haines, P.W. 2007. The geology of Darwin Crater. *Earth and Planetary Science Letters* 260:328-339.

Kemp, E.M. and Harris, W.K. 1975. The vegetation of Tertiary islands on the Ninetyeast Ridge. *Nature* 258:303-307.

Kershaw, A.P. and Sluiter, I.R. 1982. Late Cenozoic pollen spectra from the Atherton Tableland, north-eastern Australia. *Australian Journal of Botany* 30:279-295.

Lo, C.H., Howard, K.T., Chung, S.L. and Meffre, S. 2002. Laser fusion-40/argon-39 ages of Darwin impact glass. *Meteoritics and Planetary Science* 37:1555-1562.

Macphail, M.K. 1999. Palynostratigraphy of the Murray Basin, inland southeastern Australia. *Palynology* 23:199-242.

Macphail, M.K. 2007. *Australian Palaeoclimates Cretaceous to Tertiary. A review of palaeobotanical and related evidence to the year 2000.* CRC-LEME Open File Report 151 (Special Volume ISBN 1 921039 75 2).

Macphail, M.K. and Colhoun, E.A. 1982. "Tarn Shelf", Tyndall Ranges, West Coast Tasmania. *Department of Geography, Faculty of Military Studies, University of New South Wales Occasional Paper* 33:107-108.

Macphail, M.K., Hill, R.S., Forsyth, S.M. and Wells, P.M. 1991. A Late Oligocene-Early Miocene cool climate flora in Tasmania. *Alcheringa* 15:87-106.

Macphail, M.K., Jordan, G.J. and Hill, R.S. 1993a. Key periods in the flora and vegetation in western Tasmania 1. The early-middle Pleistocene. *Australian Journal of Botany* 41:673-707.

Macphail, M.K., Colhoun, E.A., Kiernan, K. and Hannan, D. 1993b. Glacial climates in the Antarctic region during the late Paleogene: evidence from northwest Tasmania, Australia. *Geology* 21:145-148.

Macphail, M.K., Alley, N.F., Truswell, E.M. and Sluiter, I.R.K. 1994. Early Tertiary vegetation: evidence from spores and pollen. In: Hill, R.S. (ed), *History of the Australian Vegetation: Cretaceous to Recent*, pp. 189-261. Cambridge: Cambridge University Press.

Macphail, M.K. and Hill, R.S. 1994. K-Ar dated palynofloras in Tasmania 1: Early Oligocene *Proteacidites tuberculatus* Zone sediments, Wilmot Dam, northwestern Tasmania. *Papers and Proceedings of the Royal Society of Tasmania* 128:1-15.

Macphail, M.K., Colhoun, E.A. and Fitzsimons, S.J. 1995. Key periods in the evolution of the Cenozoic vegetation and flora in western Tasmania: the Late Pliocene. *Australian Journal of Botany* 43:505-526.

Macphail, M.K. and Truswell, E.M. 2004. Palynology of Site 1166, Prydz Bay, East Antarctica. *Proceedings of the Ocean Drilling Program. Scientific results* 188:1-29.

Macphail, M.K. and Cantrill, D.J. 2006. Age and implications of the Forest Bed, Falkland Islands, southwest Atlantic Ocean: evidence from fossil pollen and spores. *Palaeogeography, Palaeoclimatology, Palaeoecology* 240:602-629.

Martin, H.A. 1978. Evolution of the Australian flora and vegetation through the Tertiary: evidence from pollen. *Alcheringa* 2:181-202.

Mildenhall, D.C. 1978. *Cranwellia costata* n.sp. and *Podosporites erugatus* n.sp. from Middle Pliocene (?Early Pleistocene) sediments, South Island, New Zealand. *Journal of the Royal Society of New Zealand* 8:253-274.

Mildenhall, D.C. and Pocknall, D.T. 1989. Miocene-Pleistocene spores and pollen from central Otago, South Island, New Zealand. *New Zealand Geological Survey Palaeontological Bulletin* 59:1-128.

Moar, N.T. 1993. *Pollen grains of New Zealand Dicotyledonous Plants.* Lincoln: Manaaki Whenua Press.

Muller, J. 1981. Fossil pollen records of extant angiosperms. *Botanical Review* 47:1-142.

Nickrent, D.L., Malécot, V., Vidal-Russell, R. and Der, J.P. 2010. A revised classification of the Santalales. *Taxon* 59: 538-558.

Ogg, J.G., Ogg, G. and Gradstein, F.M. 2008. *The Concise Geologic Time Scale.* Cambridge: Cambridge University Press.

Paterson, S.J. 1965. Pleistocene drift in the Mersey and Forth Valleys – probability of two glacial stages. *Papers and Proceedings of the Royal Society of Tasmania* 99:115-124.

Paterson, S.J., Duigan, S.L. and Joplin, G.A. 1967. Notes on Pleistocene deposits at Lemonthyme Creek in the Forth Valley. *Papers and Proceedings of the Royal Society of Tasmania* 101:221-225.

Partridge, A.D., Trigg, K., Montgomerie, N.R. and Blevin, J. 2002. Review and compilation of open file micropalaeontology and palynology data from offshore Tasmania. *Geoscience*

Record 2002 [CD 3 of 7].

Pocknall, D.T. and Mildenhall, D.C. 1984. Late Oligocene-early Miocene spores and pollen from Southland, New Zealand *New Zealand Geological Survey Palaeontological Bulletin* 51:1-66.

Reese, C.A., Liu, K-b and Mountain, K.R. 2003. Pollen dispersal and deposition on the ice cap of Volcan Parinacota, southwestern Bolivia. *Arctic, Antarctic and Alpine Research* 35:469-474.

Song, Z, Wang, W. and Huan, F. 2004. Fossil pollen records of extant angiosperms in China. *Botanical Review* 70:425-458.

Stover, L.E. and Partridge, A.D. 1973. Tertiary and Late Cretaceous spores and pollen from the Gippsland Basin, southeastern Australia. *Proceedings of the Royal Society of Victoria* 85:237-286.

Truswell, E.M. and Owen, J.A. 1988. Eocene pollen from Bungonia, New South Wales. *Memoir Association of Australasian Palaeontologists* 5:259-284.

Truswell, E.M. and Macphail, M.K. 2008. Polar forests on the edge of extinction: what does the fossil spore and pollen evidence from East Antarctica say? *Australian Systematic Botany* 22:57-106.

Vidal-Russell, R. and Nickrent, D.L. 2008a. The first mistletoes: origins of aerial parasitism in Santalales. *Molecular Phylogenies and Evolution* 47:523-537.

Vidal-Russell, R. and Nickrent, D.L. 2008b. Evolutionary relationships in the showy mistletoe family (Loranthaceae). *American Journal of Botany* 95:1015-1029.

Wei, W. 1991. Evidence for an earliest Oligocene abrupt cooling in the surface waters of the Southern Ocean. *Geology* 19:780-783.

13

Wind v water: Glacial maximum records from the Willandra Lakes

Jim M. Bowler
School of Earth Sciences, University of Melbourne, Victoria
jbowler@unimelb.edu.au

Richard Gillespie
The University of Wollongong, NSW, and The Australian National University, Canberra, ACT

Harvey Johnston
New South Wales Office of Environment and Heritage, Buronga, NSW

Katarina Boljkovac
The Australian National University, Canberra, ACT

Introduction

Using lakes and dry basins for discerning the patterns of climatic change faces a number of challenges. Study of the Willandra basins (Figures 1 and 2) involves reconstruction of their environmental history and its relationship to controlling climatic change. The various methods for data interpretation and hydrologic reconstruction have been discussed elsewhere (Bowler 1971, 1998). In early evaluation, the history of the Willandra Lakes was summarised in terms of three major stratigraphic units, each related to a major cycle of hydrologic change. The units Golgol, Mungo and Zanci were designated as responsible for the major stratigraphic events in the history of the system (Bowler 1971). The Zanci drying phase was directly related to the Last Glacial Maximum (LGM), the period of maximum ice extent in the Northern Hemisphere (Clark et al. 2009), glacial ice on Kosciuszko (Barrows et al. 2001) and lowest sea level (Lambeck and Chappell 2001). In later revisions, Bowler (1998) defined an Arumpo unit between Mungo and Zanci, and a final Mulurulu unit to account for evidence of late-stage filling especially in Lake Mulurulu. Coincident with the LGM (ca. 25,000-17,000 years ago), the Zanci phase, and the assumed aridity it represents, is of

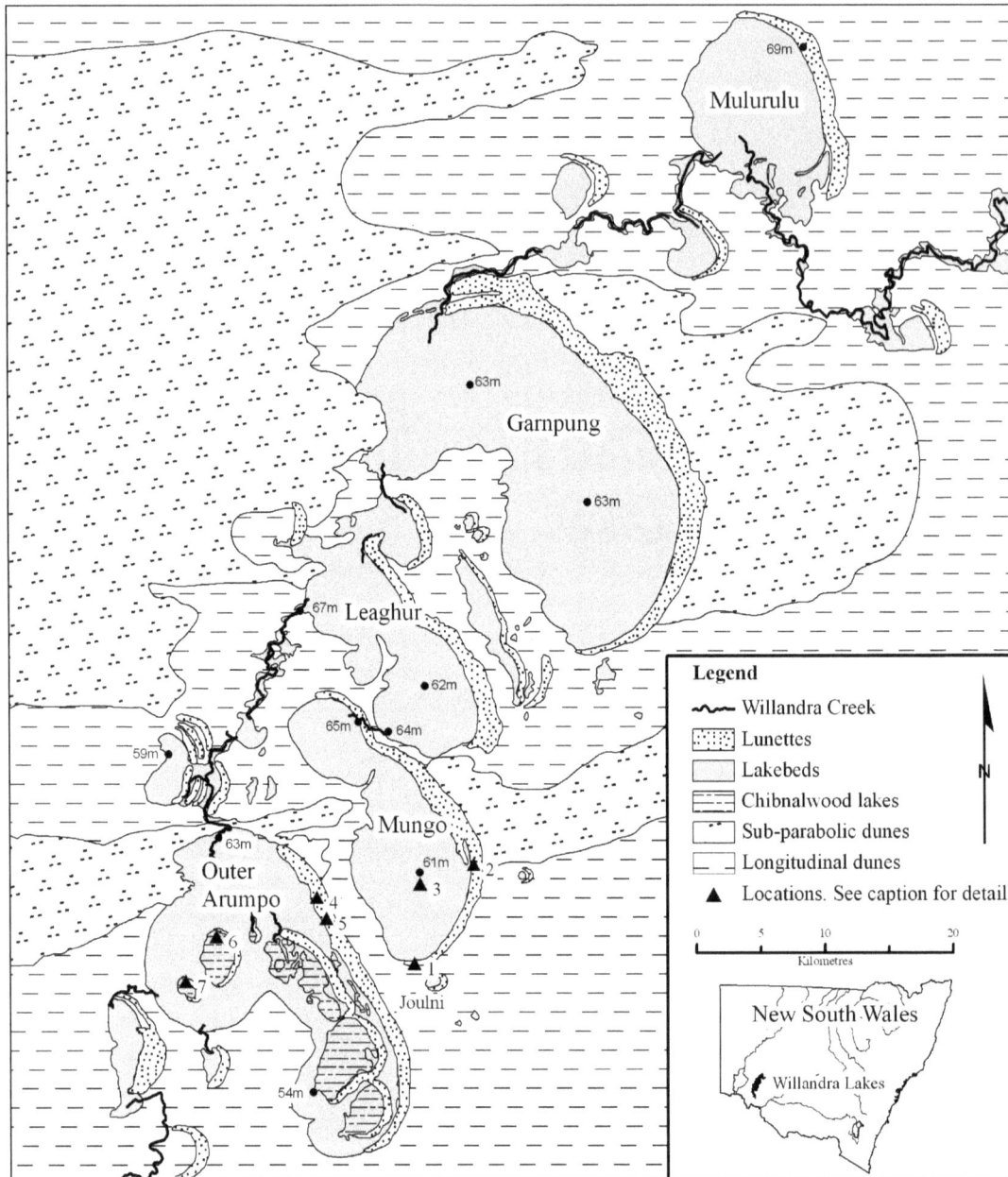

Figure 1. The Willandra Lakes (excluding Prungle), with inset location map. The location of sites mentioned in the text are shown with the following numbers: 1: Mungo I and III burial site. 2: Walls of China Visitors Area. 3: Shell Tank, Mungo lake bed. 4: Long Waterhole Gully, Outer Arumpo lunette. 5: Top Hut 1 and 3, Outer Arumpo lunette. 6: Lake Arumpo. 7: Lake Bulbugaroo. Spot heights are shown in metres (m) above Australian Height Datum (AHD).

special significance for the glacial age environmental history of inland Australia.

Questions of age, although of great importance, frequently remain difficult to assess because age estimates are constrained by limitations resulting from difficulties in field settings, and analytical constraints often with inevitably large errors (Bowler et al. 1970; Gillespie 1997; Bowler 1998). The application of radiocarbon dating to organic carbon remains frequently failed to provide consistent and reliable results, whereas dating freshwater mussel-shell carbonate proved much more reliable (Bowler 1998; Gillespie 1998). With the advent of accelerator dating enabling the analysis of very small samples, new opportunities arose; application to fish otoliths (ear stones) has provided new levels of information, shedding light particularly on the latter part of the chronology relating to the last glacial period (Gillespie 1997; Kalish et al. 1997; Boljkovac 2009). This information provides a new understanding of the patterns of lake-dune behaviour at this important time of changing climate regimes.

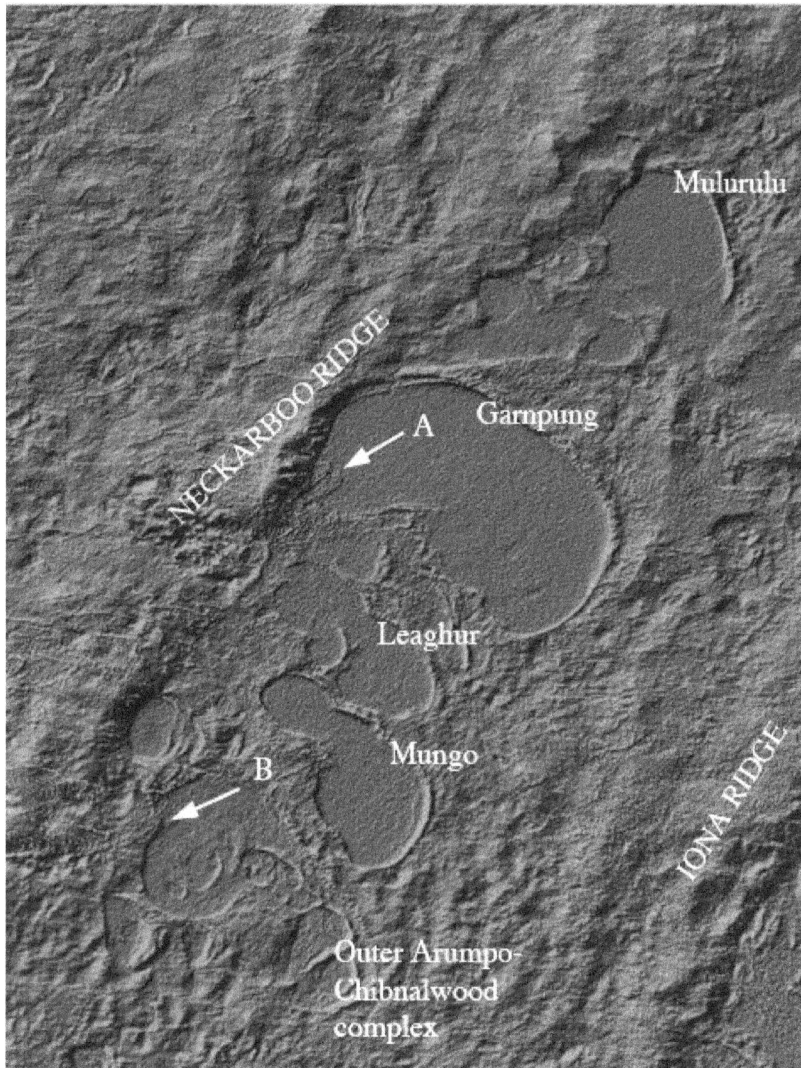

Figure 2. Digital radar image of Willandra basins (Figure 1) in the trough between tectonic Neckarboo and Iona ridges, with cover field of Mallee sand dunes. The arrows indicate mallee dune transgression across lake floors at Garnpung (A) and Outer Arumpo (B). Data from Australian Landform Atlas, courtesy Mike Sandiford. *http://jaeger.earthsci. unimelb.edu.au/Images/ Landform/landform.html*

Despite the importance of age evaluation in assessing field data, that evaluation must be consistent with field evidence. In the Willandra chain of inter-connected basins, a response to any particular climatic effect involving change in the hydrologic balance will be felt and expressed differently across different basins in the same region. Responses will vary between basins, depending on shape, size and especially position in the drainage chain (see Figure 3).

Basins close to catchments will retain much more water for longer periods than those in more distant regions further down the chain. Basins close to uplands, regions of most available water, or supplied by major rivers remained full even under conditions of glacial maximum aridity. As examples, the lake basins of Urana (Page et al. 1994), Kanyapella on the Murray River (Bowler 1967, 1978) and Nekeeya near the Grampians (Bowler 1999) all remained at high levels during the glacial maximum due entirely to the efficient nature of the catchments which supplied them. Similarly, glacially wet conditions pertained in glacial age catchments in the Flinders Ranges (Williams et al. 2001). Basins more distant from effective water supply, between or away from streams, responded more sensitively to the onset of aridity (Nanson et al. 1992; Kershaw and Nanson 1993; Magee et al. 1995; Hesse et al. 2004).

As an inter-connected but descending basin chain, the Willandra Lakes can be explained by a 'stairway' analogy, in which relative positions impose limitations on hydrologic behaviour. Basins adjacent to each other and especially those connected by overflow, one supplying the other, will vary according to individual thresholds of wetting and drying. The application of these principles to basins in the Willandra chain of lakes has not been previously addressed.

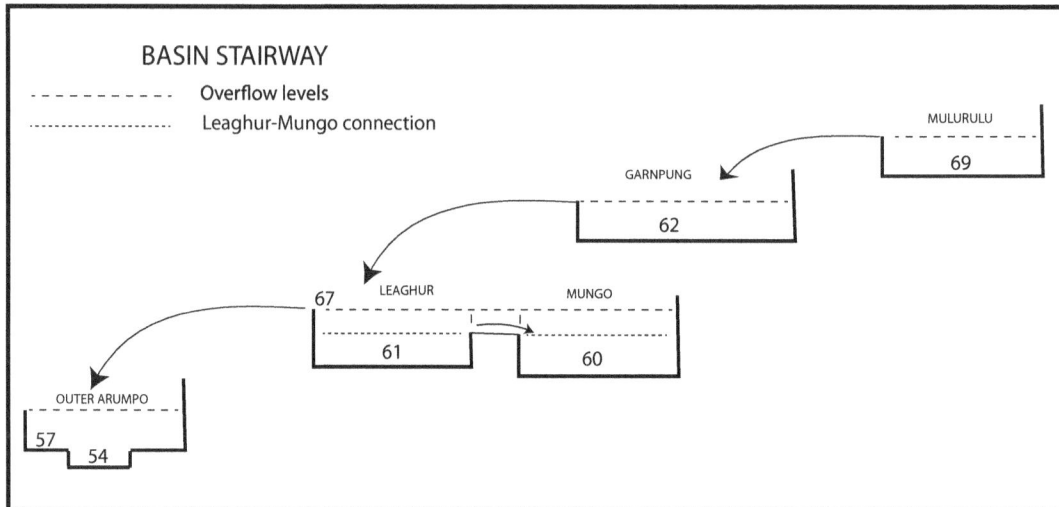

Figure 3. Diagrammatic representation of Willandra basins in 'stairway', relative to each other. Approximate levels of basin floors in metres AHD, from digital elevation and local surveys.

This paper reviews new evidence and revises previous interpretations, with special relevance to a new understanding of glacial maximum environments and their expression within the Willandra system.

Principles of lake-dune response

The simple rise and fall of a major wet to dry cycle is complicated by fluctuations at intermediate levels in the rising and falling stages. These are particularly important in providing opportunity for episodic or intermittent periods of lake floor exposure resulting from seasonal or secular discharge variability (see Figure 4A).

When the flood trend reaches an overflow level, short-term oscillations disappear. Maximum level permits generation of gravels and beach sands, providing material for foreshore dune development (quartz sand dunes, QSD). On the drying trend, the basin acts as a terminal system and salinity increases. The threshold for clay dune development (PCD) is reached when oscillations in the shallowing lake expose the dry lake floor. Basin floor 1, highest in the drainage line, first receives the floodwaters and is last to dry. Basin floor 2, further down the drainage line, receives waters later and has a much shorter duration of lake-full environment. While the long wavelength water's rise and fall defines the major cycle, smaller high-frequency oscillations of a seasonal or secular nature impose important complications. Critical elements are twofold.

Firstly, the maximum water level in the Willandra basins is controlled by the overflow outlet level connecting each basin to the next in the chain (Figure 3). At that point, production of high wave energy, with consequent deposition of beach sands and shoreline gravels, defines this important status in the lake-full regime. At overflow level, it remains in through-flow status; any accumulating salts are flushed downstream. Simultaneously, under conditions of sufficient sand supply, substantial sandy beaches provide the materials on which wind action operates, with the construction of foreshore dunes derived from well-sorted beach sands (Bowler 1973). Identified as quartz sand dunes (QSD), although often with a significant shell component, these identify a process dramatically different from the aeolian mantles composed of pelletal clay materials, the pelletal clay dune facies (PCD). As long as the level remains at overflow, seasonal-secular oscillations are suppressed by annual recharge.

Secondly, as soon as water level falls below the outlet, the basin begins to act as a terminal system. Evaporation exceeds combined precipitation and inflow, resulting in increasing salt

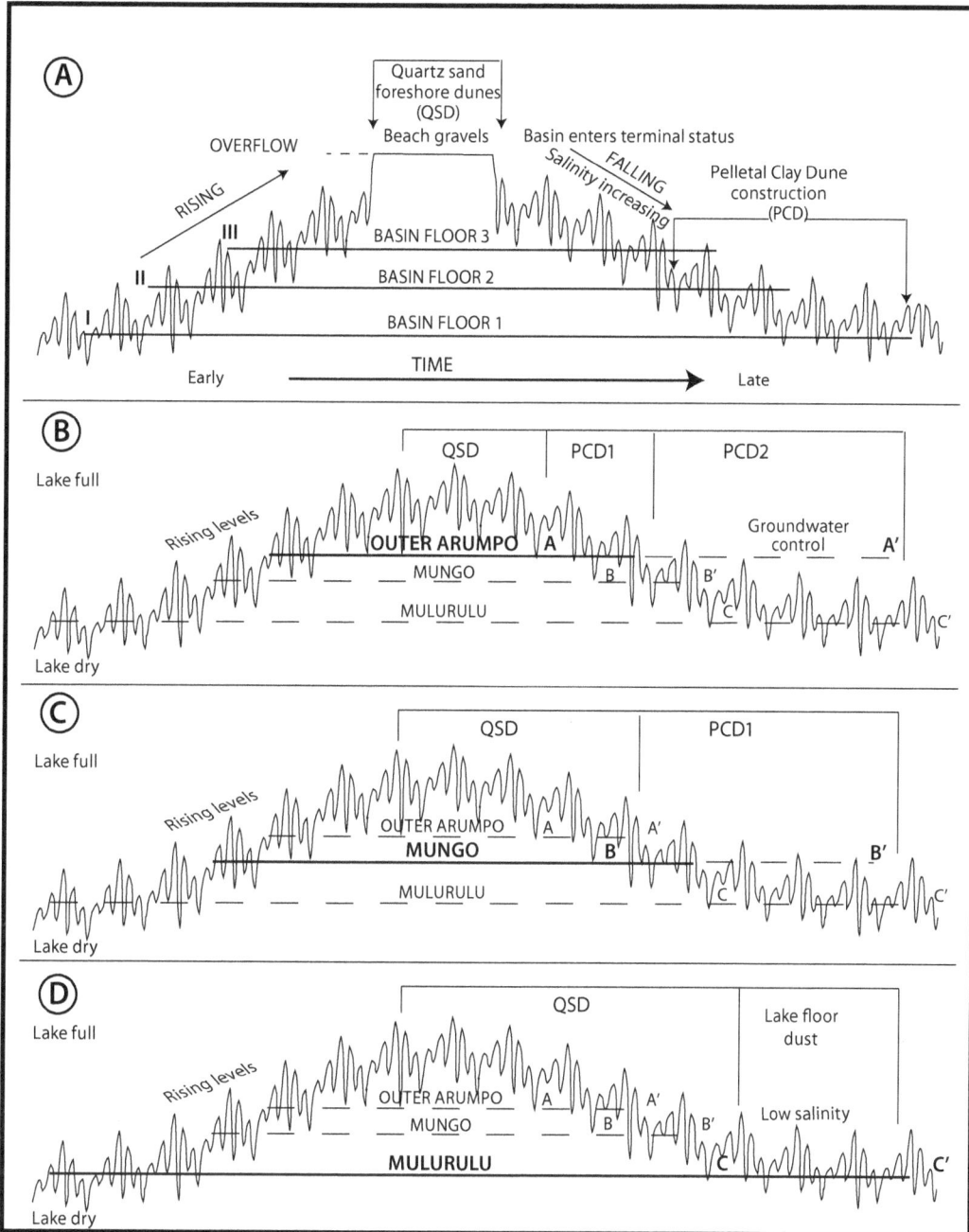

Figure 4. Detailed evaluation of different responses to lake level changes between three interconnected basins.

(A) Idealised major wet to dry lake level oscillation, showing differences in response between basins depending on locations on drainage chain. Second-order seasonal to secular oscillations are superimposed on major rise and fall of levels. Horizontal lines depict arrival times on the basin floors, showing duration of the flood events for three basins (1 Mulurulu, 2 Mungo and 3 Outer Arumpo). On the rising trend, items I, II and III depict water arrival time at each basin. Foreshore quartz dune development (QSD) may follow arrival and stabilisation at maximum water level. On falling trend, A-A', B-B' and C-C' define times of lake floor drying, setting the scene for major clay dune development (PCD).

(B) Outer Arumpo: Pattern illustrates relatively short-term lake-full stage. Quartz dunes generated from high lake level. On falling stage, early exposure of lake floor (A) permits development of clay dunes (PCD1). With further development of drying (A to A'), clay dune deposition accelerates. As saline surface waters contract, groundwaters nourish pools of higher salinity within inner lake floor depressions. Crystallisation of salts generates next phase of gypseous clay dunes inside outer basin perimeter (PCD2), with generation of Chibnalwood, Inner Arumpo and Bulbugaroo lunettes (Figure 5).

(C) Mungo: Longer period of permanent water permits development of beach sands and foreshore dune facies (QSD). In falling stage (B-B'), period of clay dune development follows evolution of water to moderate salinity, no gypsum. This develops simultaneously with PCD2 dune growth at Chibnalwood.

(D) Mulurulu: Longer period of water retention permits extensive development of beach sands and foreshore dunes. Frequent flushing by overflow prevents build-up of salts. Final drying under conditions of low salinity permits deflation of dust (clay and silt) from lake floor, but lacks effects of salt crystallisation characteristic of other basins when in terminal phase.

concentrations as water levels progressively fall. With shallowing, the situation is reached when seasonal oscillations expose at least part of the lake floor to drying. That immediately opens the possibility for wind action to begin the process of erosion, with deposition of sand-sized clay pellets (or silt-clay components) on the lunette. Thereafter, the drying process is dominated by production of the clay dune facies. This may progress to a high-salinity-groundwater phase in which seasonal groundwater discharge affects the lower parts of the drying system. Such saline groundwaters may then produce high gypsum dune content, in contrast to earlier lower-salinity deflation clays. This occurred to the inner gypseous basins in the contracting phase of Outer Arumpo. Clay dune deposition was then terminated by either a successive return of lake water, or a fall in groundwater levels, followed by vegetation recolonisation.

These idealised events take on additional significance when applied to different basins on the drainage chain, situations in which differences in the sequence dictate different levels and different response times. We take examples from three basins, examined in a two-stage process:

1.   Evaluate responses to a specific wetting and drying event.

2.   Extend to the larger region over the long-term stratigraphic record.

Major wetting and drying event

A general pattern of different basin responses to changes in controlling discharge is summarised diagrammatically in Figure 4. The example chosen relates to the final major episode of water in the Willandra system, that which preceded and led to final drying. Responses of three basins (Outer Arumpo, Mungo and Mulurulu) to a major phase of lake filling and drying have been chosen as representatives to illustrate the nature of changes that applied throughout the system. On the drying trend, critical levels A-A', B-B' and C-C' are levels at which basin floors begin exposure, while lengths of basin floor levels represent the relative duration of waters in Mulurulu, Mungo and Outer Arumpo respectively.

The Outer Arumpo basin (Figures 1, 3 and 4B), farthest from Mulurulu, is last to fill but first to dry. With a reduction in water level, short-term oscillations would provide early exposure there of the lake floor, permitting development of clay dune building.

Basin flood phase

At the onset of a high lake phase, until water reaches the basin overflow level, little distinctive sedimentary record is preserved until stabilisation at overflow level results in the deposition of high-energy beach sands and gravels. Depending on the sand supply, this sets the scene for the production of foreshore quartz dunes, QSD. The duration of this phase varies between lakes. It is shortest in the downstream system of Outer Arumpo compared with those upstream.

Drying phase

With progressive reduction in discharge associated with regional drying, when the level of the supply basin (Leaghur in this case) falls below outflow level, its downstream neighbour Outer Arumpo begins to act as a terminal system. Deprived of inflow with reduced levels, the terminal basin will eventually permit seasonal oscillations to expose the lake floor. This triggers the onset of local clay dune production on the outer ridge, the location of the former shoreline. Meanwhile, the upstream basins Mungo and Mulurulu remain within the quartz dune production phase.

In Outer Arumpo, as water level falls and salinity progresses to the salt accumulation phase, an important change occurs. Local depressions on the basin floor become sites for the next dune production phase. With final loss of surface water, saline groundwater outcrops on the lowest levels where salt crystallisation breaks clays, forming a soft 'fluffy' pelletal surface. Strong winds do the rest, resulting in the production of a new generation of saline clay dunes with high gypsum content. This is exactly the sequence in the Outer Arumpo-Chibnalwood lunette

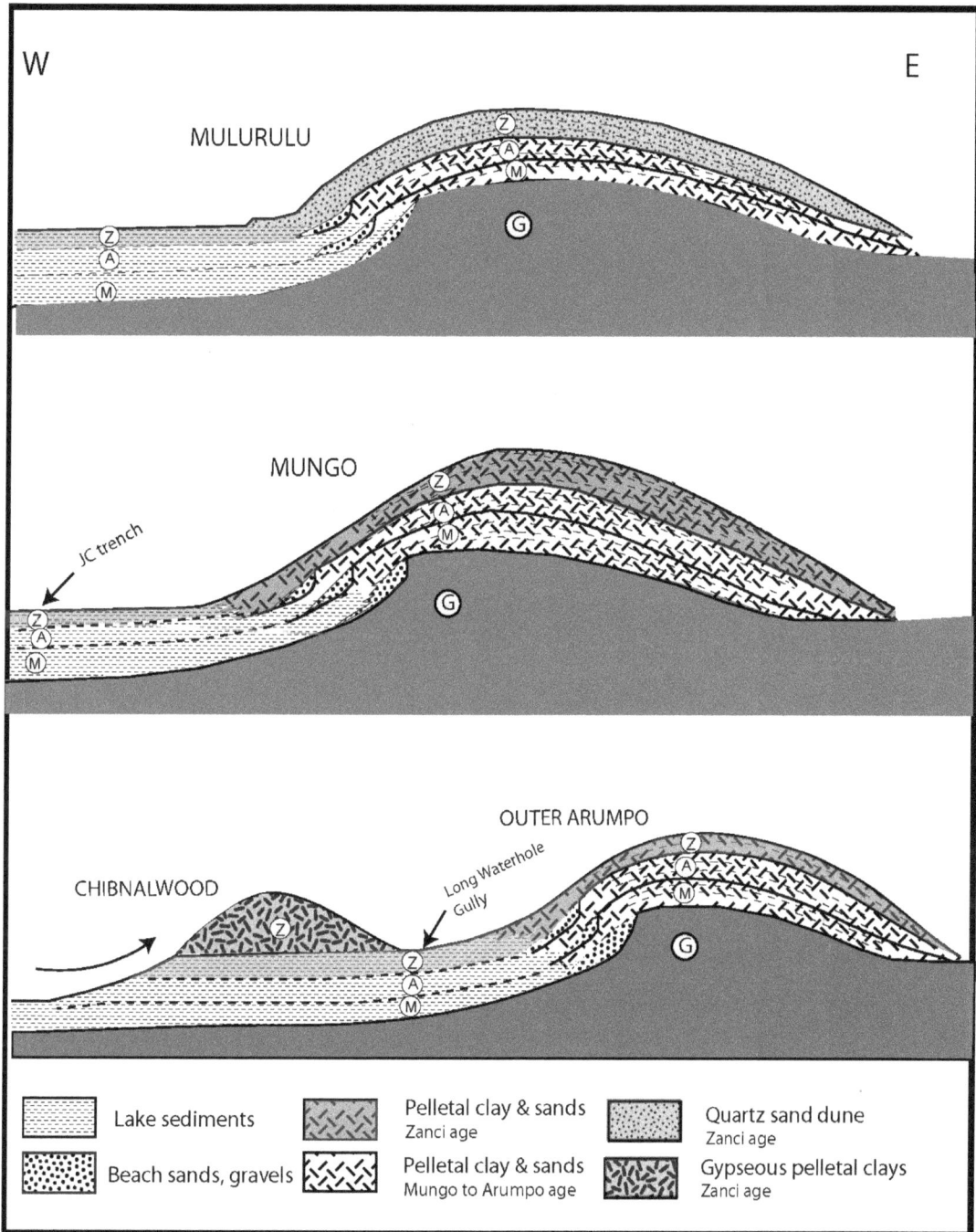

Figure 5. Schematic diagram to show relationships of LGM Zanci phase lake-dune units in three basins, Outer Arumpo, Mungo and Mulurulu. The diagram emphasises contrasts in the nature of final Zanci aeolian units between groundwater control at Chibnalwood, thin pelletal clay on Outer Arumpo, thicker at Mungo, in contrast to dominantly quartz sand control at Mulurulu. Lake bed units are discriminated on the basis of disconformities on the floor of Outer Arumpo (Long Waterhole) and Mungo (JC trench). Relatively uniform clay on Mulurulu floor does not permit direct unit discrimination.
Stratigraphic units: **G:** Golgol **M:** Mungo **A:** Arumpo **Z:** Zanci.

complex, where final groundwater control was isolated to within depressions on the basin floor within the confines of the outer shoreline, and the resultant gypseous ridges accumulated on the margins of those depressions. The inner basin lunettes of Chibnalwood, Arumpo and Bulbugaroo (Figure 1) are the result of that final wind-controlled groundwater deflation phase. A second generation of gypseous dune building followed the earlier drying trend expressed in clay dunes on the Outer Arumpo shoreline, thus defining the basin-in-basin pattern of the Outer Arumpo-Chibnalwood complex (Figures 1, 2 and 5).

At Mungo (Figures 4C and 5), where the dry threshold was not crossed until later, the trigger phase for clay dune development was reached after the initial clay dune phase at Outer Arumpo had already begun. It would have occurred contemporaneously with the gypsum dune development of the inner basins at Chibnalwood and Arumpo. Meanwhile, Mulurulu upstream (Figures 4D and 5) remained under water influence for much longer, the high lake preserving longer conditions for quartz dune formation.

Chronology and stratigraphic revision

Shell and otolith ages

A comprehensive list of radiocarbon ages for Willandra Lakes lacustrine shells and fish otolith (see Figures 7 and 8) is presented in Table 1. These details were accumulated by RG from a range of published (including Bowler et al. 1970; Clark 1987; Gillespie 1997; Kalish et al. 1997; Bowler 1998; Johnston and Clark 1998; Boljkovac 2009) and unpublished sources, brought together for the first time in the context of specific basin origin. Some samples are from controlled archaeological excavations, for example by Harry Allen and Wilfred Shawcross; many others were collected from the surface of actively eroding lunettes. Among the shell ages (n = 39), many are based on fragmented multi-shell samples, pretreatment was usually simple

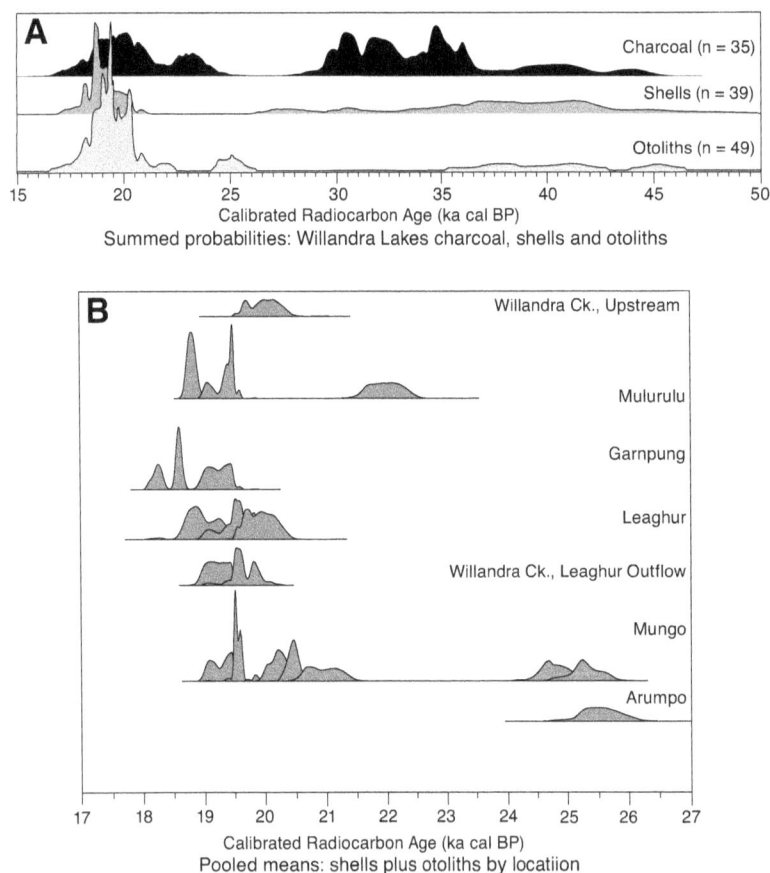

Figure 6. Radiocarbon ages for shells and otoliths.
A. Summed probability distributions (2σ calibrated) for 39 shell and 49 fish otolith radiocarbon ages, calculated with the atmospheric data in Reimer et al. (2009), using OxCal 4.1 (Bronk Ramsay 2009).
B. Location-based probability distributions of calibrated pooled mean ages for groups of shells and otoliths through the LGM. Peak height is approximately proportional to the number of samples (1-10) in each group, lake-bed shells excluded.

Figure 7. Shell-midden collection of mussel, Velesunio ambiguus, dated to near 18,000 years cal BP. L. Mulurulu. Photo from Bowler's LAKE MUNGO CD: *Window to Australia's Past*.

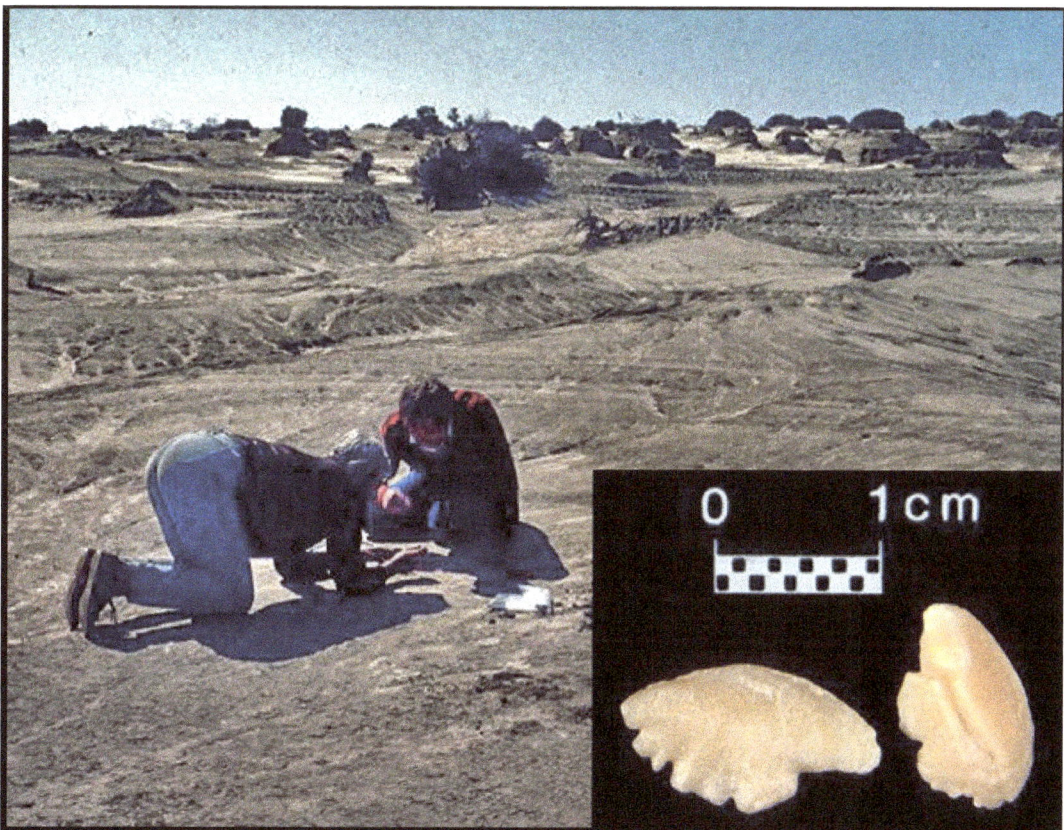

Figure 8. Lunette erosion, L. Mungo Tourist Site (Location 2, Figure 1). Keryn Walshe (left) and the late Peter Clark examine extensive fish hearth, with more than 500 otoliths on a single bedding plane in pelletal sandy clays. Inset: Golden Perch otoliths. Organic carbon date >39,000 cal BP. Photo from Bowler's LAKE MUNGO CD: *Window to Australia's Past*.

Table 1. Radiocarbon ages for shell and fish otolith carbonate from the Willandra lakes, with location of samples (dash = unknown). Calibration of raw ages using IntCal09 atmospheric data from Reimer et al. (2009), showing median probability and one sigma uncertainty from OxCal 4.1 (Bronk Ramsay, 2009). Otolith number (brackets) in Material column from Kalish et al. (1997). Laboratory codes: ANU (radiometric, ANU); N (radiometric, RIKEN); SANU (SSAMS, ANU); OZ (AMS, ANSTO).

Basin Location	Site	Material dated (carbonate)	Lab. No.	¹⁴C Age (BP)	1 sigma (years)	IntCal09 Age Cal BP (median)	1 sigma (years)
Mulurulu	ME1	shell	ANU-880A	15,120	235	18,290	260
Mulurulu	-	shell	N-2036	15,400	205	18,630	250
Mulurulu	ME1	shell	ANU-880B	15,450	240	18,670	290
Mulurulu	OT-04	otolith	SANU-8823	15,500	90	18,710	85
Mulurulu	ME3	shell	N-2035	15,500	205	18,720	240
Mulurulu	ME3	shell	ANU-948B	15,560	240	18,770	280
Mulurulu	3	otolith (7)	OZB-608	15,700	340	18,920	350
Mulurulu	OT-12	otolith	SANU-8819	15,720	90	18,860	160
Mulurulu	OT-11	otolith	SANU-8820	15,750	90	18,890	170
Mulurulu	1	otolith (23)	OZB-624	16,150	120	19,260	180
Mulurulu	1	otolith (11)	OZB-612	16,200	120	19,320	190
Mulurulu	1	otolith (9)	OZB-610	16,250	130	19,400	210
Mulurulu	ME3 m3/3	otolith (29)	OZB-630	18,350	220	21,890	270
Mulurulu	ME8	shell	ANU-1914	29,360	620	33,860	710
Garnpung	-	otolith (5)	OZB-606	14,500	600	17,640	710
Garnpung	-	otolith (4)	OZB-605	14,800	370	17,990	430
Garnpung	-	otolith (1)	OZB-602	14,900	130	18,200	220
Garnpung	-	otolith (6)	OZB-607	15,050	400	18,220	440
Garnpung	-	otolith (3)	OZB-604	15,150	230	18,310	250
Garnpung	-	otolith (30)	OZB-631	15,250	100	18,530	190
Garnpung Intensive	-	otolith (8)	OZB-609	15,400	160	18,640	200
Garnpung East	GS4	shell	ANU-266	15,400	210	18,630	260
Garnpung/ Leaghur	GL1	shell	ANU-373B	15,480	210	18,700	250
Garnpung	-	otolith (2)	OZB-603	15,600	270	18,810	300
Garnpung/ Leaghur	GL13	shell	ANU-2210	16,050	200	19,170	210
Garnpung/ Leaghur	GL13	shell	ANU-1913	16,100	220	19,210	240
Garnpung	GS10	shell	ANU-2969	25,600	500	30,370	460
Garnpung/ Leaghur	GL24	shell	ANU-2206	30,300	800	34,890	920
Garnpung	GS1	shell	ANU-2163	31,600	850	36,110	930
Garnpung/ Gogolo	GG19	shell	ANU-2164	32,100	950	36,720	1090
Leaghur	LN7	shell	ANU-461	15,690	235	18,900	250
Leaghur West	LW20	shell	ANU-2769	16,360	200	19,520	270
Leaghur West	LW5	shell	ANU-2203	16,700	180	19,860	220
Leaghur Peninsula	LP3	shell	ANU-372B	27,160	900	31,690	910
Leaghur Peninsula	-	shell	OZJ716	41,100	1200	44,740	920
Willandra Ck	TH B1, Leaghur Outflow	otolith (26)	OZB-627	16,000	130	19,150	160

Table 1: *Continued*

Basin Location	Site	Material dated (carbonate)	Lab. No.	¹⁴C Age (BP)	1 sigma (years)	IntCal09 Age Cal BP (median)	1 sigma (years)
Willandra Ck	TH B1, Leaghur Outflow	otolith (27)	OZB-628	16,400	150	19,550	210
Willandra Ck	WCU13, Upstream	shell	ANU-2238	16,800	180	19,950	240
Mungo Lakebed	Shell Tank, 1.5m	shell	N-1656	14,500	290	17,640	380
Mungo	WOC7(152)	shell	ANU-2967	16,000	180	19,140	190
Mungo	BMLM-158	otolith	SANU-8817	16,070	90	19,190	150
Mungo	LAC-9009	otolith	SANU-8816	16,090	90	19,210	160
Mungo	WOC1, surface	otolith (22)	OZB-623	16,100	150	19,200	180
Mungo	WOC1, surface	otolith (13)	OZB-614	16,150	140	19,260	190
Mungo	WOC1, surface	otolith (16)	OZB-617	16,150	140	19,260	190
Mungo	WOC1, surface	otolith (15)	OZB-616	16,200	220	19,310	270
Mungo	WOC1, surface	otolith (18)	OZB-619	16,200	120	19,320	190
Mungo	WOC1, surface	otolith (20)	OZB-621	16,250	170	19,380	240
Mungo	WOC1, surface	otolith (10)	OZB-611	16,350	150	19,510	220
Mungo	WOC1, surface	otolith (21)	OZB-622	16,450	170	19,630	220
Mungo	WOC1, surface	otolith (17)	OZB-618	16,550	120	19,720	180
Mungo	WOC1, surface	otolith (19)	OZB-620	16,600	120	19,750	180
Mungo	BMLM-011	otolith	SANU8825	16,710	90	19,860	180
Mungo	WOC1, surface	otolith (24)	OZB-625	16,800	130	19,960	210
Mungo	LAC-9004	otolith	SANU-8813	16,820	90	19,990	190
Mungo	BMLM-008	otolith	SANU-8821	16,980	90	20,170	160
Mungo	WOC1, surface, square D19	otolith (25)	OZB-626	17,000	120	20,180	200
Mungo	BMLM-010	otolith	SANU-8824	17,020	90	20,220	160
Mungo	BMLM-007	otolith	SANU-8805	17,060	90	20,260	170
Mungo	BMLM-211	otolith	SANU-8806	17,120	90	20,320	190
Mungo	MN	shell	OZJ372	17,160	150	20,390	290
Mungo	LAC-9001	otolith	SANU-8811	17,190	90	20,400	230
Mungo	BMLM-121	otolith	SANU-8807	17,240	90	20,460	240
Mungo	LAC-9008	otolith	SANU-8814	17,510	100	20,840	240
Mungo	LMB1-01	otolith	SANU-8810	20,580	130	24,580	210
Mungo	BMLM-156	otolith	SANU-8818	21,030	130	25,090	240
Mungo	WOC7(151)	shell	ANU-2205	32,400	900	37,080	1060
Mungo	WOC7(152)	shell	ANU-331	32,750	1250	37,510	1410
Mungo	LMB3-B1	otolith	SANU-8809	33,090	270	37,790	480
Mungo	WOC1, surface	otolith (14)	OZB-615	33,200	950	37,940	1120
Mungo	WOC7(151)	shell	ANU-2204	33,700	1200	38,510	1370
Mungo	Mungo B Shawcross Pit 1.1m	otolith (28)	OZB-629	35,900	800	40,900	800
Mungo	WOC1, surface, square D20	otolith (12)	OZB-613	36,100	950	41,030	910
Mungo Lakebed	ML 21	shell	ANU-4134	37,200	1000	41,960	820
Mungo	LAC-9002	otolith	SANU8812	41,740	650	45,190	470

Table 1: *Continued*

Basin Location	Site	Material dated (carbonate)	Lab. No.	¹⁴C Age (BP)	1 sigma (years)	IntCal09 Age Cal BP (median)	1 sigma (years)
Outer Arumpo	TH1 H3 area, surface, square B6	otolith	CAMS-1930	21,190	190	25,330	290
Outer Arumpo	TH1 (OA4)	shell	N-1664	22,600	430	27,230	570
Outer Arumpo	OA1	shell	ANU-2207	23,600	610	28,400	720
Outer Arumpo	TH2 (OA5)	shell	ANU 2586	33,100	850	37,820	1000
Outer Arumpo	TH1 (OA4)	shell	ANU-1692	34,450	1500	39,280	1560
Outer Arumpo	TH3 (OA6)	shell	N-2032	35,100	1800	39,810	1760
Outer Arumpo	TH3 (OA6)	shell	CAMS-1925	35,590	990	40,530	990
Outer Arumpo	TH3 (OA6)	shell	N-1665	35,600	1650	40,310	1600
Outer Arumpo	TH1 (OA4)	shell	CAMS-2038	36,170	1110	41,040	1050
Outer Arumpo	TH1 (OA4)	shell	ANU-1470	36,200	3825	40,380	3510
Outer Arumpo	TH3 (OA6)	shell	ANU-1697	36,800	1800	41,410	1660
Outer Arumpo Lakebed	TH3 (OA6)	shell	ANU-306	38,500	2550	42,740	2240
Prungle South Lakebed	-	shell	OZJ717	44,100	1400	47,380	1260
Non-finite dates							
Mungo Lakebed	Shell Tank, basal	shell	N-1657	>37,800			
Mungo Lakebed	-	shell	N-2033	>37,800			
Mungo Lakebed	-	shell	N-1658	>37,800			
Outer Arumpo Lakebed	-	shell	N-1663	>37,800			
Prungle South Lakebed	-	shell	N-1662	>37,800			

hand-scraping with weak acid etching, and the oldest results were close to maximum age limits for the available Liquid Scintillation Counting (LSC) technology. All fish otolith ages (n=49) were measured using Accelerator Mass Spectrometry (AMS) on single otolith samples, often with better decontamination procedures, stable isotope correction and smaller uncertainties.

We assume here that all shell and otolith radiocarbon ages listed in Table 1 are Conventional Radiocarbon Ages as defined in Stuiver and Polach (1977); they were calibrated with OxCal 4.1 (Bronk Ramsay 2009) using IntCal09 atmospheric data from Reimer et al. (2009). Two sigma calibrated age range (95.4% confidence interval, rounded to 10 years) and mean calibrated age with two sigma uncertainty (rounded to the nearest 100 years) are given in Table 1; non-finite ages were not calibrated. The very similar patterns of summed probability distributions for all shell ages and all otolith ages are shown in Figure 6A.

Several groups of shell and otolith ages in our date list, particularly at the younger end of the age range, were found to be statistically identical using the two sigma significance test in CALIB 6.0 (Stuiver et al. 2010). Pooled mean ages for these groups were calibrated as above, and these location-based probability distributions are shown in Figure 6B, where peak height is approximately proportional to the number of samples (1-10) in each group. The predominance of results at the younger end of the age range should be seen as a consequence of taphonomy, site destruction and opportunistic sample collection (Surovell et al. 2009; Johnson and Brook 2011).

Where measurements are available from both scintillation counting and AMS for midden shells, all results overlap at one standard deviation, for example from Outer Arumpo: Top

Hut 1 (OA4) midden has three dates with pooled mean age 40,600 ± 900 cal BP, and Top Hut 3 (OA6) midden has four dates with pooled mean age 40,700 ± 600 cal BP. There are otoliths (and emu eggshells) from the Mungo lunette with ages similar to these Outer Arumpo middens, and luminescence dates for the Lower Mungo to Upper Mungo transition and the Mungo I and III human burials (Bowler et al. 2003) in the same age range (40,000 ± 2000 cal BP). These observations suggest that the oldest Outer Arumpo and Mungo shell midden results may not be significantly in error, while the 47,300 cal BP lake-bed shells from Prungle South probably give some indication of the true age for shell samples previously reported as >37,800 BP (Bowler 1998).

Although not always from stratigraphically controlled contexts, the association of the samples with particular basins provides new information on the presence of water at particular times. The pattern of lake changes requires that we discriminate between longer climatic change events and short-term high variability flood events (as in Figure 4). Major changes in climate will register phases of long duration compared with changes in flood regime. By reason of the time required to establish the fish and shell populations, most aquatic samples will be associated with the former. In the distribution of age samples, the cluster around a modal peak helps discriminate between long-term and short-term changes in water level. Lake-full events of longer duration are more likely to generate clusters of related ages. Short-term flood events may be represented by outlier ages from main clusters.

Within the LGM timeframe, ca. 25,000-17,000 years ago, relationships between ages and source basins assist in defining patterns of change. Age patterns between Outer Arumpo, Mungo and Mulurulu are taken here as key indicators. Leaghur and Garnpung, falling between Mungo and Mulurulu, add little that is not already evident in the other two. The modal clustering of age groups relevant to the LGM period is particularly important – they provide the basis for a new data set defining this important period.

In addition to their value as reliable 'clocks', otoliths act as the most valuable archives retaining the isotopic signature of water chemistry and temperature during the life of the fish. Its secretion of a daily bank of aragonite permits high-resolution chronology within the life of the fish, separable on a seasonal, monthly or even daily basis (Kalish et al. 1997). Preliminary results of in situ $\delta^{18}O$, Sr/Ca and $^{87}Sr/^{86}Sr$ measurements on *Macquaria ambigua* (Golden Perch) otoliths (Boljkovac 2009) have already begun to yield such data, shown in Figure 9.

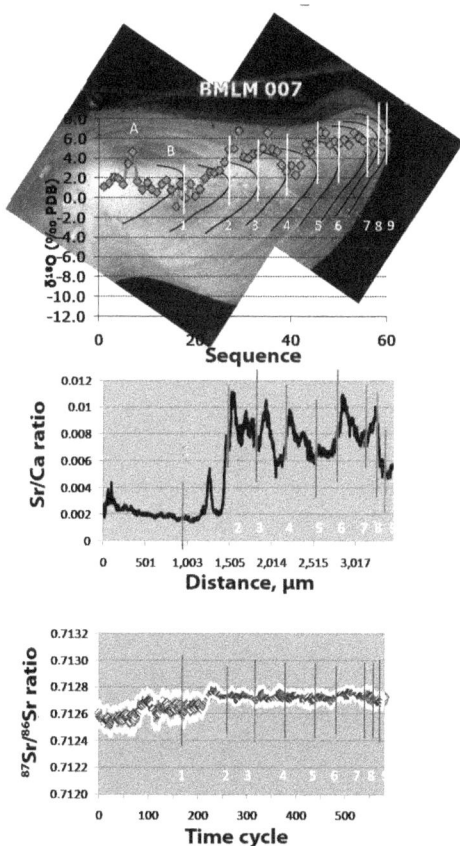

Figure 9. Measurements of $\delta^{18}O$, Sr/Ca and $^{87}Sr/^{86}Sr$ ratios in a single otolith from Mungo, showing annual variations (Boljkovac 2009).

Stratigraphic revision

Sampling constraints impose controls on the overall age distribution patterns. While many dating samples originate on erosion surfaces, ages from stratigraphically controlled sites inform some aspects of origins. Older samples, such as the otolith (OZB629, 40,900 years ago) from the Mungo B excavation (Shawcross 1998) and an organic sample from a substantial fish hearth at the Walls of China visitor area (ANU-2964, >34,500 BP) are important in the early sequence. While the latter represents meal remains, the results of human harvesting, many otoliths represent assemblages from fish mortality. Indeed, there is often a correlation between the two. In the Walls of China fish hearth example (Figure 8), hundreds of fish remains lie on, and were covered by, a blanket of dune pelletal clay. That fish feast was the result of increased salinity in the drying lake, and was an opportunistic recovery of hundreds of fish by people living on the shoreline at that time, providing a prototype example of human response to drying conditions as described earlier (Bowler 1998).

The relevance of otolith origins from mass mortality events is further enhanced by differences in modal age patterns between Mungo, 19,200-20,500 years ago, and Mulurulu, 18,800-19,300 years ago (Figure 6). If dominantly the result of human harvesting, similar ages might be expected between basins experiencing mutually similar aquatic environments. The evidence points more towards otolith age patterns reflecting later drying in the upstream basin.

In this sense, a strong bias exists between human behaviour and expressions of environmental change. The onset of drying with diminishment of regional resources goes hand in hand with increased pressure on diminishing aquatic resources. In an opportunistic sense, the occasion of fish populations made groggy by effects of rising salinity would immediately focus human harvesting on that occurrence.

A further link between otolith abundance and a period of increased salinity is suggested by overlapping ages in both otoliths and PCD development. Mungo otolith ages in the 19,200-20,400 year ago range are synchronous with ages for PCD development at that time, dated on charcoal or carbonates (ANU-312, 19,700 years ago and ANU-319, 19,400 years ago from Mungo; ANU-320, 19,900 years ago from Chibnalwood). Lacking fine time-scale resolution (to decades or even centuries) we cannot discriminate between ages of these mutually exclusive events. Within the limitations of dating accuracy, the disappearance of fish ca.19,000 years ago is entirely consistent with, and supportive of, evidence for increased salinity leading to dune building commencing about that time.

Arumpo

Detailed construction of wet-dry oscillations during Arumpo time remain poorly constrained both in terms of accurate ages and the detailed nature of oscillations. However, the reality of the main transgression is clearly defined by high-level gravels at Joulni disconformably overlying sediments of Upper Mungo age (Figures 7 and 10 in Bowler 1998), and in Outer Arumpo by layered sands in Long Waterhole, characterised there by frequent limnaeid shells, algal tubules and organic staining (Level D, Figure 11 in Bowler 1998). These three identical components occur in sandy clays of equivalent age at the toe of the Mungo lunette at the Walls of China Tourist Area, evidence of high carbonate precipitation, with molluscs grazing on microbial communities on aquatic plant surfaces.

The Arumpo lake phase is registered in Outer Arumpo by dates in the 27,600-25,300 year ago range (ANU-2207, N-1664, CAMS-1930). The absence of aquatic ages between 24,500 and 22,000 years ago is consistent with clay dune accretion on the Outer Arumpo dune at that time. This break in lake deposition is represented by the disconformity above the limnaeid-rich shell bed with microbial (algal) carbonates in Long Waterhole gully (Level E, Figure 11 in Bowler 1998).

High carbonate production would favour relatively warm conditions preceding descent into later cold conditions of the glacial maximum. Ages in the 24,500-27,000 year ago range are consistent with otolith and shell samples from Mungo and Outer Arumpo (Table 1, Figure 6B). This phase is tentatively assigned to 34,000-24,500 years ago, followed by subsequent drying and clay dune accretion both at Mungo and on the main lunette at Outer Arumpo. The return of water heralded the next wet phase (Lower Zanci).

Zanci

The onset of Zanci flooding occurred before 20,000 years ago (OZB626, OZJ372). This event is represented at Long Waterhole by the final transgressing, with beach sands containing *Sphaerium* shells (Level F, Figure 11 in Bowler 1998) disconformably overlying Arumpo deposits. Equivalent events are registered on the floor of Lake Mungo, where sands with *Sphaerium* overlie soil-cracked clays 1 m below the present surface, evident in a 3 m deep trench (JC Trench, Figure 5). Shells dated from Shell Tank on the lake floor (N-1656, 17,600 years ago) probably suffer from soil carbonate contamination, providing a minimum age only. Unionid shell dates from Tysons Lake near Hatfield (N-1655, 20,300 years ago; N-2034, 22,000 years ago) help define the last arrival of freshwater in the wider region ca. 22,000-20,000 years ago. The filling stage of the Zanci cycle at Mungo is consistent with that chronology.

Dune building

While ages of lacustrine environments are defined by dates from fish and shellfish, dune-building ages must rely on organic or carbonate radiocarbon ages supported by OSL, as defined from earlier publications (Bowler 1971, 1998; Bowler et al. 1972, 2003). Important independent confirmatory evidence is now appearing from otolith isotope analyses. Results of in situ $\delta^{18}O$, Sr/Ca and $^{87}Sr/^{86}Sr$ measurements on five Golden Perch otoliths recovered from both surface middens and the lake floor suggest frequent drying out of Lake Mungo 19,200-20,400 years ago, and a major $\delta^{18}O$ increase of 5‰ over nine years (Figure 9) indicates a rapid increase in salinity on a drying trend (Boljkovac 2009).

The close of the Zanci lacustral phase set the trigger for the next phase of pelletal clay dune building (PCD). A midden on the northern Mungo lunette (OZJ372, 20,400 years ago) represents shell harvesting approaching the closing stage of the lake (Figure 10). The overlying clay dune layer is stratigraphically continuous with the uppermost unit at Joulni, dated there on charcoal to 19,700 years ago (ANU-312). That estimate is further consistent with the shell/otolith age clusters in the Mungo record (19,200-20,400 years ago), where out of 26 samples, only the probably contaminated Shell Tank date is younger than 19,000 years ago. A simultaneous date on carbonate from the gypseous dune at Chibnalwood (ANU-320,

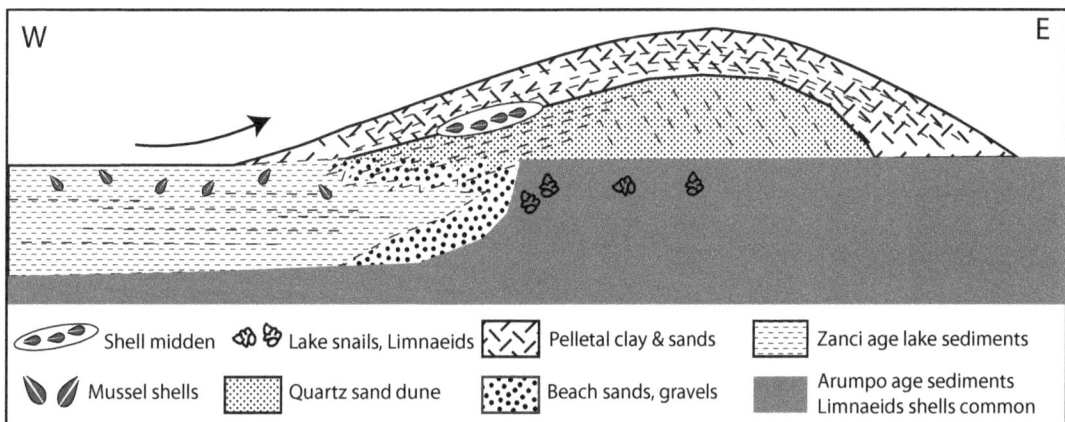

Figure 10. Schematic diagram showing notional association between LGM Zanci age disconformably cut into Arumpo unit, with frequent limnaeid shells and microbial carbonates. Lower Zanci quartz dune deflated from beach sands (20,000 years ago) overlain by pelletal clay drying phase (19,000 years ago), as identified at Mungo. Mussel shell (midden) and fish harvesting appear closely associated with the transition, wet to dry at close of lake-full phase.

19,800 years ago) would confirm contemporaneous dune formation between basins at that time (Figures 5 and 10).

Upstream, freshwater remained longer, leading to final drying of Mulurulu at 18,800 years ago. Deflation of sands from the eastern beach contributed to deposition of a final quartz sand cover contemporaneously with pelletal clay at Mungo and gypseous pelletal clay at Chibnalwood, Arumpo and Bulbugaroo, the inner basins within the Outer Arumpo system.

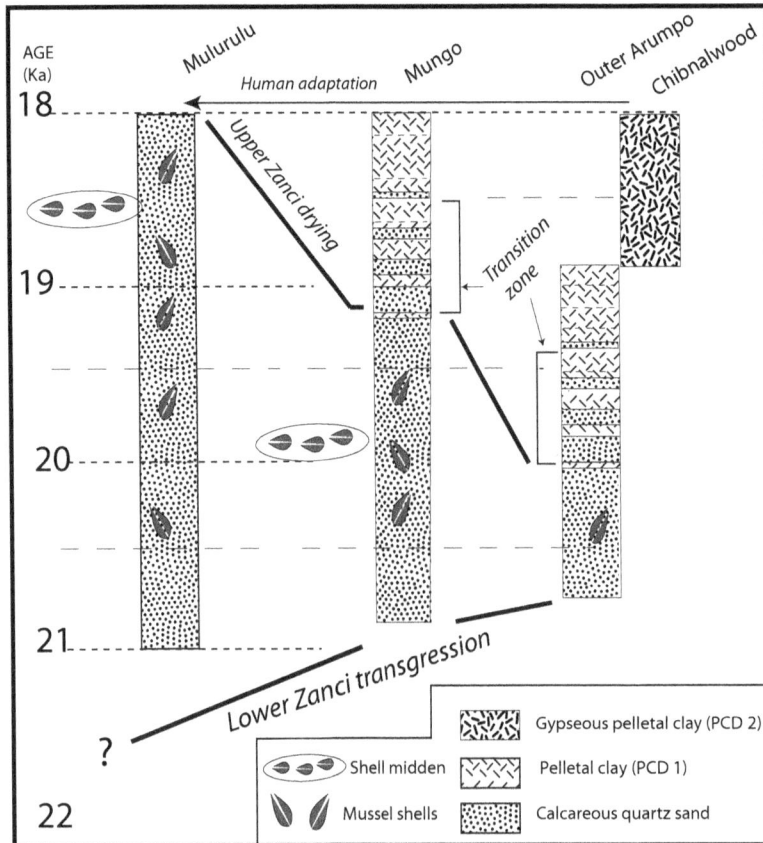

Figure 11. Simplified stratigraphic diagram depicting the Lower to Upper Zanci transition between lakes, as on Figure 4. Interbedded sands and pelletal clays developed during lake level oscillations (A-A', B-B', C-C') pass up at Mungo and Outer Arumpo into dominantly pelletal clays. Contraction to the inner basin permits saline groundwater to accelerate erosion, with deposition of gypseous clay dune at Chibnalwood (see Figure 4). At Mulurulu, waters continue to nourish sandy beaches, feeding quartz sand dunes into final lunette phase. Shell harvesting at Mungo (20,000 years ago) is continued upstream at Mulurulu after Mungo dried (19,000 years ago). People moved in response to changing water availability. Despite peak aridity, fish and shellfish continued to provided sustenance under diminishing resources.

Lacking salts necessary for pelletal clay formation, the final drying at Mulurulu did not produce the pelletal clay cover more characteristic of other lakes.

Diachroneity

The distribution of shell and otolith dates suggests drying of Lake Mungo at 19,000 years ago, compared with Garnpung at 18,400 years ago, with Lake Mulurulu and Leaghur in between. The drying relationship between Outer Arumpo, Mungo and Mulurulu is shown diagrammatically in Figure 11. The new data set permits a revision of the previous account, with construction of a new water level curve; the critical element involves the disappearance of water in Mungo by 19,000 years ago, coincident with dune building there, while water was clearly retained in Garnpung and Mulurulu until considerably later. The implications are two-fold: Firstly, this requires revision of the water level-dune building environmental synthesis curve. Secondly, nomenclature previously applied to events at this time requires modification.

Water level curve

This evidence requires amendment to the previous version of the Willandra Lakes water level oscillation curve. The general pattern of this curve, modified from Bowler (1998) and shown in Figure 12, retains the Mungo, Arumpo and Zanci major cyclic events. The style of diachronic response identified here for the final drying cycle may well have had expression

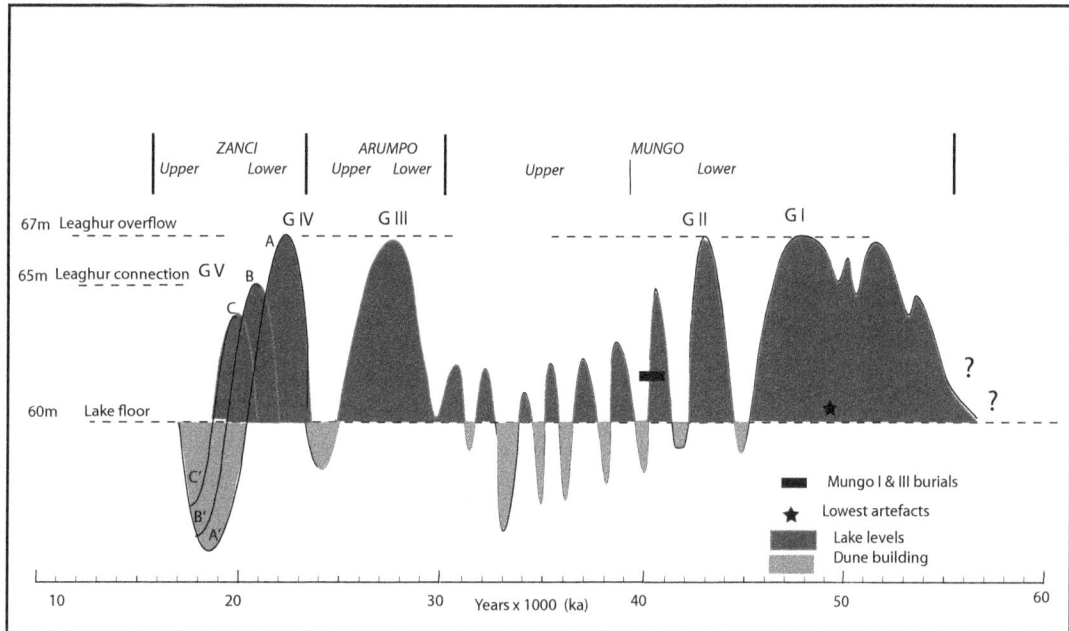

Figure 12. Lake level and dune formation summarising data mainly from Lake Mungo. Highest levels associated with gravels developed during high lake stands. LGM drying diachronous between basins A, B and C, representing progressive reduction in levels from Outer Arumpo to Mungo to Mulurulu respectively. A', B' and C' deflation events from progressively drying basin floors. Clay dune A' includes initial deposition on Outer Arumpo, with continued gypseous clay deflation from inner basins Chiblanwood, Inner Arumpo and Bulbugaroo.

between basins during earlier cyclic changes. However, the current resolution of field and age data do not permit specification of those changes. The emphasis here remains on clearer understanding of glacial maximum conditions.

The G notation in Figure 12 identifies the presence of specific gravels, indicators of high lake stand long enough to accumulate significant deposits. New stratigraphic analyses modify earlier usage in Bowler et al. (2003), the revision suggesting only a brief high lake phase coincident with Mungo I and III burials, an event insufficient to deposit an identifiable gravel signature. The event identified as G3 in Bowler et al. (2003) is now incorporated into the latter part of the G2 high lake phase.

In the construction of that curve, ages of early elements Lower to Upper Mungo were established by data drawn from earlier publications (Bowler 1971; Bowler et al. 1972). Some details of the younger Arumpo phase are clarified by the new data set. The main differences lie in two areas.

Firstly, in the time range 40,000-25,000 years ago, uncertainties remain in dating events of the Outer Arumpo lake and dune phases. In the version presented here, onset of the high lake phase represented by Level D in Long Waterhole gully (Figure 11 in Bowler 1998) and in the clays with carbonate tubules and limnaeids at Mungo Tourist site is moved towards the 30,000-28,000 year ago range. Until further work resolves these dating uncertainties, this correlation remains tentative.

Secondly, the more important revision to the curve involves the representation of the now complex and diachronic behaviour between basins in the final drying stage. This involves modification of the Zanci-Mulurulu anomaly. The tighter chronology presented here permits resolution of that period into discrete basin phases, with clay dune building progressing upstream, earliest at Outer Arumpo and later at Mulurulu (Figure 12). Although separated in time, these remain expressions of the same hydrologic event.

Nomenclature revision

The availability of new dates, especially from a range of otolith sites, has clarified an important aspect of the chronology. In so doing, it defines a new interpretation of events coinciding with, and to a large extent controlled by, climates of the Last Glacial Maximum.

In earlier interpretations, the drying of events at Mungo and Outer Arumpo were relegated to the Zanci unit, the period of major and final lunette construction. Dated to near 20,000 years ago, this event predated ages of midden shells at Mulurulu, implying a later lake filling event in that basin. Designated the *Mulurulu Unit* (Bowler 1998), that interpretation implied a complex response to glacial maximum conditions, major drying at Mungo on the one hand, followed immediately by filling of Mulurulu on the other, perhaps an unlikely combination of events during that climatic phase.

New dates now resolve the dilemma; there are no longer two discrete wet-dry events. Instead, they form part of a single complex, one in which controls are exerted as much by variations between basins as in the overall trend towards aridity.

In the Willandra system, the nature of the connected chain of lakes imposes a differential time control of filling and drying of each basin. This results in a diachronic relationship between specific events (Figure 11`). The diagrammatic water level curve (Figure 4) with seasonal-secular oscillations affects the three basins in different ways at different times.

The early Zanci high lake phase is followed by drying and clay dune development, the original Zanci dune building. In this revised interpretation, the two-stage wet-dry environment previously attributed to the Zanci dry and Mulurulu wet events are seen as evolutionary aspects of a single evolving phase. The reality now defines a single lateral change proceeding upstream, with different timing between basins. Thus, while Mulurulu was still full, the drying dune generation at Outer Arumpo occurred simultaneously with upstream flooding. Meanwhile, at Mungo, midway between the two, equivalent events occurred at intermediate times (Figure 12).

In this context, the former definition of a separate Mulurulu wet phase post-dating Zanci drying is no longer relevant. The wet-dry event is now seen as a continuum of a single cyclic change, a change coincident with events of the Last Glacial Maximum. The processes of drying and dune building were diachronic between basins rather than separate discrete events, as previously interpreted. A revision of the Zanci terminology is offered as the most appropriate amendment to the stratigraphic system (see Figure 12). The validity of the sandy Mulurulu Unit as a discrete mappable deposit lapses in favour of its incorporation as a member of the wider Zanci Unit, which, by precedence, is retained.

Climatic implications

A number of constraints limit the value of the Willandra data as a basis for quantitative climatic reconstructions (Williams et al. 1991). Firstly, this was an overflow system, thereby unable to define actual discharge volumes. Secondly, the Willandra Creek is just one of several distributary channels of the Lachlan downstream from Hillston. Despite that qualification, to simply maintain lake levels at overflow stage (an area exceeding 1000 km^2) at present evaporation rates would require diversion of the entire Lachlan River. Although providing minimum discharge records only, the Willandra flood history provides the best record of changes from major streams draining the eastern highlands, an area critical to our understanding of southeastern Australia's hydrologic and climatic change. In drying trends, episodes of lunette building, if correlating with regional linear dune expansion, have climatic implications beyond the confines of the Willandra system.

Aridity paradox: Lakes v desert dunes

The paradox of flooding in the northern basins simultaneously with clay dune building involving massive transfer of dust and salt to the east raises important questions on the nature of glacial age aridity (Williams et al. 2001, 2009). While evidence of locally permanent water bodies is secure, the Willandra also registers regional evidence for simultaneous aridity. Two lines of evidence register the response of surrounding dunefields to major changes in water balance in regions away from major drainage systems. The first is the record of desert dust, *Wüstenquarz* – records visible in microscopic analyses of Mungo lunette sediments. The second involves advances of Mallee dunefields and their encroachment across lake margins on to basin floors, advances that not only involve reactivation of the now fossil dunefields, but equally, the loss of water in the lakes necessary to permit dune advance.

In published records from the Joulni area, the sequence of lake drying and dune development records a substantial increase in the Wüstenquarz component near 45,000 years ago (Bowler et al. 2003). Wüstenquarz is virtually absent from sediments in the 60,000-45,000 years ago range, remaining uniformly high from 45,000 years ago through Upper Mungo time, to near 32,000 years ago. This would indicate at least some reactivation of desert dunes upwind coincident with the reduction of water availability in the drying basins. In that sense, two independent measures of aridity are consistent, lending confidence to a picture of substantial environmental change at that time.

Approaching the glacial maximum, the evidence for accelerated dune activation becomes more visible. In the regional setting, large areas of mallee dunes extend from the Darling River in the west to lake shorelines in the east. As extensions of inland arid landforms, evidence of past expansion exists today on the western shoreline of several Willandra basins.

On the northwestern margin of Outer Arumpo, a substantial lobe of mallee dunes has transgressed across the cliffed margin to extend some 2.5 km on to the lake floor (Figure 2). The leading edge retains its irregular advance outline and shows no sign of trimming by waters. This event clearly occurred after drying of Outer Arumpo, an event associated with the Upper Zanci drying near 20,000 years ago. The advance of mallee dunes was then synchronous with the major clay and salt deflation involved in the production of the Chibnalwood, Inner Arumpo and Bulbugaroo lunettes (Figure 1). Simultaneously, on the western shoreline of Lake Garnpung, another lobe of broad dunefield also encroached a similar distance (ca. 2.5 km) across the shoreline but, unlike the Outer Arumpo example, this advance has been trimmed and cut back, a function of water in the basin. Dune encroachment across the western shorelines implies at least a low water level or dry basins. The trimming of advancing dunes in Garnpung offers two alternative explanations.

Firstly, short alternations of dry to wet lake conditions are consistent with the progressive but intermittent drying after 20,000 years ago. A brief dry interval with dune advance may then have been interrupted by a final pulse of water while Mulurulu was in final drying phase.

Alternatively, dune advance after final drying of Garnpung may have been trimmed by much later return of restricted floodwaters in post-glacial times.

In neither case did the Garnpung waters extend south to Outer Arumpo levels, where dune lobes retain their original advancing forms. In both cases, dune advance has been arrested on drying lake floors. Water and wind have combined to preserve this interaction between competing processes.

In summary, the lakes' history provides a measure of time control on regional dune activity. Dune advance involves the loss of woodland vegetation cover, permitting massive erosion in mobile sands. In topographic form, the dunefield west of the Willandra, like those south of the Murray in Big Desert, retains patterns of advancing forms, now preserved as a virtually fossil sand sea. Now entirely stabilised by mallee vegetation, the genuine desert nature of these glacial maximum environments remain difficult for us to visualise today.

LGM paradox: Wind v water

While defining relative ages of lake and dunefield changes on the one hand, the data appear to be in conflict on the other. Desert dunes and saline gypseous lunettes advanced at 20,000-19,000 years ago while people upstream feasted on fish and shellfish. That paradox requires further analysis of controlling factors.

The two main actors evident on this LGM climatic stage, *wind* and *water* were operating, both modified by the further controlling agent *temperature*. While the role of wind and water is clearly evident in lake and dune behaviour, details of local temperature remain contentious.

Coincidence of maximum ice volume with lowest sea level would argue for major temperature reductions, a reconstruction consistent with 19,000 years ago Kosciuszko glacial advance (Barrows et al. 2004). Meanwhile, the abundance of fish and shellfish at Mungo (20,000 years ago) and Mulurulu (19,000 years ago) remains paradoxically inconsistent with locally cold temperatures.

In pointing to temperature constraints on fish spawning, Allen (1972) identified populations of Golden Perch surviving through the LGM as evidence that otolith abundance would seem to favour at least seasonally warm conditions. Coincidentally, distribution of mussel shells (*Velesunio ambiguus*) is dependent on transport by fish at the embryonic or glochidial stage. Isotopic analyses of 19,000 years ago Mulurulu mussel shells (Bowler in Shackleton's lab, Cambridge, 1975) and by Douglas (1996) defined average changes of 3-4‰ between Mulurulu and modern shells of the same species collected upstream from Lake Cargelligo. If due to temperature alone, that magnitude would involve a temperature depression of some 9-12°C. Until the effects of glacial-age lake water composition can be uncoupled from the temperature effect, such isotopic data remain to be validated. From evidence at Lake George near Canberra, Galloway (1965) proposed glacial-age summer temperature reduction of as much as 9°C below present, an estimate in line with later analyses using eggshell racemization (Miller et al. 1997) approximating that implied by the isotopic evidence.

On the other hand, given the high Southern Hemisphere insolation at LGM time (Berger and Loutre 1991), the prospect of relatively warm inland temperatures coincident with glacial and periglacial environments in southern high altitudes remains a possibility inviting further evaluation. High carbonate production, as in the microbial carbonates of Arumpo time (30,000-25,000 years ago), would favour relatively high temperatures despite proximity to the LGM.

Decreased temperature, especially on montane catchments, exerted major controls on complexities of the LGM stage. Its influence on periglacial highland extent, increased catchment efficiency, amplified by vegetation loss and by its controls on evaporation and soil moisture balance had widespread effects. Competing agents, wind and water, operated under its overall influence. Clarification of that influence poses a major challenge for future workers.

Basin drying and occupation: Age differences

The diachronous relationship in drying between the archaeologically barren Chibnalwood lunette and that of Mungo is further reflected in two different aspects of sedimentology.

Firstly, episodic variations in exposures of lake floors during drying leads to the development of the pelletal clay phases (PCD 1 and PCD 2 in Figure 4). In Lake Mungo, the Upper Zanci dune unit consists of highly variable laminated sands and sandy clays with less frequent clay-rich laminae. This variability reflects expression of the transitional nature from lake-full to dry basin, a progressive rather than abrupt change in hydrologic status (Figure 11). Such variability reflects the shallow oscillations in A-A' field of the water level model, Figure 4A.

Secondly, a major difference is reflected in the variation between the aeolian units of Mungo and Chibnalwood. While Mungo continued to receive pulses of surface inflow, Chibnalwood, isolated within the confines of the already dry Outer Arumpo, was subject to discharge by saline groundwaters only.

The upper dune at Mungo is highly variable in mineralogy and texture. By contrast, the Chibnalwood sediments are dominantly gypseous clay, with a relatively low percentage of quartz sand, and maintain relatively uniform textures throughout. Isolated from the high variability of surface inflow, this is a reflection of the final, relative uniform nature of groundwater control.

The thickness of the Chibnalwood gypseous clay is additionally important in consideration of rates and times between events. With total relief from floor to crest of 30 m, the thickness on the depositional margin of the pelletal mantle exceeds 10 m (Figure 5). Assuming a depositional rate of 1 cm per year, a relatively rapid rate of erosion and deposition, the 10 m deposit would involve some 1000 years. Although only approximate, this is consistent with the inferred interval between drying of Mungo and Mulurulu. Thus, while the Chibnalwood and Mungo dunes were building, people continued to harvest on the shores of Mulurulu throughout that 1000-year interval, from near 19,000 to 18,000 years ago (Figure 11). This provides at least a notional estimate of the time involved in changing human patterns of occupation and exploitation of the final Willandra resources.

The dramatic emergence of human footprints on Garnpung's western shore (Webb et al. 2005) has added new perspectives to the LGM occupation of the Willandra. Here, a small group of people left their footprints running and walking through saline mud ca. 19,000-23,000 years ago. The new picture reflects a regional movement, a progressive adaptation in which people may have followed the retreat of freshwaters up the drainage chain. This adaptation to the progressive drying trend reflects an earlier expression, with parallels in the records of the Mungo burials some 20,000 years earlier. The association there with increasing occupational density at the close of the Lower Mungo freshwater phase was interpreted as evidence for increased reliance on aquatic harvesting at times of regionally reduced surface waters (Bowler 1998). Corresponding to times of shallow water and increased salinity, these were precisely the environments favouring ease of fish and shellfish harvesting. This is the situation represented by the midden and otolith abundance in the closing stages of the Willandra lake environments.

Human Adaption

The people of the Willandra area occupied a region of natural system complexity, unparalleled by comparison with any other across southern Australia. The dominance of westerly winds, together with the array of dunes, lunettes and salt deflation they controlled, stood in direct competitive interaction with the dynamics of large stream discharge emanating from the snow-fed highland catchments in the southeast. Changing climatic balance at one time favoured massive stream discharge. At another, the balance gave way to colder drier climates and strong westerlies. If glacial maximum temperatures were reduced to anything like the 9°C suggested earlier, the effects on ecosystems and humans would be devastating. With much-reduced local rainfall, cold climates would be associated with extensive frost, and a much-reduced growing season, with impacts on both aquatic and terrestrial ecosystems. Golden Perch, the most frequent source of otoliths, favouring temperatures greater than 23°C (Lake 1967), would have been near their lowest limits of tolerance.

Human occupants, exposed to near freezing winters associated with strong winds in areas of little protective cover, were subject to very high levels of stress. Protective clothing and shelter construction would have been essential. Such harsh living conditions were perhaps ameliorated only by access to permanent water. It implies an adaptive ability difficult for us today to comprehend.

The new evidence presented here qualifies some previous interpretations of human adaptation to the peak of Zanci aridity. The disappearance of water was formerly interpreted as cause for people abandoning the area and moving towards rivers as a more permanent water source (Bowler 1998). The absence of any archaeological material in the dozens of gullies on Chibnalwood lunette supported movement away from locally inhospitable deflating basins; however, the absence of occupation dates between 18,000 years ago and 21,000 years ago was also interpreted as limited sampling rather than an abandonment of the region (Johnston and Clark 1998).

The ^{14}C dates presented demonstrate an absence of archaeological evidence associated with freshwater exploitation between 20,500 years ago and 25,000 years ago, followed by renewed fishing and mussel collection from 20,500 years ago to the final drying stages of the Willandra at 17,500 years ago. Locally, comparisons with the archaeological record from lakes associated with the Darling River (Balme and Hope 1990; Balme 1995), ca. 100-150 km to the west and northwest of the Willandra Lakes, are revealing. Similar although somewhat earlier patterns of commencement and abandonment of lacustrine resources have been documented at Lake Tandou, where small freshwater shell middens date from 26,500 to 31,300 years ago. At the Teryawynia Lakes, Balme and Hope (1990) found Holocene middens 6800 to 8400 years ago followed a similar pattern to that now established for the Willandra: middens at the south of the drainage system were older than those upstream. Occupation of the lakes moved upstream as the system progressively dried. The concept of opportunistic exploitation of systems under stress finds support from independent lacustrine environments.

In other areas of Australia, the LGM witnessed archaeological patterns that indicate a restriction of foraging ranges, avoidance of high-risk environments and concentration on locally available resources within refuges (Smith 1989; Veth 1989; Hiscock 2008): territorial reorganisation (Hiscock 2008) was essential in the face of changing resource availability. In this national context, the Willandra Lakes may be viewed as a complex and variable refuge rather than a high-risk and abandoned environment; it offered watered environments that contracted north over time, particularly through the latter part of the LGM. Fishing people and mussel collectors were forced to follow the contracting waters. While we may speculate that people completely abandoned the dry and deflating lakes and their surrounds, this is difficult to confirm. Further investigation of the non-lacustrine archaeological record is required.

We have now a new and somewhat paradoxical cameo of people in a rapidly changing environment, one in which glacial cold was associated with increasing aridity. It was one in which downstream drying with massive deflation blasted clouds of dust and salt into the atmosphere. Meanwhile, simultaneously further upstream, fish and shellfish were being caught and eaten. It provides a new picture of glacial maximum environments, one in which wind and water interacted to display a new dynamic of environmental change, one in which land and people were involved in a whole mosaic of new accommodation. People moved upstream responding to progressive drying. They were adapting to changing patterns of ice-age environments with their feet.

Conclusions

1. The new account of flooding and drying, refined by new levels of dating precision, demonstrates the continuity of change across hydrologic gradients within this interconnected basin system. In so doing, the justification for the previous two-stage climatic subdivision (Zanci, Mulurulu) is resolved into a single Zanci stage where differential lag effects in a single drying cycle affect different basins at different times.

2. The data significantly amplify the nature of LGM environments. Widely regarded as the time of maximum aridity, that view must be modified to include two critical observations:

    (a) In southeastern Australia, montane catchments affected by glacial or periglacial processes, despite reduction in rainfall, carried significant water volumes even through the maximum phase of glacially reduced temperatures. Meanwhile, areas distant from or in dry downstream reaches of such streams underwent major instability, with intensive reactivation of both continental and lacustrine dune systems.

    (b) Despite the major reduction in available surface waters, groundwater levels remained high. This acted as a trigger accentuating erosive instability, especially in low-lying areas where seasonal groundwater discharge fuelled processes of salt crystallisation and instability.

3. The Willandra evidence provides a new window into human response to the adversity of diminishing resources. The picture is one of selective adjustment, small migrations upstream following the pattern of disappearing lakes. It further suggests opportunistic focus on those pockets, where falling water levels, associated with increased salinity, provided increased availability to shell and fish harvesting. Throughout the at least 1000 years that people survived upstream, massive erosion with clouds of dust and salt were blowing from downstream basins of Chibnalwood, Arumpo, Bulbugaroo and their equivalents elsewhere across huge regions of southern Australia.

4. Contemporaneously with glacial age depressed temperatures, accentuated by clouds of dust and salt from westerly winds, the fishing men and women of Mungo to Mulurulu stand as exemplary survivors of massive environmental change, a cameo not without relevance to humanity today.

Acknowledgements

The ground surface levelling presented in this paper was prepared by Dick Kernebone, surveyor for the NSW Western Lands Commission. Permission to quote unpublished radiocarbon ages on Willandra shells and otoliths was generously provided by John Kalish, Stewart Fallon, Fiona Bertuch and the late John Head. We acknowledge assistance of two anonymous referees, one of whom offered detailed constructive criticism providing improvements to the paper. Simon Haberle and Bruno David have been patient and supporting editors in the paper's submission. In this dedication to A.P.K., we recall the delightful Pickwickian style he brought to entertain, enlighten and enliven a sometimes sleepy academia. We are grateful for the absence here of any pollen diagram!

Bibliography

Allen, H.A. 1972. *Where the crow flies backwards: man and land in the Darling Basin*. Unpublished PhD thesis, Department of Prehistory, Australian National University.

Balme, J. 1995. 30,000 years of fishery in western New South Wales. *Archaeology in Oceania* 30:1-21.

Balme, J. and Hope, J. 1990. Radiocarbon dates from midden sites in the lower Darling River area of western New South Wales. *Archaeology in Oceania* 32:85-101.

Barrows, T.T., Stone, J.O., Fifield, L.K. and Cresswell, R.G. 2001. Late Pleistocene glaciation of the Kosciuszko Massif, Snowy Mountains, Australia. *Quaternary Research* 55:179-189.

Barrows, T.T., Stone, J.O. and Fifield, L.K. 2004. Exposure ages for Pleistocene periglacial deposits in Australia. *Quaternary Science Reviews* 23:697-708.

Berger, A. and Loutre, M.F. 1991. Insolation values for the climate of the last 10 million years. *Quaternary Science Reviews* 10:297-317.

Boljkovac, K. 2009. *In situ SHRIMP δ18O and laser ablation ICP-MS Sr/Ca and 87Sr/86Sr measurements in fossil otoliths for palaeoclimate reconstructions at the Willandra Lakes World Heritage Area.* Unpublished BSc (Honours) thesis, Research School of Earth Sciences, Australian National University.

Bowler, J.M. 1967. Quaternary chronology of Goulburn Valley sediments and their correlation in southeastern Australia. *Journal of the Geological Society of Australia* 14(2):287-292.

Bowler, J.M. 1971. Pleistocene salinities and climatic change: Evidence from lakes and lunettes in southeastern Australia. In: Mulvaney, D.J. and Golson, J. (eds), *Aboriginal man and environment in Australia,* pp. 47-65. Canberra: ANU Press.

Bowler, J.M. 1973. Clay Dunes: Their occurrence, formation and environmental significance. *Earth-science Reviews* 9:315-338.

Bowler, J.M. 1978. Quaternary climate and tectonics in the evolution of the Riverine Plain, Southeastern Australia. In: Davies, J.L. and Williams, M.A.J. (eds), *Landform Evolution in Australasia*, pp. 70-112. Canberra: ANU Press.

Bowler, J.M. 1998. Willandra Lakes revisited: environmental framework for human occupation. *Archaeology in Oceania* 33:156-168.

Bowler, J.M. and Price, D.M. 1998. Luminscence dates and stratigraphic analysis at Lake Mungo: review and new perspectives. *Archaeology in Oceania* 33:120-155.

Bowler, J.M. 1999. Nekeeya Lunette Assessment Study: Geomorphology, origins and evolution of the Nekeeya sand lunette. *Report to Aboriginal Affairs Victoria.*

Bowler, J.M., Thorne, A.G. and Polach, H.A. 1972. Pleistocene man in Australia: age and significance of the Mungo skeleton. *Nature* 240:48-50.

Bowler, J.M., Jones, R., Allen, H. and Thorne, A.G. 1970. Pleistocene human remains from Australia: a living site and human cremation from Lake Mungo, western New South Wales. *World Archaeology* 2:39-60.

Bowler, J.M., Johnston, H., Olley, J.M., Prescott, J.R., Roberts, R.G., Shawcross, W. and Spooner, N.A. 2003. New ages for human occupation and climatic change at Lake Mungo, Australia. *Nature* 421:837-840.

Bourman, R.P., Prescott, J.R., Banerjee, D., Alley, N.F. and Buckman, S. 2010. Age and origin of alluvial sediments within and flanking the Mt Lofty Ranges, southern South Australia: a Late Quaternary archive of climate and environmental change. *Australian Journal of Earth Sciences* 57(2):175-192.

Bronk Ramsey, C. 2009. Bayesian analysis of radiocarbon dates. *Radiocarbon,* 51(1):337-360.

Clark, P.M. 1987. Willandra Lakes World Heritage Area archaeological resources study. Unpublished report, Western Lands Commission, New South Wales Department of Planning, Sydney.

Clark, P.U., Dyke, A.S., Shakun, J.D., Carlson, A.E., Clark, J., Wohlfarth, B., Mitrovica, J.X., Hostetler, S.W. and McCabe, A.M. 2009. The Last Glacial Maximum. *Science* 325:710-714.

Douglas, K. 1996. *Land systems and stratigraphy of Lake Mulurulu: Examination of Quaternary palaeoenvironments.* Unpublished BSc (Honours) thesis, School of Earth Sciences, University of Melbourne.

Galloway, R.W. 1965. Late Quaternary climates in Australia. *Journal of Geology* 73:603-18.

Gillespie, R. 1997. Burnt and Unburnt Carbon: dating charcoal and burnt bone from the

Willandra Lakes, Australia. *Radiocarbon* 39(3):239-250.

Gillespie, R. 1998. Alternative timescales: a critical review of Willandra Lakes dating, Australia. *Archaeology in Oceania* 33:169-182.

Gillespie, R., Bertuch, F. and Nicholas, W.A. (no date). Australian freshwater shells: further radiocarbon and amino acid racemization studies. Mss in preparation.

Hesse, P., Magee, J.W. and van der Kaars, S. 2004. Late Quaternary climates of the Australian arid zone: a review. *Quaternary International* 118/119:87-102.

Hiscock, P. 2008. *Archaeology of Ancient Australia*. Abingdon: Routledge.

Johnson, C.N. and Brook, B.W. 2011. Reconstructing the dynamics of ancient human populations from radiocarbon dates: 10,000 years of population growth in Australia. *Proceedings of the Royal Society B* doi:10.1098/rspb.2011.0343.

Johnston, H and Clark, P. 1998. Willandra Lakes archaeological investigations 1968-98. *Archaeology in Oceania* 33:105-119.

Kalish, J.M., Miller, G.H., Tuniz, C., Pritchard, J.C., Rosewater, A. and Lawson, E. 1997. Otoliths as recorders of palaeoenvironments: comparison of radiocarbon age and isoleucine epimerization in Pleistocene golden perch *Macquaria ambigua* otoliths from Willandra Lakes. In: *Conference Handbook, Sixth Australasian Archaeometry Conference*, Australian Museum, Sydney, 10-13 February 1997.

Kershaw, A.P. and Nanson, G.C. 1993. The last full glacial cycle in the Australian region. *Global and Planetary Change* 7:1-9.

Lake, J.S. 1967. Rearing experiments with five species of Australian freshwater fishes. 1. Inducement to spawning. Australian Journal of Marine & Freshwater Research 18:137-153.

Lambeck, K. and Chappell, J. 2001. Sea level change through the last glacial cycle. *Science* 292:679-686.

Magee, J.W., Bowler, J.M., Miller, G.H. and Williams, D.L.G. 1995. Stratigraphy, sedimentology, chronology and palaeohydrology of Quaternary lacustrine deposits at Madigan Gulf, Lake Eyre, South Australia. *Palaeogeography, Palaeoclimatology, Palaeoecology* 113:3-42.

Miller, G.H., Magee, J.W. and Jull, A.J.T. 1997. Low-latitude glacial cooling in the Southern Hemisphere from amino-acid racemization in emu eggshells. *Nature* 385:241-244.

Nanson, G.C., Price, D.M. and Short, S.A. 1992. Wetting and drying of Australia over the past 300 ka. *Geology* 20:791-794.

Page, K., Dare-Edwards, A., Nanson, G. and Price, D. 1994. Late Quaternary evolution of Lake Urana, New South Wales, Australia. *Journal of Quaternary Science* 9:44-57.

Reimer, P.J., Baillie, M.G.L., Bard, E., Bayliss, A., Beck, J.W., Blackwell, P.G., Bronk Ramsey, C., Buck, C.E., Burr, G.S., Edwards, R.L., Friedrich, M., Grootes, P.M., Guilderson, T.P., Hajdas, I., Heaton, T.J., Hogg, A.G., Hughen, K.A., Kaiser, K.F., Kromer, B., McCormac, F.G., Manning, S.W., Reimer, R.W., Richards, D.A., Southon, J.R., Talamo, S., Turney, C.S.M., van der Plicht, J. and Weyhenmeyer, C.E. 2009. IntCal09 and Marine09 radiocarbon age calibration curves, 0-50,000 years cal BP. *Radiocarbon* 51(4):1111-50.

Shawcross, W. 1998. Archaeological excavations at Mungo. *Archaeology in Oceania* 33:183-200.

Smith, M.A. 1989. The case for a resident human population in the Central Australian ranges during full glacial aridity. *Archaeology in Oceania* 24:93-105.

Stuiver, M. and Polach, H.A. 1977. Discussion: Reporting of 14C data. *Radiocarbon* 19:253-258.

Stuiver, M., Reimer, P.J. and Reimer, R. 2010. CALIB Radiocarbon Calibration Version 6.0 (http://calib.qub.ac.uk/calib/ accessed May/June 2010).

Surovell, T.A., Finley, J.B., Smith, G.M., Brantingham, P.J. and Kelly, R. 2009. Correcting temporal frequency distributions for taphonomic bias. *Journal of Archaeological Science* 36:1715-1724.

Veth, P. 1989. Islands in the interior: a model for the colonization of Australia's arid zone. *Archaeology in Oceania* 24:81-92.

Webb, S., Cupper, M.L. and Robin, R. 2006. Pleistocene human footprints from Willandra Lakes, southeastern Australia. *Journal of Human Evolution* 50:405-413.

Williams, M.A.J., Cook, E., van der Kaars, S., Barrows, T., Shulmeister, J. and Kershaw, P. 2009. Glacial and deglacial climatic patterns in Australia and surrounding regions from 35,000 to 10,000 years ago reconstructed from terrestrial and near-shore proxy data. *Quaternary Science Reviews* 28:2398-2419.

Williams, M.A.J., De Deckker, P. Adamson, D.A. and Talbot, M.R. 1991. Episodic fluviatile, lacustrine and aeolian sedimentation in a late Quaternary desert margin system, central western New South Wales. In: Willams, M.A.J., de Deckker, P., Kershaw, A.P. (eds), *The Cainozoic in Australia: A re-appraisal of the evidence*, Special Publication No. 18, Geological Society of Australia Incorporated.

Williams, M.A.J., Prescott, J.R., Chappell, J., Adamson, D., Cock, B., Walker, K. and Gell, P. 2001. The enigma of a late Pleistocene wetland in the Flinders Ranges, South Australia. *Quaternary International* 83/85:129-144.

14

Late-Quaternary vegetation history of Tasmania from pollen records

Eric A. Colhoun

School of Environmental and Life Sciences, University of Newcastle, Newcastle, NSW
Eric.Colhoun@newcastle.edu.au

Peter W. Shimeld

University of Tasmania, Hobart, Tasmania

Introduction

Vegetation forms the major living characteristic of a landscape that solicits inquiry into the history of its changes during the late Quaternary and the major factors that have influenced the changes. Early studies considered ecological factors would cause vegetation to develop until a stable climatic climax formation was attained (Clements 1936). The concept of an area developing a potential natural vegetation in the absence of humans was similar (Tüxen 1956). Both ideas held that the vegetation of an area would develop to a stable condition that would change little. However, the vegetation of a region never remains in stasis, but develops dynamically through time, influenced by changing dominant factors (Chiarucci et al. 2010).

The structure of a major vegetation formation is usually dominated by a limited number of taxa of similar physiognomy. Although many taxa are identified at most sites studied for pollen in Tasmania, the major percentages in the records are represented by fewer than 10 pollen taxa. These are widely dispersed taxa, local taxa usually being under-represented in the records (Macphail 1975). The structures of fossil pollen-vegetation formations are interpreted with regard to modern vegetation even though abiotic and biotic conditions rarely remain the same through time, and identical replication is not expected. During the late Quaternary in Tasmania, the most important abiotic changes affecting vegetation were temperature and precipitation, and the most important biotic change was the impact of Aboriginals using their major cultural tool, fire. The advent of people to a region adds another dimension to palaeoecological reconstructions and frequently reveals inconsistencies between the expected vegetation before and the extant vegetation after human occupation (Willis and Birks 2006).

During the past 35 years, pollen records have been obtained from many lake and swamp deposits located mainly in western Tasmania and more sparsely in eastern Tasmania. Until recently, the results have been used to interpret vegetation history with reference to present extensive vegetation formations defined by the major pollen components represented in site diagrams, from which former climate changes have been inferred (Table 1). Published records refer to one or at best a few sites, except for the early work of Macphail (1979), which gave a regional representation for western Tasmania, and maps by Kirkpatrick and Fowler (1998), who used pollen records to model vegetation distribution at the Last Glacial Maximum (LGM). During the past 10-15 years, pollen records at several sites have revealed that humans prevented the development of 'climax' forest during the postglacial period and produced cultural disclimax vegetation associations, especially in southwest Tasmania (Fletcher and Thomas 2007a, 2010a).

The density of analysed pollen sites in Tasmania is greater than for other mid-latitude Southern Hemisphere areas. This provides an opportunity to reconstruct the palaeoecology of major vegetation formations and associations at different times, and to evaluate the results in relation to inferred climate changes and human impacts (Table 2). In this paper we:

1. Use eight regionally distributed pollen taxa to represent broad-scale vegetation patterns on a series of time-slice maps using relative pollen data from 52 sites.

2. Discuss changes in late-Quaternary vegetation from 125,000 years ago to 1000 years ago, shown by the patterns on the maps and reference to original publications.

3. Consider the influence of climate changes on the vegetation plus human modifications not evident from the patterns on the maps.

Table 1. Referenced and acknowledged sources of pollen data.

1 Adamsons Peak (Macphail 1979)	27 Lake Vera (Macphail 1979)
2 Beatties Tarn (Macphail 1979)	28 Mathinna Plains (Thomas 1996)
3 Big Heathy Swamp (Thomas 1992)	29 Melaleuca Inlet (Thomas 1995)
4 Blakes Opening (Colhoun and Goede 1979)	30 Mowbray Swamp (van de Geer et al. 1986)
5 Broadmeadows Swamp (van de Geer et al. 1986)	31 Newall Creek (van de Geer et al. 1989)
6 Brown Marsh (Macphail 1979)	32 Newton Creek (Colhoun et al. 1993)
7 Camerons Lagoon (Thomas and Hope 1994)	33 Ooze Lake (Macphail and Colhoun 1985)
8 Coal Head (A. Fowler 1993 pers. comm.)	34 Pedder Pond (Fletcher and Thomas 2007a)
9 Crotty Road (Colhoun and van de Geer 1987)	35 Pieman Dam (Augustinus and Colhoun 1986)
10 Crown Lagoon (Sigleo and Colhoun 1981)	36 Pipe Clay Lagoon (Colhoun 1977a)
11 Dante Rivulet (Gibson et al. 1987)	37 Poets Hill Lake (Colhoun 1992)
12 Darwin Crater (Colhoun and van de Geer 1988)	38 Pulbeena Swamp (Colhoun et al. 1982)
13 Den Plain A, B, C (Moss et al. 2007)	39 Remarkable Cave (Colhoun 1977b)
14 Dublin Bog (Colhoun et al. 1991b)	40 Rocky Cape (Colhoun 1977c)
15 Eagle Tarn (Macphail 1979)	41 Smelter Creek (Colhoun et al. 1992)
16 Forester Marsh (Thomas 1996)	42 Snow Hill Marshes (C. Becker 2000 pers. comm.)
17 Governor Bog (Colhoun et al. 1991a)	43 SO36-7SL (van de Geer et al. 1994)
18 Hazards Lagoon (Mackenzie 2010)	44 Stoney Lagoon (Jones 2008)
19 Henty Bridge (Colhoun 1985a)	45 Tarn Shelf Mt Field (Macphail 1979)
20 King River (van de Geer et al. 1991)	46 Tarraleah (Macphail 1984)
21 Lake Dove (Dyson 1995)	47 Tullabardine (Colhoun and van de Geer 1986)
22 Lake Fidler (K. Harle 1993 pers. comm.)	48 Tyndall Range Tarn Shelf (Macphail 1986)
23 Lake Johnson (Anker et al. 2001)	49 Upper Lake Wurawina (Macphail 1986)
24 Lake St Clair (Hopf et al. 2000)	50 Upper Timk Lake (Harle 1989)
25 Lake Selina (Colhoun et al. 1999)	51 Waterhouse Marsh (Thomas 1996)
26 Lake Tiberias (Macphail and Jackson 1978)	52 Yarlington Tier (Harle et al. 1993)

Tasmanian environment and vegetation

Tasmania, located at 40-43°S and 144-149°E, has a complex topography. The west coast is backed by a low sloping plateau that rises from around 100 m near the coast to over 500 m at the foot of the West Coast Range. Rivers are deeply incised into the low coastal plateau. Inland of the 1000 m-high West Coast Range, north-south trending mountain ranges attain 1000-1200 m altitude and consist of Precambrian and Palaeozoic siliceous rocks with limestone in many deep valley floors. The mountains reach 1300-1500 m in the Central Highlands, where numerous peaks have caps of Jurassic dolerite that overlie siliceous rocks. The dolerites extend eastwards as a high Central Plateau that descends from around 1200 m on its western margin to 1000-900 m on its eastern and southeastern margins. Midland Tasmania is a rift valley that connects the northern coastal plains and Tamar Trough with the lower Derwent Valley of the southeast. Within the rift, late Palaeozoic and Mesozoic mudstones and sandstones underlie Tertiary sediments and basalts. Eastern Tasmania is an area of dissected hills and plateaus formed of similar siliceous sediments capped by Jurassic dolerite that attain 600-900 m. The northeast is dominated by the 1200-1500 m-high dolerite plateau of Ben Lomond and by extensive coastal plains mantled with sandy Quaternary-age sediments (Figure 1).

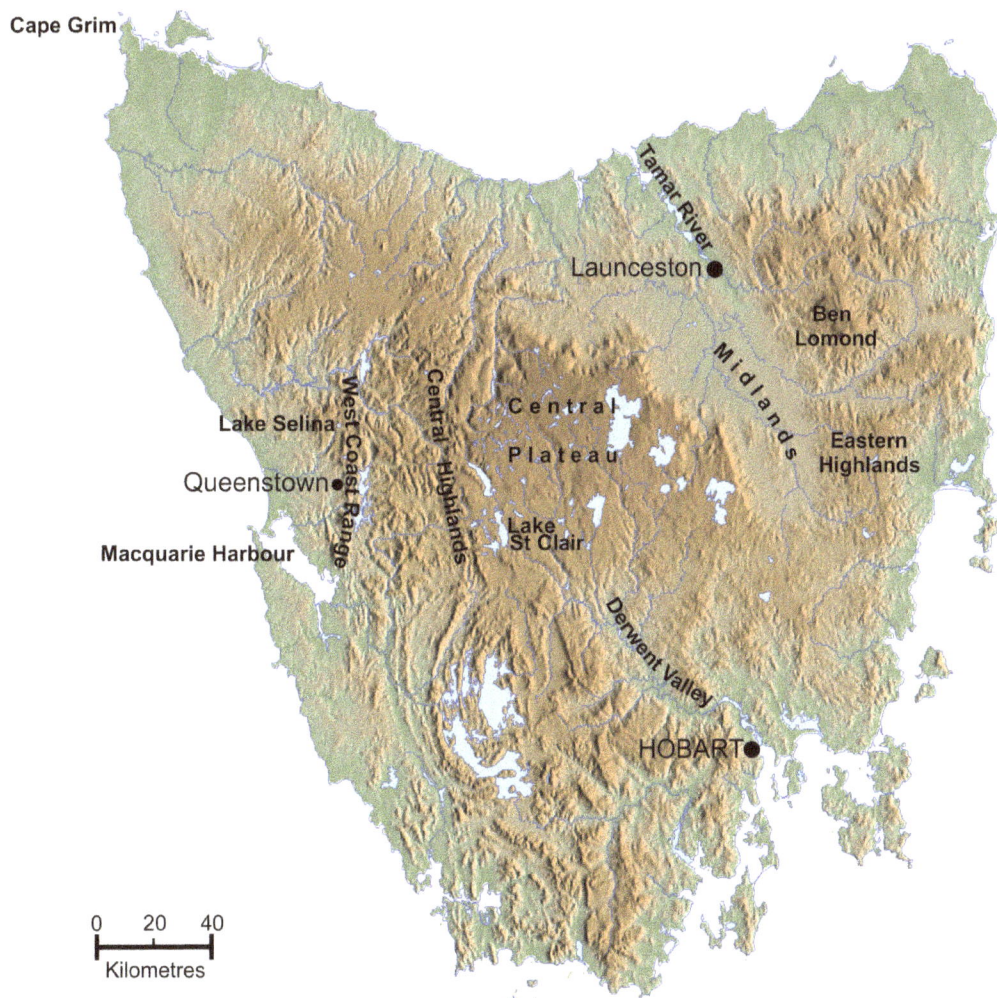

Figure 1. Topography of Tasmania.

Situated between the Australian continental high-pressure system and the Southern Polar Front, Tasmania experiences much changeability between sub-tropical continental and cool temperate oceanic weather systems. Strong continental effects occur during summer, while oceanic effects can occur in all seasons. The maritime airflows provide 3600-1200 mm precipitation annually to the western mountains. Precipitation decreases eastwards and northeastwards across the Central Plateau, which receives 1800-900 mm. The Midland Valley lies in a rain shadow and receives only 700-550 mm, while the Eastern Uplands receive 800 to more than 1200 mm, with a significant proportion coming from the east.

In winter, frost and snow are frequent above 500 m, particularly in the centre and west. Snow only blankets the terrain for short periods. Coastal areas are mild and generally snow-free, with mean temperatures of 11°C at Queenstown in the west and 12.4°C at Hobart in the southeast. Central highland Tasmania is relatively cold, with a mean temperature of 8°C at Lake St Clair. Most of central and northeastern alpine Tasmania has July mean minimum temperatures below 0°C. Nocturnal temperatures as low as –20°C may occur on high peaks and severe glazing storms occur frequently in winter. In summer, adiabatically warmed airflows descend from the plateau, bringing warm dry winds and extreme temperatures of 35-40°C, particularly to the Midlands and southeast.

During glacial times when sea level was 60-120 m lower, Tasmania was connected to Victoria, and at maximum lowering the exposed land area was double its present size. The increase in continentality reduced precipitation in the central, eastern and northern areas leeward of the mountains, and mean temperatures throughout Tasmania were also reduced (Colhoun 1991).

The vegetation of Tasmania is determined by steep environmental gradients associated with precipitation, temperature, altitude, geology and soils (nutrient availability), and by fire (Bowman and Jackson 1981; Jackson 1981a, b, c; Kirkpatrick 1982; Kirkpatrick and Dickinson 1984; Reid et al. 1999; Harris and Kitchener 2005). These factors have resulted in complex patterns of major vegetation formations and communities, as determined by the Tasmanian Vegetation Mapping Program (Figure 2). Key characteristics of the formations are described by Harris and Kitchener (2005: See for details at the community level). The modern vegetation patterns in Tasmania differ in detail from the broader patterns shown by our pollen synthesis because of:

1.   The more detailed classification used for the TASVEG Mapping Program,

2.   The generalised patterns resolved by our selection of regionally important pollen taxa for the time periods mapped, and

3.   The impact of Aboriginal occupation and burning on the vegetation during the past 35,000 years, which has altered/prevented Holocene forest recovery in the southwest (Allen 1996; Fletcher and Thomas 2007a, 2010a, b).

The local variability of climatic influence on the vegetation is indicated by the treeline, which approximates a mean temperature of 10°C for January. The treeline rises from 750 m in the southwestern mountains to 1400 m in the northeast. However, due to local ecological and environmental factors such as topographic situation, exposure to frost, effects of wind, and lack of protection from fire, the treeline is fragmented and varies over about 200-300 m altitude. Above the treeline the vegetation consists of alpine heaths, herbfields and coniferous shrubberies with numerous endemic taxa, notably the shrub conifers *Microstrobos niphophilus*, *Microcachrys tetragona* and *Podocarpus lawrenceii*.

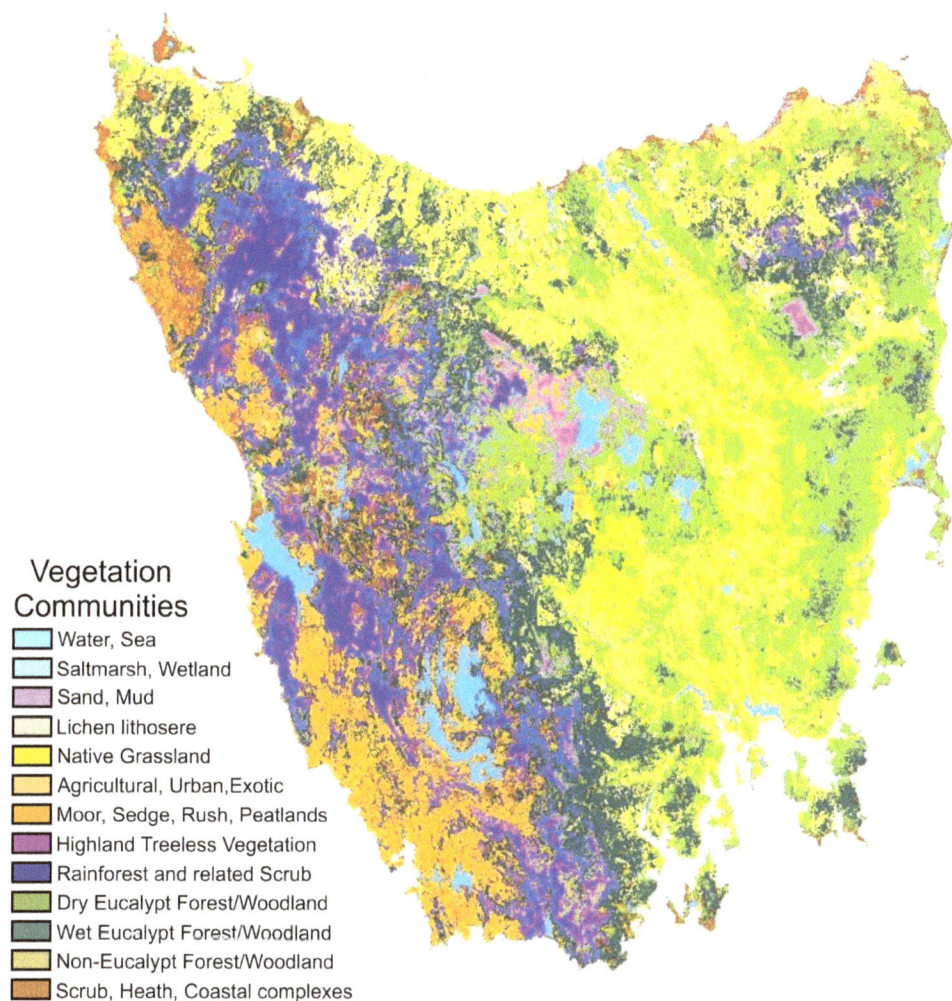

Figure 2. Vegetation map of Tasmania.

Western Tasmania is climatically suitable for cool temperate rainforest (cf. rainforest and related scrub, Figure 2), but its distribution is limited to less than half its potential area. Rainforest is dominated by the Southern Beech *Nothofagus cunninghamii* and Celery-top Pine *Phyllocladus aspleniifolius*, with lesser amounts of *Atherosperma moschatum* and *Eucryphia lucida*, plus the native conifers Huon Pine *Lagarostrobos franklinii* and King Billy Pine *Athrotaxis selaginoides* (Jarman et al. 1999). However, in many areas, the rainforest is impure or absent because of soils with poor nutrient status, bad drainage or burning. Over extensive areas, rainforest taxa and *Eucalyptus* spp. combine to form wet mixed forests. At altitude, rainforest and wet mixed forests become diminutive in form and diverse in associated species, including some species distinctive of subalpine environments such as *Nothofagus gunnii*. Rainforest also occurs extensively in valleys and on the mid slopes of mountains, as in northeastern Tasmania surrounding Ben Lomond Plateau below the zone of alpine vegetation.

Extensive areas of western Tasmania are poorly drained, have acid soils (pH 4-5.5) and have vegetation that has been extensively and frequently burned. The vegetation of drier sites is dominated by epacridaceous heathlands and locally by regenerating myrtaceous shrublands, but on wet sites lowland peatlands are dominated by the tussock-forming buttongrass sedge *Gymnoschoenus sphaerocephalus* and cord rushes Restionaceae. Blanket moorland comprised of these species may extend upslope on to ridges and plateaus (Figure 2).

The ecotonal zone between the rainforests of the west and the dry sclerophyll forest and woodland of much of eastern Tasmania (cf. dry eucalypt forest/woodland, Figure 2) is dominated by wet sclerophyll *Eucalyptus* spp. forest and woodland (cf. wet eucalypt forest/woodland, Figure 2), which occurs in a belt extending from west-northwest to east-southeast across west-central Tasmania as far as the southeast coast, and on the slopes of the northeastern highlands. Wet sclerophyll forests are largely the product of burning during the Holocene, which favoured the dominance of *Eucalyptus* spp. Their understoreys are characterised by regenerating rainforest taxa and several mesic broadleaved shrub and small tree taxa, including notably *Pomaderris apetala* (Jackson 1981c). Above about 700 m the wet sclerophyll *Eucalyptus* forests become subalpine *Eucalyptus* woodlands that include a diversity of small tree and shrub taxa.

Leeward, in the rain shadow of the western mountains, dry sclerophyll *Eucalyptus* spp. forest and woodland is dominant. The formation is extensive between Lake St Clair and the eastern margin of the Central Plateau, and throughout southeastern and eastern Tasmania. Species of *Eucalyptus* are dominant, with the greatest diversity in the southeast. Understoreys consist of drought-tolerant shrubs, grasses and sedges in nutrient-poor, nutrient-rich and poorly drained areas (Duncan and Brown 1985). Sclerophyll forest and woodland also extended along much of the northwest coast region before European land clearance, with isolated areas extending as far as Cape Grim.

Native grassland occurred mainly on the driest southeastern lower parts of the Central Plateau, middle Derwent Valley and Midlands before European settlement. Lowland grasslands comprise species of *Poa, Themeda, Austrodanthonia* and *Austrostipa*. Highland grassland occurs on plain areas and valley floors on the Central Plateau, and extends northwest to Middlesex Plains north of Cradle Mountain National Park. The dominant grasses *Poa gunnii* and *Poa labillardière* form tussocks. Much of the lowland native grassland was associated with sparse trees that would have given a savanna-like or parkland aspect to the environment before their removal on settlement (Harris and Kitchener 2005).

Late-Quaternary vegetation map reconstruction

The vegetation maps have been reconstructed using pollen records from 52 sites (Table 1 and Figure 3a-i). The maps reflect the broad-scale patterns of vegetation formations and associations within Tasmania during the past 125,000 years, as shown for the time-slices oxygen isotope stages (OIS=MIS) 5e, 4, 3 and 2, and for 12,000, 9000, 6000 and 1000 radiocarbon years. The time slices 12,000, 9000 and 6000 would be slightly older than shown on the maps, with calibrated ages of approximately 13,500, 10,000 and 6800 years ago respectively. The plotted data on the maps represent what the climatic climax potential natural vegetation was prior to 35,000 BP or would have been afterwards in the absence of Aboriginal impact. Some known areas with disclimax vegetation associations due to human impact and reflected in the pollen diagrams are discussed in the text.

Map reconstruction is based on eight pollen taxa that are the major components in the regional pollen rain which best represent regional vegetation formations (Macphail 1975). The taxa selected are: *Nothofagus cunninghamii, Phyllocladus aspleniifolius, Lagarostrobus franklinii, Pomaderris apetala, Eucalyptus* spp., *Allocasuarina* spp., Asteraceae (tubuliflorae) and Poaceae. These taxa account for much of the pollen represented in full pollen diagrams, and most are widely distributed. Full pollen counts at the majority of sites used a sum of around 300 grains of tree, shrub and herb taxa sufficient for identification of vegetation associations using the modern analogue technique (MAT) in which a limited number of major taxa combined with a sum of 150 grains is considered adequate (Lytle and Wahl 2005). The pollen counts of the eight taxa extracted from the full counts have been normalised to 100% for classification of the pollen-vegetation formations and asssociations. Four major Vegetation Formations can be defined (the headings in Table 2). The limited number of taxa used, though biased against local

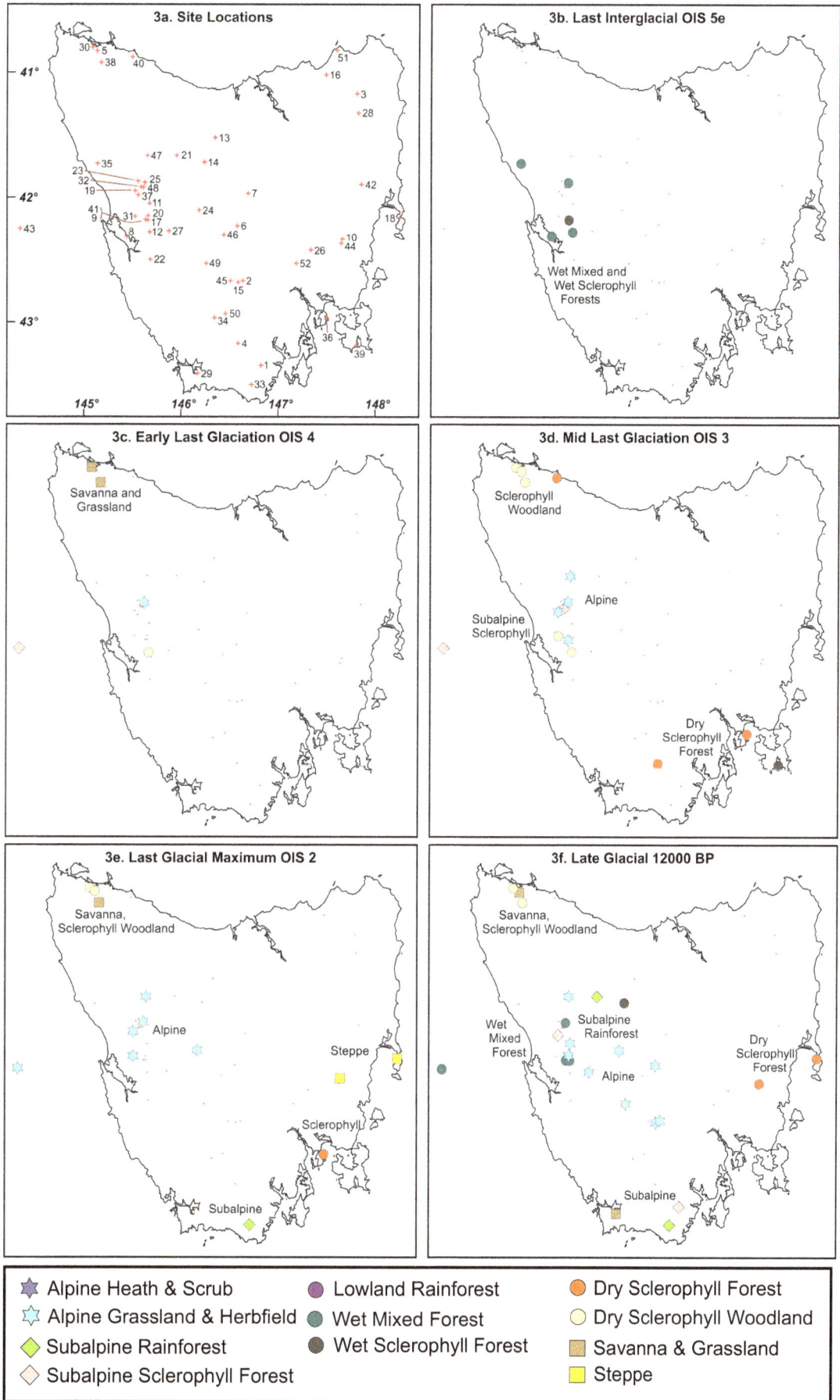

Figure 3. Late-Quaternary vegetation maps.

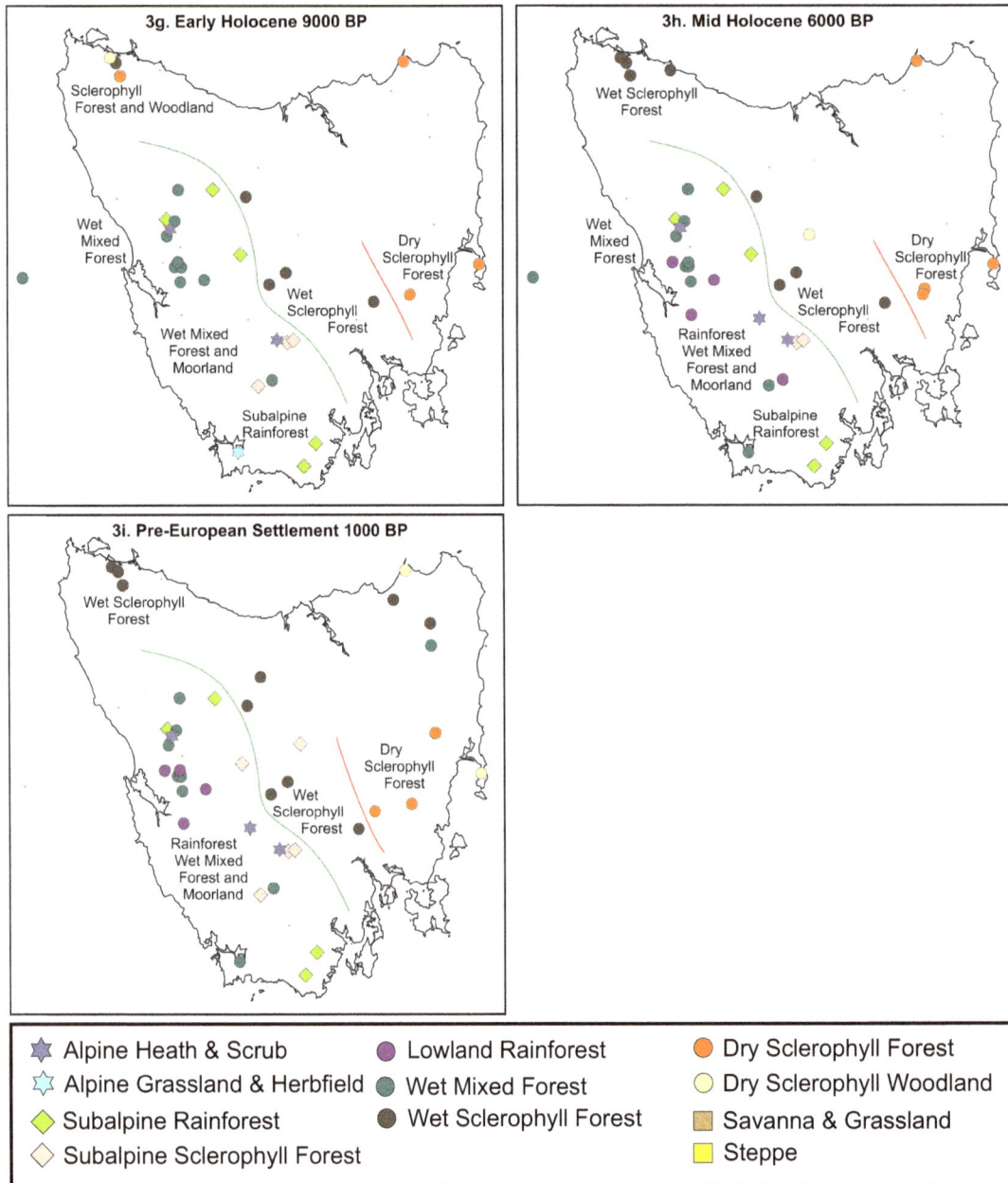

Figure 3. *Continued*

taxa of limited distribution, is suitable for highlighting the major composition of much of the regional vegetation.

The formations are subdivided into 12 associations using a combination of the normalised percentage representation of the regionally important taxa, indicator species for the associations, reference to full spectra of dated and relevant age-interpolated horizons on pollen diagrams, and ecological knowledge. The mapped pollen-defined vegetation associations are thus a broad-scale interpretive model of late-Quaternary vegetation that can be compared with modern vegetation (Table 2). However, a few caveats are necessary. First, pollen transport is generally from west to east across Tasmania, and wet forest taxa, especially *N. cunninghamii* and *P. apetala*, appear consistently in small quantities in records from Midland and northeastern Tasmania and may cause the association to be classified as wet sclerophyll forest or woodland when other evidence clearly indicates a dry sclerophyll association. In such cases, the long-distance transport component has been deleted before classification. Second, at locations where the major pollen input is from locally dispersed taxa, the classification will be biased against revealing the local vegetation association. This has occurred mainly where Aboriginals have extensively burned the vegetation and is qualified in the text. Third, it is possible that palaeo-associations recognised by application of the MAT may not exactly represent present vegetation associations.

Interpretation of the vegetation associations from the pollen record requires recognition

Table 2. Pollen-defined late-Quaternary vegetation associations for Tasmania.

Vegetation associations	Major criteria	Indicators
Wet Forests: Rainforest taxa (*N. cunninghamii* + *P. aspleniifolius* + *L. franklinii*) >30%		
1. Lowland rainforest	Rf taxa >70%, *Eucalyptus* <5%	*Atherosperma moschatum, Eucryphia lucida, Anodopetalum biglandulosum,* treeferns
2. Wet mixed forest	Rf taxa >70%, *Eucalyptus* >20%	*Allocasuarina, Dicksonia antarctica*
3. Subalpine rainforest	Rf taxa >70%	*Nothofagus gunnii,* ± *Athrotaxis* spp.
4. Subalpine sclerophyll forest	Rf taxa 30-70%, *Eucalyptus* >30%	*Allocasuarina, P. aspleniifolius, Nothofagus gunnii, Microstrobos niphilus*
Sclerophyll Forests and Woodlands: *Eucalyptus* >30% (forest), 10-30% (woodland)		
5. Wet sclerophyll forest	Rf taxa 5-30%, *P. apetala* >2% (*Eucalyptus* >30%)	*Dicksonia antarctica*
6. Wet sclerophyll woodland	Rf taxa 5-30%, *P. apetala* >2% (*Eucalyptus* 10-30%)	*Dicksonia antarctica*
7. Dry sclerophyll forest	Rf taxa <5%, *P. apetala* <2% (*Eucalyptus* >30%)	Dry indicator taxa (e.g. *Dodonaea viscosa*)
8. Dry sclerophyll woodland	Rf taxa <5%, *P. apetala* <2% (*Eucalyptus* 10-30%)	Dry indicator taxa
Grasslands: Non-woody taxa (*Poaceae* + *Asteraceae*) >80%		
9. Savanna and grassland	*Eucalyptus* >10% (savanna) *Eucalyptus* <10% (grassland)	
10. Steppe	*Eucalyptus* <10% and Chenopodiaceae >10%	*Plantago*
Alpine: Alpine shrub and herb taxa		
11. Alpine heath and scrub	(Poaceae + Asteraceae) <50%	*Microcachrys tetragona, Microstrobos niphophilus, Diselma archeri, Athrotaxis* spp., Epacridaceae
12. Alpine grassland and herbfield	(Poaceae + Asteraceae) >50%	*Astelia alpina, Plantago* and alpine herbs

not only of the taxa contributing to the regional pollen rain, but also the degree to which they are over-, proportionately- or under-represented. Fletcher and Thomas (2007b) have analysed modern pollen from western Tasmania and shown that of the eight taxa used in this study, *N. cunninghamii, P. aspleniifolius* and *P. apetala* are over-represented, *Eucalyptus,* Poaceae and *Allocasuarina* are proportionately-represented, and Asteraceae and *L. franklinii* under-represented. They have also been able to differentiate rainforest, moorland and alpine vegetation from a limited number of major pollen taxa. They show that rainforest (cf. Association 1) is characterised by *N. cunninghamii* and *P. aspleniifolius,* and frequently contains *L. franklinii* and *Eucryphia* spp. They also show that most species in the alpine zone (cf. Associations 1 and 2), except for Poaceae, are under-represented. These include the herb *Astelia alpina* and the coniferous shrubs *Microcachrys tetragona, Microstrobos niphophilus*, the deciduous beech *Nothofagus. gunnii* and Epacridaceae. Thirdly, they have identified moorland, which is and was widespread in southwest Tasmania at least during the Holocene (*vide infra*), but is not represented in our broad-scale classification of associations. The moorland is shown to be identifiable from a combination of the under-represented taxa Ericaceae, buttongrass *Gymnoschoenus sphaeocephalus, Melaleuca* and *Leptospermum*, but it also includes some pollen of well-represented Poaceae, *Eucalyptus* and over-represented *N. cunninghamii* and *Phyllocladus*.

Some of Tasmania's pollen sites occur in alpine sites above the modern treeline in southwestern Tasmania. The records contain significant quantities of *N. cunninghamii* and *P. aspleniifolius*, which have wide dispersal ability (Fletcher and Thomas 2007b). Other species are much less widely dispersed. A study of modern *Eucalyptus* pollen transport from a sharp woodland-edge eastwards across Liawenee Moor on the Central Plateau shows that *Eucalyptus* accounted for 50-70% of total pollen beneath the woodland canopy and concentrations of only 5% outside the woodland on the treeless Moor (Shimeld and Colhoun 2001).

Before interpreting the mapped data, it is necessary to comment on the major taxa represented in the modern vegetation and how the pollen is likely to be represented in the analogue associations outlined in Table 2. Of particular importance is the dispersal of the pollen quantitatively assessed by Fletcher and Thomas (2007b), but also qualitatively apparent from the representation of pollen in many pollen diagrams obtained from diverse geographic locations. The time slices represented on the maps are also of limited precision and the time periods they represent need qualification.

Southern Beech *N. cunninghamii* is the dominant species of lowland rainforest. Celery-top Pine *P. aspleniifolius* is also widespread, while Huon Pine *L. franklinii* occurs locally in river valleys and as subalpine mountain stands. Each taxon is an abundant pollen producer. In mountainous western Tasmania, *N. cunninghamii* and *Phyllocladus* can be transported in quantity upslope into adjacent alpine areas (Macphail 1975, 1979). In addition, given the prevalent westerly winds, the pollen can be transported in small quantities (2-3%) eastwards across the entire island. *Lagarostrobos* is much less widely dispersed, though occasional grains do travel far. Spores of the main treefern *Dicksonia antarctica* can be widely distributed, especially by water, and may occur in abundance at riparian sites.

In contrast, Leatherwood *Eucryphia lucida*, Sassafras *Atherosperma moschatum* and Horizontal Scrub *Anodopetalum biglandulosum* are generally sub-canopy trees of wet forests that are insect pollinated. *Atherosperma* and *Anodopetalum* do not flower abundantly and the pollen is deposited within the forest. *Eucryphia* may occur extensively in riparian situations, where it flowers abundantly, but the pollen is also deposited locally. Hence, these indicators demonstrate presence of wet forest, and with a predominance of *N. cunninghamii* and *Phyllocladus* and <5% *Eucalyptus*, indicate lowland rainforest. Pure rainforest without *Eucalyptus* is of limited occurrence in Tasmania. The Forestry Commission recognises that rainforest can contain a *Eucalyptus* component of up to 5% (Hickey pers. comm. 2003). Allowing for other trace pollen of regional origin, we define the lowland rainforest as having >70% *Nothofagus* + *Phyllocladus* + *Lagarostrobos* pollen. With more pollen of *Eucalyptus* and less of rainforest taxa, the forest is defined as wet mixed forest. *Pomaderris apetala* forms an understorey tree in wet forests and is especially evident where the forest has been periodically burnt. The pollen can be widely dispersed in small amounts (ca. 1%) but where it occurs in larger amounts, usually considerably exceeding 2%, it indicates wet sclerophyll *Eucalyptus* forest. At altitudes above 500-700 m, subalpine rainforest may contain significant quantities of Native Pines, *Athrotaxis* spp. and the dwarf Deciduous Beech, *N. gunnii*. Both species can produce relatively abundant pollen, with that of the pines being more widely dispersed than the beech, which is deposited locally.

Eucalyptus is insect pollinated and there appears to be a close relationship between tree cover and quantity of pollen produced. A division has been made between the subalpine, wet and dry sclerophyll forests and woodlands where normalised *Eucalyptus* pollen values between 10% and 30% infer regional woodlands and greater than 30% infer regional forests. This approximates Specht's (1970) woodland and forest structural forms. Unlike *Eucalyptus*, the pollen of *Allocasuarina* spp. is wind dispersed. It can vary considerably in abundance, be transported widely and occur in small quantities at sites far beyond its source area. It can also occur abundantly in association with coastal communities.

The vegetation of non-wooded environments is dominated by pollen of Poaceae and Asteraceae. Pollen from isolated *Eucalyptus* trees, which can be locally abundant, plus up to 5% of other long-distance transported pollen, make separation of the dry savanna-like vegetation and grassland difficult, which at the broad scale would probably form a mosaic. Here we use >10% *Eucalyptus* pollen as indicating savanna-like vegetation.

Chenopodiaceae are recorded by a few pollen grains in many spectra at Tasmanian sites. Macphail (1979) suggested the Chenopodiaceae pollen may have been transported from southern mainland Australia, but high values associated with native *Plantago* spp. suggest they are likely to indicate local steppe vegetation, especially during drier conditions in the last glaciation.

The vegetation of alpine areas consists either of alpine heath and scrub or alpine grassland and herbfield often in a complex mosaic pattern. In addition to Poaceae and Asteraceae, the heath may contain one or more species of coniferous shrub taxa and numerous species of Epacridaceae. In areas of alpine grassland and herbland, pollen of grass genera (not differentiated) is abundant and is probably over-represented in the pollen assemblage relative to its source plants occurrence in the field. Pollen of Asteraceae is also abundant, and although in Tasmania there is an abundance of alpine Asteraceae shrub spp., it is not possible to separate the pollen of herbs and shrubs. The Pineapple Grass *Astelia alpina* is a consistent indicator of wet alpine vegetation, while pollen of native *Plantago* spp. is consistently represented in alpine herbaceous vegetation.

The time slices on the maps represent broad but not overlapping periods that have been selected to detect major temporal changes in the vegetation. The radiocarbon dating of pollen-sediment sequences in Tasmania has been undertaken over several decades. Assays have been made by various laboratories and until recently reported only in radiocarbon years. Many of the sequences have been taken from alpine and subalpine lakes where sediments particularly of pre-Holocene age are low in organic carbon. In order to obtain dates, some of the core samples from which carbon has been extracted are 5-10 cm long. Other dates mainly of Holocene age have been obtained from individual wood and charcoal fragments, and from small samples of organic lake mud and peat. Much of the dating is not precise and standard error values can be large. In addition, residual traces of humic acid have affected some of the older pre-Holocene samples, making their ages appear younger than they really are. Due to constraints on the precision of the dates, original determinations have been used rather than calibrated ages, except where the latter are specified. It is estimated that errors may be up to around ± 1000 years for Holocene dates, but may be greater for older dates that are here allocated broadly to isotope stages as reflected by the pollen curves.

Unfortunately, limited financial resources have resulted in most Tasmanian pollen diagrams not being closely dated, and the chronologies of many depend on only a few dates. Thus, in this reconstruction, the ages of many pollen spectra are linearly interpolated between dated samples assuming uniform sedimentation rates, which is rarely the case. In addition, two to three pollen spectra may be combined to represent the interpreted vegetation assemblage of designated time slices. The time slices 1000, 6000, 9000 and 12,000 BP were selected to represent pre-European, mid-Holocene, early-Holocene and late-glacial-Holocene transition vegetation, which they reasonably do. Radiocarbon calibration indicates the 6000 and 9000 BP time slices are approximately 1000 years older (6800 and 10,000 cal BP) and the 12,000 BP slice 1500 older (13,500 cal BP), but given the overall limitations of the dating, the difference is of little significance.

The vegetation associations represented in Figure 3e for OIS 2 are derived from pollen spectra dated to or interpolated to have occurred around the LGM between ca. 18,000 and 25,000 BP. Those on Figure 3d for OIS 3 are derived from average pollen values from spectra over the period interpolated as 30,000-55,000 BP, neglecting the fact that OIS 3 has had short warmer and colder phases of climate. Several marked fluctuations of vegetation that may reflect these climate changes have been averaged out for the map. The vegetation maps Figures 3c and 3b for OIS 4 and OIS 5e have no absolute dating but are derived from pollen spectra obtained from sediment sequences attributed stratigraphically to these isotope stages.

The late-Quaternary vegetation maps

Only a limited number of pollen-vegetation records exist before the Holocene (Figures 3b-3f), whereas many records have been obtained for the Holocene (Figures 3g-3i).

Last interglacial

Figure 3b represents the last-interglacial vegetation for central-western Tasmania probably during OI Substage 5e. Only one site at Lake Selina (Figure 4) has a complete vegetation record for Substages 5e to 5a (Colhoun et al. 1999). Records from other sites are attributed to Substage 5e on palynological grounds, but some might belong to interstadial Substages 5a or 5c. Assuming their attribution to Substage 5e is correct, then the last-interglacial vegetation in the west-coast mountain region of Tasmania consisted predominantly of wet mixed forest in which *N. cunninghamii* was dominant, with *Phyllocladus, Allocasuarina, Lagarostrobos* and *Eucalyptus* present in quantity.

The break in the Lake Selina record indicates the section obtained from a short surface

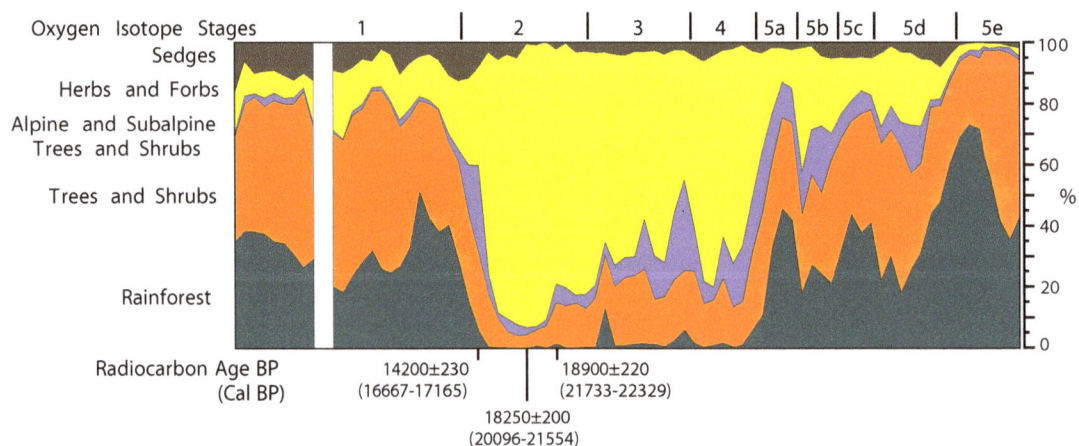

Figure 4. Summary pollen diagram for Lake Selina.

core and the section from the longer main core. There is no time break in the pollen sequence. The radiocarbon dates have been calibrated using Calib Rev 6.0 (Stuvier and Reimer 1993).

In the north, the vegetation at Pieman Damsite (Reece Dam) closely resembled modern cool temperate *Nothofagus* rainforest, though there was a strong riparian element represented by *Eucryphia* and *Anodopetalum. Eucalyptus* values averaged 5% (Colhoun 1980). At Lake Selina, rainforest taxa peaked during OI Substages 5e, c and a, but were reduced by around two-thirds during OI Substages 5b and d (Figure 4). *Eucalyptus* was scarcely represented, though *Allocasuarina* varied from 10% to 15%, which may reflect local presence around the lake. Lake Selina is the only site in Tasmania with pollen records for OI Substages 5d and 5b. Both are characterised by greater amounts of Epacridaceae, Poaceae, Asteraceae and *Microstrobos* than 5e, 5c and 5a, which points to the occurrence of heath and herbaceous vegetation, and colder conditions.

Further south in the King Valley, the wet sclerophyll forest of the interglacial deposits at Smelter Creek had 58% *Allocasuarina*, which probably reflects a riparian rather than regional aspect of the vegetation (Colhoun et al. 1992). In the adjacent Andrew Valley, the last-interglacial vegetation at Darwin Crater had abundant *N. cunninghamii* and *Phyllocladus*, with *Lagarostrobos*, which is still abundant adjacent to the crater (Colhoun and van de Geer 1998).

North of Tasmania, a site at Yarra Creek on southeast King Island contains lagoonal peat and organic sand beds within 1.6 m of the modern cobble beach. The sandy peat 30-50 cm above the beach was dated by thermoluminescence to younger than 120,000 ± 7000 BP, and the organic deposit which extended to 1.7 m above the beach has been suggested to be of Substage 5e age. The vegetation was dominated by cool temperate rainforest and wet sclerophyll forest taxa, with 57-67% *Phyllocladus* and 12-19% *Eucalyptus. N. cunninghamii, Cyathea* and *Dicksonia* are well represented. The data are consistent with wet forest vegetation extending northwards to King Island during the last interglacial (Porch et al. 2009), but the deposit occurs well below the level attained by the last interglacial marine transgression on King Island of 20-21 m (Jennings 1959), and must have been formed after sea level began to retreat. It thus must be of late 5e age or belong to Substage 5c or 5a.

Of particular biological, temporal and probable stratigraphic significance is that at Pieman Damsite and Darwin Crater, the spores of *Cyathea australis* were more abundant during the last interglacial than those of *Dicksonia Antarctica*. Throughout the Holocene, *D. Antarctica* is the more dominant in wet forests.

Early last glaciation

Figure 3c represents the early last glaciation vegetation of OIS 4, which is recorded at five sites. At Lake Selina, abundant pollen of Poaceae, Asteraceae, Apiaceae, Chenopodiaceae and *Microstrobos* indicates alpine grassland and herbfield with alpine shrubs occurred around 516 m altitude on the northwestern mountains (Colhoun et al. 1999). Further south, at an altitude of 180-200 m at Darwin Crater, the vegetation is marginally classified as dry sclerophyll (subalpine) woodland. However, the abundance of pollen of Epacridaceae, Gramineae, Asteraceae, *Astelia*, *Plantago*, Ranunculaceae, Scrophulariaceae and Apiaceae in the full spectra showed that the sparse subalpine woodland of *Eucalyptus* and *Allocasuarina* species probably occurred locally within or adjacent to alpine grassland and herbfield vegetation, the latter of which was extensive in mountainous western Tasmania (Colhoun and van de Geer 1998).

The pollen in marine core SO36-7SL west of Macquarie Harbour indicates the extensive occurrence of subalpine sclerophyll forest in central-western Tasmania during OIS 4. This site would have received pollen from many vegetation communities in the catchment of the Gordon River and on the mountains of the southern West Coast Range. Apart from 6-7% pollen of *Eucalyptus* and of *Allocasuarina*, most pollen is Asteraceae, Poaceae and Chenopodiaceae, indicating widespread herbaceous vegetation, but there are also alpine shrub and herb species (van de Geer et al. 1994).

Pollen diagrams from Pulbeena and Mowbray swamps in northwest Tasmania are dominated by Poaceae with Asteraceae and Cyperaceae, and contain only 5% *Eucalyptus* pollen. The regional vegetation had the structure of a savanna or grassland, with sparse trees, while the swamps were covered with sedges (Colhoun et al. 1982; van de Geer et al. 1986).

Mid last glaciation

The mid last glaciation of OIS 3 extended from approximately 59,000 to 24,000 BP (Martinson et al. 1987). During this period, there was some variation in the vegetation at several sites. The vegetation associations represented in Figure 3d are based on average values for spectra in the age range 30,000-55,000 BP.

The most complete sequence for western Tasmania is from Lake Selina, where alpine grassland and herbfield occurred during most of OIS 3. The vegetation consisted predominantly of Asteraceae, Poaceae and Apiaceae, with 10% *Microstrobos*. The vegetation for most of the period was closer in composition to the glacial-age vegetation of OIS 4 and OIS 2 than to interglacial wet forests. Of particular interest are three fluctuations, with the largest peak at the beginning of OIS 3 when significant amounts of *Microstrobos* and *N. gunnii* indicate an alpine heath component, and seem to indicate that vegetation and climatic conditions varied between alpine and subalpine.

Subalpine sclerophyll woodland was probably widely distributed. At Newton Creek in the West Coast Range at 550 m altitude, a sequence is dated basally to 34,000 BP. The pollen record for the lower part comprised high quantities of *Athrotaxis-Diselma* (similar pollen forms) and some *Astelia*, indicating subalpine-alpine vegetation, but the upper part was dominated by the sclerophyll taxa *Allocasuarina* and *Eucalyptus*, and the rainforest taxon *Phyllocladus* (Colhoun et al. 1993). Offshore in Marine Core SO36-7SL, *Allocasuarina* with *Eucalyptus* and *Phyllocladus* are the most important taxa, but *Athrotaxis*, *Microstrobos* and *Astelia* confirm extensive subalpine-alpine vegetation in west-central Tasmania (van de Geer et al. 1986).

Further south in western Tasmania, the vegetation at Newall Creek and at Darwin Crater is classified as dry sclerophyll woodland, although the climate was probably wet. In both cases, the vegetation was likely to have been subalpine in composition. At Newall Creek, in addition to dominant Poaceae and Asteraceae, *Eucalyptus* averaged 30% and pollen of rainforest species was negligible. Several alpine indicator taxa were present, the most important being *Astelia* and the bolster plant *Donatia novae-zelandiae*. At Darwin Crater, the record is difficult to interpret due to a bed of gravel causing a break in pollen sedimentation. The record indicates a co-dominance of Poaceae and Asteraceae, with around 10% *Eucalyptus*, little pollen of rainforest taxa, and significant quantities of *Astelia*, *Plantago* and Apiaceae, probably indicating alpine conditions (van de Geer et al. 1994; Colhoun and van de Geer 1998).

During OIS 3, the vegetation of the northwestern plains was dry sclerophyll woodland and forest. *Eucalyptus* with abundant *Leptospermum* (probably on the surface or adjacent to the swamps) and lesser amounts of *Melaleuca* and Poaceae were the most important taxa. Alpine herb and shrub taxa were notably absent at Pulbeena and Mowbray swamps (Colhoun et al. 1982; van de Geer et al. 1986). At Rocky Cape, further east on the northwest coast, organic horizons within alluvial fan gravels mainly contained pollen of *Eucalyptus* and *Allocasuarina* with Poaceae and Asteraceae, suggesting open dry sclerophyll forest (Colhoun 1977a). These northwestern sclerophyll woodlands and forests reflected their more continental location, the result of glacial lowering of sea level and draining of Bass Strait.

In southeastern Tasmania, the vegetation consisted of dry sclerophyll forest at Blakes Opening in the middle Huon Valley, where *Eucalyptus* was the most important taxon for most of the period (Colhoun and Goede 1979). Similarly, *Eucalyptus* forest occurred at Pipe Clay Lagoon in the 5000-year period that preceded the LGM (Colhoun 1977c). Wet sclerophyll forest occurred on the southern part of Tasman Peninsula, where in a gully at Remarkable Cave the dominant *Eucalyptus* pollen is accompanied by 1.4-13.2% *Pomaderris* and by up to 5% *N. cunninghamii* and 4% *Phyllocladus* (Colhoun 1977b).

Last Glacial Maximum

The Last Glacial Maximum vegetation of Figure 3e has been reconstructed for the period 18,000-24,000 BP during OIS 2. Alpine grassland and herbfields dominated all valley and mountain sites from Tullabardine Creek to Newall Creek in the west and extended as far as Lake St Clair in the interior. Poaceae and Asteraceae pollen are dominant. There is usually less than 10% *Eucalyptus* and negligible amounts of rainforest taxa, but traces of alpine taxa occur widely. Offshore, Marine Core SO 36-7SL indicates the widespread presence of alpine grassland and herbfields, with very high quantities of Poaceae, Asteraceae and Chenopodiaceae pollen. *Eucalyptus* and *Allocasuarina* are about 10% each, while *Astelia, Athrotaxis* and *Microstrobos* are well represented (Colhoun 1985a; Colhoun and van de Geer 1986; van de Geer et al. 1989, 1994; Colhoun et al. 1999; Hopf pers. comm. 2011).

At Dante Rivulet in the upper King Valley, bolsters of *Donatia novae-zelandiae* occur on the surface of a fossil alpine soil at 230 m altitude some 750 m below the modern treeline. The bolster is dated to 18,800 BP (22,300 cal BP) and is buried by outwash sediments deposited by the LGM ice advance. Asteraceae and Poaceae are the most important pollen types, but the bolster also contains numerous local Cyperaceae and Epacridaceae, spores of *Gleichenia*, plus small amounts of a wide range of alpine herbs and subalpine shrubs (Gibson et al. 1987; Colhoun et al. 2010).

At Ooze Lake cirque in the highly oceanic mountains of southern Tasmania, the vegetation comprised subalpine rainforest of *Lagarostrobos, N. cunninghamii* and *Phyllocladus* that extended to 880 m around 18,000 BP immediately after the cirque glacier had melted. Most of the *Lagarostrobos* pollen is immature, indicating severe environmental stress (Macphail and Colhoun 1985).

In the lowlands of northwestern Tasmania, the vegetation was savanna and dry sclerophyll woodland. At Mowbray Swamp, the vegetation was almost exclusively Poaceae, with small amounts of *Eucalyptus* and *Leptospermum*. Pollen of rainforest taxa is negligible and none is present for alpine-subalpine taxa. The vegetation was similar at Broadmeadows Swamp, except that *Melaleuca* as well as *Leptospermum* was well represented adjacent to or on the swamps (van de Geer et al. 1986). The vegetation at Pulbeena Swamp was a savanna, with Poaceae and Asteraceae very abundant, and less than 5% *Eucalyptus*. No rainforest, subalpine or alpine pollen are present (Colhoun et al. 1982). On Hunter Island adjacent to northwest Tasmania, pollen from the archaeological site at Cave Bay Cave for the period dating 23,000-14,750 BP contains abundant Poaceae and Asteraceae, with *Eucalyptus* being the only significant tree species represented. Such a cave site preferentially represents regional over local pollen, and the assemblage is compatible with that at Pulbeena, indicating the vegetation of northwest Tasmania at and following the LGM was savannah-like grassland, which probably extended from the Adelaide region to Bass Strait and represented colder, drier conditions than present (Hope 1978).

The vegetation history of Midland and coastal eastern Tasmania during OIS 2 is restricted to records from three sites.

At Crown Lagoon in the dry eastern Midlands, a 2 m sediment core, though undated, is thought to extend from earlier than 25,000 BP until it was drained during European settlement. The pollen record suggests the vegetation varied from savanna or open woodland with *Eucalyptus* and *Allocasuarina* before OIS 2, to grassland or steppe at the peak of glacial dryness, with Poaceae, Asteraceae, abundant Chenopodiaceae and less than 5% *Eucalyptus* (Sigleo and Colhoun 1981).

On the east coast at Freycinet Peninsula, an old glacial-age deflation hollow that now forms the coastal Hazards Lagoon provided a 157 cm sediment record. The record extends to earlier than 18,000 BP (21,000 cal BP), and includes the peak of the LGM. At that time, the vegetation comprised a steppe to grassland vegetation, with Poaceae and Asteraceae dominant. Chenopodiaceae and Epacridaceae pollen were also abundant (Mackenzie 2010).

In coastal southeast Tasmania, peaty sediments formed in a pond on the floor of a deflation hollow at Pipe Clay Lagoon are dated to 20,000-22,000 BP. The pollen record indicates the occurrence of dry sclerophyll forest of *Eucalyptus*, with Poaceae and Asteraceae (Colhoun 1977c).

At the three Midland and east-coast sites, pollen of rainforest taxa is negligible and none is present for alpine-subalpine taxa. There was a sharp north-northwest to south-southeast-trending boundary between the wet alpine and subalpine associations of the west and the dry sclerophyll woodland/forest-grassland-steppe associations of the east during the LGM. Similarly, the vegetation of northwestern Tasmania, then connected to Victoria by a reduction in sea level and exposure of Bass Strait, consisted of dry sclerophyll woodland, savanna and grassland.

Late last glaciation

The vegetation shown in Figure 3f represents the later part of Termination 1 when glaciers finally melted from the highlands of western Tasmania around 15,000-14,000 BP and early changes in the vegetation from glacial to interglacial conditions occurred (Colhoun et al. 2010). The vegetation can be regarded as of late-glacial age up to 12,000 BP but age calibration would indicate that the changes at some sites commenced a few millennia earlier. During this late-glacial period, alpine vegetation was still very extensive in upland western Tasmania and wet forests were restricted to lower altitudes.

Marine Core SO 36-7SL and sites at 180 m at Governor Bog and 200 m at Smelter Creek in the King Valley show that a transition from non-forest to wet mixed forest occurred in the valleys and on the lowlands, and extended to over 516 m at Lake Selina during the period 14,000-10,000 BP (earliest at Lake Selina cal age 16,700-17,100 BP, Figure 4) (Colhoun et al. 1991a, 1992, 1999; van de Geer et al. 1994). However, wet mixed forest was not widely distributed in the mountains, as alpine grassland and herbfield still remained at higher altitudes, together with alpine shrubs, as at Poets Hill in the West Coast Range, Lake Vera, Lake Wurawina and Mt Field before post-glacial expansion of forest vegetation. At Lake St Clair in the centre and Dove Lake at Cradle Mountain further north, subalpine rainforest occurred, while further east at Brown Marsh the vegetation was mainly tussock grassland with alpine herbs. At Dublin Bog in the Mersey Valley, *Eucalyptus* wet sclerophyll forest replaced grassland by 13,000 BP (Macphail 1979, 1986; Colhoun et al. 1991b; Colhoun 1992).

In mountainous southeastern Tasmania, subalpine rainforest had already expanded to 880 m at Ooze Lake. However, at 960 m at nearby Adamsons Peak the vegetation was subalpine sclerophyll forest with slightly over 30% *Eucalyptus* pollen, and significant quantities of pollen of alpine herbs and shrubs, notably *Astelia* and *N. gunnii* (Macphail and Colhoun 1985; Macphail 1986).

On the northwestern plains, as far as dating and pollen zone correlation allow, the vegetation was savanna at Broadmeadows Swamp and dry sclerophyll woodland at Mowbray and Pulbeena swamps (Colhoun et al. 1982; van de Geer et al. 1986).

In eastern Tasmania, steppe-grassland vegetation was still present in the dry Midlands at Crown Lagoon, with high values for Poaceae, Asteraceae and Chenopodiaceae, and under 10% *Eucalyptus* (Sigleo and Colhoun 1981). At nearby Stoney Lagoon, dry sclerophyll forest with 81% *Eucalyptus* pollen plus mainly Poaceae and Asteraceae is recorded first around 12,000 BP. At Hazards Lagoon on the east coast, *Eucalyptus* pollen had increased from the LGM to around 40-50%, and Poaceae and Asteraceae had decreased to below 15%. The vegetation was dry sclerophyll forest with a grassy understorey (Jones 2008; Mackenzie 2010).

Early Holocene

The early-Holocene vegetation pattern shown in Figure 3g dating to around 9000 BP represents the time post-dating the last glaciation when forest expansion was occurring in much of Tasmania. Wet mixed forests of *N. cunninghamii, Phyllocladus, Eucryphia,* around 10% *Eucalyptus,* and pollen of mesic shrubs occurred throughout the West Coast Ranges except at high altitude. Wet mixed forest also occurred at Upper Timk Lake in the southeast, where pollen of Poaceae, Asteraceae and alpine taxa were negligible. However, at high altitude, as at Lake Johnston in the west, Lake Dove in the northwest, Lake St Clair in the centre, and Adamsons Peak and Ooze Lake in the south, the vegetation was subalpine rainforest located close to treeline, which in addition to a dominance of rainforest taxa also contained indicator alpine shrub and some herb taxa, *N. gunnii* and *Astelia* being present at all sites. At around 1000 m at Beatties and Eagle tarns at Mt Field, *Eucalyptus* and *Pomaderris* pollen were abundant, indicating that wet sclerophyll forest was well established on the mountains below the tarns which occurred in a subalpine environment close to treeline. At higher altitude (1158 m) on Tarn Shelf, alpine grassland and herbfield still occurred (Macphail 1979; Macphail and Colhoun 1985; Harle 1989; Dyson 1995; Hopf et al. 2000; Anker et al. 2001).

Further east in west-central Tasmania, wet sclerophyll forests began to be established at lower altitudes. At 440 m in a deep river valley at Tarraleah, *Eucalyptus* increased strongly to 60-90%, with an accompanying rise in the main indicator taxon *Pomaderris.* Small quantities of pollen from *Ziera arborescens, Phebalium squameum, Monotoca glauca* and *Bauera rubioides* typically associated with wet forests, and *Bursaria spinosa* and *Dodonaea viscosa* associated with dry forests and woodlands suggest that the site was located towards the eastern part of the wet sclerophyll forest zone. Still further east at an altitude of 750 m at Brown Marsh, *Eucalyptus* was increasing with *Pomaderris,* while Poaceae and alpine herb and shrub taxa were decreasing, and the forest on the lower southeastern part of the Central Plateau was still relatively open (Macphail 1979, 1984).

At 650 m on Yarlington Tier west of Colebrook, a similar stand of *Eucalyptus* wet sclerophyll forest was established, with Poaceae and Asteraceae decreasing and *Pomaderris* increasing. Of particular significance is the occurrence of 6-11% pollen of *N. cunninghamii,* which is more than expected by transport from western forests. Local occurrence is confirmed by a small stand of *Nothofagus* on the site. Around 9000 BP, an increase in *Pomaderris* with *Atherosperma, Phyllocladus* and *Dicksonia* indicates change from dry to wet sclerophyll forest at the end of the glacial period. The presence of *Nothofagus* and *Atherosperma* at Yarlington raises the question of whether these species had expanded further eastwards during the early Holocene than they occur today, or whether they survived throughout the last glacial despite regionally dry and cold conditions by virtue of being located in a suitable topographic and hydrologic habitat. The latter explanation has been preferred (Harle et al. 1993).

Further north at Dublin Bog in the Mersey Valley, wet sclerophyll forest of *Eucalyptus* with *Pomaderris* had been established by 13,000 BP, with very few ancillary shrub and herb taxa except for a small rainforest component of *N. cunninghamii* and *Phyllocladus* and the treefern *Dicksonia.* The forest remained similar in composition throughout the early Holocene (Colhoun et al. 1991b).

In northwest Tasmania, the savannah and dry sclerophyll forests of 12,000 years ago were now largely replaced by wet sclerophyll forest at Mowbray and Broadmeadows swamps, though Pulbeena Swamp was little different but classified as dry sclerophyll forest. At all sites, *Eucalyptus* was the major regional component, with abundant *Melaleuca* and *Leptospermum* locally adjacent to or on the swamps (Colhoun et al. 1982; van de Geer et al. 1986).

The vegetation of Midland and eastern Tasmania during the early Holocene is represented by three sites. At Lake Tiberias in the southern Midlands, the vegetation was dry sclerophyll forest with 75% *Eucalyptus*, around 5% *Allocasuarina* and 10% Poaceae. *Pomaderris* attained 5-15% and represents abundant transport from western forests, along with 5% *N. cunninghamii*. At nearby Stoney Lagoon, dry sclerophyll forest is dominated by 66% *Eucalyptus* with 12% Poaceae and 4% Asteraceae, and by 9% *Pomaderris* and 5% *Phyllocladus*, which, like at Lake Tiberias, was transported from the west.

At Hazards Lagoon, the vegetation was also dry sclerophyll forest with 50% *Eucalyptus* and 30% *Allocasuarina*. Rhamnaceae (*Pomaderris*) was 5-10% and Poaceae and Asteraceae both less than 5%. *N. cunninghamii* pollen was negligible. The transport of pollen types from western forests was less than in the Midlands and the *Allocasuarina* probably reflects near coastal influences (Macphail and Jackson 1978; Jones 2008; Mackenzie 2010).

In northeast Tasmania, a coastal site at Waterhouse Marsh has 20-40% *Eucalyptus* and 10-15% *Allocasuarina*, plus 20-30% Poaceae and 10% Asteraceae. There is a significant wet forest component of *Pomaderris*, *N. cunninghamii* and *Phyllocladus* and spores of *Dicksonia* and *Cyathea* that would have been derived from wet forests in the valleys and on the mountain slopes of the uplands to the south. This rainforest component results in the site being classified marginally as wet sclerophyll forest, though the local vegetation of the plain was almost certainly dry sclerophyll forest (Thomas 1996).

In southwest Tasmania, a sediment core taken at Pedder Pond on the outwash plains east of Lake Pedder and west of the foothills of Mt Anne showed that from the beginning of the Holocene record at 10,350 BP until after 9000 BP, the most important taxa were moorland species that included *Gymnoschoenus sphaerocephalus*, which is usually very under-represented by pollen but abundant in the vegetation, Restionaceae and Epacridaceae. Pollen of *Eucalyptus* slightly exceeds 10% and Poaceae attains about 10%. Pollen of *N. cunninghamii* averages about 5% and *Phyllocladus* 5-10%. The high amount of pollen of moorland taxa and low amounts of rainforest taxa combined with high quantities of charcoal led Fletcher and Thomas (2007a) to conclude that during the postglacial period moorland vegetation had always occupied the Lake Pedder area and that rainforest had not colonised it. Here, the vegetation is classified by the regional pollen types as subalpine sclerophyll forest, which differs from that based on local taxa, which would indicate presence of moorland-heathland.

Mid Holocene

The mid-Holocene vegetation pattern of about 6000 BP (Figure 3h) represents postglacial optimum forest development (Macphail 1979), though maximum rainforest developed at different times in different locations (Colhoun 1996). At the regional scale, there is little difference between the 9000 and 6000 BP patterns in western Tasmania because the major divide between wet forest vegetation in much of the west, and dry forest vegetation over most of the Midlands and east had been established by 9000 BP (Figure 3g).

In western Tasmania, regionally distributed pollen types indicate that at 6000 BP, wet mixed forest dominated by *N. cunninghamii* and *Phyllocladus* with 5-10% *Eucalyptus*, plus *Bauera rubioides* and *Dicksonia* occurred at most sites. Cool temperate lowland rainforest occurred only at a few sites adjacent to major rivers as at Newell Creek and Lake Fidler, or at higher altitude as at Lake Vera or Upper Lake Timk where the montane rainforest was protected from fire (Macphail 1979; Harle 1989; van de Geer et al. 1989; Harle et al. 1999).

At several sites where regional pollen representation classes the site as wet mixed forest as at Poets Hill west of Lake Margaret, King River Railway Bridge, Governor Bog and Smelter Creek in the King Valley, abundant local taxa including Epacridaceae, *Leptospermum*, *Melaleuca* and Restionaceae indicate high inputs from either local bog surfaces or mosaics of vegetation communities (Colhoun et al. 1991a, 1992; van de Geer et al. 1991; Colhoun 1992).

In the centre at Lake St Clair and towards the north at Lake Johnston and Lake Dove, subalpine rainforest persisted. In addition, subalpine rainforest persisted at high altitude at Adamsons Peak and Ooze Lake in the southeast (Macphail 1979; Macphail and Colhoun 1985; Dyson 1995; Hopf et al. 2000; Anker et al. 2001).

Further east in central Tasmania, wet sclerophyll forest occurred at Tarraleah. However, at higher altitude, around 1000 m at Eagle Tarn and Beatties Tarn at Mt Field, the vegetation comprised subalpine sclerophyll forest in which around 30-50% *Eucalyptus*, 1-10% *Pomaderris* and 1-10% *N. gunnii* complemented *N. cunninghamii* and *Phyllocladus*. At higher altitude (1158 m) on Tarn Shelf, *Eucalyptus* decreased to 20% and alpine taxa including *N. gunnii*, *Microcachrys* and *Astelia* amounted to 5-10% each. The vegetation was alpine heath, and scrub and forest did not expand to Tarn Shelf during the Holocene. Similar vegetation occurred at Upper Lake Wurawina at 1040 m in the Denison Range, with 20-30% *Eucalyptus*, 10% each for *N. gunnii* and *Athrotaxis-Diselma*, and 10-30% *Astelia* (Macphail 1979, 1986).

At Pulbeena, Mowbray and Broadmeadows swamps in lowland northwest Tasmania, the regional vegetation at 6000 BP was wet sclerophyll forest and there was also widespread swamp forest. At Pulbeena, there was a marked increase in *Eucalyptus* plus *Melaleuca*, very small increases in *Monotoca*, Rhamnaceae (*Pomaderris*) and traces of rainforest taxa, indicating change from dry to wet sclerophyll forest between 9000 and 6000 BP. At Mowbray, *Eucalyptus* and *Melaleuca* increased and small quantities of *Monotoca* and *Acacia* (probably Blackwood *Acacia melanoxylon*) occurred. At Broadmeadows, *Eucalyptus* and *Melaleuca* also increased and there was more *Monotoca* than at Mowbray (Colhoun et al. 1982; van de Geer et al. 1986).

At Yarlington Tier (650 m altitude) west of Colebrook adjacent to the southern Midlands, there was very little change in the *Eucalyptus* wet sclerophyll forest between 9000 and 6000 BP, with only a slight reduction in Poaceae and increase in *Pomaderris* (Harle et al. 1993).

On the dry eastern part of the Central Plateau at Camerons Lagoon (1100 m) on Liawenee Moor, the regional vegetation at 6000 BP was dry sclerophyll woodland, with about 29% *Eucalyptus*, 40% Poaceae and 15% Asteraceae. The *Eucalyptus* would have grown on the surrounding dolerite ridges, while the Poaceae and Asteraceae would have covered a grassy upland plain.

At Lake Tiberias in the eastern Midlands, about 10% of both *N. cunninghamii* and *Pomaderris* transported from wet forests to the west is present. *Eucalyptus* (70-75%) is the dominant taxon and the forest was dry sclerophyll forest with 10% Poaceae and 5% *Allocasuarina*. Similar dry sclerophyll forest occurred at nearby Stoney Lagoon. Dry sclerophyll forest also extended to Hazards Lagoon in the Freycinet Peninsula, which by the middle Holocene was a coastal site and had around 25-30% pollen of *Allocasuarina* (Macphail and Jackson 1978; Jones 2008; Mackenzie 2010).

In northeast Tasmania, the regional vegetation at Waterhouse Marsh remained dry sclerophyll forest when the long-distance-transported rainforest component is excluded. Forester Marsh (1000 m) in the upper Forester River catchment is dated to 4400 BP and shows the mountain vegetation contained the same rainforest and treefern taxa as recorded at Waterhouse Marsh, but here it was local. *Eucalyptus* was the dominant taxon, with over 30% pollen, and the vegetation was wet sclerophyll forest (Thomas 1996).

In southwest Tasmania at Pedder Pond, the regional pollen gives a maximum signal for rainforest taxa, with about 10% *Eucalyptus* at 6000 BP, which classes the vegetation as wet mixed forest in Figure 3h. However, high local inputs of Epacridaceae and Restionaceae plus other shrubs and the buttongrass *Gymnoschoenus* indicates the vegetation was moorland rather than forest (Fletcher and Thomas 2007a). At Melaleuca Inlet, virtually no pollen of rainforest taxa and *Eucalyptus* is recorded, but local pollen of Epacridaceae, *Melaleuca squamea*, *Monotoca* and Restionaceae are abundant and indicate shrubby moorland occupied this most southwesterly corner of Tasmania (Macphail et al. 1999). Similar results from Thomas (1995) show that

Gymoschoenus moorland and wet scrub occupied the area for at least the past 12,000 BP, and the continuous presence of charcoal indicates the vegetation association was maintained by the occurrence of frequent burning.

Pre-European settlement

The pre-European settlement vegetation pattern in Figure 3i is represented by pollen spectra of about 1000 BP age. The regional pollen indicates no significant change for lowland rainforest and wet mixed forest in central and northwestern Tasmania. The distribution of lowland rainforest is analogous to that mapped for modern vegetation (Kirkpatrick and Dickinson 1984; Harris and Kitchener 2005; Figure 2). Subalpine rainforest occurs at altitude in the western mountains and extends to the southern mountains. Alpine vegetation is confined to higher than 1000 m. A sharp north-northwest south-southeast-trending divide occurs between northwestern and central Tasmania, where the dominant forest was wet sclerophyll forest with subalpine sclerophyll woodland at higher altitude. This boundary has not moved westwards since it was established before 9000 BP, and there was no change in the wet sclerophyll forest of northwest Tasmania between 6000 and 1000 BP (Figures 3h and 3i).

Wet mixed forest and wet sclerophyll forest expanded in the highlands of northeastern Tasmania during the Holocene, as indicated at Forester Marsh and Mathinna Plains (950 m) where abundant pollen of *N. cunninghamii*, *Phyllocladus* and treeferns, and over 10% *Eucalyptus* indicates that during the past millennium wet forests were widespread (Thomas 1996). Chloroplast DNA studies show that *N. cunninghamii* survived within northeast Tasmania and was not dispersed from western Tasmania across the relatively dry northern Midlands after deglaciation. Although one haplotype (C1) is the most common in western Tasmania, another (NE1) is only found in the uplands of northeastern Tasmania (Worth et al. 2009). It is likely other wet-forest species survived locally.

There is a marked contrast between Figures 3h and 3i with the expansion of dry sclerophyll forest and contraction of wet sclerophyll forest southwestward during the late Holocene. Macphail (1979) first observed from sites at Mt Field and Adamsons Peak that after 6000 BP floristically simple *N. cunninghamii* rainforests and scrubs were replaced by open subalpine *Eucalyptus* woodlands and alpine communities. He attributed the retreat of the montane rainforest communities from their alpine limits to increases in drought and frost. Structural change also occurred in the wet sclerophyll forest at Tarraleah during the late Holocene, with strong decreases in *N. cunninghamii* and *Pomaderris* and an increase in *Eucalyptus* from less than 30% to more than 50% (Macphail 1984).

Near the eastern boundary of the wet sclerophyll forest there is a decrease in *Pomaderris* and an increase in *Allocasuarina* and Poaceae at Yarlington Tier, suggesting drier conditions. A marked rise in *Allocasuarina* at Lake Tiberias also indicates increased dryness. Further north at Camerons Lagoon on the Central Plateau, *N. cunninghamii* and *Pomaderris* decreases and *Eucalyptus* increases. The pollen data indicate westward decreasing precipitation, which also extended to Den Plain in the Mersey Valley during the late Holocene. In the eastern Midlands, *Eucalyptus* and *Allocasuarina* increases at Stoney Lagoon, while at Hazards Lagoon there is a very strong increase of *Allocasuarina* in the dry sclerophyll forest, which reflects its coastal location (Macphail and Jackson 1978; Harle et al. 1993; Thomas and Hope 1994; Moss et al. 2007; Jones 2008; Mackenzie 2010).

In southwest Tasmania, there is a marked contrast between the vegetation of the deeply incised river valleys, the inland basins and the lowland plains. In the lower Gordon Valley, lowland rainforest, with *N. cunninghamii*, *Phyllocladus*, *Lagarostrobos*, *Eucryphia* and *Anodopetalum*, is well developed. In contrast, at Pedder Pond on Huon Plains adjacent to the eastern end of Lake Pedder, any rainforest that may have developed has been supplanted by moorland dominated by Restionaceae, Epacridaceae and *Gymnoschoenus*, with shrubs of *Leptospermum* and *Melaleuca*. Similar shrubby-sedge moorland occurs around Melaleuca Inlet in the far southwest (Thomas 1995; Harle et al. 1999; Macphail et al. 1999; Fletcher and Thomas 2007a).

The influences of climate and people on the vegetation changes

The major driving influence on late-Quaternary vegetation changes was climate with its two main components, temperature and precipitation (Jackson 1968; Macphail 1980; Colhoun 2000). When the Tasmanians crossed Bass Strait 34,000-35,000 BP (38,000-39,000 cal BP), new pressures were exerted on the vegetation by the hunting of Red-necked wallaby *Macropus rufogriseus* and extensive use of fire, which caused the development of disclimax associations detectable in numerous pollen records (Cosgrove et al. 1990; Allen 1996; Cosgrove 1999; Fletcher and Thomas 2010a).

Western Tasmania receives 3500 to 1800 mm precipitation per annum and sustains rainforest where there is more than 1200 mm per annum and more than 50 mm in all months. Wet sclerophyll forests occur where there is more than 1000 mm per annum, with more than 25 mm in the driest month (Jackson 1983). During the late Quaternary, only temperature reduction and altitude would have limited the wet forest associations unless summers were much drier. This is unlikely because the southward expansion of the Australian continental high-pressure system and northern extension of Antarctic sea ice would have compressed and strengthened onshore westerly winds, bringing more moisture to western Tasmania than at present. Thus, western Tasmanian late-Quaternary vegetation changes were primarily controlled by temperature.

Estimating temperature change is difficult on land because the modern analogue technique (MAT) of comparing fossil pollen-vegetation associations with modern vegetation from one site may not fully represent regional limiting values for the associations. Snowline estimates in complex mountain topography are also limited by relatively large errors but are an independent proxy (Table 3; Colhoun 1985b; Colhoun et al. 1999).

Nevertheless, the Lake Selina record (Figure 4) is currently the best resolved late-Quaternary pollen sequence of vegetation changes for western Tasmania, and closely compares with the sequence of δ Deuterium changes in the Vostok ice core, indicating that Lake Selina records both regional and hemispheric climate signals (Colhoun et al. 1999; Petit et al. 1999).

First estimates of temperature changes for western Tasmania were based on calculations of reduction of the Stage 2 snowline of the West Coast Range glacial system from the mean atmospheric freezing level, using data from western meteorological stations and a lapse rate of 0.63°C/100 m, determined from a transect between sea level at Hobart and the summit of Mt Wellington (Nunez and Colhoun 1986) (Table 3). A value of 6.5°C for mean temperature depression was obtained. A higher value has been determined inland using mean summer freezing level for Stage 2 at Mt Field of 7.4°C. Unfortunately, mean lapse rate is site specific and makes determination of regional lapse rates and temperature comparisons difficult. The variations in values in rows 1 and 2 of Table 3 reflect variations between maximum values based on snowline estimates and minimum values of the pollen-vegetation associations at Lake Selina using the MAT (Colhoun 1985b; Nunez 1988; Colhoun et al. 1999, 2010; Mackintosh et al. 2006).

A regional picture that provides lower average values has been obtained using 26 pollen sites from western Tasmania plus a transfer function model (Fletcher and Thomas 2010a). These results in row 3 of Table 3 are comparable with results from marine sediment cores using alkenones and faunal assemblages on the East Tasman Plateau and South Tasman Rise, shown in rows 4 and 5. Together, they indicate temperature changes during the late Quaternary were

Table 3. Temperature estimates for Tasmania. **a:** Mean 4 fluctuations **b:** STF was south of Core GCO7 in early Holocene.

Temperature depression based on	5e	5d	5c	5b	5a	4	3	2	1	Sources
Estimated from present by variation in amplitude of inferred treeline curve in Lake Selina summary pollen diagram calibrated to 6.5°C temperature depression at LGM based on glacial snowline estimates	+1-2	4	2-3	4	2-3	>5	3-5a	6.5	+1-2 early 0 late	Colhoun 1985b; Colhoun et al. 2010
Minimum temperature depression from present inferred from the pollen-vegetation zones in the Lake Selina summary curve	0	2.2->3.5	0.6-2.2	2.2->3.5	0.6-2.2	>3.5	2.8->3.5	>3.5	<0.6 > 0 (?) 0 late	Colhoun et al. 1999; Table 3
Modern pollen data from 26 sites in western Tasmania calibrated by transfer function model for mean annual temperature reductions	+1	2	1	2	1	4	0-91-3.9	3.7-4.2	+0.4-0.84 at 7200-8300 BP 0 late	Fletcher and Thomas 2010a
Alkenone palaeothermometry of Core FRI/94-GC3 (44° 15'S, 149° 59'E) East Tasman Plateau	+2.6	1.1	+1	1.1	+1	3.1	1.5	3.1	+0.9 early 0 late	Pelejero et al. 2006
Alkenone palaeothermometry for Core GCO7(45° 09'S, 146° 17'E) on South Tasman Rise	+2-3					3.9-5.2	0.4-2.4	4.6-5.4	+4 at 11,000 BPb 0 late	Sikes et al. 2009

equivalent on land and sea (Pelejero et al. 2006; Sikes et al. 2009).

Some features of the temperature records in Table 3 should be highlighted. The mean temperature for Substage 5e when both rainforest and wet mixed forest developed in western Tasmania was 1-3°C warmer than during the late Holocene and slightly greater than values representing the early-Holocene thermal maximum. The mean temperature of 1.9°C below present and greater seasonal range recorded for Yarra Creek on King Island (Porch et al. 2009) seems not to support a Substage 5e age for the site, and a 5c or 5a age is more likely.

The sequence of Substages 5a to 5d is only recorded by pollen at Lake Selina. The estimates of temperature change are slightly greater than those derived by the Fletcher and Thomas model, while the values in the marine record of Pelejero et al. (2006) for Substages 5a and 5c are less, and are as warm as the Holocene. At Lake Selina, Substages 5a and 5c were cooler than 5e, as indicated by montane rainforest with *Phyllocladus* more important than *Nothofagus*. Substages 5b and 5d were significantly colder than 5a and 5c, as indicated by the mosaic of subalpine rainforest, shrubland and heathland.

All temperature estimates suggest Stage 4 was almost as cold as Stage 2 and the values compare with sea surface temperatures (SSTs) obtained for the southern Tasman Sea, the South Tasman Rise and west of Tasmania of a decrease of between 2°C and 5°C for the LGM (Barrows et al. 2000, 2007; Pelejero et al. 2006; Sikes et al. 2009). The vegetation at Lake Selina was alpine grassland and herbfield during both stages. That a temperature depression of at least 4.5-4.7°C (based on lapse rates of 0.6 and 0.63°C/100 m altitude) occurred immediately preceding the peak of the LGM is supported by the fossil *Donatia* bolster at 230 m at Dante Rivulet, which today occurs at over 1000 m (Gibson et al. 1987; Colhoun et al. 1999).

During Stage 3, temperature depression was more variable due to the occurrence of three marked fluctuations in Stage 3 and a fourth at the beginning of Stage 2, as the overall trend was downward towards the LGM (Table 3). Some fluctuations were sufficiently cold for the formation of glaciers in Stage 3 as well as Stage 2. The climate fluctuations were also not local, as similar changes occur in marine records as far apart as Core SO 136-GC3 taken off the west coast of South Island, New Zealand, and others in the South Atlantic (Kanfoush et al. 2000; Mackintosh et al. 2006; Pelejero et al. 2006; Colhoun et al. 2010).

At Lake Selina, the vegetation of Stage 3 was predominantly herbaceous including some alpine herb taxa, but there were also considerable shrub taxa including abundant *Microstrobos*. Pollen of rainforest was negligible. The pollen-vegetation assemblage suggests climate during most of the stage was closer to glacial than interglacial.

During Stage 2, the massive reduction of rainforest pollen over western Tasmania indicates little rainforest or wet mixed forest was present. As modern treeline in the West Coast Range occurs at 1100 m and snowline was reduced by 1000 m, there would have been little continuous forest higher than 100 m above sea level at the LGM. Such an inferred lowering of treeline reflects a 6°C reduction in mean temperature compared with today during Stage 2, even though greater than values in the broader regional and marine records. Areas of wet forest and woodland would have survived in scattered local refugia in protected sites from which they expanded during the deglaciation period. The process is represented in the West Coast Range, where Huon Pine *Lagarostrobos* is thought to have survived Stage 2 near Lake Johnston and expanded to an altitude of 1040 m immediately succeeding melting of the cirque glacier (Anker et al. 2001).

The most notable factor that drove re-expansion of wet forests throughout much of western Tasmania during the late glacial and early Holocene was a marked rise in temperature. Radiocarbon dates from marine Core GC07 which occurs on the South Tasman Rise close to southern Tasmania indicate the LGM ended at 18,700 cal BP. Alkenone data show SST rose gradually from 19,000 cal BP until 16,000 cal BP, when it exceeded late-Holocene values by about 1°C, after which it fell by 1°C between 16,000 cal BP and 14,000 cal BP during the Antarctic Cold Reversal. Temperature then rose to a maximum between 13,400 cal BP and 11,000 cal BP, when it increased by 1-2.8°C, before it rapidly declined by 4°C as the Sub Tropical Front (STF) moved north across the South Tasman Rise (4°C is the difference between the SST of subantarctic water south of and subtropical water north of the STF). GC07 occurs north of the STF in summer and the present alkenone SST is 16-17°C (Sikes et al. 2009).

A study of chironomids at Platypus and Eagle tarns at Mt Field also indicates deglaciation by 15,700-15,200 cal BP and rise in temperature of the summer quarter to 0.7°C above modern between 15,000 cal BP and 13,000 cal BP, followed by a decline of 1.8-2.5°C to a minimum between 11,100 cal BP and 10,000 cal BP. The earliest forest expansion in western Tasmanian pollen diagrams generally commences around 13,000 BP (14,700 cal BP), but becomes widespread after 11,000 BP. The forest expansion has been suggested to relate to increases in winter temperature and precipitation rather than summer temperature (Markgraf et al. 1986; Rees and Cwynar 2010).

Maximum climatic warmth in the early Holocene occurred between around 11,000 and 7500 BP and was followed by a late-Holocene cooling. This is consistent with widespread survival of alpine and subalpine vegetation associations in central and northernwestern Tasmania until 12,000 BP (13,500 cal BP), as shown on Figure 3f and regional development of wet mixed forest, as shown on Figure 3g by 9000 BP (10,000 cal BP). Although the alkenone data indicate a marked SST downturn of 4°C after 11,000 cal BP as the STF moved northwards, it did not inhibit expansion of lowland rainforest of *Nothofagus* over central-western Tasmania during the early to middle Holocene (Figure 3g). The chironomid data from Mt Field indicate there was a sharp increase in temperature to a maximum of 10°C by 9300 cal BP at Eagle Tarn and 9.4°C by 9800 cal BP at Platypus Tarn (modern temperature of warmest quarter 9.5°C). Macphail (1979) recognised a mid-Holocene thermal maximum from 8000 BP to 5000 BP (8900-5700 cal BP), after which a retreat of rainforest and expansion of sclerophyll vegetation occurred at high altitude, probably due to the combined effects of cooling, drying and burning. Chironomid-inferred temperatures at Mt Field also generally decrease during the late Holocene (Rees and Cwynar 2010).

During the late Quaternary, only valley heads and high slopes of the mountains of northeast Tasmania preserved extensive areas of *Nothofagus* temperate rainforest. Elsewhere, wet-forest vegetation was confined to gullies and sites where water was concentrated to provide local wet habitats. East of the western mountains, the lower eastern Central Plateau was relatively dry, the Midlands were heavily rain-shadowed and very dry, and the Eastern Highlands were also drier than western Tasmania. The vegetation of this area was dominated initially by grassland and steppe during the termination of Stage 2, after which sclerophyll woodland and forest expanded over most of the area, as represented on Figures 3e-3i and already discussed.

While variations in the vegetation associations were partly (*vide infra*) a response to drought under the cold glacial conditions, there is little evidence that permits quantification of how dry it was. Most evidence is based on geomorphology rather than palaeoecology. At Newdegate Cave in southeast Tasmania on the margin of the wet climate of southwest Tasmania, stalagmite growth was reduced from 16.1 mm/1000 years to 0.3 mm/1000 years at 116,700 BP, signalling a very rapid change to aridity between Substages 5e and 5d (Zhao et al. 2001). Though there is no record of vegetation in eastern Tasmania at this time, the rapidity and apparent strength of this change may be reflected in the rapidity of change from wet mixed forest to subalpine forest at Lake Selina (Figure 3b).

Eastern Tasmania was very much drier during the LGM than present, as indicated by extensive linear dunes dated by OSL to 23,800-16,800 BP on the northeastern coastal plain. In addition, a crescentic dune, the Dunlin Dune, was formed by two phases of aeolian activity that occurred from before 29,000 cal BP until after 14,500 cal BP, with an interval between 21,000 cal BP and 16,000 cal BP when a podzolised palaeosol was formed. Aeolian activity during and immediately after the LGM extended to southern Tasmania, where at Southwood Road a 1.9 m-thick dune was deposited around 19,100-18,700 cal BP. Extensive lagoon and lunette systems also occur in the northeast, Midlands and southeast, and one lunette at Rushy Lagoon in the northeast was formed after 9600-9300 cal BP. There are also extensive source-bordering river dunes along the Derwent River in which the sediment was derived from streams loaded with glacial outwash (Nicolls 1958; Cosgrove 1985; Duller and Augustinus 2006; McIntosh et al. 2009). In addition to this widespread evidence for aeolian activity during and succeeding the LGM, McIntosh et al. (2009) have presented extensive geomorphological evidence of landscape instability and erosion below 600 m, which they convincingly demonstrate largely resulted from Aboriginal burning of vegetation after 35,000 BP. Thus, the development of vegetation throughout eastern Tasmania during at least the period of occupation in the middle to late Holocene after postglacial sea level rise (Brown 1986, 1991; Kee 1990, 1991) must have been strongly influenced by aboriginal burning as well as climate change.

When the first Tasmanians crossed the Bass Plain around 40,000 years ago, they moved southwards into the northern valleys, into the central Florentine Valley, throughout the southwestern valleys from the Weld River in the southeast to the Franklin and Mackintosh rivers in the west, and into the southeastern Central Plateau and Derwent Valley. Their major pursuit was hunting the Red-necked wallaby using their main cultural tool, fire, which would have had a major impact on the vegetation (Murray et al. 1980; Kiernan et al. 1983; Stern and Marshall 1993; Cosgrove 1995, 1996; Allen 1996; Allen and Porch 1996; Stone and Everett

2010).

Numerous pollen diagrams dating to the period after about 35,000 BP reveal the impact of aboriginal burning on the vegetation.

In northern Tasmania, the Forth Valley was occupied from 34,000 BP, as indicated by the archaeological site at Parmeener Meethaner and the adjacent Mersey Valley to the east from before 10,900 BP at Warragarra Rockshelter (Cosgrove 1995; Allen and Porch 1996). The pollen record at Dublin Bog in the Mersey Valley reveals that during the past 13,100 BP, charcoal was abundant and the vegetation was *Eucalyptus* forest. Climatically, the valley should have had rainforest during the Holocene, yet rainforest pollen occurs only in small amounts that can be regarded as background values. It seems highly probable that burning of the vegetation by Aboriginals favoured establishment of sclerophyll forest and prevented the postglacial expansion of rainforest. Further north at Den Plain, *Eucalyptus* forest was dominant throughout the past 3000 years until European settlement 200 years ago when grass and herbs became more abundant (Colhoun et al. 1991b; Moss et al. 2007).

No pollen sites have been recorded from the Florentine Valley by which to judge aboriginal impact on the vegetation. However, at Mt Field to the east, Macphail (1979) attributed the decrease of rainforest and expansion of sclerophyll woodland at high altitude partly to the effects of aboriginal burning. Further south at Ooze Lake, a high-altitude community of grassland and sedgeland-heath probably resulted from aboriginal burning between 16,500 and 13,500 BP, before the expansion of rainforest and wet forest vegetation (Macphail and Colhoun 1985).

In western Tasmania, the core region of lowland cool temperate rainforest and wet mixed forest as discussed, it is notable that at some sites wet forests expanded much less than would have been expected during the Holocene. At Governor Bog in the King Valley, a sequence shows that after 13,000 BP, small trees and shrubs comprised ca. 60% of the total pollen sum and included Epacridaceae, *Melaleuca*, *Leptospermum*, *Monotoca* and *Bauera* and only 30% of rainforest species. Charcoal is abundant throughout the sequence. A mosaic vegetation of heath, shrub and forest was suggested. Similar sequences with abundant charcoal occurred at Smelter Creek and King River Railway Bridge (Colhoun et al. 1991a, 1992; van de Geer et al. 1991). It is suggested that the occurrence of heath-scrub vegetation in the King Valley rather than *Nothofagus* rainforest is a result of aboriginal occupation and burning (Thomas 1995).

Figure 2 shows the vegetation of large areas of far southwestern and parts of northwestern Tasmania are moorland that consist mainly of sedges including the buttongrass *Gymnoschoenus* and heath with many small Epacridaceae and Myrtaceae shrubs. The pollen records from Pedder Pond and Melaleuca Inlet indicate that much of this region has been moorland throughout the Holocene and the area was not extensively invaded by the postglacial expansion of rainforest. The modern vegetation is regarded as a cultural artefact resulting from aboriginal burning of the landscape that maintained moorland vegetation established in this superhumid lowland landscape during glacial times (Fletcher and Thomas 2007a, 2010a, b). The moorland vegetation in this region may have been particularly shaped by the cold wet climate as well as human impact, as suggested by both macrofossil and microfossil evidence from Melaleuca Inlet for predominantly wet scrub and sedgeland-heath vegetation of at least 38,800 years' age (Jordan et al. 1991).

While the evidence for modification of vegetation by aboriginal burning during the Holocene in southwestern Tasmania is clear, the question of to what extent they may have modified late-Pleistocene vegetation is more difficult to assess due to few available sites and more rigorous temperature conditions acting on the vegetation. It has been indicated here that Stage 4 was almost as cold as Stage 2 and that during both, the vegetation of the mountains of western Tasmania was mainly alpine grassland and herbfield. The archaeological evidence indicates that the region could not have been occupied until the final part of Stage 3 and Stage

2, so the alpine grassland and herbfields of Stage 4 and most of Stage 3 would have reflected climatic influences alone.

Four sites at Tullabardine Creek, Henty Bridge, Newell Creek and Lake Selina all show strong Poaceae peaks during and succeeding the peak of the LGM, with associated alpine herbs and shrubs plus heathland shrubs and sedges. Charcoal was only counted at Lake Selina. It is equally abundant in Stage 2 as in Stage 4, is greater than during Stage 3, but is much less abundant than during the Holocene. It is not possible to determine whether any of the herbaceous peaks during Stage 2 at these sites were produced by aboriginal burning rather than representing the vegetation of the cold glacial climate, but it is a possibility.

At Lake Selina, a whole-core NRM analysis showed that for all colder periods (5d, 5b, 4, late 3 and 2), the NRM values were higher than at other times, especially during Stages 5e and 1 (see Colhoun et al. 1999: Figure 5). The NRM values reflect the amount of minerogenic sediment in the core *vis a vis* the organic sediment, and represent the amount of catchment erosion. As Stage 2 had three times the NRM values of Stage 4 and climatic conditions were similar, there appears to have been much more erosion during the LGM than during Stage 4. If Aboriginals had inhabited the region during this period, it is possible they contributed to the erosion by burning of the catchment, but there is no independent evidence.

Conclusion

During the late Quaternary, there were major changes to the vegetation of Tasmania that were primarily climatically driven. The changes occurred mainly in response to variations in temperature in the west and to temperature and precipitation in the east. After the arrival of the Tasmanians around 35,000 BP, their hunter-gatherer mode of life utilising fire impacted strongly on the vegetation to produce disclimax communities which can be detected in numerous pollen diagrams. Cultural modification of the vegetation is most noticeable in southwest Tasmania, where maintenance of moorland from glacial times prevented Holocene expansion of *Nothofagus* cool temperate rainforest.

Acknowledgements

The authors thank colleagues indicated in Table 1 who have willingly contributed pollen data for the mapping. They thank Stephen Harris and Anne Kitchener, and the Department of Primary Industries Parks Water and Environment of Tasmania for permission to publish Figure 2. They thank Penny Jones and Lydia Mackenzie for supplying unpublished pollen data. They acknowledge the University of Newcastle, the University of Tasmania, the Australian Research Council, and the Australian Nuclear Science and Technology Organisation for supporting the work. They also thank Ian Thomas and Michael Shawn-Fletcher for reviewing the paper and providing many constructive comments and help.

References

Allen, J. (ed) 1996. *Report of the Southern Forests Archaeological Project*. Volume 1. School of Archaeology La Trobe University. Bundoora, Victoria 3083, Australia.

Allen, J. and Porch, N. 1996. Warragarra Rockshelter. Chapter 10. In: Allen, J. (ed), *Report of the Southern Forests Archaeological Project* Volume 1. pp. 195-217. School of Archaeology,

La Trobe University, Bundoora, Victoria 3083, Australia.

Anker, S.A., Colhoun, E.A., Barton, C.E., Petersen, M. and Barbetti, M. 2001. Holocene vegetation and palaeoclimatic and palaeomagnetic history from Lake Johnston, Tasmania. *Quaternary Research* 56:264-274.

Augustinus, P.C. and Colhoun, E.A. 1986. Glacial history of the upper Pieman and Boco valleys, western Tasmania. *Australian Journal of Earth Sciences* 33:181–191.

Barrows, T.T., Juggins, S., De Deckker, P., Thiede, J. and Martinez, J.I. 2000. Sea-surface temperatures of the southwest Pacific Ocean during the Last Glacial Maximum. *Paleoceanography* 15:95-109.

Barrows, T.T., Juggins, S., De Deckker, P. and Calvo, E. 2007. Long-term sea surface temperature and climate change in the Australian-New Zealand region. *Paleoceanography* 22:1-17.

Bowman, D.J.M.S. and Jackson, W.D. 1981. Vegetation succession in south-west Tasmania. *Search* 12:358-362.

Brown, S.H. 1986. *Aboriginal Archaeological Resources in South East Tasmania*. Occasional Paper No. 12. National Parks and Wildlife Service Tasmania.

Brown, S.H. 1991. *Aboriginal Archaeological Sites in Eastern Tasmania*. Occasional Paper No. 31, National Parks and Wildlife Service Tasmania.

Chiarucci, A., Araújo, M.B., Decorq, G., Beierkuhnlein, C. and Fernández-Palacios, J.M. 2010. The concept of potential natural vegetation: An epitaph? *Journal of Vegetation Science* 21:1172-1178.

Clements, F.E. 1936. Nature and Structure of the Climax. *Journal of Ecology* 24:252-284.

Colhoun, E.A. 1977a. Late Quaternary fan gravels and slope deposits at Rocky Cape, northwestern Tasmania: Their palaeoenvironmental significance. *Papers and Proceedings of the Royal Society of Tasmania* 111:13-27.

Colhoun, E.A. 1977b. The Remarkable Cave, southeastern Tasmania: Its geomorphological development and environmental history. *Papers and Proceedings of the Royal Society of Tasmania* 111:29-39.

Colhoun, E.A. 1977c. A sequence of Late Quaternary deposits at Pipe Clay Lagoon, southeastern Tasmania. *Papers and Proceedings of the Royal Society of Tasmania* 111:1-12.

Colhoun, E.A. 1980. Quaternary fluviatile deposits from the Pieman Dam site, western Tasmania. *Proceedings of the Royal Society of London* B207:355-384.

Colhoun, E.A. 1985a. Pre-Last Glaciation Maximum vegetation history at Henty Bridge, western Tasmania. *New Phytologist* 100:681-690.

Colhoun, E.A. 1985b. Glaciations of the West Coast Range, Tasmania. *Quaternary Research* 24:39-59.

Colhoun, E.A. 1991. Climate during the Last Glacial Maximum in Australia and New Guinea. Evidence inferred from biogeographical and geomorphological data. *Australian and New Zealand Geomorphology Group* Special Publication No. 2. pp. 1-71.

Colhoun, E.A. 1992. Late Glacial and Holocene vegetation history at Poets Hill Lake, western Tasmania. *Australian Geographer* 23(1):11-23.

Colhoun, E.A. 1996. Application of Iversen's Glacial-Interglacial Cycle to interpretation of the late Last Glacial and Holocene vegetation history of western Tasmania. *Quaternary Science Reviews* 15:557-580.

Colhoun, E.A. 2000. Vegetation and climate change during the Last Interglacial-Glacial cycle in western Tasmania, Australia. *Palaeogeography, Palaeoclimatology, Palaeoecology* 55(1-2):195-209.

Colhoun, E.A. and Goede, A. 1979. The Late Quaternary deposits of Blakes Opening and the Middle Huon Valley, Tasmania. *Philosophical Transactions of the Royal Society of London* 286B:371-395.

Colhoun, E.A. and van de Geer, G. 1986. Holocene to Middle Last Glaciation vegetation history at Tullabardine Dam, western Tasmania. *Proceedings of the Royal Society of London* 229B:177-207.

Colhoun, E.A. and van de Geer, G. 1987. Vegetation history and climate before the maximum of the Last Glaciation at Crotty, western Tasmania. *Papers and Proceedings of the Royal Society of Tasmania* 121:69–74.

Colhoun, E.A. and van de Geer, G. 1998. Pollen analysis of 0-20m at Darwin Crater, western Tasmania, Australia. *International Project on Paleolimnology and Late Cenozoic Climate, Newsletter* 11:68-89.

Colhoun, E.A., Benger, S.N., Fitzsimons, S.J., van de Geer, G. and Hill, R.S. 1993. Quaternary organic deposit from Newton Creek Valley, western Tasmania. *Australian Geographical Studies* 31(1):26-38.

Colhoun, E.A., Kiernan, K., Barrows, T. and Goede, A. 2010. Advances in Quaternary studies in Tasmania. In: Bishop, P. and Pillans, B. (eds), *Australian Landscapes*. Geological Society, London, Special Publications 2010, v.346. pp. 165-183.

Colhoun, E.A., Pola, J.S., Barton, C.E. and Heijnis, H. 1999. Late Pleistocene vegetation and climate history of Lake Selina, western Tasmania. *Quaternary International* 57/58:5-23.

Colhoun, E.A., van de Geer, G. and Fitzsimons, S.J. 1991a. Late glacial and Holocene vegetation history at Governor Bog, King Valley, western Tasmania, Australia. *Journal of Quaternary Science* 6(1):55-66.

Colhoun, E.A., van de Geer, G. and Fitzsimons, S.J. 1992. Late Quaternary organic deposits at Smelter Creek and vegetation history of the Middle King Valley, Western Tasmania *Journal of Biogeography* 19:217-227.

Colhoun, E.A., van de Geer, G. and Hannan, D. 1991b. Late Glacial and Holocene Vegetation History at Dublin Bog, north-central Tasmania. *Australian Geographical Studies* 29(2):337-354.

Colhoun, E.A., van de Geer, G. and Mook, W. 1982. Stratigraphy, pollen analysis and palaeoclimatic interpretation of Pulbeena Swamp, northwestern Tasmania. *Quaternary Research* 19(1):108-126.

Cosgrove, R. 1985. New evidence for Early Holocene Aboriginal occupation in northeast Tasmania, *Australian Archaeology* 21:19-36.

Cosgrove, R. 1995. Late Pleistocene behavioural variation and time trends: the case from Tasmania. *Archaeology in Oceania*, 30:83-104.

Cosgrove, R. 1996. ORS 7 Rockshelter. Chapter 5. In: Allen, J. (ed), *Report of the Southern Forests Archaeological Project* Volume 1. pp. 69-89. School of Archaeology, La Trobe University, Bundoora, Victoria 3083, Australia.

Cosgrove, R. 1999. Forty-two degrees South: The archaeology of Late Pleistocene Tasmania. *Journal of World Prehistory* 13(4):357-402.

Cosgrove, R., Allen, J. and Marshall, B. 1990. Palaeoecology and Pleistocene human occupation in south central Tasmania. *Antiquity* 64:59-78.

Duller, G.A.T. and Augustinus, P.C. 2006. Reassessment of the record of linear dune activity in Tasmania using optical dating. *Quaternary Science Reviews* 25:2608-2618.

Duncan, F. and Brown, M.J. 1985. Dry sclerophyll vegetation in Tasmania. *Wildlife Division Technical Report 85/1* National Parks and Wildlife Service, Tasmania.

Dyson, W.D. 1995. *A pollen and vegetation history from Lake Dove, Western Tasmania*. B. Honours thesis, Department of Geography, University of Newcastle, Australia.

Fletcher, M-S. and Thomas, I. 2007a. Holocene vegetation and climate change from near Lake Pedder, south-west Tasmania, Australia. *Journal of Biogeography* 34:665-677.

Fletcher, M-S. and Thomas, I. 2007b. Modern pollen-vegetation relationships in western

Tasmania, Australia. *Review of Palaeobotany and Palynology* 146:146-168.

Fletcher, M-S. and Thomas, I. 2010a. The origin and temporal development of an ancient cultural landscape. *Journal of Biogeography* 37:2183-2196.

Fletcher, M-S. and Thomas, I. 2010b. A Holocene record of sea level, vegetation, people and fire from western Tasmania, Australia. *The Holocene* 20:351-361.

Gibson, N., Kiernan, K.W. and Macphail, M.K. 1987. A fossil bolster plant from the King River, Tasmania. *Papers and Proceedings of the Royal Society of Tasmania* 121:35-42.

Harle, K.J. 1989. Palaeoenvironments of the Mt. Anne massif region in south-western Tasmania. Honours thesis, Monash University, Melbourne.

Harle, K.J., Hodgson, D.A. and Tyler, P.A. 1999. Palynological evidence for Holocene palaeoenvironments from the lower Gordon River valley, in the World Heritage Area of southwest Tasmania. *The Holocene* 9:149-162.

Harle, K.J., Kershaw, A.P., Macphail, M.K. and Neyland, M.G. 1993. Palaeocological analysis of an isolated stand of *Nothofagus cunninghamii* (Hook.) Oerst. in eastern Tasmania. *Australian Journal of Ecology* 18:161-170.

Harris, S. and Kitchener, A. (eds) 2005. *From Forest to Fjaeldmark.* Department of Primary Industries, Water and Environment, GPO Box 44, Hobart.

Hopf, F.V.L., Colhoun, E.A. and Barton, C.E. 2000. Late-glacial and Holocene record of vegetation and climate from Cynthia Bay, Lake St Clair, Tasmania. *Journal of Quaternary Science* 15(7):725-732.

Hope, G.S. 1978. The Late Pleistocene and Holocene vegetational history of Hunter Island, north-western Tasmania *Australian Journal of Botany* 26:493-514.

Jackson, W.D. 1968. Fire, air, water and earth – An elemental ecology of Tasmania. *Proceedings of the Ecological Society of Australia* 3:9-16.

Jackson, W.D. 1981a. Vegetation. In: Jackson, W.D. (ed), *The Vegetation of Tasmania.* pp. 11-16. Australian Academy of Science, Canberra.

Jackson, W.D. 1981b. Fire – patterned vegetation. In: Jackson, W.D. (ed), *The Vegetation of Tasmania.* pp. 17-35. Australian Academy of Science, Canberra.

Jackson, W.D. 1981c. Wet sclerophyll. In: Jackson, W.D. (ed), *The Vegetation of Tasmania.* pp. 68-72. Australian Academy of Science, Canberra.

Jackson, W.D. 1983. Tasmanian rainforest. In: Blakers, R. and Robertson, P. (eds), *Tasmania's Rainforests. What Future?* pp. 9-39. Australian Conservation Foundation, Hobart.

Jarman, S.J., Kantvilas, G. and Brown, M.J. 1999. Floristic Composition of Cool Temperate Rainforest. Chapter 7. In: Reid, J.B. Hill, R.S., Brown, M.J. and Hovenden, J. (eds), *Vegetation of Tasmania,* pp. 145-159. Monotone Art Printers, Tasmania.

Jennings, J.N. 1959. The coastal geomorphology of King Island, Bass Strait, in relation to changes in the relative level of land and sea. *Records of the Queen Victoria Museum, Launceston* New Series, No. 11:1-39.

Jones, P. 2008. The palaeoecology of the Midlands, Tasmania. Honours thesis, University of Melbourne.

Jordan, G.J., Carpenter, R.J. and Hill, R.S. 1991. Late Pleistocene vegetation and climate near Melaleuca Inlet, south-western Tasmania. *Australian Journal of Botany* 39:315-333.

Kanfoush, S.L., Hodell, D.A., Charles, C.D., Thomas, P.G., Mortyn, P.G. and Ninnemann, U.S. 2000. Millenial-scale instability of the Antarctic ice sheet during the Last Glaciation. *Science* 288:1815-1818.

Kee, S. 1990. *Midland Archaeological Site Survey.* Occasional Paper No. 26. Department of Parks, Wildlife and Heritage, Hobart, Tasmania.

Kee, S. 1991. *Aboriginal Archaeological Sites in North East Tasmania.* Occasional Paper No. 28. Department of Parks, Wildlife and Heritage, Hobart, Tasmania.

Kiernan, K., Jones, R. and Ranson, D. 1983. New evidence from Fraser Cave for glacial age man in Southwest Tasmania. *Nature* 301:28-32.

Kirkpatrick, J.B. 1982. Phytogeographical analysis of Tasmanian alpine floras. *Journal of Biogeography* 9:255-271.

Kirkpatrick, J.B. and Dickinson, K.J.M. 1984. *1:500,000 Vegetation Map of Tasmania*, Forestry Commission, Hobart.

Kirkpatrick, J.B. and Fowler, M. 1998. Locating likely glacial forest refugia in Tasmania using palynological and ecological information to test alternative climatic models. *Biological Conservation* 85:171-182.

Lytle, D.E. and Wahl, E.R. 2005. Palaeoenvironmental reconstructions using the modern analogue technique: the effects of sample size and decision rules. *The Holocene* 15(4):551-566.

Mackenzie, L. 2010. *Late Quaternary environments of Freycinet Peninsula, Eastern Tasmania.* B. Honours Thesis. School of Geography, Planning and Environmental Management, University of Queensland.

Mackintosh, A.N., Barrows, T.T., Colhoun, E.A. and Fifield, L.K. 2006. Exposure dating and glacial reconstruction at Mt Field, Tasmania, Australia, identifies MIS3 and MIS2 glacial advances and climatic variability. *Journal of Quaternary Science* 21:363-376.

Macphail, M.K. 1975. Late Pleistocene environments in Tasmania. *Search* 6:295-300.

Macphail, M.K. 1979. Vegetation and climates in southern Tasmania since the Last Glaciation. *Quaternary Research* 11:306-341.

Macphail, M.K. 1980. Regeneration processes in Tasmanian forests: A long-term perspective based on pollen analysis. *Search* 11:84-190.

Macphail, M.K. 1984. Small-scale dynamics in an early Holocene wet sclerophyll forest in Tasmania. *New Phytologist* 96:131-147.

Macphail, M.K. 1986. "Over the Top": Pollen-based reconstructions of past alpine floras and vegetation in Tasmania. Chapter 11. In: Barlow, B.A. (ed), *Flora and Fauna of Alpine Australasia.* pp. 173-204. CSIRO, Melbourne.

Macphail, M.K. and Jackson, W.D. 1978. The late Pleistocene and Holocene history of the Midlands of Tasmania: Pollen evidence from Lake Tiberias. *Proceedings of the Royal Society of Victoria* 90:287-300.

Macphail, M.K. and Colhoun, E.A. 1985. Late Last Glacial vegetation, climates and fire activity in southwest Tasmania *Search* 16(1-2):43-45.

Macphail, M.K., Pemberton, M. and Jacobson, G. 1999. Peat mounds of southwest Tasmania: Possible origins. *Australian Journal of Earth Sciences* 46:667-677.

Markgraf, V., Bradbury, J.P. and Busby, J.R. 1986. Paleoclimates in southwestern Tasmania during the last 13,000 years. *Palaios* 1:368-380.

Martinson, D.G., Pisias, N.G., Hays, J.D., Imbrie, J., Moore, T.C. and Shackleton, N.J. 1987. Age dating and the orbital theory of the ice ages: Development of a high resolution 0 to 300,000-year chronostratigraphy. *Quaternary Research* 27:1-29.

McIntosh, P.D., Price, D.M., Eberhard, R. and Slee, A.J. 2009. Late Quaternary erosion events in lowland and mid-altitude Tasmania in relation to climate change and first human arrival. *Quaternary Science Reviews* 28:850-872.

Moss, P.T., Thomas, I. and Macphail, M. 2007. Late Holocene vegetation and environments of the Mersey Valley. *Australian Journal of Botany* 55:74-82.

Murray, P., Goede, A. and Bada, J.L. 1980. Pleistocene human occupation at Beginners Luck Cave, Florentine Valley, Tasmania. *Archaeology & Physical Anthropology in Oceania* 15:142-152.

Nicolls, K.D. 1958. Aeolian deposits in river valleys in Tasmania. *Australian Journal of Science*

21:50-51.

Nunez, M. 1988. A regional lapse rate for Tasmania. *Papers and Proceedings of the Royal Society of Tasmania* 122:53-57.

Nunez, M. and Colhoun, E.A. 1986. A note on air temperature lapse rates on Mount Wellington. *Papers and Proceedings of the Royal Society of Tasmania* 120:11-15.

Pelejero, C., Calvo, E., Barrows, T.T., Logan, G.A. and De Deckker, P. 2006. South Tasman Sea alkenone palaeothermometry over the last four glacial/interglacial cycles. *Marine Geology* 230:73-86.

Petit, J.R., Jouzel, J., Raynaud, D., Barkov, N.L., Barnola, J.M., Basile, I., Bender, M., Chappellaz, J., Davis, J., Delayguye, G., Delmotte, M., Kotlyakov, V.M., Legrand, M., Lipenkov, V., Lorius, C., Pépin, L., Ritz, C., Saltzman, E. and Stievenard, M. 1999. Climate and atmospheric history of the past 420,000 years from the Vostok ice core, Antarctica. *Nature* 399:429-436.

Porch, N., Jordan, G.J., Price, D.M., Barnes, R.W., Macphail, M.K. and Pemberton, M. 2009. Last interglacial climates of south-eastern Australia: Plant and beetle-based reconstructions from Yarra Creek, King Island, Tasmania. *Quaternary Science Reviews* 28:3197-3210.

Rees, A.B.H. and Cwynar, L.C. 2010. Evidence for early postglacial warming in Mount Field National Park, Tasmania. *Quaternary Science Reviews* 29:443-454.

Reid, J.B., Hill, R.S., Brown, M.J. and Hovenden, M.J. (eds) 1999. *Vegetation of Tasmania. Flora of Australia Supplementary Series Number 8*. Australian Biological Resources Study.

Shimeld, P.W. and Colhoun, E.A. 2001. *Eucalyptus* spp. pollen transport across Liawenee Moor on the Central Plateau of Tasmania. *Papers and Proceedings of the Royal Society of Tasmania* 135:51-55.

Sigleo, W.R. and Colhoun, E.A. 1981. A short pollen diagram from Crown Lagoon in the Midlands of Tasmania. *Papers and Proceedings of the Royal Society of Tasmania* 115:181-188.

Sikes, E.L., Howard, W.R., Samson, C.R., Mahan, T.S., Robertson, I.G. and Volkman, J.K. 2009. Southern Ocean seasonal temperature and Subtropical Front movement on the South Tasman Rise in the late Quaternary. *Palaeoceanography* 24:PA2201, doi:10.1029/2008/PA001659.

Specht, R.L. 1970. Vegetation. In: Leeper, G.W. (ed), *The Australian Environment* (4th edition). CSIRO, Melbourne University Press.

Stern, N. and Marshall, B. 1993. Excavations at Mackintosh 90/1: A discussion of stratigraphy, chronology and site information. *Archaeology in Oceania* 28:183-192.

Stone, T. and Everett, A. 2010. Interim report on the Jordan River levee excavation. pp. 1-75. CHMA, 22 Queen Street, Sandy Bay, Tasmania 7005.

Stuvier M. and Reimer P.J. 1993. Extended 14C data base and revised CALIB 3.0 14C age calibration. *Radiocarbon* 35:215-230.

Thomas, I. 1992. The Holocene archaeology and palaeoecology of northeastern Tasmania, Australia. Unpublished PhD thesis, University of Tasmania, Hobart.

Thomas, I. 1995. Where have all the forests gone? New pollen evidence from Melaleuca Inlet in southwestern Tasmania. In: Dixon, G. and Aitkin, D. (eds), *Institute of Australian Geographers 1993 Conference Proceedings*. pp. 295-301.

Thomas, I. 1996. Environmental changes in northeast Tasmania during the Holocene. *Records of the Queen Victoria Museum and Art Gallery, Launceston* 103:73-78.

Thomas, I. and Hope, G. 1994. An example of Holocene vegetation stability from Camerons Lagoon, a near treeline site on the Central Plateau, Tasmania. *Australian Journal of Ecology* 19:150-158.

Tüxen, R. 1956. Die heutige potentielle natürliche Vegetation als Gegenstand der

Vegetationskartierung. *Angewandte Pflanzensoziologie (Stolenau)* 13:4-42.

van de Geer, G., Colhoun, E.A. and Mook, W.G. 1986. Stratigraphy, pollen analysis and palaeoclimatic interpretation of Mowbray and Broadmeadows Swamps, north western Tasmania. *Australian Geographer* 17:121-133.

van de Geer, G., Fitzsimons, S.J. and Colhoun, E.A. 1989. Holocene to middle Last Glaciation vegetation history at Newell Creek, western Tasmania. *New Phytologist* 111:549-558.

van de Geer, G., Fitzsimons, S.J. and Colhoun, E.A. 1991. Holocene vegetation history from King River Railway Bridge, western Tasmania. *Papers and Proceedings of the Royal Society of Tasmania* 125:73-77.

van de Geer, G., Heusser, L.E., Lynch-Stieglitz, J. and Charles, C.C. 1994. Palaeoenvironments of Tasmania inferred from a 5-75 ka marine pollen record *Palynology* 18:33-40.

Willis, K.J. and Birks, H.J.B. 2006. What is natural? The need for a long-term perspective in biodiversity conservation. *Science* 314:1261-1265.

Worth, J.R.P., Jordan, G.J., McKinnon, G.E. and Vaillancourt, R.E. 2009. The major Australian cool temperate rainforest tree *Nothofagus cunninghamii* withstood Pleistocene glacial aridity within multiple regions: Evidence from the chloroplast. *New Phytologist*, 182:519-532.

Zhao, J., Xia, Q. and Collerson, K.D. 2001. Timing and duration of the Last Interglacial inferred from high resolution U-series chronology of stalagmite growth in Southern Hemisphere. *Earth and Planetary Science Letters* 184:635-644.

15

Holocene environments of the sclerophyll woodlands of the Wet Tropics of northeastern Australia

Patrick T. Moss
Climate Research Group, School of Geography, Planning and Environmental Management, The University of Queensland, St Lucia, Queensland
patrick.moss@uq.edu.au

Richard Cosgrove
La Trobe University, Bundoora, Victoria

Åsa Ferrier
La Trobe University, Bundoora, Victoria

Simon G. Haberle
The Australian National University, Canberra, ACT

Introduction

The Wet Tropics region of northeastern Australia has been the focus of palynological research into the late Quaternary history of climate, vegetation and human environmental impact for a number of years (Moss and Kershaw 2000, 2007; Kershaw et al. 2007, 2003a, 2003b; Kershaw 1994, 1986). Numerous palynological records covering the Holocene period have been examined, but they have either been concentrated within the core rainforest area due to the availability of volcanic crater sites on the Atherton Tableland (e.g. Kershaw 1983, 1975, 1971, 1970; Walker and Chen 1987; Chen 1988; Walker 2007); and/or situated in coastal areas where successional processes in mangroves have tended to mask more regional signals (e.g. Grindrod and Rhodes 1984; Grindrod 1985; Crowley et al. 1990; Gagan et al. 1994; Crowley and Gagan 1995). Recently, there has been a focus on high-resolution records that investigate

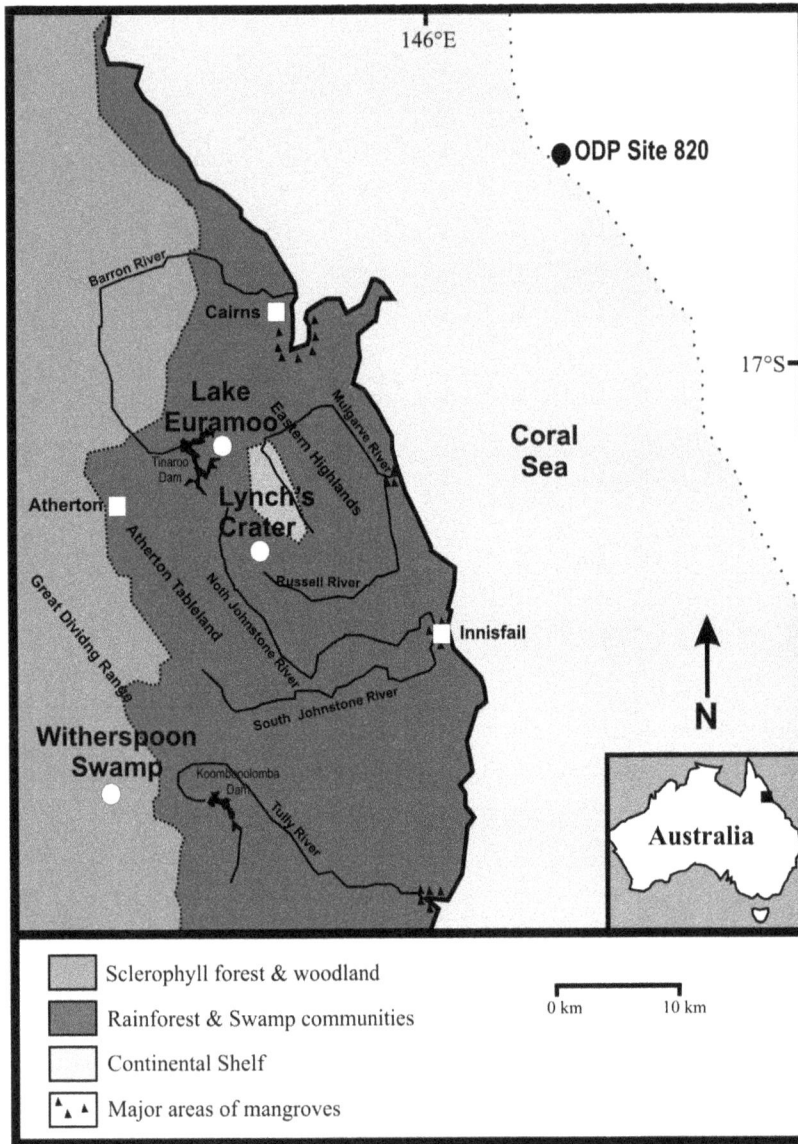

Figure 1. Location of the Witherspoon Swamp site in relation to major environmental features of the Wet Tropics region, as well as the Lynch's Crater, Lake Euramoo sites and ODP 820 sites.

the response of the Wet Tropics ecosystem to rapid climate change and human impact for the Late Quaternary period. A key record, located within the modern mesophyll rainforests of the Atherton Tableland region, was taken from Lake Euramoo (Haberle 2005; Figure 1). This 8.4 m core covers the past 23,000 years and provides a high-resolution record (i.e. centennial to decadal scale) of vegetation change and fire history for the region. Five pollen zones were identified, with dry sclerophyll woodland dominating from 23,000 to 16,800 cal BP; changing to wet sclerophyll forest with patches of mesophyll rainforest from 16,800 to 8600 cal BP; followed by expansion of mesophyll rainforest from 8600 to 5000 cal BP; slight contraction in mesophyll rainforest from 5000 to 70 cal BP; and degraded mesophyll rainforest with an increase in fire and invasive species from 70 cal BP to the present. The process of rainforest expansion was thought to be at least partly controlled by changes in insolation (dominated by the precessional component at equatorial latitudes), although local factors (i.e. biomass change associated with forest development) and human impact, particularly fire use, have been suggested to also have played significant roles (Haberle 2005).

Archaeological research within the Wet Tropics suggests that people have occupied the region for at least the past 40,000 to 30,000 years (Kershaw 1986; Cosgrove et al. 2007). However, it appears that human occupation occurred at a very low intensity until around 2500 cal BP, peaking after 1500 cal BP, when the exploitation of toxic nut varieties and the development of permanent occupation in the rainforests of this region occurred (Haberle and David 2004; Cosgrove et al. 2007). It has been suggested that fire, El Niño Southern Oscillation (ENSO) activity and shifting vegetation played a significant role in the history of permanent human occupation within the region (Haberle and David 2004; Cosgrove et al. 2007). This implies significant linkages between environmental alterations from the region and human cultural responses.

A broader picture of environmental change for the Wet Tropics region is provided through palynological analysis of sediment cores taken from Witherspoon Swamp. These records are some of the first to examine a significant portion of the Holocene period (i.e. past 8000 years) from the scelerophyll communities located to the west of the Wet Tropics region, and as such, provide a complementary site to the existing records taken from the rainforest communities of the region. In particular, the Witherspoon Swamp records provide an assessment of changes in the rainforest-sclerophyll boundary, as well as possible insight into human impacts on the sclerophyll communities of the Wet Tropics region.

Regional setting and methods

A key characteristic of the Wet Tropics region is the substantial topographic variation that has a significant influence on the area's climate and vegetation (Tracey 1982; Figure 1). Marked climatic gradients are associated with the variable physiography, with mean annual temperatures exceeding 24°C along the coast, falling below 21°C on the Atherton Tableland, and to below 17°C on the highest peaks. The southeasterly trades dominate the region's weather patterns and produce much of the rain, although this is supplemented by monsoonal northwesterlies and occasional cyclones. Most precipitation falls in summer (November to April) but almost continuous cloud cover maintains moist conditions throughout the year on the higher ranges. The region is also highly sensitive to ENSO variability, with significant reductions in rainfall during El Niño years when the activity of the trade winds is much reduced, and significant increases in precipitation during La Niña years when there are stronger southeasterlies, which is also augmented by the more southerly movement of the Inter Tropical Convergence Zone, as well as increased cyclone activity (Nichols 1992). Vegetation distribution is closely linked to the climatic and topographic variability. Rainforest, including floristically similar swamp communities on the poorly drained coastal lowlands, occupy the high rainfall [>1500 mm Mean Annual Precipitation (MAP)] parts of the region, which extend from the coast to the western edge of the Atherton Tableland. Sclerophyll communities, most frequently dominated by eucalypts, generally replace rainforest in areas that receive fewer than 1500 mm of MAP, which include the western areas of the study region, as well as a large patch in the rain shadow to the west of the Eastern Highlands (see Figure 1). Mangrove communities are found in low-energy coastal embayments, such as Trinity Inlet near Cairns, Mutchero Inlet at the mouth of the Russell and Mulgrave Rivers, at the mouth of the North and South Johnstone Rivers and at the mouth of the Tully River.

Witherspoon Swamp (17°49'S, 145°24'E, elevation 652 m) is located in the southwestern corner of the Atherton Tableland (commonly referred to as the Evelyn Tableland) within the sclerophyll woodland communities of the Wet Tropics region, and is around 5 km to 10 km from the nearest rainforest communities (Figure 1). The swamp itself is situated within a small depression and consists of an area of relatively permanent water with numerous reeds, sedges and other aquatic plants growing on the swamp surface (Figure 2). It is surrounded by a large area of *Melaleuca* forest to the north and east, and the regional vegetation is characterised by open eucalypt forest with a grassy understorey.

Precipitation is around 800 mm to 1000 mm per year and the dominant land-use at present

Figure 2. Photograph of the Witherspoon Swamp site.

is cattle grazing. The cores obtained from this site are the first ones to be taken from a modern sclerophyll community from the region. Two cores were obtained from the site, the first (WS 1) was collected in October 2004 from the edge of the swamp and is 1.4 m in length, while the second (WS 2) was collected in November 2006 from the centre of the swamp and is 1.9 m long.

The Witherspoon Swamp samples were prepared for pollen analysis using the technique developed by van der Kaars (1991). This involved using sodium pyrophosphate to disaggregate the sediments, which were then further processed by using sodium polytungstate (specific gravity of 2.0) to float the lighter organic fraction from the heavier minerogenic component. The samples then underwent acetolysis to darken the pollen grains, remove extraneous organic matter and improve their visibility under a light microscope. All samples were mounted in glycerol. Pollen identification and counting was undertaken using a light microscope at x 400 magnification and the pollen sum consisted of a minimum of 300 dryland pollen taxa. Charcoal analysis involved counting all black angular particles above 5 mm in diameter as carbonised particles across three transects, and exotic *Lycopodium* spores were also counted to allow for the calculation of charcoal concentrations or carbonised particles per cubic cm (Wang et al. 1999).

Pollen diagrams were produced using TGView (Grimm 2004) and the pollen diagrams for WS 1 and WS 2 include both pollen and charcoal counts (see Figures 3 and 4). The pollen diagrams are divided into zones based on the results of a stratigraphically constrained classification undertaken by CONISS (Grimm 1987, 2004) on taxa contained within the pollen sum. Age control was based on AMS radiocarbon dating of bulk sediments from both cores, with two radiocarbon dates from WS 1 (determined by the University of Waikato radiocarbon laboratory) and two radiocarbon dates from WS 2 (undertaken by the Australian Nuclear Science and Technology Organisation). The Hughen et al. (2006) chronology was used to provide calibrated ages for both records and all dates are presented in Table 1. In addition, sediment samples for Witherspoon WS 2 underwent Loss-on-Ignition analysis, which involved heating the samples at 490°C for 12 hours to remove the organic fraction from the samples. Figure 4 presents the results of the analyses (as percentage ash content or inorganic fraction per dry weight).

Table 1. Radiocarbon dates for the WS 1 and WS 2 cores.

Record	Top (m)	Bottom (m)	Age BP (^{14}C years)	Cal BP (median)
WS 1	0.48	0.50	1709 ± 122	1974
WS 1	1.38	1.40	4259 ± 35	4919
WS 2	0.60	0.65	1980 ± 40	2287
WS 2	1.85	1.95	6830 ± 60	7889

The Witherspoon Swamp pollen records

Both of the cores (WS 1 and WS 2) taken from Witherspoon Swamp are dominated by sclerophyll taxa and relatively low rainforest gymnosperm and angiosperm values are recorded (Figures 3 and 4). In addition, both cores provide a record of vegetation change that extends back to the early to mid-Holocene period. The base (140 cm) of WS 1 dates to ca. 5000 cal BP, with an age of ca. 2000 cal BP at 45 cm, while the base (190 cm) of the longer WS 2 core dates to ca. 7800 cal BP, with an age of 2300 cal BP at 60 cm. The key difference between the cores is in terms of representation of wetland taxa, with higher aquatic values, particularly Cyperaceae in WS 2, while WS 1 shows higher *Selaginella* values from 140 cm to 100 cm. This difference between the two pollen records may reflect the location of the cores within the swamp. WS 1 is located closer to the swamp edge and the higher representation of *Selaginella* may therefore reflect this taxa growing on the relatively exposed swamp edge. In contrast, WS 2 was taken in the centre of the lake and the higher aquatic values represent sedges and other aquatic taxa (e.g. *Polygonum* and *Nymphoides*) that are growing within the swamp itself.

The classification of the pollen undertaken on the cores identified six zones in both records. The zones are categorised by the depth values as it was thought that two radiocarbon ages per record were insufficient for the development of a reliable age model. The six zones in both records are described in more detail below:

Core WS 1

Zone WS 1A (140 cm to 130 cm)

This zone is characterised by the highest representation of rainforest gymnosperm and angiosperm values in the core (particularly at 140 cm). Poaceae dominate the dryland taxa and Casuarinaceae is the most significant sclerophyll arboreal taxon. *Selaginella* occurs in low abundances, while Cyperaceae, *Polygonum* and *Nymphoides* are the dominant aquatic taxa. Charcoal values maintain a low representation in this zone.

Zone WS 1B (120 cm to 100 cm)

A significant increase in Poaceae is observed in this zone, while sclerophyll arboreal taxa, rainforest gymnosperms and rainforest angiosperms decrease. Aquatic taxa and pteridophytes, particularly the fern ally *Selaginella*, reach their highest abundances in this zone. The highest charcoal values in the record are observed in this zone at 100 cm.

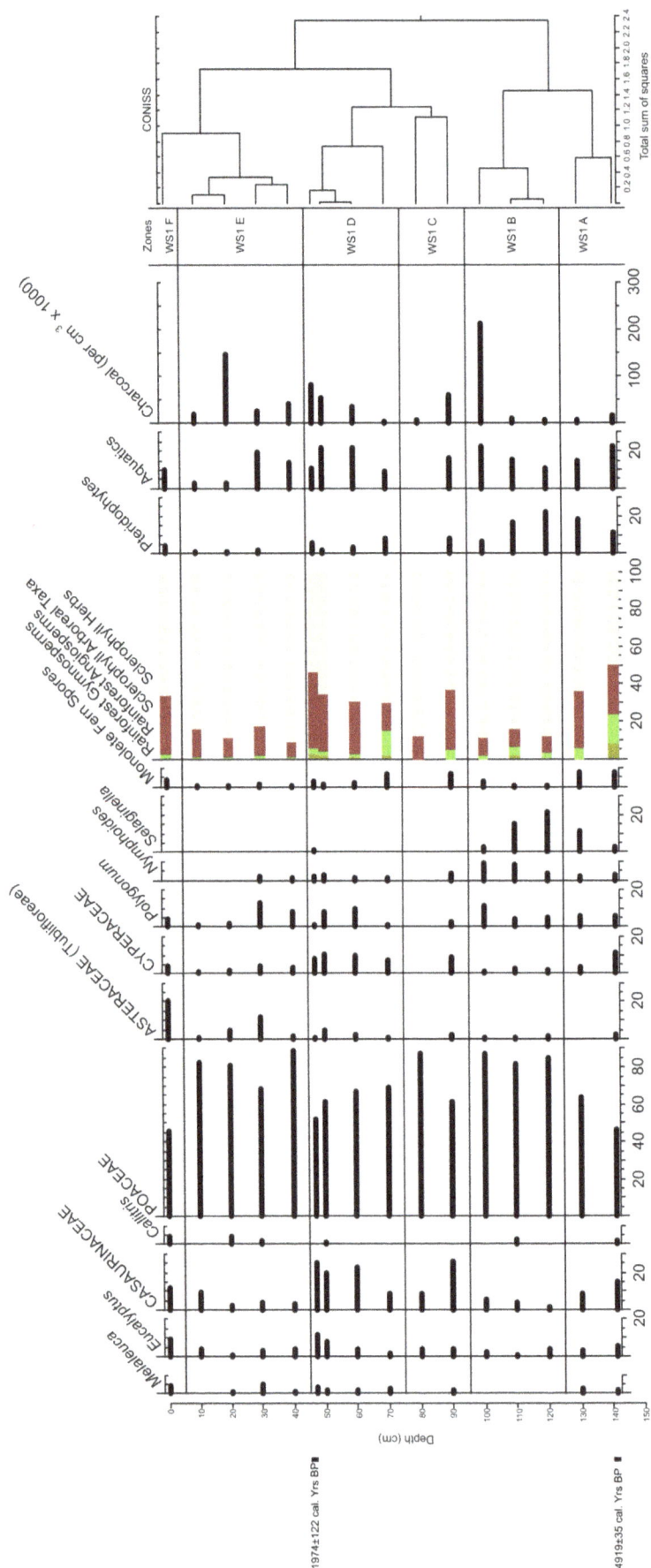

Figure 3. The Witherspoon Swamp core (WS 1) taken from the edge of the swamp. The solid lines reflect the vegetation zones derived from the classification of the pollen taxa.

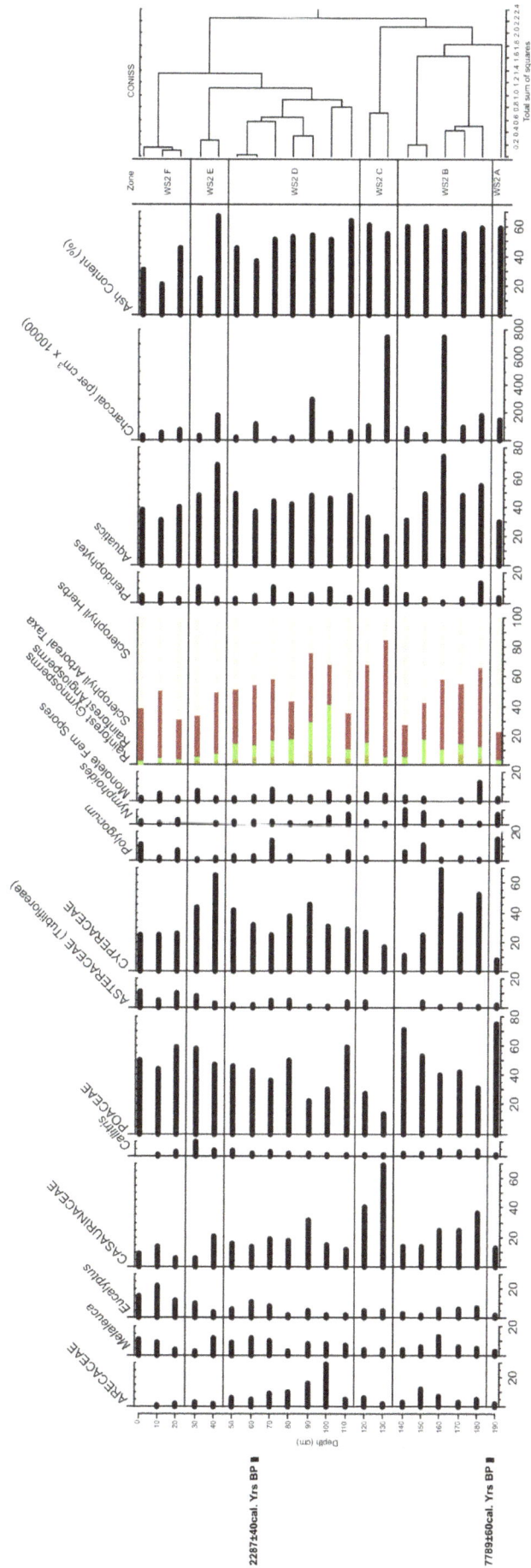

Figure 4. The Witherspoon Swamp core (WS 2) taken from the centre of the swamp. The solid lines reflect the vegetation zones derived from the classification of the pollen taxa.

Zone WS 1C (90 cm to 80 cm)

This zone is characterised by an increased representation of sclerophyll arboreal taxa, particularly Casuarinaceae. Poaceae values initially decrease at 90 cm but increase again at 80 cm, mainly at the expense of rainforest gymnosperms and rainforest angiosperms, which completely disappeared from the record at this depth. Aquatic and pteridophyte values, particularly *Selaginella* (which do not return in any significant values in the rest of the record) sharply decrease and are also absent from the 80 cm depth. Charcoal values decrease sharply in this zone.

Zone WS 1D (70 cm to 45 cm)

A decrease in Poaceae values is seen in this zone, while there is an increase in sclerophyll arboreal taxa, particularly Casuarinaceae, *Eucalyptus* and *Melaleuca*. In fact, the Myrtaceous taxa reach their highest values in the record at 45 cm. Rainforest gymnosperm and rainforest angiosperm values increase, as do aquatic and pteridophyte values. Charcoal values increase towards the top of this zone.

Zone WS 1E (40 cm to 10 cm)

This zone is characterised by an increased representation of Poaceae, while sclerophyll arboreal taxa, rainforest gymnosperm and rainforest angiosperm values decrease sharply. Pteridophyte abundances decline, while aquatic values, particularly *Polygonum*, initially increase from 40 cm to 30 cm and then sharply decline. Charcoal values are consistent with the previous zone, but there is a peak (the second highest in the record) at 20 cm.

Zone WS 1F (0 cm)

Sclerophyll arboreal taxa abundances increase sharply in this zone. Poaceae values decline and there is a sharp increase in representation of Asteraceae (Tubifloreae) pollen. In addition, there is an increased representation of rainforest angiosperm, ferns and aquatics in this zone, but very low charcoal values are observed.

Core WS 2

Zone WS 2A (190 cm)

Poaceae dominates in this zone, with low representation of rainforest taxa and sclerophyll arboreal taxa. Relatively low aquatic values are seen in this zone, with *Polygonum* being the dominant taxa. Charcoal values are low and ash content is ca. 60% and maintains this value until the top of zone WS 2D (60 cm).

Zone WS 2B (180 cm to 140 cm)

This zone is characterised by a significant increase in rainforest gymnosperm, rainforest angiosperm (particularly Arecaceae) and sclerophyll arboreal taxa, particularly Casuarinaceae. Poaceae values initially decline but then increase towards the top of the zone. Aquatic values increase, with Cyperaceae dominating from 180 cm to 160 cm, then decrease from 150 cm to 140 cm, being replaced by *Polygonum* and *Nymphoides*. There is a peak in monolete fern spores at 180 cm, as well as a peak in charcoal at 160 cm.

Zone WS 2C (130 cm to 120 cm)

Poaceae values sharply decline in this zone and there is a significant increase in Casurainaceae representation. There is a peak in rainforest gymnosperm and angiosperm values at 120 cm. Aquatic values decline in this zone and there is a peak in charcoal at 130 cm.

Zone WS 2D (110 cm to 50 cm)

This zone observes a significant increase in Poaceae representation, a sharp decline in Casuarinaceae values and a peak in Arecaceae values, particularly at 100 cm. There is also increased representation of *Melaleuca* and *Eucalyptus* at 70 cm to 50 cm. Aquatic values, particularly Cyperaceae, increase in this zone and there is a peak in *Polygonum* and monolete fern spores at 70 cm. Charcoal maintains relatively low representation in this zone, except for a peak at 90 cm. Ash content values are around 50% to 60% from 110 cm to 70 cm and then decline to 40% at 60 cm, suggesting an increase in organic content at this depth.

Zone WS 2E (40 to 30 cm)

Poaceae abundances increase in this zone and there is a decrease in Casuarinaceae and Arecaceae values. *Melaleuca* has a peak at 40 cm and there is a peak in *Eucalyptus* at 30 cm. *Callitris* and Cyperaceae observe their highest values in the record in this zone and there is a peak in monolete fern spores at 30 cm. Charcoal values are relatively low in this zone as well as the next one (WS 2F). There is an initial peak at 40 cm in ash content (ca. 70%), which sharply declines to ca. 20% at 30 cm.

Zone WS 2F (20 cm to 0 cm)

This zone is characterised by continued high values of Poaceae, as well as an increase in the sclerophyll arboreal taxa. There is also a sharp increase in Asteraceae (Tublifloreae) abundances and the disappearance of *Callitris* in the top sample. Cyperaceae values decrease and there is a peak in *Polygonum* values at 20 cm and 0 cm. Ash content is lowest in this zone ranging from 50% to 20%.

Holocene rainforest-sclerophyll boundary stability

One of the key features of the Witherspoon Swamp records is the relative stability in the rainforest-sclerophyll boundary in the Wet Tropics region during the Holocene period. In general, both the WS 1 and WS 2 cores are dominated by sclerophyll trees, particularly Casuarinaceae, and grasses. Peaks in rainforest gymnosperms and angiosperms may reflect minor expansions in the rainforest boundary (which is within 5 km to 10 km of the site). Significant rainforest taxa in both records (e.g. palms) can be typical of swamp communities rather than a reflection of the presence of extensive mesophyll rainforest (i.e. as would be high values of Cunoniaceae and *Elaeocarpus*). In addition, Witherspoon Swamp is fed by the Blunder Creek catchment, which originates from the nearby rainforest-covered escarpment to the east of the swamp and probably responsible for the long-distance transport of rainforest pollen to the site during flood events. A key consequence of these results is that there have been no significant climatic alterations (e.g. significantly wetter phases) that have facilitated the further expansion westward of rainforest. That is, once fully established between 9500 and 7500 cal BP (Haberle 2005), the rainforest-sclerophyll boundary was highly stable, with only minor fluctuations at the fringe of this boundary. These alterations at the fringe could be associated with slightly wetter and drier phases linked to ENSO variability, particularly during the late Holocene, as well as human use of fire influencing the boundary over the length of the Holocene period (Ash 1988).

Although it does not reveal a significant expansion of rainforest during the Holocene period, the Witherspoon Swamp records do suggest some significant climatic alterations (generally associated with precipitation changes) that reflect broader regional Holocene patterns. The WS 2 core suggests a drier environment around 8000 cal BP, possibly indicating that mesophyll rainforest did not reach its modern-day distribution until ca. 7500 cal BP, as suggested by the

Lake Euramoo record (Haberle 2005). In fact, a wetter phase, characterised by an increase in sedges and sclerophyll arboreal taxa and a decline in grass, is seen in the WS 2 record from 180 cm to 140 cm. This suggests higher precipitation during this early to mid-Holocene period, which is further supported by relatively high rainforest values in the WS 1 core at the base of this record (ca. 5000 cal BP). In addition, a general drying trend is observed in the WS 1 record with the decline and virtual disappearance of *Selaginella*, as well as lower representation of *Nymphoides* at 100 cm, probably reflecting a shift to lower water levels in the swamp. These results are similar to findings by Shulmeister and Lees (1995) from Groote Eylandt, Northern Territory, and by Luly et al. (2006) for the eastern shores of Cape York, Queensland, which suggests that these drying and wetting trends are a general feature of the sclerophyll forests and woodlands of northern Australia.

Some climatic variability is suggested by periodic increases in grass representation and charcoal values, alternating with increased values of rainforest taxa and sclerophyll arboreal taxa in both records. This vegetation alteration may be linked to the onset of ENSO conditions in the region ca. 5000 cal BP (i.e. a combination of El Niño and La Niña events) (Haberle and David 2004; Shulmeister and Lees 1995; Donders et al. 2007). A key change in the record appears to occur sometime after ca. 2000 cal BP (ca. 60 cm to 40 cm in both WS 1 and WS 2), with a return to generally drier conditions, seen through the expansion of grass values at the expense of the sclerophyll arboreal taxa. There is an increase in ash content variability (values ranging from 30% to 70%) from 60 cm in the WS 2 record, and this may reflect increased climatic variability beginning around 2000 cal BP. These alterations in ash content may represent changes in fluvial inputs of lithogenic sediments from the surrounding areas, with higher rainfall increasing inorganic content, and lower precipitation increasing organic values through less mineral matter transported to the swamp. Finally, a return to wetter conditions, lower grass values and higher sclerophyll arboreal representation is observed at the top of both records (0 cm in WS1 and 20 cm in WS 2), although this may also be associated with European disturbance. Generally, these alterations are consistent with similar changes in the Lake Euramoo record (Haberle 2005), which suggest that climatic variability associated with ENSO activity affected rainforests, sclerophyll forests and woodlands in a comparable fashion.

Implications for human occupation of the sclerophyll communities of the Wet Tropics region

The Witherspoon Swamp record has a number of implications for understanding potential human impacts on the sclerophyll communities of the Wet Tropics region of northern Australia. The presence of extensive swamp deposits extending back to at least the past 8000 years from Witherspoon Swamp suggests that this site may have formed a good camping area, with a source of permanent water and extensive food resources located within and around the swamp. To date, no archaeological surveys have been conducted around Witherspoon Swamp and further research should be undertaken to investigate whether this site did form an important resource for Aboriginal communities during the Holocene period. However, the pollen and charcoal record does not provide any clear evidence of Aboriginal impacts on the surrounding landscape through anthropogenic burning. Nearby records from the rainforest communities, including Lake Euramoo (Haberle 2005) and Lynch's Crater (Hiscock and Kershaw 1992), suggest – through increased burning, expansion of disturbance indicators (i.e. *Macaranga/ Mallotus*) and destruction of swamp forest at Lynch's Crater – that greater occupation of the region occurred over the past 5000 years. There are slight peaks in charcoal in the Witherspoon Swamp records at ca. 2000 cal BP that may reflect increased human impacts, but as suggested by Haberle and David (2004), it is extremely difficult to disentangle the relative roles of human impact and climate in landscape change for the humid tropics region.

Increased climatic variability through increased ENSO intensity and frequency around 2000 years ago may have facilitated a significant change in Aboriginal subsistence strategies. Gagan et al. (2004) suggest that the most intense period of ENSO activity occurred from 2500 to 1700 years ago, with a reduction in rainfall in the order of 20% to 40%, as well as precipitation becoming highly seasonal. Cosgrove et al. (2007) suggest that this significant climatic event made the surrounding sclerophyll woodlands and forests more marginal for human subsistence and facilitated the development of the Aboriginal rainforest culture based on toxic nut cultivation and processing. As discussed previously, the WS 2 record observes an increase in ash content variability from ca. 2000 years ago and this may reflect increased climatic variability associated with increased ENSO activity. In addition, the WS 1 record indicated an increase in rainforest and sclerophyll arboreal taxa around 50 cm to 45 cm, which then declines significantly at 40 cm, associated with a corresponding increase in grass. This decline has been directly dated to around 2000 years and also suggest highly variable climates associated with increased ENSO activity. These results support Cosgrove et al.'s (2007) suggestions that alterations in ENSO activity may be a key driver of subsistence changes for the Aboriginal people of the Wet Tropics region. However, further research is required to verify these results in terms of high-resolution palynological analysis of the Witherspoon Swamp record and examination of other sites within the sclerophyll communities of the region, as well as direct evidence, through artefacts, of human occupation changes around the margin of the swamp. Finally, both the WS 1 and WS 2 records provide evidence of European occupation (Zones WS 1F and WS 2F), with an increase in Asteraceae (Tublifloreae) pollen, decreased representation of grass, increased sclerophyll arboreal taxa values and decreased burning. These alterations most likely reflect activities associated with cattle grazing, particularly the imposition of fire suppression, and support similar findings in the Lake Euramoo record (Haberle 2005).

Conclusion

Landscape changes observed within both the sclerophyll and rainforest communities of the Wet Tropics region of northern Australia during the Holocene period suggest a significant amount of environmental variability. Key alterations include: a dry early Holocene; an early to mid-Holocene climatic optimum phase; and onset of drier, more variable environments, with greatest variability occurring around 2500 to 1700 cal BP, most likely associated with enhanced ENSO activity from the mid-Holocene (from 5000 years ago onwards). These natural climatic alterations may have had a profound impact on human activity within the region, and as suggested by Cosgrove et al. (2007), this increased climatic variability from 2500 to 1700 cal BP may have played a key role in the development of the unique indigenous rainforest culture of the region. However, the Witherspoon Swamp record, along with other sites from the modern rainforest communities of the region (particularly Lake Euramoo, Haberle 2005), also suggest that once established, the rainforest-sclerophyll boundary was highly stable during the Holocene period and that landscape alterations occurred within each community, rather than resulting in a significant expansion/contraction of these communities. These results also suggest that complex interactions between natural climate change and human impacts played a key role in shaping the northern Australian Wet Tropics environment during the Holocene period.

Acknowledgements

We thank Mr Trevor Austin, of Blunder Park, for permission to core Witherspoon Swamp and the Jirrbal people for their support and permission to work in their traditional country. Radiocarbon dating of the WS 2 samples was supported by AINSE Grant AINGRA08049. Research funding for this project was provided by a research grant from the School of Geography, Planning and Environmental Management, the University of Queensland and by an Australian Research Council Discovery Project grant (DP0986579). The authors would like to thank two anonymous reviewers for constructive comments on the manuscript.

References

Ash, J. 1988. The location and stability of rainforest boundaries in north-eastern Queensland, Australia. *Journal of Biogeography* 15:619-630.

Chen, Y. 1988. Early Holocene population expansion of some rainforest trees at Lake Barrine basin, Queensland. *Australian Journal of Ecology* 13:225-233.

Cosgrove, R., Field, J. and Ferrier, Å 2007. The archaeology of Australia's tropical rainforests. *Palaeogeography, Palaeoclimatology, Palaeoecology* 251:150-173.

Crowley, G.M. and Gagan, M.K. 1995. Holocene evolution of coastal wetlands in wet-tropical northeastern Australia. *Holocene* 5:385-399.

Crowley, G.M., Anderson, P., Kershaw, A.P. and Grindrod, J. 1990. Palynology of a Holocene marine transgressive sequence, lower Mulgrave River valley, north-east Queensland. *Australian Journal of Ecology* 15:231-40.

Donders, T.H., Haberle, S.G., Hope, G., Wagner, F. and Visscher, H. 2007. Pollen evidence for the transition of the Eastern Australian climate system from the post-glacial to the present-day ENSO mode. *Quaternary Science Reviews* 26:1621-1637.

Gagan, M.K., Johnson, D.P., and Crowley, G.M. 1994. Sea level control of stacked late Quaternary coastal sequences, central Great Barrier Reef. *Sedimentology* 41:329-351.

Gagan, M.K., Hendy, E.J., Haberle, S.G. and Hantoro, W.S. 2004. Post-glacial evolution of the Indo-Pacific Warm Pool and El Niño-southern Oscillation. *Quaternary International* 118-119:127-143.

Grimm, E.C. 1987 CONISS: a FORTRAN 77 program for stratigraphically constrained cluster analysis by the method of incremental sum of squares. *Computers and Geoscience* 13:13-35.

Grimm, E.C. 2004. *TGView Version 2.0.2*, Springfield, IL, USA: Illinois State Museum.

Grindrod, J. 1985. The palynology on a prograded shore, Princess-Charlotte Bay, North Queensland, Australia. *Journal of Biogeography* 12:323-348.

Grindrod, J. and Rhodes, E.G. 1984. Holocene sea-level history of a tropical estuary: Missionary Bay, north Queensland. In: Thom, B.G. (ed), *Coastal Geomorphology in Australia*, pp. 151-177. Sydney: Academic Press.

Haberle, S.G. 2005. A 23,000-yr pollen record from Lake Euramoo, Wet Tropics of NE Queensland, Australia. *Quaternary Research* 64:343-356.

Haberle, S.G. and David, B. 2004. Climates of change: human dimensions of Holocene environmental change in low latitudes of the PEPII transect. *Quaternary International* 118-119:165-179.

Hiscock, P. and Kershaw, A.P. 1992. Palaeoenvironments and prehistory of Australia's Top End. In: Dodson, J. (ed), *The Naive Lands: Prehistory and Environmental Change in Australia and the Southwest Pacific*, pp. 43-71. Melbourne: Longman Cheshire.

Hughen, K., Southon, J., Lehman, S., Bertrand, C. and Turnbull, J. 2006. Marine-derived 14C calibration and activity record for the past 50,000 years updated from the Cariaco Basin. *Quaternary Science Reviews* 25:3216-3227.

Kershaw, A.P. 1970. A pollen diagram from Lake Euramoo, North-East Queensland, Australia.

New Phytologist 69:785-805.

Kershaw, A.P. 1971. A pollen record from Quincan Crater, north-east Queensland, Australia. *New Phytologist* 70:669-805.

Kershaw, A.P. 1975. Stratigraphy and pollen analysis of Bromfield Swamp, north-eastern Australia. *New Phytologist* 75:173.

Kershaw, A.P. 1983. A Holocene pollen record from Lynch's Crater, north-eastern Queensland, Australia. *New Phytologist* 94:399-412.

Kershaw, A.P. 1986. The last two glacial-interglacial cycle from northeastern Australia: implication for climate change and Aboriginal burning. *Nature* 322:47-49.

Kershaw, A.P. 1994. Pleistocene vegetation of the humid tropics of northeastern Queensland, Australia. *Palaeogeography, Palaeoclimatology, Palaeoecology* 109: 399-412.

Kershaw, A.P., van der Kaars, S. and Moss, P.T. 2003a. Late Quaternary Milankovitch-scale climate change and variability and its impact on monsoonal Australia. *Marine Geology* 201:81-95.

Kershaw, A.P., Moss, P.T. and van der Kaars, S. 2003b. Causes and consequences of long-term climatic variability on the Australian continent. *Freshwater Biology* 48:1274-1283.

Kershaw, A.P., Bretherton, S.C. and van der Kaars, S. 2007. A complete pollen record of the last 230 ka from Lynch's Crater, north-eastern Australia *Palaeogeography, Palaeoclimatology, Palaeoecology* 251:23-45.

Luly, J.G., Grindrod, J. and Penny, D. 2006. Holocene palaeoenvironments and change at Three-Quarter Mile Lake, Silver Plains Station, Cape York Peninsula, Australia. *The Holocene* 16:1085-1094.

Moss, P.T. and Kershaw, A.P. 2000. The last glacial cycle from the humid tropics of northeastern Australia: comparison of a terrestrial and a marine record. *Palaeogeography, Palaeoclimatology, Palaeoecology* 155:155-176.

Moss, P.T. and Kershaw, A.P. 2007. A late Quaternary marine palynological record (oxygen isotope stages 1 to 7) for the humid tropics of northeastern Australia based on ODP site 820. *Palaeogeography, Palaeoclimatology, Palaeoecology* 251:4-22.

Nichols, N. 1992. Historical El Niño/Southern Oscillation variability in the Australasian region. In: Diaz, H.F. and Markgraf, V. (eds), *El Niño: Historical and Paleoclimatic Aspects of the Southern Oscillation*, pp. 167-179. Cambridge: Cambridge University Press.

Shulmeister, J. and Lees, B.G. 1995. Pollen evidence from tropical Australia for the onset of an ENSO-dominated climate at *c*. 4000 BP. *The Holocene* 5:10-18.

Tracey, J.G. 1982. *The vegetation of the humid tropical region of north Queensland*. Melbourne: CSIRO.

van der Kaars, W.A. 1991. Palynology of eastern Indonesian marine piston-cores: A late Quaternary vegetational and climatic record for Australasia. *Palaeogeography, Palaeoclimatology, Palaeoecology* 85:239-302.

Walker, D. 2007. Holocene sediments of Lake Barrine, north-east Australia, and their implications for the history of the lake and catchment environments. *Palaeogeography, Palaeoclimatology, Palaeoecology* 251:57-82.

Walker, D and Chen, Y. 1987. Palynological light on tropical rainforest dynamics. *Quaternary Science Reviews* 6:77-92.

Wang, X., van der Kaars, S., Kershaw, P., Bird, M. and Jansen, F. 1999. A record of fire, vegetation and climate through the last three glacial cycles from Lombok Ridge core G6-4, eastern Indian Ocean, Indonesia. *Palaeogeography, Palaeoclimatology, Palaeoecology* 147:241-256.

16

Holocene vegetation change at treeline, Cropp Valley, Southern Alps, New Zealand

Matt S. McGlone
Landcare Research, Lincoln, New Zealand
mcglonem@landcareresearch.co.nz

Les Basher
Landcare Research, Nelson, New Zealand

Introduction

New Zealand treelines have been well studied over the past few decades. Peter Wardle carried out extensive observational and experimental studies of their ecology and suggested that the length and warmth of summer was critical in permitting alpine trees to make sufficient growth to survive winter (Wardle 1985a). He also showed that New Zealand treelines were low when compared with global treelines, in particular with those of southern South America, and that exotic *Pinus contorta* could grow up to 300 m above the indigenous treeline (Wardle 1985b, 2008). A later global study demonstrated that warmth of the growing season was the one universal factor controlling position of treeline (Körner and Paulsen 2004). Both Wardle (2008) and Körner and Paulsen (2004) suggested New Zealand alpine trees were incapable of persisting under the cool growing season temperatures typical of treeline sites in most other regions, a possible result of having insufficient time to evolve cool-climate adaptations.

The 20th century trend towards warmer temperatures has intensified interest in treelines, as, being temperature-sensitive, they should be responding to the ca. 1°C increase in mean annual temperatures since 1900 AD. New Zealand investigations have shown there has been little or no response (Wardle and Coleman 1992; Cullen et al. 2001), and a global synthesis (Harsch et al. 2009) indicated that, while many treelines were rising, almost as many were static.

Despite this interest, and the large extent of alpine terrain in New Zealand, relatively little palaeoecological work has been reported from sites above, at, or close to treeline. In contrast, Tasmania and southeastern Australia, in part because of Peter Kershaw's long-standing interest, have many intensively investigated subalpine to alpine sites and a wealth of information on their treeline history (e.g. Kershaw and Strickland 1988; McKenzie 1997, 2002; Kershaw et al. 2007).

Central western districts of the Southern Alps are of great interest from both ecological and biogeographic perspectives because they lie in the western 'beech' gap of the Southern Alps. This extensive region lacks *Nothofagus*, arguably because of its exclusion by ice during the Last Glacial Maximum (McGlone et al. 1996). As the vast majority of New Zealand treelines are of *Nothofagus*, the history of a non-*Nothofagus* treeline gives a valuable additional perspective. The work presented here consists of several pollen profiles from a west-central Southern Alps alpine site. A summary diagram from this site with a line of accompanying text was included in a review of New Zealand postglacial vegetation history (McGlone 1988). Given the inherent interest in treelines, the renewed importance of understanding them in the context of climate changes, and the paucity of palaeoecological sites at treeline in New Zealand, it was thought appropriate to fully document the site and discuss it in the context of Australasian alpine results.

Environmental setting

Topology and geology

The Cropp River drains an alpine-montane basin of some 28.5 km², centred on 43°05'S, 170°58'E, on the western flanks of the central Southern Alps. This section of the Southern Alps has extremely high annual precipitation and high erosion rates. The basin is 8 km by 3.5 km and has a predominantly east-west orientation, ranging in elevation from 2140 m (Mount Beaumont) to 240 m at the confluence of the Cropp and Whitcombe Rivers (Figure 1).

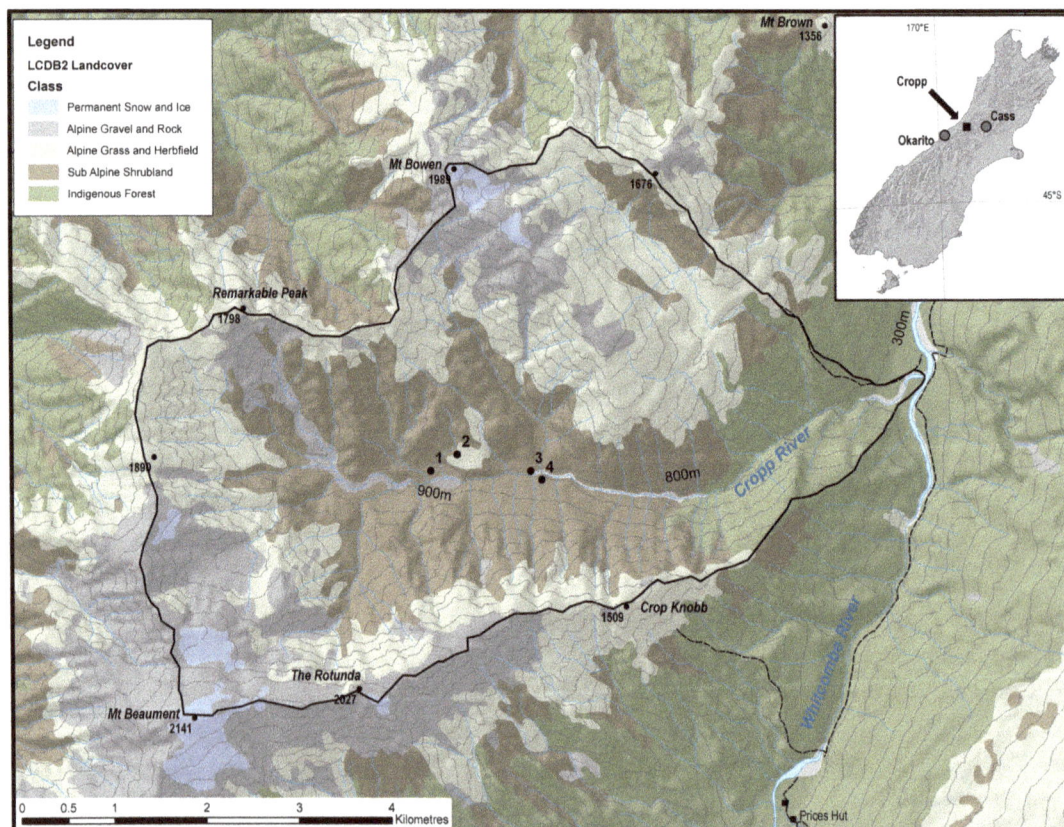

Figure 1. Location map, Cropp Valley, west-central Southern Alps, New Zealand. Inset: South Island, New Zealand.

The basin is composed of high-grade metamorphic rocks of green schist to amphibolite facies. The rocks are highly fissile and commonly intensely fractured. As it lies 5-8 km southeast of the Alpine Fault, the basin is close to the maximum uplift zone of the central Southern Alps, with an estimated uplift rate of 12 mm a^{-1} (Little et al. 2005).

Regional glacial chronology indicates that the Cropp Basin was filled with ice until about 16,000 cal BP (calibrated calendar years before 1950 AD; Gellatly et al. 1988). More recent dating of tills on the eastern side of the Southern Alps shows ice retreated after ca. 13,000 cal BP, following a late-glacial cooling during the Antarctic cold reversal between 14,900 and 12,900 cal BP (Kaplan et al. 2010). Till dated at 11,986 ± ca. 12,1000 cal BP from the middle reaches of the basin is interpreted as a subsequent late-glacial readvance (Basher and McSaveney 1989) and gives a maximum age for the postglacial surfaces. U-shaped cirques are present at the head of the Cropp River and its larger tributaries, but most of the glaciated landscape has been modified by postglacial fluvial and mass movement erosion. Slopes are steep (30-60°), valley profiles v-shaped, and stream grades steep and uneven. Regolith is thin (generally ≤2m). A wide range of soils occurs, including yellow-brown earths, podzols, gley podozols, gley recent soils, gley soils and organic soils. Soils are generally shallow, strongly leached and infertile.

Climate

The western front ranges of the Southern Alps are characterised by frequent, heavy rainfall (>100 mm/day), annual precipitation in excess of 10,000 mm and persistent cloudiness. Mean annual rainfall of the Cropp Basin is estimated at 10,800 mm a^{-1} (Griffiths and McSaveney 1983). Rainfall is usually well distributed throughout the year. Snow falls are frequent and may occur at any time during the year. Snow cover persists for about three months at treeline. Mean annual air temperature at Cropp Hut (865 m above sea level; 1982-1985; Tonkin and Basher 2001) is estimated at 5.5 ± 0.5°C, with a mean monthly range of 0-10.5°C, and a mean annual soil temperature (at 0.5 m depth) of 6.0 ± 0.5°C. Mean daily temperature ranges from 15°C to –5.5°C, while the absolute range is from 23.4°C to –10.1°C. Frosts may occur at any time of the year but are most common and severe in June and July.

Vegetation

Land cover of the basin ranges from permanent snow and ice to dense montane forest, and the high, often torrential, rainfall promotes extensive landslides and soil erosion in the upper reaches. Vegetation descriptions of the valley follow Norton (1983) and our own observations (Basher 1986). Below 350 m to 400 m asl, dense, mixed conifer-angiosperm forests predominate, with tall conifers (*Dacrydium cupressinum*, *Prumnopitys ferruginea* and *Podocarpus hallii*) emergent over an angiosperm canopy dominated by *Weinmannia racemosa*. From 400 m, the tall conifers gradually decline and *Weinmannia racemosa* and *Metrosideros umbellata* low forest extends up to 600 m. A *Metrosideros umbellata* belt with emergent *Libocedrus bidwillii* continues to 800 m, grading into a *Libocedrus bidwillii* forest without *Metrosideros umbellata* that interfingers with a low subalpine forest dominated by tall asterads (*Olearia colensoi*, *O. ilicifolia*, *O. lacunosa*,) and heaths (*Archeria traversii*, *D. longifolium*, *Dracophyllum traversii*) and other angiosperm small trees and shrubs (*Coprosma pseudocuneata*, *Griselinia littoralis*, *Myrsine divaricata*, *Pseudopanax colensoi*). Tall conical emergents of *Libocedrus bidwillii* and lower stature *Podocarpus hallii*, *Phyllocladus alpinus* and *Halocarpus biformis* occur in this association. This subalpine forest grades into alpine shrubland between 900 m and 1000 m asl, with a range of asterad and heath-dominated communities, *Halocarpus biformis* on old, leached soils, and *Hoheria glabrata* on young debris soils. Alpine grasslands replace shrubland at elevations between 900 m and 1400 m asl, depending on aspect, exposure and soil depth and age. *Chionochloa* spp. tussock grasses dominate well-drained sites, and species-rich communities of sedges, reeds, herbs and prostrate shrubs and cushion plants occur on poorly drained sites. These include the creeping podocarp *Lepidothamnus laxifolium* and low-growing *Podocarpus nivalis*.

There are numerous definitions of treeline in the literature but here we use the definition of Körner and Paulsen (2004): "the connecting line between the uppermost groups of upright trees of at least 3 m in height". By this criterion, treeline in Cropp Valley lies at about 1000 m asl. The regional treeline in this section of the western Southern Alps lies between 1200 m and 1150 m, with annual average temperatures of 5.8-6.2°C (Ellen Cieraad pers. comm. 2011). The Cropp Valley treeline is therefore somewhat lower and colder than the regional treeline, probably because of cold-air drainage from the high reaches of the basin.

Sites and methods

Site locations are given in Figure 1, site stratigraphies in Figure 2, and radiocarbon date information in Table 1. All sites were sampled from exposed faces in pits or stream exposures.

Table 1. Sample locations and radiocarbon date details. Radiocarbon calibrations were calculated using the software CALIB v 5.0.1 (1) using the Southern Hemisphere Calibration dataset SHCal04 (2) (Stuiver and Reimer 1993; McCormac et al. 2004).

Site	Grid ref. (NZMS1)	Material	Cm below surface	^{14}C age BP	Cal BP (median)	^{14}C Lab code
Site 1. Cropp Hut: 895 m	J34/439903	Peat	25	1053 ± 87	908	NZ6881
		Peat	55	2851 ± 85	2917	NZ6878
		Peat	88-90	7065 ± 90	7837	NZ6115
Site 2. Tarkus Knob: 950 m	J34/444904	Peat	175	1748 ± 38	1600	NZ5367
	J34/441905	Peat	160-162	7381 ± 70	8134	NZ5369
		Peat	170-180	8390 ± 76	9337	NZ5368
Site 3. Steadman Creek: 820 m	J34/446902	Wood	See text	10295 ± 120	12098	NZ6576

Site 1: Cropp Hut (field sample number: X8404)

A concave basin peat has formed in the distal zone of a prominent high-level alluvial cone of the north bank of Cropp River slightly upstream of the site of the former Cropp Hut (destroyed by a debris flow). The site has an elevation of 895 m and is on a forest-free subalpine valley floor. The plant communities of the general area are grassland (*Chionochloa pallens* and *C. rubra*) or grassland-low shrub on gently sloping surfaces, while steeply sloping surfaces are covered with subalpine scrub (*Coprosma* spp., *Dracophyllum uniflorum*, *Olearia lacunosa*, *Phyllocladus alpinus*, *Podocarpus nivalis*, *Senecio bennettii*) and forest (*Archeria traversii*, *Dracophyllum longifolium*, *Libocedrus bidwillii*, *Halocarpus biformis*). The peat surface is covered with *Chionochloa rubra*, *Carpha alpina*, *Celmisia glandulosa*, *Donatia novae-zelandiae* and *Oreobolus pectinatus*, and with substantial stands of the creeping podocarp, *Lepidothamnus laxifolium*. Ten samples were taken for pollen analysis and pollen results are given in Figures 3a and b.

Site 2: Tarkus Knob (field sample number X8207)

Opposite the former Cropp Hut is a prominent ice-smoothed bedrock promontory informally named Tarkus Knob (elevation at site: 950 m). A hollow along the northern margin of this feature has been in-filled by alluvium which contains two buried peat layers and is capped by a peaty soil. Three radiocarbon-dated samples for pollen analysis have been taken from two exposures along the hollow (Figure 2). The present vegetation is *Chionochloa rubra* grassland-cushion bog (*Carpha alpina*, *Celmisia glandulosa*, *Donatia novae-zelandiae*, *Lepidothamnus laxifolium, Schoenus pauciflorus*). Surrounding vegetation on well-drained slopes is as described for the Cropp Hut site.

Site 3: Terminal moraine, Steadman Creek (field sample X8512)

This site (grid reference NZMS2690 J34/446902; altitude 820 m) is described in Basher and McSaveney (1989). A 30-40 m high terrace of the Cropp River consists of angular to sub-rounded boulders of schist, some with surface striations, in a sandy matrix. Decomposed twigs and logs (up to 20 cm in diameter) were found tightly embedded in a matrix of angular clasts towards the base of the section just above the bed of the river. A piece of wood was dated (10,295 ± 120 BP) and the matrix surrounding the wood sampled for pollen analysis.

Site 4: Surface sample sites (field samples X8621a, b and c)

Surface samples were taken from the surface of the Tarkus and Cropp Hut sites and are included as the upper levels in those pollen diagrams. A further surface sample was taken as far down the valley as could be reached on foot, at the confluence of the Cropp River and Reckless

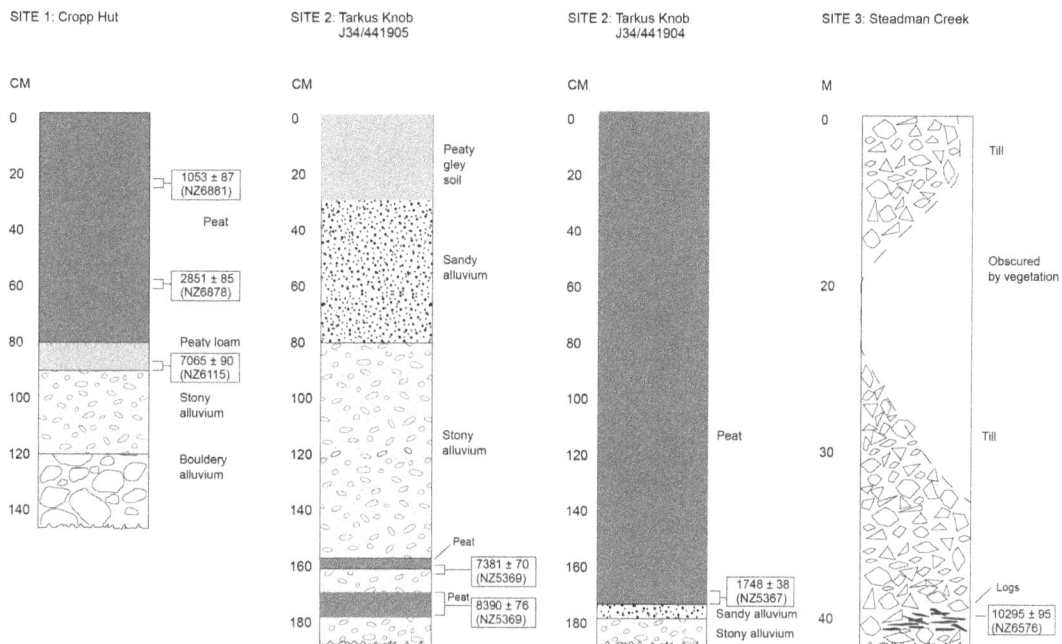

Figure 2. Site stratigraphies, Cropp Basin. Conventional radiocarbon ages in text boxes. The two Tarkus Knob exposures are less than 100 m apart. Note change of scale between Sites 1 and 2 and Site 3.

Torrent at an altitude of 820 m. The site was a patch of wet *Chionochloa rubra* grassland within subalpine scrub approximately 1.5 km from tall montane forest.

Pollen and spore analysis followed standard extraction procedures (dehumification in 10% KOH, digestion in 40% HF, followed by acetolysis and mounting in glycerine jelly for light microscope examination). A terrestrial pollen sum of >250 grains was used and results are presented as a percentage of the terrestrial pollen sum.

Results and discussion

Modern pollen representation (see Figures 3 and 4)

Alpine and subalpine sites derive a large proportion of their pollen rain from wind-pollinated trees in lower altitude forests (Moar 1970; McGlone 1982; Pocknall 1982). The two uppermost sites sampled from subalpine grasslands derive ca. 35% of their pollen rain from lowland tree podocarps (*Dacrycarpus dacrydioides*, *Dacrydium cupressinum*, *Prumnopitys ferruginea*, *P. taxifolia*), whereas the lower surface sample site (Site 4, Reckless Torrent), in dense, low subalpine forest, derives only 21% of its pollen from that source. *Metrosideros* and *Weinmannia*, the primary canopy species of the high montane forest in the lower reaches of the basin, make up <5% of the pollen rain. Although virtually absent from the upper altitude forests, two trees, *Ascarina lucida* and *Quintinia serrata*, and a liana, *Ripogonum scandens*, are also registered. Grass and subalpine scrub species (asterads, *Coprosma*, *Dracophyllum*, *Halocarpus*, *Myrsine*, *Phyllocladus*) average ca. 30% for the two upper sites, but ca. 65% for the lower. Although never making up a substantial proportion of the pollen rain, upland herbs, cushion plants and prostrate shrubs are represented. Although *Libocedrus bidwillii* is prominent in the upland low forest of the basin and wind-pollinated, it contributes surprisingly little to the pollen rain, averaging only ca. 1%.

Steadman Creek, till exposure (Figure 4)

The Cropp till at 850 m is attributed to a late-glacial advance at ca. 12,100 cal BP, during which a glacier occupied most of the upper basin (Basher and McSaveney 1989). The wood was incorporated into the advance as the ice moved down valley. The pollen is derived from silt associated with the wood fragments and, as it was reworked by or deposited directly on the ice, is not directly comparable to that from the peat sequences. Exceptionally high levels of *Hoheria* (22%) and abundant *Coprosma*, asterads and grass indicate an open low subalpine forest-shrubland, typical of what might be expected on disturbed sites within the basin at the present day. *Hoheria glabrata* currently does not extend more than about 200 m above the altitude of the till exposure and we conclude that subalpine low forest must have been close to its current limit during the advance. Nevertheless, a number of present-day components of the basin vegetation are missing from the pollen spectrum, including *Dracophyllum*, *Halocarpus* and *Libocedrus*, and lowland conifer forest trees make up only 6% of the pollen rain, rather than the 30% average for the Holocene sites. *Metrosideros* and *Weinmania* are at trace levels only. Two broad conclusions can be made.

First, the Cropp glacier must have advanced down valley at a time when mean annual temperatures were close to those of the present. *Hoheria glabrata* forest does not extend higher than 1050 m in central Westland (Wardle 1991) and thus the maximum depression in mean annual temperature relative to the present can have only been of the order of 1.2°C. However, as the wood was included within the till and not over-ridden, it is likely the *Hoheria* trees grew up-slope of the site. Given the steep gradient of the river, the *Hoheria* forest is likely to have grown at 900-1000 m altitude, that is, at its current limit.

The second conclusion is that the lowland to montane forest conifer emergents that provide more than 30% of the pollen rain at present must have been very much reduced in extent. This inference is supported by the pollen results from the extensive Okarito Pakihi mire at 70 m asl approximately 63 km to the southwest of the Cropp Basin (Newnham et al. 2007), where emergent conifers made up ca. 30% of the pollen rain at 12,000 cal BP, versus 95% during the Holocene. We therefore have the somewhat paradoxical results that woody vegetation was growing close to or at its present limit, while the upper limit to montane forest appears to have been lower and conifer forest in general much less abundant than now.

Cropp Hut–Tarkus Knob peat sequences (Figures 3, 4 and 5)

Cropp Hut and Tarkus Knob peats are in close proximity and both are currently in grassland vegetation, differing in altitude by 50 m. Their pollen percentages at matched time intervals are

Cropp Hut (Site 1): Trees and tree ferns

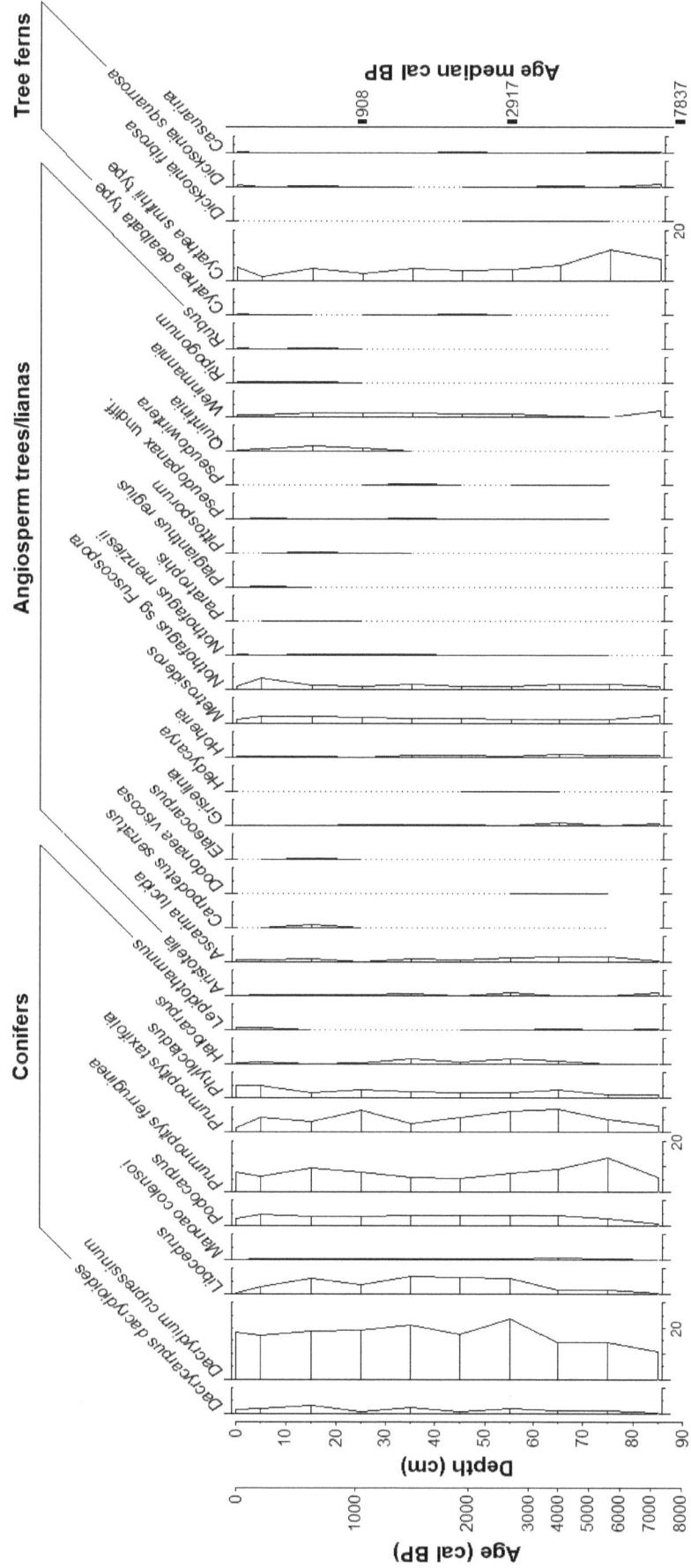

Figure 3a. Cropp Hut (Site 1). Percentage pollen and spore diagram: trees and tree ferns. Pollen sum: dryland trees, shrubs and herbs.

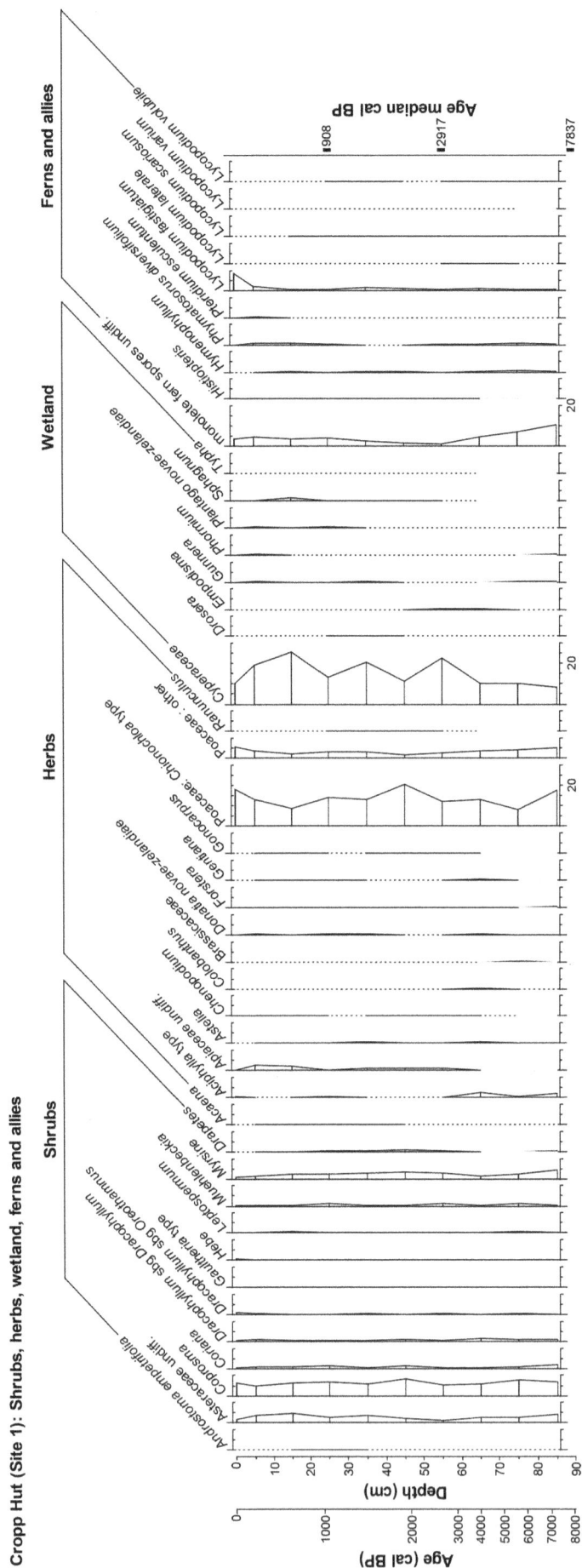

Figure 3b. Cropp Hut (Site 1). Percentage pollen and spore diagram: shrubs, herbs, fern and fern allies. Pollen sum: dryland trees, shrubs and herbs.

Tarkus Knob (Site 1), Steadman Creek (Site 3) & Reckless Torrent surface sample (Site 4)

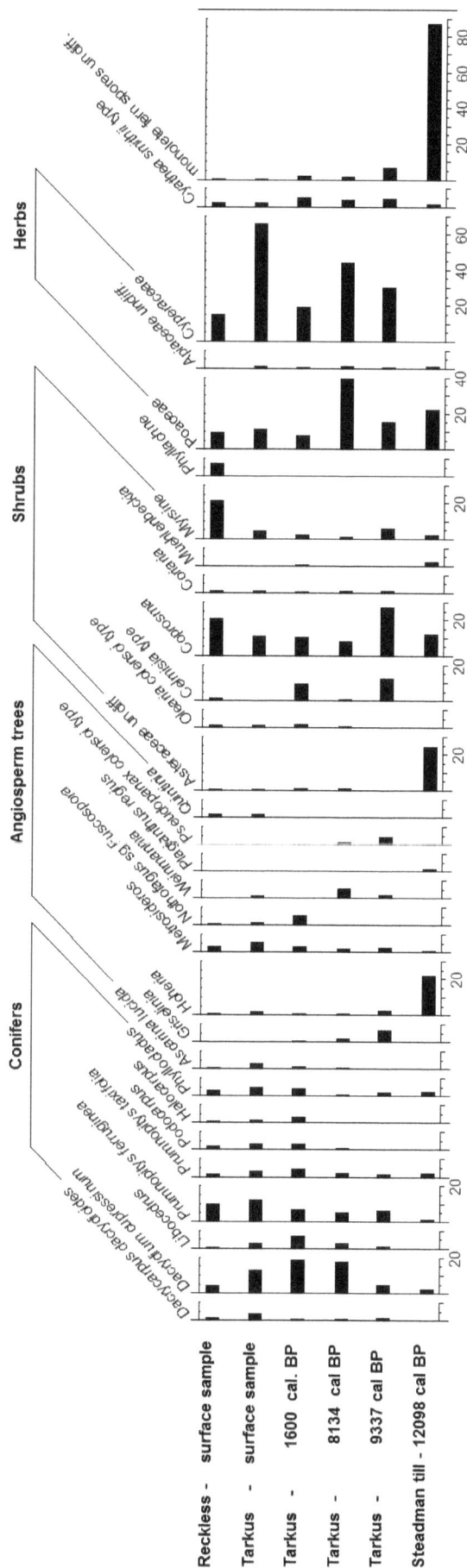

Figure 4. Surface samples: Reckless Torrent, (820 m); Tarkus Knob (950 m). Three dated peat samples: Tarkus Knob (Site 2) buried peats; Steadman Till (Site 3), silt associated with *Hoheria* wood fragments. Percentage pollen and spore diagram (major types only). Pollen sum: dryland trees, shrubs and herbs.

close, and they can be for most purposes treated as a single record. To get a generalised view of changes, a composite diagram for the arboreal pollen has been constructed from the Cropp Hut, Tarkus Knob and Steadman sequences (Figure 5).

Local vegetation cover appears to have been remarkably uniform throughout the Holocene. Grass pollen, dominated by *Chionochloa*, asterads, *Coprosma Dracophyllum*, *Myrsine*, and sedges and alpine herbs are consistently present at levels suggesting little or no change in the local vegetation structure from the current day (Figures 3 and 4).

The oldest sample (Tarkus Knob site; 9337 cal BP) is ca. 2500 years younger than the till. Independent confirmation of the age of this sample is provided by the trace levels of *Nothofagus* versus the 1-5% levels for those younger than 9000 cal BP. *Nothofagus* became the dominant forest cover between 8200 and 7500 cal BP in the Cass Basin, 70 km to the east (McGlone et al. 2004), and this is the only likely source for this pollen type. The basal sample from the Cropp Hut peat section is younger than 8000 cal BP and has a significant *Nothofagus* representation.

The contribution of lowland conifer pollen rose from 13% to 20-25% by 7500 cal BP and reached present day levels of 30-40% by 6000 cal BP (Figure 5). High-altitude montane

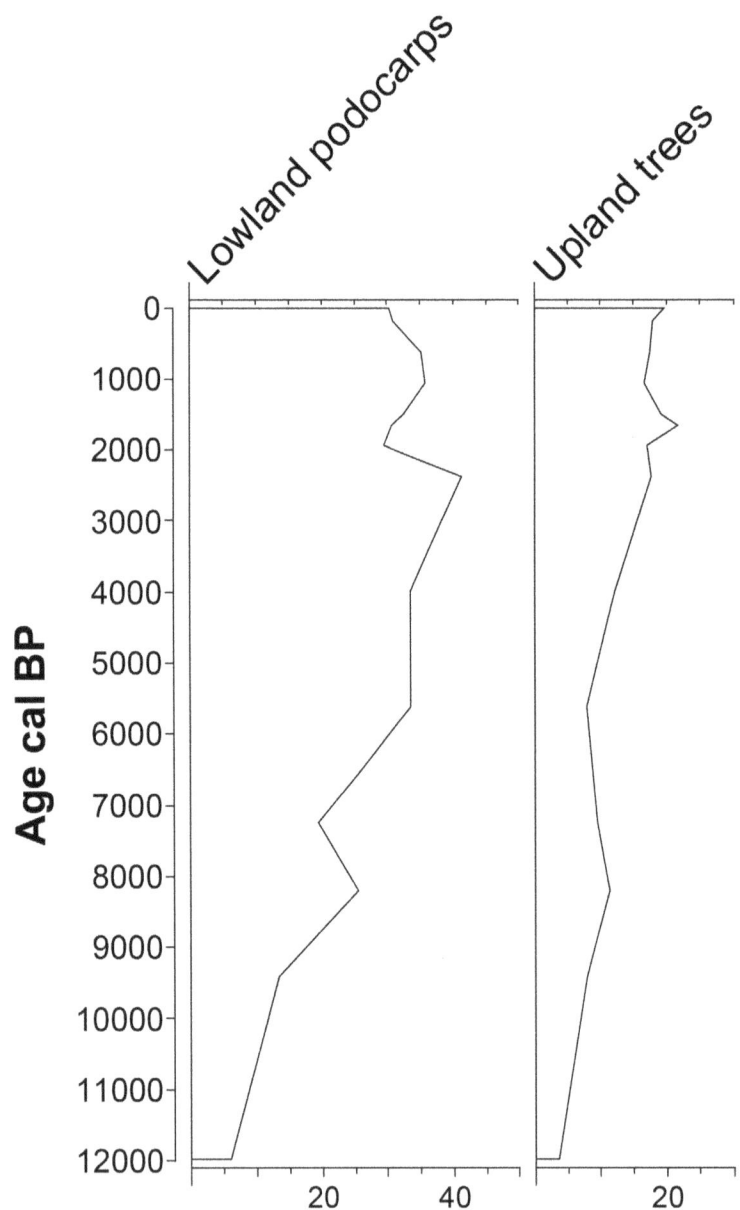

Figure 5. Summary composite percentage pollen diagram (Cropp, Tarkus, Steadman). Lowland pollen types: *Dacrycarpus dacrydioides, Dacrydium cupressinum* and *Prumnopitys* spp. Upland pollen types: *Libocedrus, Phyllocladus, Podocarpus, Halocarpus, Metrosideros, Weinmannia* and *Quintinia*. Note: lowland pollen types are those from emergent wind-pollinated trees that dominate the lowlands; the upland types are those that are known to occur either near the Cropp sites or in the upper montane forest (regardless of whether they also occur further downslope).

forest types expanded from 8% to 12% between 9500 and 4000 cal BP, but did not reach current levels of 18% until 2400 cal BP, which is when *Libocedrus bidwillii* became abundant within the basin. Tree-fern spores are dominated by *Cyathea smithii* type (which includes the short subalpine *C. colensoi*) at levels which suggest wind dispersal from montane forests but its percentage representation does not change in the course of the Holocene.

Quintinia is the last of the montane forest species to be registered at the site (between 1600 and 900 cal BP). This canopy tree, characteristic of infertile soils in central Westland, is confined to scattered patches in the south of the district, and does not occur in south Westland. Wardle (1988) suggested that its distribution reflects a slow movement south from refugia in north Westland at the end of the last glaciation. However, it was present at Bell Hill, north Westland, before 10,400 cal BP (Moar 1971), but even there did not become abundant until ca. 3500 cal BP (Pocknall 1980). A 7000-year migration lag for local spread seems improbable. It is far more likely that *Quintinia* was present throughout its current range at the end of the Last Glacial Maximum, as suggested by its trace occurrences then at Okarito Pakihi (Newnham et al. 2007), but remained uncommon until some aspect of the climate or soils became more favourable for the tree in the course of the late Holocene.

Climate inferences

As discussed in regard to *Quintinia*, there is no reason to suspect any of the dominant tree species were affected by migration lags: *Weinmannia*, *Metrosideros*, *Quintinia* and *Libocedrus* have small, wind-dispersed seeds and the podocarp trees have bird-dispersed fleshy fruits. While soil requirements may have been a constraint for some species, under the mild, high-rainfall environment of Cropp Basin soils mature and become podozolized extraordinarily quickly, and landslides caused by heavy rain or earthquakes are a constant source of rejuvenated surfaces within the basin (Tonkin and Basher 2001). We therefore argue that the most probable explanation for the Holocene vegetation changes is the direct or indirect consequences of changing climate. Given that the Cropp Basin is directly in the path of the rain-bearing westerlies and has an enormous surplus of well-distributed precipitation far beyond the growth requirements of the vegetation, it further seems unlikely that changes in rainfall would have major effects on the vegetation cover. For that reason, we focus on changes in temperature and seasonality.

A marine-core alkenone sea-surface temperature record 200 km west of the Cropp Basin shows that temperatures peaked in the later late glacial, and were about 1.5°C warmer than present at 12,000 cal BP (Barrows et al. 2007). Pollen-based transfer functions show mean annual temperatures in South Island sites to have been warmer than now from about 12,000 cal BP (Wilmshurst et al. 2007). The occurrence of tree-sized *Hoheria glabrata* high within the Cropp catchment at 12,000 cal BP, close to its upper limit, implies that temperatures during the growing season were within 1°C of those of the present. Warmer than present climates in the Cropp Basin by 12,000 cal BP therefore seem highly probable.

Nevertheless, some observations do not fit a scenario of summer temperatures at or above current values. In the cloudy, high-precipitation, mild climates of the western flank of the Southern Alps, winter cold is not a significant factor except in valley bottoms exposed to cold-air drainage (Leathwick et al. 2003). It follows that regional alpine treelines are controlled by growing-season warmth, as demonstrated by Körner and Paulsen (2004). If summer temperatures were 1.5°C warmer than now, vegetation zones should have been some 250 m higher and the upper limit to closed montane forest at ca. 850-1000 m asl. Forest would have surrounded the Cropp sites. The absence of montane forest trees, and conifers in particular, from the catchment in the early Holocene demonstrates this did not happen. New Zealand glaciers in the central Southern Alps are much more responsive to summer ablation than winter precipitation (Anderson and Mackintosh 2006). The fact that a glacier advanced beyond Holocene limits to fill the upper Cropp Valley and produce the Cropp till is therefore also incompatible with summer temperatures warmer than those of the present.

Cool summers but higher than present mean annual temperatures necessarily imply much warmer and shorter winters. In temperate latitudes, warm winters are important in the lowlands as they permit frost-sensitive species to flourish. In contrast, at treeline, warm winters are largely irrelevant or even detrimental, as warm temperatures may break dormancy and increase frost. Cool summers and warm winters create an intensely oceanic environment which is understood to further limit tree growth through restricting photosynthesis in summer and increasing metabolic energy expenditure in winter (Stevens and Fox 1991; Crawford 2000). Tall podocarps were less common than now during this intensely oceanic phase in the nearby lowlands at Okarito (Newnham et al. 2007) and it is possible that they were also restricted by cool summers. We conclude that in the late glacial and early Holocene, a warm-winter/ cool-summer regime depressed altitudinal forest zones and may have had a disproportionate influence on tall conifer trees.

The Holocene vegetation sequence within the Cropp Valley (Figure 5) is best interpreted as a response to steady increases in summer temperature. By 8200 cal BP, upper montane forests were present in the lower reaches of the valley. By 2500 cal BP, *Libocedrus bidwillii* had occupied the upper part of valley, and lowland-forest pollen rain peaked about the same time, suggesting that upslope movement of forest zones had not reached current altitudes until then.

Other New Zealand sites near treeline or in montane settings support the scenario that summer temperatures have been rising in the course of the Holocene. Pollen and macrofossil evidence from subantarctic Auckland and Campbell Islands indicates that forest did not reach current extent until 6000 cal BP, and may have continued to extend upslope into the late Holocene (McGlone 2002; McGlone et al. 2010). In the Garvie Mountains (Central Otago), high-altitude grassland was abruptly replaced by coniferous forest at 8300 cal BP (McGlone et al. 1995). Pollen sites close to treeline in northwest Nelson suggest late-Holocene spread of subalpine *Nothofagus* forest into low woody vegetation and grassland (Shulmeister et al. 2003). In the North Island, in an altitudinal sequence of organic soil pollen sites across treeline (1140-1470 m asl) on the isolated massif of Mount Hauhungatahi, Tongariro, low conifer forest rose to dominance at ca. 9700 cal BP, but the current tall *Nothofagus-Libocedrus* forest spread only from ca. 8000 cal BP, and was not dominant until several thousand years later (Horrocks and Ogden 2000).

A forest ecosystem process model has been used in inverse mode to explore early Holocene climate scenarios based on pollen and macrofossil data from Cass, an intermontane site 70 km east of the Cropp Valley at 600 m altitude (McGlone et al. 2011). Best fit between early Holocene vegetation and climate was obtained when summers were no warmer than present but winters substantially warmer. If early Holocene summers were set at 0.5°C cooler than present, and compensated for by much warmer winters to achieve an overall warmer climate (+1.5°C), suitable matches could also be made. Annual ocean-surface warmth and terrestrial summer cooling in the early Holocene can thus be reconciled.

The mountain ranges of Tasmania and the southeastern coast between 43°S and 30°S provide the closest comparable sites in Australia to Cropp Valley. Changes in Australian upland regions are complicated by major alterations in rainfall patterns which result in asynchronous vegetation change between sites (Donders et al. 2007). Nevertheless, there is agreement that summer temperatures and forest cover peaked in the mid Holocene, as early recognised by Kershaw (1988). At the northern edge of the temperate forest zone in New South Wales, at high-altitude (1200-1500 m asl) sites on Barrington Top, current forest was established by 10,000 cal BP and vegetation typical of warmer and moister conditions was widespread over the plateau between 7500 and 4000 years ago (Dodson et al. 1986). In Kosciuszko National Park in New South Wales, sites at treeline had grassy alpine herbfields until ca. 12,500 cal BP and *Eucalyptus* increased to form similar communities to those at current treeline by 7500 cal BP. Subalpine woody communities reached maximum development between 6000 and 3000 cal BP (Martin 1986, 1999). It is possible that drier conditions earlier in the Holocene may have delayed this maximum. In south-central Victoria, forest reached current levels at several high-altitude sites after 11,500 cal BP (McKenzie 1997, 2002; Kershaw et al. 2007), but wet *Eucalpytus* forests and *Nothofagus* did not achieve maximum extent until ca. 7000 cal BP (McKenzie 2002). Changes after ca. 4500 cal BP are complicated by the onset of ENSO and changing precipitation, but the decline in moisture-loving *Nothofagus* may have resulted from the transpiration stress of increasing summer temperatures. In Tasmania, maximum treeline altitude and warmest growing-season temperatures appear to have been achieved by 7000 cal BP (Anker et al. 2001; Fletcher and Thomas 2007).

In summary, Australian results are consistent with those from the Cropp Valley in that they show treelines were lower than now before 11,500 cal BP and current treelines mostly formed only after 7000 cal BP.

Schaefer et al. (2009) note that Holocene glacier advances in the central Southern Alps of New Zealand diminish in volume towards the present. They suggest this is largely due to increased summer ablation, consistent with our claim that treelines have risen and summers warmed in the course of the Holocene. It follows then that ocean-surface temperatures, which fall during the mid-to-late Holocene as treelines rise (Barrows et al. 2007), must reflect substantial winter cooling.

Conclusions

The latter part of the late glacial and early Holocene in New Zealand was characterised by low treelines relative to the present. In the upper Cropp Valley, although shrubs and small trees extended up to current treeline, the current montane-alpine forest was absent, including the conifer *Libocedrus*. Similar patterns of change are seen at near-treeline sites throughout New Zealand, usually involving the late spread of *Nothofagus* forest into early Holocene tall shrubland. We conclude that highly oceanic environments during the late glacial-early Holocene period, characterised by warm winters and cool summers, disproportionately affected trees in the montane to alpine zones. There is a great deal of similarity in timing between the New Zealand montane-alpine vegetation sequences over the late glacial and Holocene, and those of the mountainous areas of Tasmania and southeastern Australia. This probably reflects a common response to a hemispheric temperature trend. Although increasing precipitation after a drier early Holocene is often suggested as a contributory factor to changing treelines at Australian sites, this is far less likely to have been an issue at New Zealand alpine sites.

We suggest that during the latter half of the late glacial and the early Holocene, cool summers, warm winters and generally drier than present conditions prevailed in southeastern Australia and New Zealand. Summers shortened but became warmer, while winters lengthened and cooled in the course of the Holocene. As a consequence, apparently contradictory trends emerged in the mid-to-late Holocene: lowland forests changed character, losing frost-sensitive components as winters became cooler, while alpine trees began to replace previous low-growing shrublands in response to warmer summers. The decline from peak summer warmth in montane-alpine regions noted in Australia after 5000-4000 cal BP is not as obvious in New Zealand records.

Acknowledgements

We thank Neville Moar and Janet Wilmshurst for their comments on a draft of the manuscript, and Rewi Newnham and Geoffrey Hope for their reviews. MSM would also like to record his gratitude to Peter Kershaw for his encouragement and support over many years.

References

Anderson, B. and Mackintosh, A. 2006. Temperature change is the major driver of late-glacial and Holocene glacier fluctuations in New Zealand. *Geology* 34:21-124.

Anker, S.A., Colhoun, E.A., Barton, C.E., Peterson, M. and Barbetti, M. 2001. Holocene vegetation and paleoclimatic and paleomagnetic history from Lake Johnston, Tasmania. *Quaternary Research* 56:264-274.

Barrows, T.T., Lehman, S.J., Fifield, L.K. and De Deckker, P. 2007. Absence of cooling in New Zealand and the adjacent ocean during the Younger Dryas Chronozone. *Science* 318:86-89.

Basher, L.R. 1986. Pedogenesis and erosion history in a high rainfall, mountainous drainage basin – Cropp River, New Zealand. Unpublished doctoral thesis, Department of Soil Science. Lincoln College, University of Canterbury.

Basher, L.R. and McSaveney, M.J. 1989. An early Aranuian glacial advance at Cropp River, central Westland, New Zealand. *Journal of the Royal Society of New Zealand* 19:263-268.

Crawford, R.M.M. 2000. Ecological hazards of oceanic environments. *New Phytologist* 147:257-281.

Cullen, L.E., Stewart, G.H., Duncan, R.P. and Palmer, J.G. 2001. Disturbance and climate warming influences on New Zealand *Nothofagus* tree-line population dynamics. *Journal of Ecology* 89:1061-1071.

Dodson, J.R., Greenwood, P.W. and Jones, R.L. 1986. Holocene Forest and Wetland Vegetation Dynamics at Barrington Tops, New South Wales. *Journal of Biogeography* 13:561-585.

Donders, T.H., Haberle, S.G., Hope, G., Wagner, F. and Visscher, H. 2007. Pollen evidence for the transition of the eastern Australian climate system from the post-glacial to the present-day ENSO mode. *Quaternary Science Reviews* 26:1621-1637.

Fletcher, M.S. and Thomas, I. 2007. Holocene vegetation and climate change from near Lake Pedder, south-west Tasmania, Australia. *Journal of Biogeography* 34:665-677.

Gellatly, A.F., Chinn, T.J.H. and Röthlisberger, F. 1988. Holocene glacier variations in New Zealand – a review. *Quaternary Science Reviews* 7:227-242.

Griffiths, G.A. and McSaveney, M.J. 1983. Hydrology of a basin with extreme rainfalls – Cropp River, New Zealand. *New Zealand Journal of Science* 26:293-306.

Harsch, M.A., Hulme, P.E., McGlone, M.S. and Duncan, R.P. 2009. Are treelines advancing? A global meta-analysis of treeline response to climate warming. *Ecology Letters* 12:1040-1049.

Horrocks, M. and Ogden, J. 2000. Evidence for Lateglacial and Holocene tree-line fluctuations from pollen diagrams from the Subalpine zone on Mt Hauhungatahi, Tongariro National Park, New Zealand. *Holocene* 10:61-73.

Kaplan, M.R., Schaefer, J.M., Denton, G.H., Barrell, D.J.A., Chinn, T.J.H., Putnam, A.E., Andersen, B.G., Finkel, R.C., Schwartz, R. and Doughty, A.M. 2010. Glacier retreat in New Zealand during the Younger Dryas stadial. *Nature* 467:194-197.

Kershaw, A.P. 1988. Australasia. In: Huntley, B. and Webb lll, T. (eds), *Vegetation History*, pp. 237-306. Dordrecht: Kluwer Academic Publishers.

Kershaw, A.P., McKenzie, G.M., Porch, N., Roberts, R.G., Brown, J., Heijnis, H., Orr, M.L.,

Jacobsen, G. and Newallt, P.R. 2007. A high-resolution record of vegetation and climate through the last glacial cycle from Caledonia Fen, southeastern highlands of Australia. *Journal of Quaternary Science* 22:481-500.

Kershaw, A.P. and Strickland, K.M. 1988. The development of alpine vegetation on the Australian mainland. In: Good, R. (ed), *The Scientific Significance of the Australian Alps: the proceedings of the first Fenner Conference on the Environment*, pp. 113-125. Canberra: Australian Alps National Parks Liaison Committee and the Australian Academy of Science.

Körner, C. and Paulsen, J. 2004. A world-wide study of high altitude treeline temperatures. *Journal of Biogeography* 31:713-732.

Leathwick, J.R., Wilson, G., Rutledge, D., Wardle, P., Morgan, F., Johnston, K., McLeod, M. and Kirkpatrick, R. 2003. *Land Environments of New Zealand*. Auckland: David Bateman.

Little, T.A., Cox, S., Vry, J.K. and Batt, G. 2005. Variations in exhumation level and uplift rate along the oblique-slip Alpine fault, central Southern Alps, New Zealand. *Geological Society of America Bulletin* 5-6:707-723.

Martin, A.R.H. 1986. Late Glacial and Holocene alpine pollen diagrams from the Kosciusko National Park, New South Wales, Australia. *Review of Palaeobotany and Palynology* 47:367-409.

Martin, A.R.H. 1999. Pollen analysis of Digger's Creek Bog, Kosciuszko National Park: Vegetation history and tree-line change. *Australian Journal of Botany* 47:725-744.

McCormac, F.G., Hogg, A.G., Blackwell, P.G., Buck, C.E., Higham, T.F.G. and Reimer, P.J. 2004. SHCal04 Southern Hemisphere calibration, 0-11.0 cal kyr BP. *Radiocarbon* 46:1087-1092.

McGlone, M.S. 1982. Modern pollen rain, Egmont National Park, New Zealand. *New Zealand Journal of Botany* 20:253-262.

McGlone, M.S. 1988. New Zealand. In: Huntley, B. and Webb III, T. (eds), *Vegetation History*, pp. 557-599. Dordrecht: Kluwer Academic Publishers.

McGlone, M.S. 2002. The late Quaternary peat, vegetation and climate history of the southern oceanic islands of New Zealand. *Quaternary Science Reviews* 21:683-707.

McGlone, M.S., Hall, G.M.J. and Wilmshurst, J.M. 2011. Seasonality in the early Holocene: Extending fossil-based estimates with a forest ecosystem process model. *The Holocene* 21:517-526.

McGlone, M.S., Mark, A.F. and Bell, D. 1995. Late Pleistocene and Holocene vegetation history, central Otago, South Island, New Zealand. *Journal of the Royal Society of New Zealand* 25:1-22.

McGlone, M.S., Mildenhall, D.C. and Pole, M.S. 1996. History and palaeoecology of New Zealand *Nothofagus* forests. In: Veblen, T.T., Hill, R.S. and Read, J. (eds), *The Ecology and Biogeography of Nothofagus forest*, pp. 83-130. New Haven: Yale University Press.

McGlone, M.S., Turney, C.S.M. and Wilmshurst, J.M. 2004. Late-glacial and Holocene vegetation and climatic history of the Cass Basin, central South Island, New Zealand. *Quaternary Research* 62:267-279.

McGlone, M.S., Turney, C.S.M., Wilmshurst, J.M., Renwick, J. and Pahnke, K. 2010. Divergent land and ocean temperature trends in the Southern Ocean over the past 18,000 years. *Nature Geoscience* 3:622-626.

McKenzie, G.M. 1997. The late Quaternary vegetation history of the south-central highlands of Victoria, Australia. 1. Sites above 900 m. *Australian Journal of Ecology* 22:19-36.

McKenzie, G.M. 2002. The late Quaternary vegetation history of the south-central highlands of Victoria, Australia. II. Sites below 900 m. *Austral Ecology* 27:32-54.

Moar, N.T. 1970. Recent pollen spectra from three localities in the South Island, New Zealand *New Zealand Journal of Botany* 8:210-221.

Moar, N.T. 1971. Contributions to the Quaternary history of the New Zealand flora. 6. Aranuian pollen diagrams from Canterbury, Nelson, and north Westland, South Island. *New Zealand Journal of Botany* 9:80-145.

Newnham, R.M., Vandergoes, M.J., Hendy, C.H., Lowe, D.J. and Preusser, F. 2007. A terrestrial palynological record for the last two glacial cycles from southwestern New Zealand. *Quaternary Science Reviews* 26:517-535.

Norton, D.A. 1983. Population dynamics of subalpine *Libocedrus bidwillii* forests in the Cropp River Valley, Westland, New Zealand. *New Zealand Journal of Botany* 21:127-134.

Pocknall, D.T. 1980. Modern pollen rain and Aranuian vegetation from Lady Lake, North Westland, New Zealand. *New Zealand Journal of Botany* 18:275-284.

Pocknall, D.T. 1982. Modern pollen spectra from mountain localities, South Island, New Zealand. *New Zealand Journal of Botany* 20:361-371.

Schaefer, J.M, Denton, G.H., Kaplan, M., Putnam, A., Finkel, R.C., Barrell, D.J.A., Andersen, B.G., Schwartz, R., Mackintosh, A., Chinn, T. and Schluchter, C. 2009. High-Frequency Holocene Glacier Fluctuations in New Zealand Differ from the Northern Signature. *Science* 324:622-625.

Shulmeister, J., McLea, W.L., Singer, C., McKay, R.M. and Hosie, C. 2003. Late Quaternary pollen records from the Lower Cobb Valley and adjacent areas, North-West Nelson, New Zealand. *New Zealand Journal of Botany* 41:503-533.

Stevens, G.C. and Fox, J.F. 1991. The causes of tree line. *Annual Review of Ecology and Systematics* 22:177-191.

Stuiver, M.P. and Reimer, P.J. 1993. Extended 14C data base and revised CALIB 3.0 14C age calibration program. *Radiocarbon* 35:215-230.

Tonkin, P.J. and Basher, L.R. 2001. Soil chronosequences in subalpine superhumid Cropp Basin, western Southern Alps, New Zealand. *New Zealand Journal of Geology and Geophysics* 44:37-45.

Wardle, P. 1985a. New Zealand timberlines. 3. A synthesis. *New Zealand Journal of Botany* 23:263-271.

Wardle, P. 1985b. New Zealand timberlines. 1. Growth and survival of native and introduced tree species in the Craigieburn Range, Canterbury. *New Zealand Journal of Botany* 23:219-234.

Wardle, P. 1988. Effects of glacial climates on floristic distribution in New Zealand. 1. A review of the evidence. *New Zealand Journal of Botany* 26:541-555.

Wardle, P. 1991. *Vegetation of New Zealand*. Cambridge: Cambridge University Press.

Wardle, P. 2008. New Zealand forest to alpine transitions in global context. *Arctic Antarctic and Alpine Research* 40:240-249.

Wardle, P. and Coleman, M.C. 1992. Evidence for rising upper limits of 4 native New Zealand forest trees. *New Zealand Journal of Botany* 30:303-314.

Wilmshurst, J.M., McGlone, M.S., Leathwick, J.R. and Newnham, R.M. 2007. A pre-deforestation pollen-climate calibration model for New Zealand and quantitative temperature reconstructions for the past 18 000 years BP. *Journal of Quaternary Science* 22:535-547.

17

Vegetation and water quality responses to Holocene climate variability in Lake Purrumbete, western Victoria

John Tibby
Geography, Environment and Population, University of Adelaide, Adelaide, South Australia
john.tibby@adelaide.edu.au

Dan Penny
The University of Sydney, Sydney, NSW

Paul Leahy
Environment Protection Authority Victoria, Macleod, Victoria

A. Peter Kershaw
Monash University, Clayton, Victoria

Introduction

Palaeoenvironmental research can provide useful perspectives about the vulnerability and resilience of ecosystems to future climate change by documenting ecosystem response to past natural and human-induced environmental change (e.g. Dearing 2008, 2011). Such information is important since instrumental records for all but a few localities are non-existent, or are temporally short relative to ecological timescales (Smol 2008) and because ecosystem changes are, or will soon be, beyond the magnitude of anything witnessed in the historical period (Hansen 2005).

Between 1997 and 2010, lakes in western Victoria responded to prolonged hydrological deficit in a range of ways. Many dried out, while the vast majority underwent substantial changes in water quality and ecology (Leahy et al. 2010). Recent research has shown that a number of lakes once thought to be permanent landscape features are now subject to drying, while many other lakes are projected to be seasonally dry by the end of the 21st century (Leahy et al. 2010). The present water quality and ecology of Western District Ramsar-listed lakes (lakes Beeac, Bookar, Colongulac, Corangamite, Cundare, Gnarput, Milangil, Murdeduke and Terangpom) bears little resemblance to their conditions when listed under the Ramsar agreement in 1982, and most have been at least seasonally dry over the past decade. Given that there has been a very substantial reduction in the number of western Victorian lakes, including those recognised as nationally and internationally significant, ecosystem preservation in the remaining lakes is paramount. Of the numerous lakes in western Victoria, a small number of deeper sites (>5 m deep) will persist throughout the 21st century, even with marked hydrological deficit (Kirono et al. 2009). Given that deep lakes will be important landscape refugia over this century, prudent management is important. Although the limnology of western Victorian lakes is relatively well studied (e.g. Timms 1976; De Deckker 1983; De Deckker and Williams 1988; Tibby and Tiller 2007), there is still considerable uncertainty about their likely future behaviour. With the aim of enhancing understanding of its possible future behaviour, this study focuses on the environmental history of one such lake: Lake Purrumbete, the deepest natural freshwater lake in Victoria.

Previous palaeoenvironmental studies at Lake Purrumbete have been undertaken by De Deckker (1982) and Yezdani (1970). However, these studies were based on records that were undated and also, in the case of Yezdani (1970), short. Yezdani (1970) briefly described the pollen and algae (including diatoms) at a coarse resolution from a core collected near the edge of the lake. All the samples were post-European, as indicated by the presence of *Pinus* pollen throughout the record. De Deckker's (1982) macrofossil record for Purrumbete was not radiocarbon dated, however the 580 cm record was suggested to cover about 6000-7000 years, an inference largely consistent with the results herein. De Deckker (1982) inferred a constantly fresh lake with water depth >35 m throughout the record. However, the proxies used by De Deckker did not allow fine-scale estimation of palaeoclimatic changes.

Lake Purrumbete has one of the longest records of continuous water-quality monitoring in Victoria (Mitchell and Collins 1987; Tibby and Tiller 2007). These data are important to the management of this and other lake systems. In particular, such data can be used to predict future responses to climate change through coupling with dynamic climate modelling such as the Kirono et al.'s (In press) modelling of salinity response to predicted climate in the western Victorian lakes Bullenmerri, Gnotuk and Keilambete. Although valuable, such exercises are limited because modelling of climate-driven biological responses is considerably more difficult than predicting physical responses (such as changes in lake depth, salinity or stratification). Climate projections for southwestern Victoria consistently predict increased temperature, while rainfall estimates are somewhat more variable (i.e. while most model predictions infer future declines in rainfall, some suggest there may be rainfall increases, CSIRO and BOM 2007). However, the combined effect of temperature increases and changes in rainfall is likely to result in reductions in effective precipitation (Kirono et al. In press). Combining future climate scenarios with the observed behaviour of lakes makes it possible to hypothesise about the response of lakes to future warming, although it is axiomatic that such models are calibrated over short time periods relative to often lengthy ecosystem responses to climate change.

Building on the earlier research of Mitchell and Collins (1987), Tibby and Tiller (2007) analysed the relationship between Lake Purrumbete water quality and climate for the period 1984-2000. In line with expectations, they demonstrated that lake salinity (measured as electrical conductivity) increased in response to reduced effective moisture and that water temperature was strongly correlated with air temperature. In addition, they also showed that there was a strong negative relationship between air temperature and nutrient concentrations (specifically total phosphorus) in the water column (r^2 = 0.61, p < 0.005, n = 165). Tibby and Tiller (2007) suggest that periods of increased temperature lead to lengthened periods of water column stratification. This, in turn, results in nutrients being depleted from the epiliminion through the uptake of phosphorus by algae, which eventually sink, sequestering nutrients to the sediments.

Based on a combination of these observations about future warming and Lake Purrumbete's relationship to measured climate, it appears that future change may be expected to increase water temperature and salinity, while decreasing average water-column nutrient concentrations. In order to assess this scenario, therefore, we utilised a long-term record of environmental change preserved in Lake Purrumbete sediments to examine the nature of the Lake's response to (past) climate change, with a view to more fully understanding the possible nature of future change. This approach arose from observations that the diatom records from this and other western Victorian lakes including Lake Surprise (Tibby et al. 2006) and Tower Hill Lake (D'Costa et al. 1989) exhibit very marked shifts during the Holocene, suggesting lacustrine conditions that are very different to those recorded in even relatively long monitoring time series.

Our analysis focuses predominantly on the pollen and diatom record derived from a 6 m Mackereth core extracted from Lake Purrumbete in the late 1990s. In addition, pollen data are derived from an associated frozen spade core, representing an undisturbed record of the most recently deposited sediments. We utilise the record of precipitation-evaporation ratio (P:E) derived from nearby Lakes Keilambete, Gnotuk and Bullenmerri (Jones et al. 1998, see Figure 1 for site locations) as a means of interpreting the response of the aquatic and terrestrial ecosystems in Lake Purrumbete and its surrounds to climate variability. Although there are differing interpretations of the precise timing of changes in these records (particularly from the most intensively studied Lake Keilambete), there is a coherent record of effective precipitation inferred by a number of authors using a variety of proxies including sediment grain size and composition, ostracod composition and shell chemistry, and pollen (Dodson 1974; Bowler 1981; De Deckker 1982; Chivas et al. 1985; Jones et al. 1998), which, importantly, have been observed in other lakes in the region (Gell 1998; Tibby et al. 2006). While it is not possible to de-couple the separate effects of precipitation and evaporation in Jones et al.'s (1998) record, it nevertheless provides a quantitative estimate of effective moisture through the Holocene. Hence, we use the P:E record as a reference point to examine how water quality and vegetation in and around Lake Purrumbete responded to arid and humid phases during the Holocene. Importantly, the inferred effective moisture history from western Victoria can be compared with the inferred effective moisture history predicted for the region (Kirono et al. 2009), allowing future likely changes to be placed in context.

Study site

Lake Purrumbete is a large, fresh, clear-water, alkaline, eutrophic maar crater (see Table 1 for summary water-quality information). It has a maximum breadth of more than 2.8 km, a maximum depth of 45 m, a surface area of more than 5.5 km^2 and a volume of 157 x 106 m^3 (Timms 1976). Vegetation surrounding the lake, as for the region in general, is heavily modified by recent land use. Remnant native vegetation can be broadly classified as grasslands or open grassy woodlands (Kershaw et al. 2004), with strong edaphic controls on vegetation apparent, particularly with respect to the distribution of soils weathered from basalt (D'Costa et al. 1989).

Ecological studies of Lake Purrumbete began in the late 1960s (Hussainy 1969; Yezdani 1970; Timms 1976). Both Hussainy (1969) and Yezdani (1970) noted the presence of spring blooms of *Melosira granulata* (=*Aulacoseira granulata*) over a period of three years. Gasse et al. (1997) showed that the dominant planktonic diatoms in the centre of Lake Purrumbete sampled in sediment traps, and to a lesser extent phytoplankton sweeps, were *Cyclotella meneghiniana*

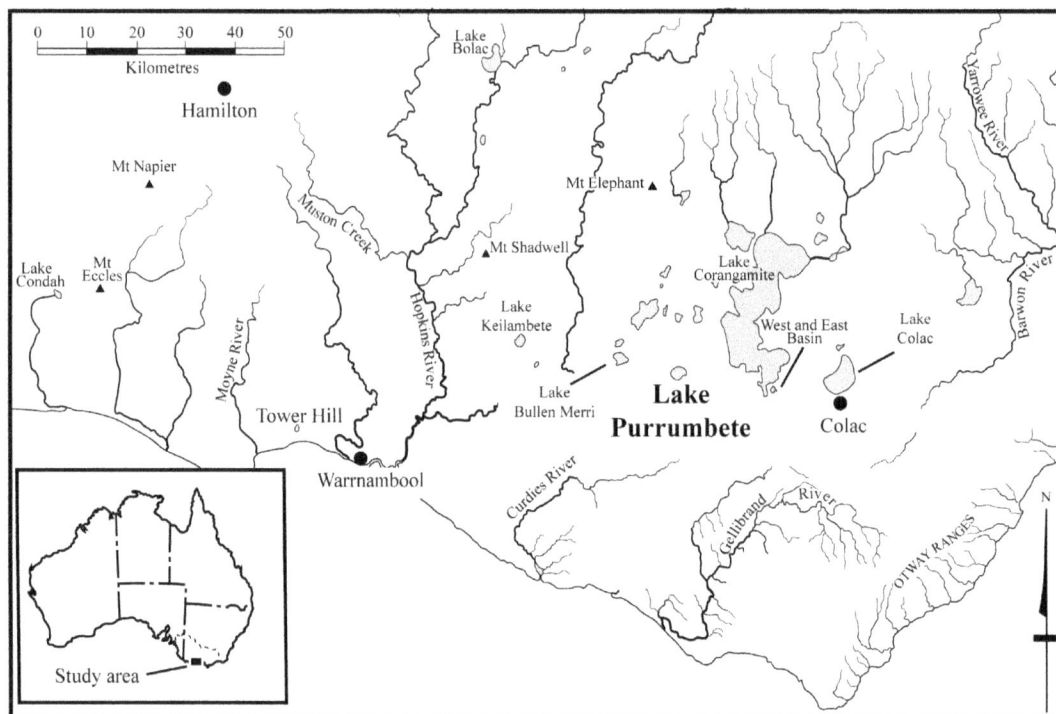

Figure 1. Location of Lake Purrumbete and other sites referred to in the text.

Table 1. Summary of water quality data from Lake Purrumbete 1984-2000. See Tibby and Tiller (2007) for details of methods.

Variable	Unit	Average	*n* determinations
Water temperature	°C	15.2	179
Electrical conductivity	µS cm^{-1}	739	180
pH	pH Units	8.6	175
Total nitrogen	mg l^{-1}	0.792	178
Total phosphorus	mg l^{-1}	0.102	178
Orthophosphate (PO$_4$)	mg l^{-1}	0.083	176
Dissolved oxygen	mg l^{-1}	9.749	180
Turbidity	NTU	1.155	148
Secchi depth	metres	4.6	81

and *Aulacoseira granulata*. Interestingly, they also noted that, in summer phytoplankton sweeps, the epiphytic diatom *Cocconeis placentula* is the most abundant diatom, despite the substantial distance of their study site from the shoreline.

Despite the relative longevity of ecological study at Lake Purrumbete, palaeoecology can provide additional information to facilitate better lake management. For example, aquatic macrophyte surveys in the late 1960s (Yezdani 1970) did not detect the current dominant macrophyte in the lake *Vallisneria americana* var. *americana* (EPA Victoria, unpub. data). In the late 1960s, *Myriophyllum* spp. were dominant (Yezdani 1970). Mitchell and Collins (1987) detected *Vallisneria gigantea* (= *Vallisneria americana* var. *americana*) in Lake Purrumbete by 1984. From the perspective of lake managers, it is not clear whether the current dominance by *Vallisneria* is related to recent shifts in climate, the lake nutrient status, or part of natural variability.

Methods

Sediment coring was undertaken from an anchored floating platform close the centre of the lake. A near 6 m long core, 5 cm in diameter, was collected with a Mackereth sampler

(Mackereth 1958). As the corer tends to disturb or fail to collect very unconsolidated surface sediments, material from the topmost 1 m was retrieved intact on a frozen spade (Neale and Walker 1996).

Samples of 1 cm³ were taken at 10 cm intervals (n=59) from the Mackereth core and prepared for diatom analysis using a modification of Battarbee (1986), where three-hour treatments in 10% HCl and 10% H_2O_2 were used to remove carbonate and organic matter, respectively. Following these treatments, samples were rinsed repeatedly in distilled water, and they were then mounted using Naphrax mounting medium. Diatoms were identified using 1000x magnification with a Zeiss Axioskop fitted with differential interference contrast, with Krammer and Lange-Bertalot (1986, 1988, 1991a, b) the main source of taxonomic information. A minimum of 300 valves were counted per sample, with diatom abundances calculated using the microspheres method described in Battarbee (1986).

Separation cells of *Aulacoseira granulata* were identified, as Gomez et al. (1995) have shown that *A. granulata* forms a larger number of separation cells during periods of stratification, so that its filament (or "chain") is shorter, rendering it easier for such chains to remain suspended in poorly mixed water columns.

Samples of 2 cm³ of wet sediment were extracted from the core and the frozen spade for pollen analysis, with sampling resolution increasing from a maximum of 0.4 m below 3.0 m depth to a minimum of 0.04 m in the frozen-spade sample, which represents the upper 0.6 m of the pollen record (n=37). The Mackereth core and the frozen-spade sample were correlated using the first appearance of introduced *Pinus* pollen in both records as a biostratigraphic marker. The samples were prepared following the method of van der Kaars et al. (2000) and pollen grains counted at 400x magnification with an Olympus CH-2 microscope. Pollen counts ranged between 133 and 406, with an average of 259 palynomorphs per sample. Taxonomy was based primarily on an extensive modern reference set in the School of Geography and Environmental Science, Monash University. Pollen and spore counts are expressed as abundance relative to a standardised dryland pollen sum for southeastern Australia (D'Costa and Kershaw 1997). *Banksia* was excluded from the standard pollen sum in order to avoid distortion in the data set due to extreme over-representation of this taxon at 0.4 m depth in the core. Pollen and diatom relative abundance data (excluding *Banksia* and taxa <1% respectively) were classified into zones using a stratigraphically constrained cluster analysis with Euclidean distance as the similarity metric (Grimm 1987). Pollen and diatom zone boundaries were established at arbitrary thresholds of 0.6 and 3.0 respectively.

Four pollen concentrates (prepared following Regnell 1992) were submitted for accelerator mass spectrometry radiocarbon dating (Fink et al. 2004). Radiocarbon ages (BP) were calibrated (cal BP) using the INTCAL09 dataset (Reimer et al. 2009) in the programme CALIB 6.0 (Stuiver and Reimer 1986). The resulting calibrated ages (Table 2) have a near-linear relationship with depth (r^2=0.992), and a simple linear regression was used as a basis for the chronological model used in Figures 2 and 3.

Results

Diatom and pollen results are presented in Figures 2 and 3, respectively. Diatom zone 4 is characterised by the highest relative abundances of non-planktonic taxa in the record, although the planktonic *Aulacoseira granulata* is the species with the highest individual relative abundance. Of the non-planktonic diatoms, the major taxa are *Fragilaria* aff. *zeilleri*, *Pseudostaurosira brevistriata*, *Staurosirella pinnata* and *Staurosira elliptica*. The relative abundance of non-planktonic diatoms, both in total and individually, is lower in Zone 3. Zone 2 is dominated by *Aulacoseira granulata* and the concentration of diatoms is lowest in this zone. In Zone 1,

Table 2. Results of AMS radiocarbon analysis on pollen concentrates. ¹⁴C ages are calibrated using the program CALIB 6.0 (Stuiver and Reimer 1986) with the INTCAL 09 dataset (Reimer et al. 2009). Ages are reported at 1σ (68.3%) and 2σ (95.4%) confidence, and the relative area of the probability distribution for each intercept of the calibration curve is given in square brackets.

Depth (cm)	Lab code	¹⁴C age BP	% modern C	Cal BP (1σ)	Cal BP (2σ)
64-65	OZG077	450 ± 35	94.55 ± 0.37	494-526 [1]	340-347 [0.013] 459-540 [0.987]
275-278	OZG076	3660 ± 40	63.38 ± 0.31	3903-3991 [0.771] 4041-4072 [0.229]	3865-4088 [1]
405-408	OZG075	4970 ± 40	53.84 ± 0.26	5652-5746 [1]	5606-5758 [0.849] 5822-5885 [0.151]
575-577	OZG074	7160 ± 50	41.03 ± 0.24	7880-7887 [0.061] 7932-8003 [0.939]	7848-8026 [1]

Discostella stelligera is the dominant diatom in every sample, although its abundance is lower in the upper four samples. The concentration of diatoms is highest in Zone 1.

The pollen record has been divided into three zones based on stratigraphically constrained cluster analysis. Zone 3 (580-170 cm depth; ca. 7600-1700 cal BP) is consistently dominated by Poaceae pollen, averaging 40% of the pollen sum. Common sclerophyll woodland taxa (*Eucalyptus*, Casuarinaceae) are strongly represented with some variability but no clear trend. Cyperaceae and *Myriophyllum* are the most commonly recorded aquatic pollen types in this zone, but their abundance is low and highly variable. The common dryland pollen types are stable throughout Zone 2 (160-36 cm; ca. 1500-150 cal BP), with a slight decline in the relative abundance of Causarinaceae pollen as depth decreases. The very high values for *Banskia* recorded at 40 cm depth are likely a result of over-representation, perhaps due to the interment of flower parts in the sediment, and are not taken as indicative of a change in catchment vegetation. Cyperaceae pollen becomes more abundant, with some variability, through Zone 2, reaching a maximum at 44 cm depth (190 cal BP). Zone 1 of the pollen record (32-0 cm depth; 130 cal BP-present) is characterised by a dramatic increase in the relative abundance of Poaceae pollen (to an average of 83% of the pollen sum), and concomitant decreases in all other dryland pollen taxa in the sum. Pollen from exotic plants (*Pinus*, *Plantago lanceolata*, Cupressaceae) appears for the first time in the record at 40 cm depth (ca. 170 cal BP) and increases dramatically to a maximum relative abundance at 16 cm depth (ca. 50 cal BP). *Myriophyllum* pollen increases at the Zone 2/Zone 1 boundary, to reach a maximum value in Zone 1 at 16 cm depth (ca. 50 cal BP), the strongest representation of that taxon for the entire record. This pattern is not apparent in any of the other aquatic plants in the record (*Triglochin*, *Ruppia*, *Potamogeton*). Cyperaceae, most probably occupying a narrow littoral margin of the steep-sided crater, is poorly and irregularly represented in Zone 1.

Discussion

Diatom and inferred water-quality response to Holocene climate variability

There have been substantial and dramatic changes to the Lake Purrumbete diatom community since 8000 cal BP, with some species completely absent from the early or late part of the record and rapid shifts in the abundance of the dominant and sub-dominant taxa. In Zone 4, from approximately 8000 to 5500 cal BP, non-planktonic diatoms, particularly taxa in the Fragiliariaceae, are more numerous in total than planktonic species, although the planktonic diatom *Aulacoseira granulata* is the single most abundant species. *Aulacoseira granulata*, although variable, declines through this period. Of the non-planktonic species, *Fragilaria* aff. *zeilleri* is the most abundant. However, the taxonomic and therefore the ecological affinity of this taxon is uncertain, apart from the likelihood that, similar to other chain-forming Fragiliariaceae

(Bennion et al. 2001; Sayer 2001), it is not planktonic. The second most abundant non-planktonic diatom is *Pseudostaurosira brevistriata*, which peaks in the middle part of this zone. *Pseudostaurosira brevistriata*, *Staurosirella pinnata* and *Staurosira elliptica* are abundant in shallow lakes (<5 m deep) in Australia (e.g. Gell et al. 2002; Tibby et al. 2007) and elsewhere (Bennion et al. 2001; Sayer 2001).

Given that Lake Purrumbete is currently >40 m deep and that De Deckker (1982) suggests that there is little evidence for marked lake level changes in the lake, the high relative abundance of non-planktonic diatoms 8000-5500 cal BP is intriguing, since it would tend to indicate, *a priori*, a lake depth considerably shallower than 40 m. This is particularly the case since other palaeoenvironmental records from the region (e.g. Bowler 1981; D'Costa et al. 1989) and Jones et al.'s (1998) P:E record indicate that this period was one of maximum water availability (with P:E up to 1.1 during this period). The abundance of *Aulacoseira granulata* in combination with the Fragilariaceae provides a possible insight into this conundrum since *Aulacoseira granulata* is a diatom that requires turbulent mixing to remain suspended in the water column (Bormans and Webster 1999). Hence, it is likely that the high relative abundance of this colonial taxon, as opposed to solitary planktonic species such as *Discostella stelligera* and *Cyclostephanos dubius*, represents times when wind-generated mixing is elevated (see Wang et al. 2008). Thus, it is possible that the relatively high representation of the Fragilariaceae may result from these taxa being transported to the central lake environment by the same mixing that advantaged *A. granulata*. Notably, Gasse et al. (1997) report that a large proportion of diatoms they inferred to be derived from the littoral zone were found in centre-lake sediment traps from Lake Purrumbete.

From approximately 5500 cal BP to 4000 cal BP (Zone 3), the dominant diatom is the planktonic species *Cyclostephanos dubius*, which generally increases in relative abundance through this zone (Figure 2). This taxon is abundant during a period of decreased effective moisture where the precipitation:evaporation ratio was between 0.89 and 0.94 (Jones et al. 1998) (see Figure 2). Indeed, the dominance of *Cyclostephanos dubius* commences concurrently with an inferred step change in regional moisture at 5500 cal BP (Jones et al. 1998). *Cyclostephanos dubius* is commonly found in lakes with elevated nutrient concentrations (Bradshaw and Anderson 2003). In northwest European lakes, it has a total phosphorus optimum of 176 µg l⁻¹ (Bradshaw and Anderson 2003), while Tibby (2004) derived an optimum of 76 µg TP l⁻¹ for *Cyclostephanos* aff. *dubius* (maximum relative abundance < 9%), which may be closely related to *C. dubius*, in southeast Australian water storages. *C. dubius* has not been recorded in modern Australian lake sediments at abundances as great as in the Lake Purrumbete record, including in western-Victorian diatom calibration data sets (Gell 1997; Barr 2010).

Given Tibby and Tiller's (2007) observation that water-column nutrient concentrations decrease in Lake Purrumbete as a result of persistent stratification during extended warming, it might be expected that the decreased effective moisture experienced from 5500 cal BP to 3500 cal BP would be associated with *decreased*, rather than increased, nutrient concentrations. Despite this, the increasing abundance of *C. dubius* in the mid Holocene mirrors the decreased moisture witnessed in Lakes Keilambete, Bullenmerri and Gnotuk. As a result, while it could be expected that nutrients would be removed from the water column as a result of longer periods of stratification associated with higher temperatures, and/or less nutrients being delivered to the lake via rainfall, epilimnetic nutrient concentrations in Lake Purrumbete remained high through this period.

There was a relatively rapid turnover of species between 4000 cal BP and 3500 cal BP, with *Cyclostephanos dubius* giving way to *Discostella stelligera* after a short (approximately 500-year) phase of *Aulacoseira granulata* dominance. The sustained abundance of *Aulacoseira granulata*, with its requirement for water column turbulence, likely indicates a period of greater wind

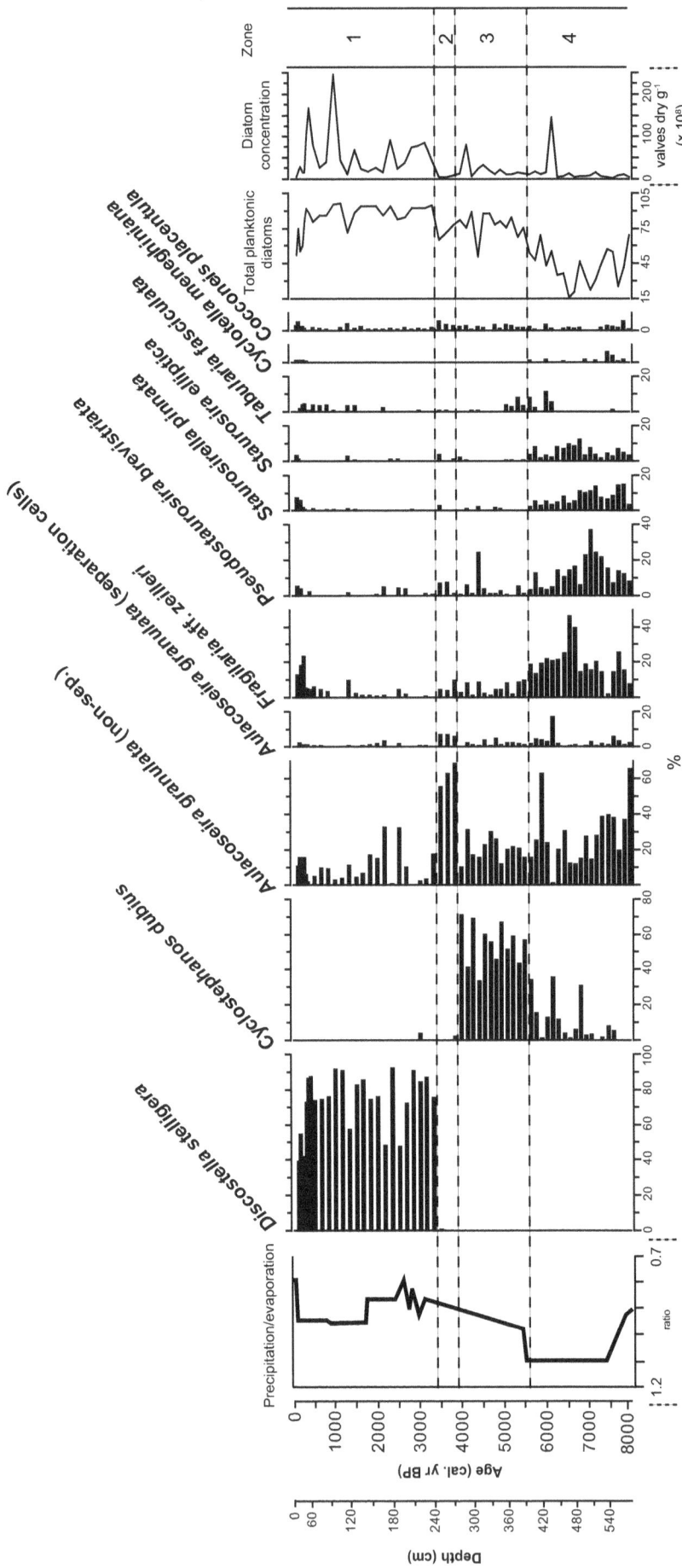

Figure 2. Diatom stratigraphy, total proportion of planktonic taxa and diatom concentration from the Lake Purrumbete Mackereth core. Only taxa with a relative abundance >5% in at least one sample were included. The inferred precipitation/evaporation (P/E) ratio derived by Jones et al. (1998) from western Victorian palaeoenvironmental data is also shown. The P/E ratio illustrated is the lower estimate of the range illustrated in Jones et al. (1998) non-sep. = the non-separation cells.

generated mixing during this time.

For approximately the past 3500 years, the diatom community in Lake Purrumbete has been dominated by *Discostella stelligera*, a freshwater planktonic species that, in high relative abundances such as observed in the Lake Purrumbete record, is indicative of oligotrophic waters (see Tibby 2004 and references therein). *Discostella stelligera*, which has a total phosphorus optimum of 16 µg l⁻¹, dominates the record at a time of substantial climatic variability and persists through relatively arid and relatively wet phases during which, for example, trees grew at the edges of lakes Keilambete and Bullenmerri and were subsequently drowned approximately 2000 years ago (Bowler 1981). Indeed, the period from 3100 cal BP to 2000 cal BP is one of substantial climate variability in western Victoria, with two >5 m oscillations in the level of Lake Keilambete recorded during this time and with the inferred P/E of 0.78 at 2500 cal BP the lowest recorded since the early Holocene (Jones et al. 1998). However, apart from this period being largely coincidental with the onset of the *Discostella stelligera* dominance in the Lake Purrumbete record, there is little to differentiate it from the period of sustained higher moisture availability between 2000 cal BP and 110 cal BP where P/E reached 0.95 (see Figure 2). Arguably, *Cyclotella stelligera* may have been advantaged relative to *Cyclostephanos dubius* from 3100 cal BP to 2000 cal BP because the conditions which resulted in increased aridity (e.g. a reduction in winter storm tracks delivering moisture to the region) advantaged this smaller, less silicified taxon in a more stratified lake environment. The persistence of this taxon following this period (i.e. post 2000 cal BP), however, is less explicable, but it appears to highlight a degree of resistance to substantial climate variability.

Vegetation response to Holocene climate variability and European impact

In contrast to the dramatic changes apparent in the diatom record, the overriding characteristic of the Lake Purrumbete pollen record is one of stability though the greater part of the Holocene. All of the common dryland pollen types maintain, with some variability, their values over time. The Casuarinaceae are perhaps the only exception, with a decrease in relative abundance from around 4000 cal BP. The apparent stability of *Eucalyptus* and Chenopodiaceae pollen throughout this period does not appear to support soil salinisation as a likely cause of the decline in Casuarinaceae in our record (*sensu* Crowley 1994) and it is not clear whether the data presented here are in fact part of the broader decline in Casuarinaceae observed at a number of sites in the western Victorian Basaltic Province (see Kershaw et al. 2004). The muted vegetation response to the climatic changes known to have occurred in the area, and particularly to the relatively arid and variable climatic conditions centred around 3000-4000 BP (Bowler and Hamada, 1971; Bowler 1981; Chivas et al. 1985, 1986; Figure 3) which are shown here to have had dramatic implications for the freshwater ecology of Lake Purrumbete, is remarkable, and has been noted elsewhere (Dodson 1974, p. 716-717, 2001, Dodson et al. 2004). We conclude that the amplitude and/or duration of climatic variability during the Holocene did not exceed the resilience (*sensu* Holling 1973) of the sclerophyll woodland flora.

As with other vegetation records from western Victoria, the impact of non-indigenous land-use practices since the early part of the 19th century (from 32 cm depth in our record) is dramatic, reflecting the expansion of grasslands at the expense of woodland and forest, and the introduction of exotic plants. It is unclear what processes are driving the increase in *Myriophyllum* pollen during this period of European settlement. Gell et al. (1993), citing Orchard's (1985) seminal work on the genus, interpreted similar recent increases in *Myriophyllum* pollen as evidence of mass-flowering following stranding of *Myriophyllum* beds. Arguably, similar patterns at other lakes from the region (Tower Hill, main lake, zone ML1; D'Costa et al. 1989; Cobrico Crater, central core; Dodson et al. 2004) imply that recent changes in the aquatic flora are perhaps indicative of some regional phenomenon, rather than site-specific variability in

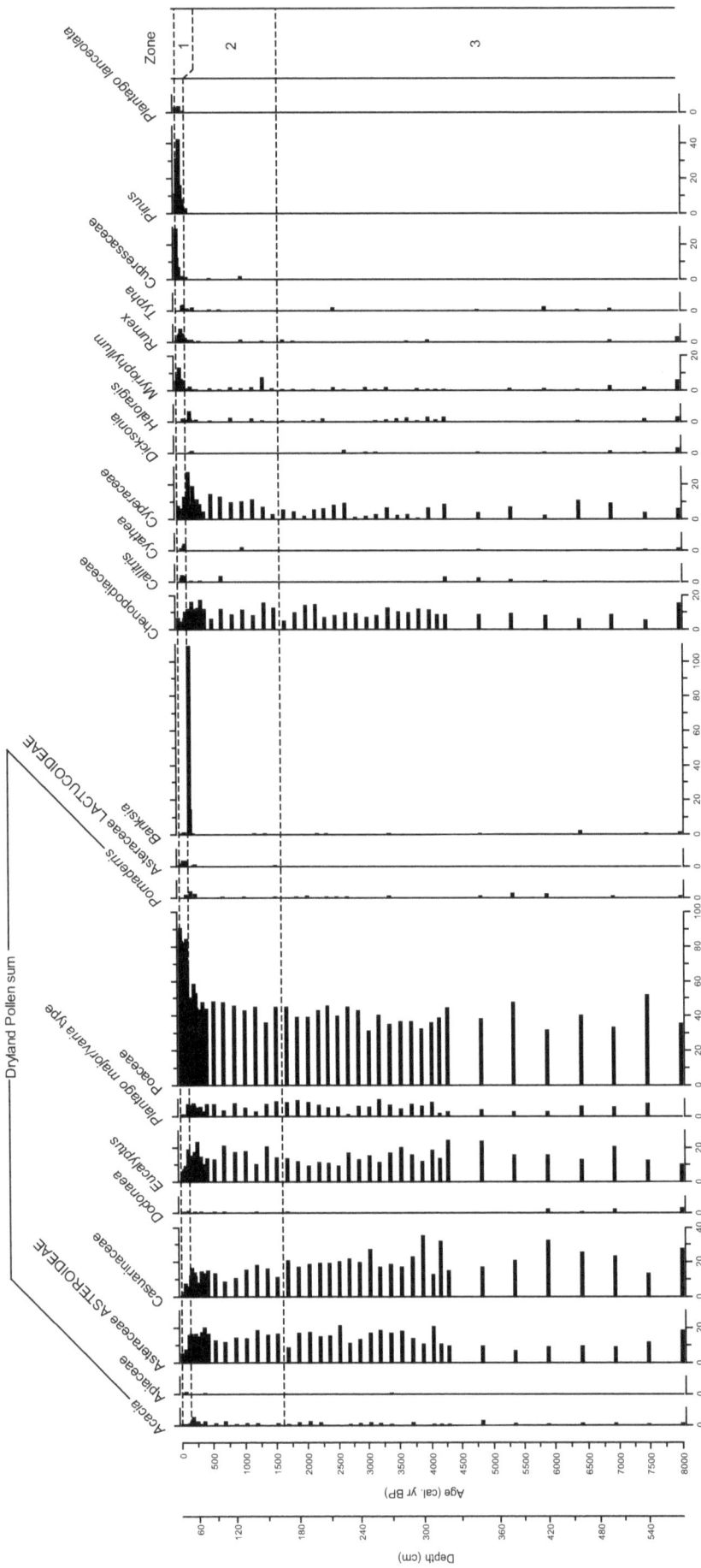

Figure 3. Pollen stratigraphy from the Lake Purrumbete core, showing variation in selected taxa against time. All taxa expressed as % relative abundance (see text). The upper 60 cm of the record is taken from a frozen spade core, and correlated to the longer Mackereth core using the first appearance of Pinus pollen as a biostratigraphic marker.

water levels.

Implications for understanding lake response to future climate change

In many Western District lakes, salinity concentrations are highly sensitive to climate (Kirono et al. 2009; Leahy et al. 2010) and recent climate-driven salinity increases have led to losses in biodiversity (Leahy et al. 2010). Future climate changes will further increase the loss of diversity in many western district Lakes. Although salinity in Lake Purrumbete is related to climate (Tibby and Tiller 2007, Yihdego 2010), modelled salinity increases over the coming decades are predicted to be moderate (Yihdego 2010) and not likely to result in large losses of diversity. Given the relative resilience of Lake Purrumbete to future climate driven salinity increases, factors such as macrophyte abundance and algal dynamics, which themselves are mediated by nutrient concentrations, are much more likely to be future drivers of diversity in Lake Purrumbete.

The Lake Purrumbete diatom record indicates that water quality, in particular nutrient status, can exhibit both marked sensitivity and apparent resilience to climate variation. The former is most amply demonstrated by the transition from dominance of a species associated with high nutrient concentrations, *Cyclostephanos dubius*, to *Discostella stelligera*, a taxon with markedly lower nutrient status. By contrast, resistance to environmental change is seen by the continued dominance of *Discostella stelligera* through lengthy periods of contrasting late-Holocene climate. Similarly, the palaeoenvironmental record shows that macrophytes like *Myriophyllum* have exhibited resilience over thousands of years, and then a late-Holocene rapid rise in abundance. This then provides lake managers with evidence that the rapid changes in *Myriophyllum* observed in the historical period have not occurred previously in the Holocene. In combination, these data suggest that while lakes such as Lake Purrumbete may 'resist' a degree of climate variability, when shifts do occur they are likely to be more abrupt, sustained and severe than can be predicted from even lengthy monitoring. From a lake-management perspective, climate-driven changes in mixing regime and nutrient cycling may propagate large and sudden changes in lake ecology. Hence, lake managers should give priority to reducing diffuse sources of nutrients to the lake to minimise the risks associated with changes in nutrient status.

Conclusion

The diatom record from Lake Purrumbete indicates that large lake systems can undergo rapid and sustained shifts in their water quality and ecology in response to climate change, even when lake levels show only minor alterations to climate perturbation (De Deckker 1982). Moreover, the limnological sensitivity of Lake Purrumbete to climatic variability is shown to be much greater than the catchment vegetation. It is clear that the limnology of Lake Purrumbete during the early to mid-Holocene was notably different to that of the late Holocene. Indeed, as with other lakes in the region, it appears that water-column nutrient concentrations in this lake were elevated, relative to the time of settlement, from 8000 to approximately 3500 years ago. As a result, the record of past environmental change in lakes such as this can provide an otherwise unobtainable insight into the behaviour of these systems.

Acknowledgements

JT, DP and PL thank Peter Kershaw for guidance and inspiration over many years. We thank Patrick De Deckker and Cameron Barr for informative discussions. The comments of two reviewers are also appreciated. This research was supported by AINSE grant 02/098P. The late Gary Swinton drew Figure 1.

References

Barr, C. 2010. Droughts and flooding rains: a fine-resolution reconstruction of climatic variability in western Victoria, Australia, over the last 1500 years. Unpublished PhD thesis, University of Adelaide.

Battarbee, R.W. 1986. Diatom Analysis. In: Berglund, B.E. (ed), *Handbook of Holocene palaeoecology and palaeohydrology*, pp. 527-570. John Wiley and Sons, Chicester.

Bennion, H., Appleby, P.G. and Phillips, G.L. 2001. Reconstructing nutrient histories in the Norfolk Broads, UK: implications for the role of diatom-total phosphorus transfer functions in shallow lake management. *Journal of Paleolimnology* 26:181-204.

Bormans, M. and Webster, I.T. 1999. Modelling the spatial and temporal variability of diatoms in the River Murray. *Journal of Plankton Research* 21:581-598.

Bowler, J.M. 1981. Australian salt lakes. A palaeohydrological approach. *Hydrobiologia* 82:431-444.

Bowler, J.M. and Hamada, T. 1971. Late Quaternary stratigraphy and radiocarbon chronology of water level fluctuations in Lake Keilambete, Victoria. *Nature* 232:330-332.

Bradshaw, E.G. and Anderson N.J. 2003. Environmental factors that control the abundance of Cyclostephanos dubius (Bacillariophyceae) in Danish lakes, from seasonal to century scale. *European Journal of Phycology* 38:265-276.

Chivas, A.R., De Deckker, P. and Shelley, J.M.G. 1985. Strontium content of ostracods indicate lacustrine palaeosalinity. *Nature* 316:251-253.

Chivas A.R., De Deckker, P., Shelley, J.M.G. 1986. Magnesium and strontium in non-marine ostracod shells as indicators of palaeosalinity and temperature. *Hydrobiologia* 143:135-142.

Commonwealth Scientific and Industrial Research Organisation (CSIRO) and Bureau of Meteorology (BOM) 2007. *Climate change in Australia*. Commonwealth Scientific and Industrial Research Organisation.

Crowley, G.M. 1994. Quaternary soil salinity events and Australian vegetation history. *Quaternary Science Reviews* 13:15-22.

D'Costa, D., Edney, P., Kershaw, A.P. and De Deckker, P. 1989. Late Quaternary palaeoecology of Tower Hill, Victoria, Australia. *Journal of Biogeography* 16:461-482.

D'Costa, D. and Kershaw, A.P. 1997. An expanded recent pollen database from south-eastern Australia and its potential for refinement of palaeoclimatic estimates. *Australian Journal of Botany* 45:583-605.

De Deckker, P. 1982. Holocene ostracods, other invertebrates and fish remains from cores of four maar lakes in southeastern Australia. *Proceedings of the Royal Society of Victoria* 94:183-219.

De Deckker, P. 1983. Australian salt lakes: their history, chemistry, and biota – a review, *Hydrobiologia* 105:231-244.

De Deckker, P. and Williams, W.D. 1988. Physico-chemical limnology of eleven, mostly saline permanent lakes in western Victoria. *Hydrobiologia* 162:275-286.

Dearing, J.A. 2008. Landscape change and resilience theory: a palaeoenvironmental assessment from Yunnan, SW China. *The Holocene* 1:117-127.

Dearing, J.A. 2011. Learning from the Past. *Global Change* 77:16-19.

Dodson, J.R. 1974. Vegetation and climatic history near Lake Keilambete, western Victoria. *Australian Journal of Botany* 22:709-717.

Dodson, J.R. 2001. Holocene vegetation change in the mediterranean-type climate regions of Australia. *The Holocene* 11: 673-680.

Dodson, J.R., Frank, K., Fromme, M., Hickson, D., McRae, V., Mooney, S. and Smith, J.D. 2004. Environmental systems and human impact at Cobrico Crater, south-western Victoria. *Australian Geographical Studies* 32:27-40.

Fink, D., Hotchkis, M., Hua, Q., Jacobsen, G., Smith, A.M., Zoppi, U., Child, D., Mifsud, C., van der Gaast, H., Williams, A. and Williams, M. 2004. The ANTARES AMS facility at ANSTO. *Nuclear Instruments and Methods in Physics Research Section B: Beam Interactions with Materials and Atoms.* 223-224:109-115.

Gasse, F., Barker, P.A., Gell, P.A., Fritz, S.C. and Chalie F. 1997. Diatom-inferred salinity of palaeolakes, an indirect tracer of climate change. *Quaternary Science Reviews* 16:547-563.

Gell P.A. 1997. The development of diatom database for inferring lake salinity, western Victoria, Australia: Towards a quantitative approach for reconstructing past climates. *Australian Journal of Botany* 45:389-423.

Gell, P.A. 1998. Quantitative reconstructions of the Holocene palaeosalinity of paired crater lakes based on a diatom transfer function. *Palaeoclimates* 3(1-3):83-96.

Gell, P.A., Sluiter, I.R. and Fluin, J. 2002. Seasonal and inter-annual variations in diatom assemblages in Murray River-connected wetlands in northwest Victoria, Australia. *Marine and Freshwater Research* 53:981-992.

Gell, P.A., Stuart, I.-M. and Smith, J.D. 1993. The response of vegetation to changing fire regimes and human activity in East Gippsland, Victoria, Australia. *The Holocene* 3:150-160.

Gomez, N., Riera, J.L. and Sabater, S. 1995. Ecology and morphological variability of *Aulacoseira granulata* (Bacillariophyceae) in Spanish reservoirs. *Journal of Plankton Research* 17(1):1-16.

Grimm, E.C. 1987. CONISS: a FORTRAN 77 program for stratigraphically constrained cluster analysis by the method of incremental sum of squares. *Computers and Geosciences* 13(1):13-35.

Hansen, J.E. 2005. A slippery slope: How much global warming constitutes "dangerous anthropogenic interference"? *Climatic Change* 68:269-279.

Holling, C.S. 1973. Resilience and the Stability of Ecological Systems. *Annual Review of Ecology and Systematics* 4:1-23.

Hussainy, S.U. 1969. *Ecological studies on some microbiota of lakes in western Victoria.* Unpublished PhD thesis. Monash University. 254 pp.

Jones, R.N., McMahon, T.A. and Bowler, J.M. 1998. A high resolution Holocene record of P/E ratio from closed lakes, western Victoria. *Palaeoclimates* 3(1-3):51-82.

Kershaw, A.P., Tibby, J. Penny, D., Yesdani, H., Walkley, R., Cook, E. and Johnston, R. 2004. Latest Pleistocene and Holocene vegetation and environmental history of the western plains of Victoria. *Proceedings, Royal Society of Victoria* 116:141-163.

Kirono, D.G.C., Kent, D.M, Jones, R.N. and Leahy, P.J. 2009. Modelling lake levels under climate change conditions: three closed lakes in Western Victoria. In: Anderssen, R.S., Braddock R.D. and Newham L.T.H. (eds), *18th World IMACS Congress and MODSIM09 International Congress on Modelling and Simulation. Modelling and Simulation Society of Australia and New Zealand and International Association for Mathematics and Computers in Simulation.* pp. 4312-4318.

Kirono, D.G.C., Kent, D.M., Jones, R.N. and Leahy, P.J. In press. Assessing climate change impacts and risks on three salt lakes in Western Victoria, Australia. *Human and Ecological Risk Assessment.*

Krammer, K. and Lange-Bertalot, H. 1986. Bacillariophyceae. 1: Teil: Naviculaceae. Gustav Fischer Verlag, Jena.

Krammer, K. and Lange-Bertalot, H. 1988. Bacillariophyceae. 2: Teil: Bacillariaceae, Epthimiaceae, Surirellaceae. Gustav Fischer Verlag, Jena.

Krammer, K. and Lange-Bertalot, H. 1991a. Bacillariophyceae. 3: Centrales, Fragilariaceae, Eunotiaceae. Gustav Fischer Verlag, Stuttgart.

Krammer, K. and Lange-Bertalot, H. 1991b. Bacillariophyceae. 4: Achnanthes, Kritische Ergänzunhen zu Navicula (Lineolatae) und Gomphonema Gesamtliteraturverzeichnis Teil 1-4. Gustav Fischer Verlag, Stuttgart.

Leahy, P.J., Robinson, D., Patten, R. and Kramer, A. 2010 *Lakes in the Western District of Victoria and Climate Change.* EPA Science Report. http://www.epa.vic.gov/publications/

Mitchell, B.D. and Collins, R. 1987. A limnological study of Lake Purrumbete. Part A. Introduction, methods and physico-chemical parameters. Report 87-3A. Centre for Aquatic Sciences, Warrnambool Institute of Advanced Education.

Mackereth, F.J.H. 1958. A portable core sampler for lake deposits. *Limnology and Oceanography* 3:181-191.

Orchard, A.E. 1985. *Myriophyllum* (Haloragaceae) in Australasia. II. The Australian Species. *Brunonia* 8:173-291.

Neale, J.L. and Walker, D. 1996. Sampling sediment under warm deep water. *Quaternary Science Reviews* 15:581-590.

Regnell, J. 1992. Preparing pollen concentrates for AMS dating – a methodological study from a hard-water lake in southern Sweden. *Boreas* 21:373-377.

Reimer, P.J., Baillie, M.G.L., Bard, E., Bayliss, A., Beck, J.W., Blackwell, P.G., Ramsey, C.B., Buck, C.E., Burr, G.S., Edwards, R.L., Friedrich, M., Grootes, P.M., Guilderson, T.P., Hajdas, I., Heaton, T.J., Hogg, A.G., Hughen, K.A., Kaiser, K.F., Kromer, B., McCormac, F.G., Manning, S.W., Reimer, R.W., Richards, D.A., Southon, J.R., Talamo, S., Turney, C.S.M., van der Plicht, J. and Weyhenmeyer, C.E. 2009. IntCal09 and Marine09 radiocarbon age calibration curves, 0-50,000 years CAL BP. *Radiocarbon* 51:1111-1150.

Sayer, C.D. 2001. Problems with the application of diatom-total phosphorus transfer functions: examples from a shallow English lake. *Freshwater Biology* 46:743-757.

Smol, J.P. 2008. *Pollution of Lakes and Rivers: A Paleoenvironmental Perspective.* Malden, MA, Blackwell.

Stuiver, M. and Reimer, P.J. 1986. A computer program for radiocarbon age calibration. *Radiocarbon* 28:1022-1030.

Tibby, J. 2004. Development of a diatom-based model for inferring total phosphorus in south-eastern Australian water storages. *Journal of Paleolimnology* 31:23-36.

Tibby, J. and Tiller, D. 2007. Climate-water quality relationships in three Western Victorian (Australia) lakes 1984-2000. *Hydrobiologia* 591:219-234.

Tibby, J., Kershaw, A.P., Builth, H., Philibert, A. and White, C. 2006. Environmental change and variability in south-western Victoria: changing constraints and opportunities for occupation and land use. In: David, B., Barker, B. and McNiven, I.J. eds), *The Social Archaeology of Australian Indigenous Societies*, pp. 254-269. Canberra, Aboriginal Studies Press.

Tibby, J., Gell, P.A., Fluin, J. and Sluiter, I. 2007. Diatom-salinity relationships in wetlands: assessing the influence of salinity variability on the development of inference models. *Hydrobiologia* 591:207-218.

Timms, B. 1976. A comparative study of the limnology of three maar lakes in western Victoria. I. Physiography and physicochemical features. *Marine and Freshwater Research* 27:35-60.

van der Kaars, S., Wang, X., Kershaw, P., Guichard, F. and Setiabudi, D.A. 2000. A late Quaternary palaeoecological record from the Banda Sea, Indonesia: patterns of vegetation, climate and biomass burning in Indonesia and northern Australia. *Palaeogeography, Palaeoclimatology, Palaeoecology* 155:135-153.

Wang, L., Lu, H., Liu, J., Gu, Z., Mingram, J., Chu, G., Li, J., Rioual, P., Negendank, J.F.W., Han, J. and Liu, T. 2008. Diatom-based inference of variations in the strength of Asian winter monsoon winds between 17,500 and 6000 calendar years B.P. *Journal Geophysical Research* 113(D21):D21101.

Yezdani, G.H. 1970. *A study of the Quaternary vegetation history in the volcanic lakes region of western Victoria.* Unpublished PhD thesis. Monash University. 570 pp.

Yihdego, Y. 2010. *Modelling of lake level and salinity for Lake Purrumbete in western Victoria.* Latrobe University unpublished report.

18

Fire on the mountain: A multi-scale, multi-proxy assessment of the resilience of cool temperate rainforest to fire in Victoria's Central Highlands

Patrick J. Baker
School of Biological Sciences, Monash University, Clayton, Victoria
patrick.baker@monash.edu

Rohan Simkin
Monash University, Clayton, Victoria

Nina Pappas
Monash University, Clayton, Victoria

Alex McLeod
Monash University, Clayton, Victoria

Merna McKenzie
Monash University, Clayton, Victoria

Introduction

A common feature of many Australian landscapes is the interdigitation of eucalypt-dominated sclerophyll forest with rainforest. In most instances, the eucalypt forests dominate the landscape, with rainforest restricted to relatively small fragments and strips that are often (but not always) associated with topographic features such as riparian zones or southeastern-facing slopes. However, these patterns reflect the current state of a dynamic system. Over several hundreds of thousands of years, the relative dominance of the rainforests and eucalypt forests has waxed and waned across these landscapes in near synchrony (Kershaw et al. 2002; Sniderman et al. 2009). During periods of relatively warm, dry conditions, the eucalypt-dominated vegetation has expanded and the rainforest contracted across the landscape. When the climate has been relatively cool and moist, the rainforests have expanded and the eucalypt forest contracted. This is, in part, thought to be a direct consequence of the ambient environmental conditions and their impact on regeneration success. However, the indirect influence of climate, in particular as a driver of fire regimes, may be as important, if not more important, in defining the structure, composition and relative abundance of rainforest and eucalypt taxa at the landscape scale. During warmer, drier periods fires occur more frequently or are more severe, whereas during cooler, wetter periods the opposite holds (Kershaw et al. 2002). Because rainforest and eucalypt species are considered to be fire-sensitive and fire-resilient, respectively, periods of more frequent or more intense fires should favour the expansion of eucalypt forests and the retreat of rainforests.

Much of the palaeoecological research examining the historical variability of climate-fire interactions and their role in modifying plant community composition has focused on very long time scales – typically millennia or longer. However, these patterns in long-term vegetation dynamics are driven by the accumulation of events that occur on much shorter time scales (i.e. annual to decadal) – scales that are relevant to management and conservation. In an attempt to link this ecological time scale with the longer-term patterns in vegetation dynamics in a temperate Australian context, Jackson (1968) proposed a probabilistic model that related forest composition – in particular, the relative abundance of rainforest and eucalypt species – to fire frequency. He suggested that where fire return intervals were <50 years, the vegetation would shift towards communities dominated by grasses and shrubs. In contrast, where fire return intervals were >350 years, the vegetation would be dominated by rainforest species. At fire return intervals between these thresholds, Jackson hypothesised that mixed forest dominated by eucalypts should persist. A key assumption of Jackson's model is that rainforest tree species are both sensitive to fire and lack resilience to fire. Thus, in the presence of fire return intervals of less than approximately 300 years, rainforest species have a low probability of persistence within a landscape. Although Jackson developed these ideas for Tasmanian forests, his conceptual model has been widely applied to other Australian landscapes where rainforest and eucalypt forest occur together (e.g. Noble and Slatyer 1980; Ash 1988; Unwin 1989; Bowman 2000) and is commonly used to describe the relationship between eucalypt and rainforest communities to the general public (Figure 1).

In the complex mountainous terrain of Victoria's Central Highlands, areas of cool temperate rainforest dominated by *Nothofagus cunninghamii* (myrtle beech) and *Atherosperma moschatum* (southern sassafras) occur along headwater streams within a landscape matrix of tall open forest dominated by *Eucalyptus regnans* (mountain ash) and *E. delegatensis* (alpine ash). One of the most striking features of the ecology of these tall open forests is the degree to which their dynamics are driven by catastrophic fires. The tall open forests of southeastern Victoria and Tasmania are subject to intense crown fires, which kill a large proportion of the trees over tens or hundreds of thousands of hectares in a single event. The energy released in these fires can be extraordinary, with fireline intensities commonly exceeding 100,000 kW m^{-1}. The eucalypts in these forests are well adapted to this type of fire regime and release seed stored in their crowns soon after a fire passes. The high light environment and heat-treated surface soils promote quick germination and vigorous growth of the new seedlings. This leads to dense, fast-growing stands of tall mountain and alpine ash and is the ecological basis of the clear-fell, burn and sow (CBS) silvicultural

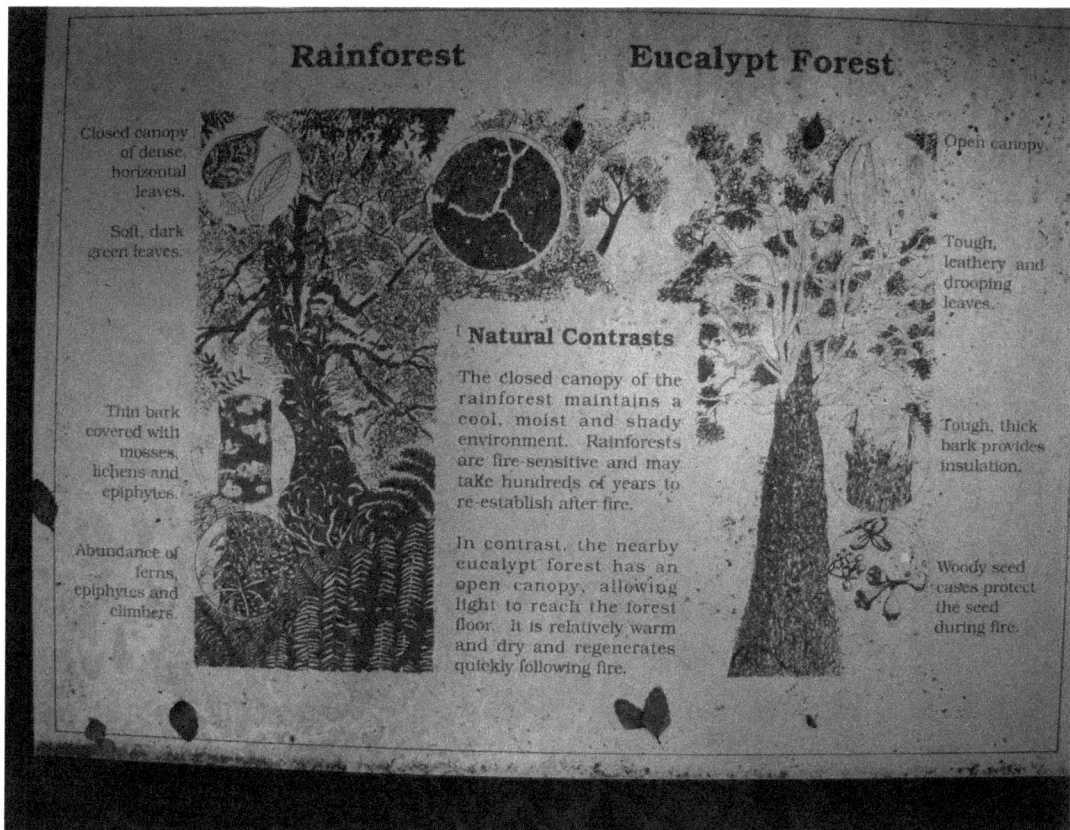

Figure 1. Interpretive signpost along a hiking trail in cool temperate rainforest near Marysville in the Central Highlands, Victoria. One of the stated 'Natural Contrasts' between the rainforest and the eucalypt forest is that 'rainforests are fire-sensitive and may take hundreds of years to re-establish after fire'. The rainforest immediately adjacent to the trail survived the 2009 fires; all *Eucalyptus regnans* trees within 100 m of the sign were killed by the fires.

system that is widely applied in the tall open forests of the Central Highlands (Attiwill 1994).

Fires are as much a part of the Central Highlands and the surrounding areas as the flora and fauna. In most years, fires occur somewhere within these landscapes. However, since European settlement, three fire seasons have stood out from the rest in terms of extent, ferocity and damage: 1851, 1939 and 2009. Each of these fires was preceded by a period of prolonged high temperature and low relative humidity; each was fanned by hot, dry winds from the arid centre of the continent; and each led to substantial loss of human life and property. From an ecological perspective, each of these fires led to widespread mortality of the forest canopy over hundreds of thousands of hectares and a subsequent landscape-wide pulse of regeneration. Given that the time between these fires was 88 years and 70 years, Jackson's model would suggest that the landscape would be predominantly eucalypt forest and that rainforest taxa would be severely disadvantaged. However, in the wake of the 2009 fires, it was evident that large tracts of cool temperate rainforest in areas subjected to high-intensity crown fires had survived more or less intact (Pappas 2010). In some instances, cool temperate rainforest stands that survived the 2009 fires were also known to have survived the 1939 and 1851 fires (Simkin and Baker 2008). If cool temperate rainforest is as fire-sensitive as widely believed, this presents a significant conundrum and seems to run counter to Jackson's model of fire frequency and forest composition. More fundamentally, it raises a simple question: How does cool temperate rainforest in the Central Highlands respond to fire? There are two key issues that must be addressed in answering this question. The first is the temporal scale of enquiry; the second is the distinction between the history of fire and the history of rainforest species on the landscape. Recent work by Whitlock et al. (2003, 2004) has highlighted the strength of using a multi-proxy approach to deal with these issues simultaneously.

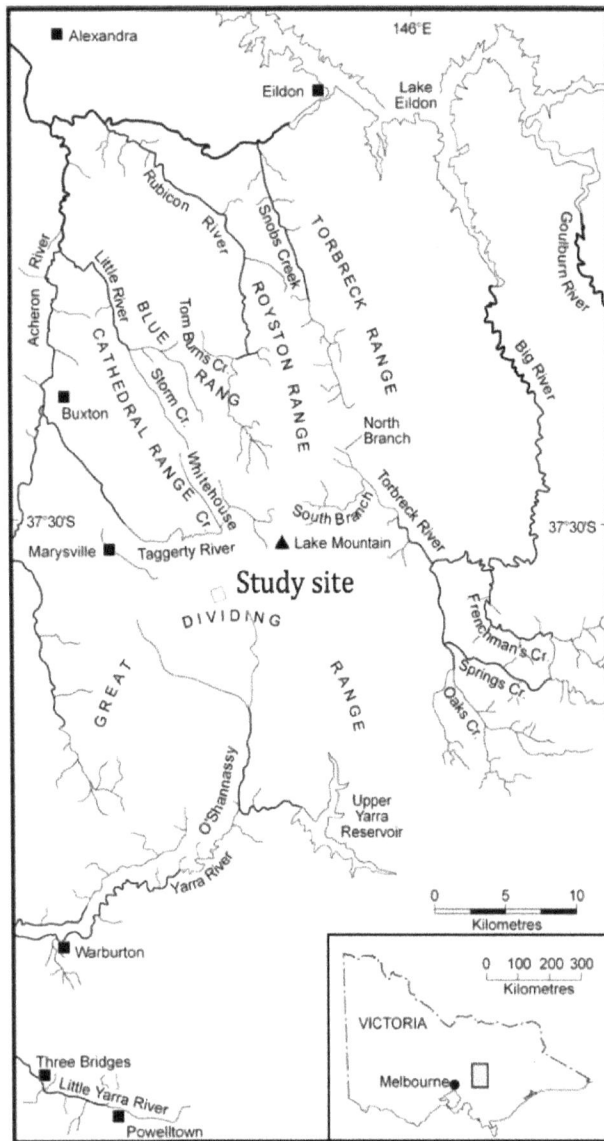

Figure 2. Location of the Bellel Creek study site from which tree ring, pollen and soil charcoal samples were collected. After McLeod (2007).

Fire ecology in southeastern Australia has benefited from research across a range of temporal and spatial scales, but much of it has focused on either direct observation and experimental manipulation or on proxy records. Inferences from these data are at very different time scales. Direct observations are relevant at annual to decadal scales, whereas most of the proxy data are relevant at the scale of millennia or tens of millennia. A multi-proxy approach allows us to bridge this scale gap and to differentiate between the history of fire and the history of rainforest. For example, soil charcoal provides data on fire occurrence, while pollen records provide data on the rainforest occurrence. Tree rings provide insights into the response (i.e. survival and recruitment) of rainforest species to specific, known fire events, while direct observations provide insights into the scale and severity of fires on rainforest, and their immediate response. The greatest impediment to this multi-proxy approach in Australia has been the lack of multiple proxies from the same area. In this chapter, we take advantage of a unique collection of proxy data sets collected from a single site, Bellel Creek, near Lake Mountain in the Central Highlands of Victoria (Figure 2), as well as other proxy data sets from southeastern Australia that provide additional local and regional context. Together with direct observations from the 2009 Victorian bushfires, we use them to critically examine the resilience of cool temperate rainforest to fire across a wide range of time scales (e.g. decadal to millennial).

Cool temperate rainforest and fire

Evidence from direct observations

In Australian plant ecology, it is almost axiomatic that rainforests are highly sensitive to fire and that rainforest tree species lack the ability to survive all but the lowest intensity fires (e.g. Jackson 1968; Noble and Slatyer 1980; Busby and Brown 1994; Bowman 2000, Figure 1). Nonetheless, the empirical evidence for these assertions is surprisingly weak and is confounded by vague or inconsistent definitions of 'rainforest' and 'sensitivity' (Hill 2000). In a continent dominated by eucalypts and acacias, which are often highly adapted to fire, rainforest species are relatively poorly adapted to fire. However, their persistence on the landscape for hundreds of thousands of years suggests that they must have some inherent capacity to either survive fire or regenerate in the wake of fire.

In southeastern Australia, there are large tracts of forest dominated by tall eucalypts such as *E. regnans* and *E. delegatensis* in which the dominant disturbance regime is one of catastrophic crown fires (Ashton 2000). Yet within these landscapes, there exist areas of cool temperate rainforest that have obviously persisted for centuries, if not millennia. Howard (1973) described the distribution and ecology of cool temperate rainforest in Victoria's Central Highlands. She showed that rainforest species, such as *N. cunninghamii* and *A. moschatum*, were typically restricted to riparian areas and cool, moist gullies that were found in the broader landscape matrix dominated by tall eucalypt species. However, she did find areas – for example, near Mt Donna Buang – in which whole hillsides were dominated by rainforest. Similarly, Lindenmayer et al. (2000) found that while *N. cunninghamii* were commonly associated with cooler, wetter microclimatic conditions, individual trees were not restricted to these sites.

The long-term maintenance of areas of rainforest within these landscapes depends on the successful regeneration of rainforest species. Howard (1973) and Read and Hill (1985) have both suggested that seedlings of the relatively shade-tolerant *N. cunninghamii* could establish under closed canopy, but that successful regeneration was more often associated with gaps in the forest canopy. These studies provided the foundation for the widely held view that cool temperate rainforest is a self-replacing 'climax' vegetation type. As canopy trees in the rainforest senesce and die, they create small gaps in the forest canopy that allow new individuals of rainforest species to establish and, in the absence of exogenous disturbance, perpetuate the rainforest on that site. In reality, however, few, if any, areas that support cool temperate rainforest are free of exogenous disturbance. The landscape matrix of tall eucalypt forests in which they occur is susceptible to high-intensity crown fires, which are likely to impact on the patches and strips of rainforest that are scattered within this matrix.

There are two potential questions that directly bear on the issue of rainforest resilience to fire and for which direct observations may provide useful insights. First, do high-intensity, landscape-scale crown fires kill rainforest trees? And, second, if they do kill rainforest trees, do rainforest species regenerate or are they replaced by eucalypt species? Because catastrophic fire events are relatively rare, there are few studies addressing either question. Recently, Pappas (2010) has documented the impacts of the 2009 fires on rainforest located in areas of the Central Highlands north of Marysville that were among the most severely burned. The first and most obvious feature of the 2009 fires is that despite their extent and intensity, the resulting pattern of burn severity was highly variable (Figure 3). Across the broader landscape, some areas of forest were completely consumed (that is, all trees were killed and the leaves and small branches immolated), while others were much less severely impacted. Post-fire surveys of other large, catastrophic fire events (e.g. 1988 Yellowstone fires, 2002 Oregon Biscuit fire) have shown that fire impacts and the resulting re-establishment of vegetation are highly heterogeneous across a range of spatial scales (Kashian et al. 2004; Schoennagel et al. 2008; Donato et al. 2009). In the cool temperate rainforests that Pappas (2010) studied, fire impacts ranged from complete stand-level mortality to small patch burns of <100 m^2 to areas that showed no evidence of fire impacts. Indeed, the most striking finding of the survey was the variability in fire impacts on rainforest, particularly given that in nearly every case adjacent stands of *E. regnans* and *E. delegatensis* suffered complete crown loss and mortality – in many cases, right up to the margins of the rainforest. So, cool temperate rainforest does appear to be relatively resilient to intense fires within the landscape. In part, this may be explained by the wetter, cooler microclimatic conditions of the riparian zones where the largest rainforest patches are found. However, the close proximity of fire-killed eucalypts to the rainforest margin (often <20 m) suggests that topographic and microclimatic differences cannot alone explain the differential impacts of mortality between the areas of rainforest and eucalypt-dominated forest.

The other issue – do rainforest species regenerate in the wake of the fire – has also received little attention due to the relative rarity of large, catastrophic fires over the past century. In a survey of two fires in Tasmania in the early 1980s, Hill and Read (1984) observed that in an area of rainforest that was burned, there was abundant rainforest regeneration (from both seed and sprout) – and no indication of eucalypt incursion. However, in an area of mixed forest, where the mature individuals of rainforest species were less abundant and there was a eucalypt overstorey, the regeneration was dominated by eucalypts. Pappas (2010) found that seedlings and vegetative sprouts of *N. cunninghamii* occur in fire-induced gaps in rainforest within 18 months of the 2009 fires; however, their distribution was highly patchy. In addition, she found that eucalypt regeneration was limited to the ecotones between the rainforest patches and the surrounding eucalypt forest and that most of the woody regeneration from seed within the rainforest was from *Acacia* species (*A. dealbata* and *A. frigrescens*).

Evidence from tree rings

Tree rings are an important source of proxy data on past climatic conditions and ecological dynamics. The strict dating control on, and annual resolution of, tree-ring chronologies allow for accurate dating of past environmental conditions and ecological events (e.g. fires, windstorms) over several centuries, and in some cases, millennia. Tree rings have been used extensively in the northern temperate zone to study the interactions between climate variability, historical fire regimes and forest dynamics (e.g. recruitment, growth and mortality; Swetnam and Betancourt 1990; Swetnam 1993; Brown and Wu 2005). These studies have provided important insights into forest dynamics because they address the time scale of decades to centuries, which is intermediate to the shorter time scales of direct observation and the longer time scales of palaeoecological studies based on pollen and charcoal abundances in lake and ocean sediments. In Australia, where most tree species do not form annual growth rings, relatively few dendrochronological studies have been conducted, limiting inference at this intermediate time scale. The few tree-ring studies that have been conducted in southeastern Australia have focused exclusively on

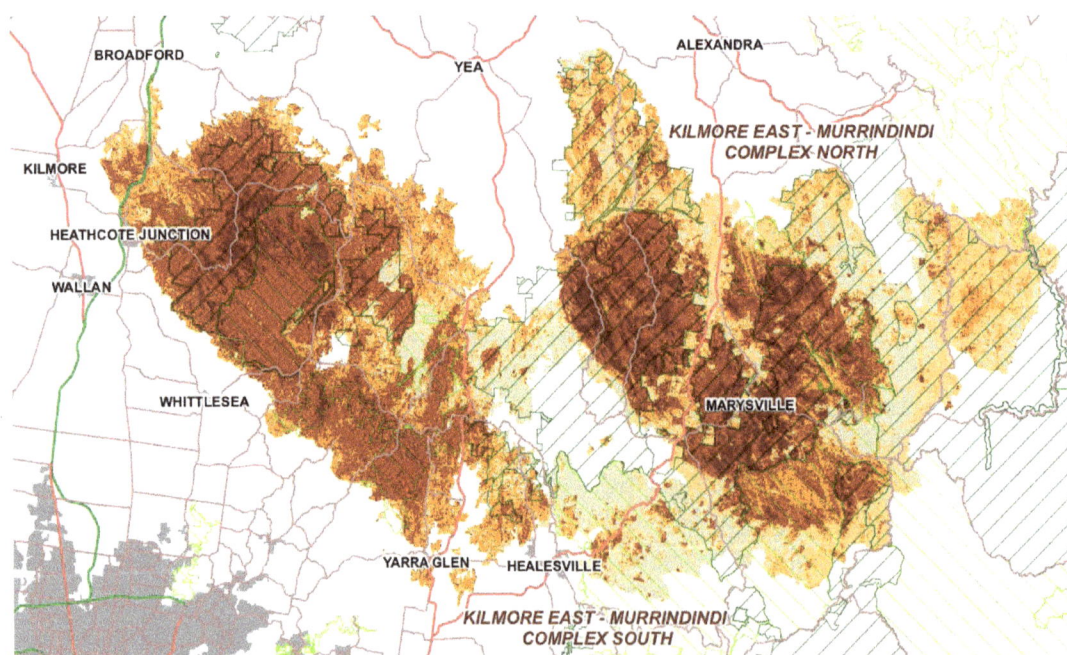

Figure 3. Variability in fire severity in the 2009 Kilmore-Murrindindi fire complex. Within the fire boundaries, darker colours represent more severe fires and lighter colours represent less severe fires. Map provided courtesy of the Department of Sustainability and Environment, Victoria.

reconstructing historical climate variability using the long-lived conifers endemic to Tasmania (Cook et al. 1991; Allen et al. 2001, In press). The best example of this is the multi-millennial reconstruction of summer temperatures in western Tasmania that was developed from huon pine (*Lagarastrobus franklinii*) (Cook et al. 1991, 2000).

Recently, however, Simkin and Baker (2008) used dendrochronological techniques to examine the role of fire on forest dynamics across an edaphic gradient in the Central Highlands of Victoria. Individual trees of several species were cored in rainforest along the riparian margins of Bellel Creek and in eucalypt forest on the mid slopes and upland sites away from the creek. The rainforest was dominated by *Nothofagus cunninghamii* and *Atherosperma moschatum* ; the eucalypt forest was dominated by *Eucalyptus regnans* and *E. delegatensis*. The mid-storey tree species, *Acacia dealbata*, was common near the transition from rainforest to eucalypt forest and in the eucalypt forest. Although *N. cunninghamii* is typically associated with riparian sites, it was relatively abundant in the mid storey of the eucalypt forest for several hundred metres away from Bellel Creek. Detailed statistical analyses of inventory plot data across the Central Highlands have shown that *N. cunninghamii* is not strictly limited to wet sites and in some instances has been found in dry sites on relatively exposed ridges (Lindenmayer et al. 2000).

The reconstructed age distributions and growth patterns derived from the tree-ring samples at Bellel Creek demonstrated that the 1939 bushfire had a dramatic impact on the mid-slope and upland sites dominated by eucalypts. In these areas, all of the sampled trees for all species on both sides of Bellel Creek had established in the years immediately after the fire. However, in the rainforest area, many of the trees that were alive at the time of sampling in 2006 were established well before the 1939 fires. Indeed, some individuals – mostly large, gnarled *N. cunninghamii*, many with old fire scars – were established more than 200 years before that fire (Simkin and Baker 2008), and thus may have also survived the other major bushfire that had swept through the Central Highlands in 1851.

The other finding of interest from these dendroecological analyses was that across all of the sites (i.e. riparian, mid slope and upland) there was a distinct pulse of *N. cunninghamii* recruitment immediately after the 1939 fire (Figure 4). Few studies have focused on the conditions that favour recruitment of rainforest species in southeastern Australia. Howard (1973) examined patterns of natural regeneration of *N. cunninghamii* and, in particular, in situ recruitment and the role of gap dynamics in creating the appropriate conditions for successful regeneration. Although a recruitment pulse in eucalypts and acacias is to be expected after a severe bushfire, only Hill and Read (1984) have previously noted (based on direct observation) that rainforest species are capable of vigorous regeneration after fire. However, at Bellel Creek, the reconstructed age distributions obtained from tree rings provide strong empirical evidence of widespread recruitment of *N. cunninghamii*, the dominant species of the rainforest canopy in these forests, in the wake of the 1939 fire (Simkin and Baker 2008). Howard (1973) provides some support for this in her earlier study of *N. cunninghamii* on Mt Donna Buang in the Central Highlands. She used ring counts (without cross-dating) of *N. cunninghamii* to age the canopy trees, and noted that they comprised a single age cohort of individuals that had established in it, which contradicts the idea that rainforest species such as *N. cunninghamii* occurring in the understorey or mid storey of mountain ash forests in some areas of the Central Highlands are the 'climax' species that will eventually replace the overstorey eucalypts as part of a relay-floristics-like succession (Jackson 1968; Howard 1973). Rather, the shorter *N. cunninghamii* established contemporaneously with the eucalypts, but due to differences in height growth rates and relative shade tolerance, have formed stratified, even-aged, mixed-species stands (see e.g. Oliver and Larson 1996).

Evidence from pollen

Pollen records are a mainstay of palaeoecological research around the world. In southeastern Australia, they have provided important insights into the tempo and mode of climatic variability and the ensuing changes in vegetational composition during the late Quaternary, and in particular the Holocene, periods. However, the pollen record in southeastern Australia has several important limitations. First, because of the geomorphological history of the region, pollen records are relatively sparse, being restricted to either calderal lakes in the broad peneplains, or perched bogs or swamps in areas of more varied topography. Second, the weak intra-annual seasonality and substantial intra-decadal variability in climatic conditions mean that annual varving of sediments is weak or absent, limiting the temporal resolution of palaeoecological reconstructions for the region. Third, the substantial changes in land-use history over the past two centuries have compromised the quality of many potential palaeoecological sites due to physical mixing of the sediment profile (e.g. from livestock trampling).

Despite these limitations, palaeoecological reconstructions of vegetation in the Central Highlands have provided important insights into historical changes in the distribution and abundance of various plant taxa for at least the past 35,000 years. This has been achieved through the development and analysis of multiple pollen-based reconstructions, which have shown several common, and consistent, patterns. McKenzie (1997, 2002) used pollen and microcharcoal to develop the most comprehensive palaeoecological reconstruction of vegetation change in the Central Highlands. The pollen record compiled by McLeod (2007) from Bellel Creek offers a point of comparison with the other proxy records (i.e. tree rings, soil charcoal) developed from the same site. Interpretation of long-term variability in rainforest dynamics is based primarily on the presence of *N. cunninghamii* pollen in the sediment samples, which show that *N. cunninghamii* has been present within the Central Highlands landscape for most of the past 35,000 years. McKenzie (1997) notes the presence of *N. cunninghamii* in the basal portions

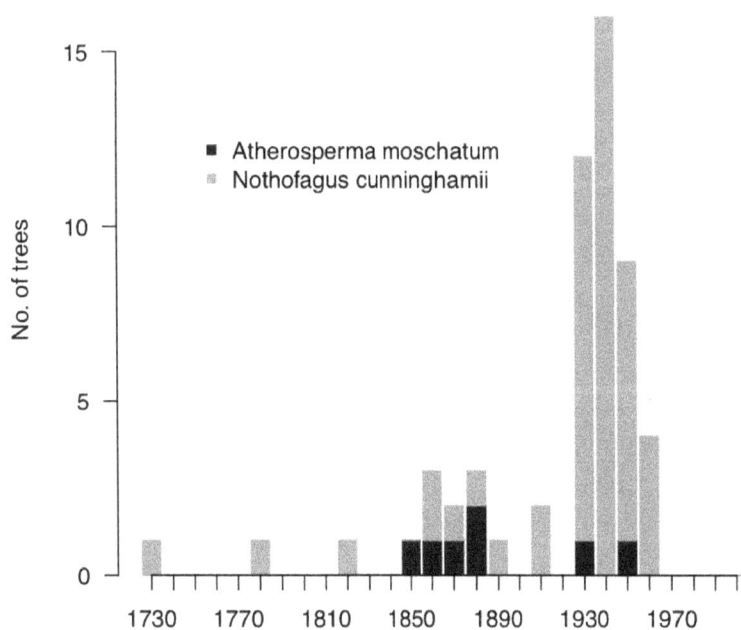

Figure 4. Age distribution of sampled rainforest trees at Bellel Creek. A distinct pulse of recruitment is evident after the 1939 fire, which burned through the area. However, as shown in the age distribution, many Nothofagus survived the 1939 fires (and several survived the 1851 fire as well). Data from Simkin and Baker (2008).

of several high-elevation (>900 m) sediment cores that have been dated to ca. 31,500 BP and again from 20,000 BP, at which point the *N. cunninghamii* pollen are found continuously until the present, albeit with fluctuating abundance. The period 31,500-20,000 BP was during the last glacial, in which climatic conditions were as much as 5-8°C below current mean annual temperatures. At high-elevation sites in the Central Highlands, the vegetation would have been limited to either woodlands of extremely cold-tolerant eucalypts (e.g. *E. pauciflora* [snow gum]) or treeless alpine vegetation (e.g. tussock grasslands), both of which are evident in the pollen record from that period and both of which are found at much higher elevations today (e.g. above 1800 m on Mt Kosciuszko to the northeast).

The early Holocene is a period of rapid change in the pollen record from the Central Highlands (McKenzie 1997, 2002; McLeod 2007). At ca. 12,000 BP, treeline was near or below 900 m and wet sclerophyll and rainforest taxa were limited to low-elevation sites. By 9500 BP, herbaceous alpine taxa had retreated upwards across the Central Highlands and were limited to representation in only the highest elevation samples (McKenzie 2002). This change was accompanied by a rapid expansion of wet sclerophyll and rainforest taxa into higher elevation sites across the Central Highlands. During this period, *N. cunninghamii* pollen is a consistent component of the pollen record, but at sufficiently low levels to indicate that it was present but not particularly abundant. By ca. 6000 BP, *N. cunninghamii* reaches its highest representation in the pollen records across the entire elevational gradient of sites (168-1440 m) in the Central Highlands, including at the Bellel Creek site (McLeod 2007). However, fossil charcoal is found in all records from this period, suggesting that the likely warmer, moister climatic conditions of this 'forest optimum' (McKenzie 2002) did not eliminate the occurrence of fire within the landscape. Rather, fires were likely restricted to periods of anomalously dry climatic conditions that were too short to be recorded in the sedimentary record. From ca. 4500 BP, the abundance of pollen from *N. cunninghamii* and wet sclerophyll taxa decrease and charcoal levels increase, suggesting the onset of warmer, drier conditions. Kershaw et al. (2002) suggest that these changes were associated with a mid-Holocene strengthening of the El Nino-Southern Oscillation.

Evidence from soil charcoal

The pollen record suggests that the abundance of rainforest taxa in the Central Highlands has fluctuated over time. These changes in abundance are typically interpreted in terms of long-term climatic variability and, in particular, the role of climate on fire occurrence. Estimates of fire occurrence in the sediment cores come from measures of fine particulate charcoal, which may be produced locally or blown in from areas outside the catchment. As such, it is difficult to disentangle local and regional influences on the abundance of fine particulate charcoal measured in most sediment cores (Clark 1988). In contrast, macroscopic charcoal particles derived from partially combusted wood are almost exclusively local in origin. Because macro-charcoal may persist in the soil for tens of thousands of years, it can be used to reconstruct millennial-scale variability in fire activity across relatively small areas. Stratigraphic analysis of charcoal within the soil profile, combined with radiocarbon dating, can be used together to develop a reconstruction of local fire activity at relatively coarse temporal scales. When compared with pollen records, which have similar dating resolution, soil charcoal records can provide a local context for interpreting the regional fire signal provided by the pollen and sediment records.

In the Central Highlands, McLeod (2007) used a network of soil pits at Bellel Creek to characterise the distribution and abundance of macro-charcoal particles in the forest soils for most of the past interglacial/glacial/interglacial cycle. Radiocarbon dating of the soil charcoal fragments revealed that fire activity at Bellel Creek has been highly variable since Oxygen Isotope Stage 3 (OIS3, ca. 50,000 BP, Figure 5). This variability is reflected in extended periods with little or no production of macro-charcoal fragments, punctuated by periods of extremely high macro-charcoal abundance. The three periods in which charcoal fragments were most abundant were 45,000-55,000 BP, 11,000-13,000 BP, and 0-2500 BP. Because the accuracy of radiocarbon dating decreases near the margins of prediction (ca. 45,000-50,000 BP), the errors associated with estimated ages are greatest for the oldest samples (1σ errors = 1700-3250 calendar years). In contrast, the estimated ^{14}C dates from 0-15,000 BP are all relatively tightly constrained (1σ errors = 40-155 calendar years).

The most notable gap in the macro-charcoal record is from 13,000 BP to 38,000 BP, which is coincident with the Last Glacial Maximum (LGM) and the accompanying cooler, drier conditions. The pollen records from Bellel Creek and other sites within the Central Highlands suggest that during this period, the vegetation of the upper Central Highlands may have been largely treeless, with plants in the families Poaceae and Asteraceae dominating the vegetation (McKenzie 1997, 2002; McLeod 2007). Anatomical identification of the charcoal fragments found that 65% were from *Eucalyptus* spp., suggesting a forest composition similar to that which currently dominates most of the Central Highlands (Table 1; McLeod 2007). Notably, only 2% of the charcoal fragments were attributed to Fagaceae (most likely *Nothofagus cunninghamii*) and all of these were restricted to the lower slope positions along the riparian margins.

The pulse of charcoal fragments at 11,000-13,000 BP occurs at the termination of the LGM, a period of rapid climatic and ecological changes across the region. Warming was occurring in the southwest Pacific as early as 17,000 BP (Turney et al. 2006), with markedly wetter conditions by 12,000-13,000 BP. Pollen from sediment cores collected above 900 m in the Central Highlands documents the transition from treeless alpine and sub-alpine vegetation to forest vegetation around this time (McKenzie 1997). The subsequent gap in soil charcoal fragments from 5000-11,000 BP is associated with forest expansion locally, regionally and globally, as the cool, dry conditions of the LGM ceded to the warmer, moister conditions of the early Holocene. Macro-charcoal fragments reappear at 4500 BP and are found in great abundance from 2500 BP to the present, suggesting a shift to a forested landscape in which fire was a more prominent disturbance or the vegetation was more flammable. As noted above, Kershaw et al. (2002) have proposed that this change was associated with a more general intensification of the El Nino-Southern Oscillation and its increasing influence on regional climate dynamics. The hotter, drier conditions would have increased the probability of fire occurrence and been more conducive to the successful establishment of eucalypts and other sclerophylous plant species.

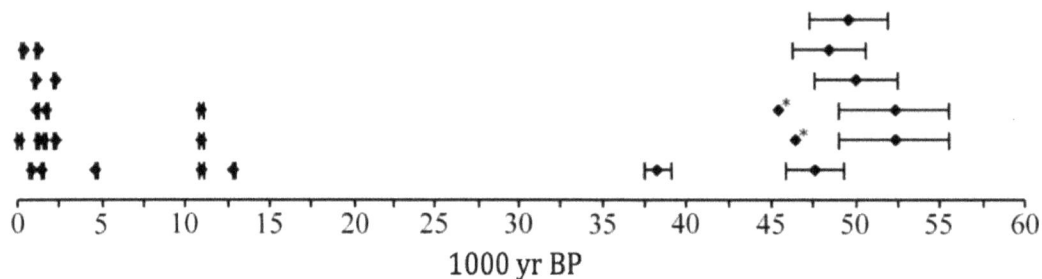

NB: (*) indicates the lower limit of an infinite age, presented in ^{14}C years

Figure 5. Temporal spread of all AMS 14C dates on charcoal fragments obtained from Bellel Creek in the Central Highlands of Victoria. Data are presented in calendar years BP with 1Ɖ error bars. An asterisk (*) indicates the lower limit of an infinite age. From McLeod (2007).

So, how resilient is cool temperate rainforest to fire?

The tree species that dominate Australian rainforests are generally considered to be sensitive to fire (e.g. Jackson 1968; Groves 1981; Bowman 2000, Figure 1), due to the absence or poor development of adaptations to fire, such as thick bark, vigorous epicormic sprouting and seritony, which are so well developed in most non-rainforest taxa (Gill and Ashton 1968; Gill 1981). Yet, across Australia areas of rainforest are commonly found within a matrix of fire-prone

Table 1. Relative abundance of taxa at family level, grouped by slope position, from 476 soil charcoal fragments from Bellel Creek in the Central Highlands, Victoria. Species in the genera Eucalyptus and Nothofagus are in the families Myrtaceae and Fagaceae, respectively. From McLeod (2007).

Family	Slope position		
	Upper	Middle	Lower
Myrtaceae	121	83	109
Leguminoseae	42	7	17
Fagaceae	0	0	11
Other (incl. unidentified)	21	15	50
Totals	**184**	**105**	**187**

sclerophyll vegetation, often dominated by Eucalyptus species. A range of hypotheses has been proposed to explain this pattern (see e.g. Bowman 2000). From first principles, however, the juxtaposition of fire-adapted sclerophyll forest and putatively fire-sensitive rainforest requires that at least one of the following hypotheses be true:

1. Fires do not transgress rainforest boundaries;

2. Rainforest species possess a degree of resilience to fire; and

3. Fire facilitates regeneration of rainforest species.

The cool temperate rainforests of the Central Highlands of southeastern Australia provide a unique opportunity to test these hypotheses for several reasons. First, cool temperate rainforest in the Central Highlands coexists with wet sclerophyll forest, where it typically occurs as scattered patches and strips within broad tracts of wet sclerophyll forest. Second, over the past two decades, palaeoecologists have developed multiple proxies – tree rings, soil charcoal and pollen – from the Central Highlands that provide a rich, multi-layered historical context for these forests that is not available in other regions of Australia. And third, in 2009 a major catastrophic fire event occurred, permitting direct observation of the impacts of an intense landscape-scale fire on cool temperate rainforest.

In the following sections, we consider the evidence for and against each of the three hypotheses described above. We focus particular attention on integrating the available palaeo-proxy data with other sources of data on the contemporary ecology of the rainforest and wet sclerophyll forest taxa. We realise that these hypotheses are not mutually exclusive and that in some instances it will be difficult to differentiate one from the other given the available evidence. For example, if a fire does not burn a patch of rainforest, is that because the fire did not reach the rainforest or because the rainforest species are resilient to the fire and thus minimally affected? In addition, we realise that we may be accused of establishing 'straw man' hypotheses that can be easily knocked down. These are fair concerns, but we believe that the three hypotheses provide a coherent framework for identifying the impact of fire on cool temperate rainforest tree taxa and considering the implications for long-term forest dynamics, which is the basis of management and conservation of these unique ecosystems.

Fires do not transgress rainforest boundaries

If we accept the assumption that rainforest taxa are not well adapted to fires, then to persist within the landscape over long periods they must not be subjected to killing fires. That is, fires of some threshold level of intensity must not burn into the rainforest, otherwise widespread mortality would occur. The proxy records from the Central Highlands provide strong evidence that rainforest tree species and fire have been present together within the landscape for long periods. The pollen records indicate that the dominant rainforest tree species, *N. cunninghamii*, has been present within the landscape, albeit at varying abundance, for most of the past 30,000 years. The micro- and macro-charcoal record also indicates that fire has occurred over much of this period, with the most recent 2000 years showing the highest levels of fire in the proxy record. In addition to the proxy records of past fire occurrence across the Central Highlands landscape, historical records and direct observations provide further evidence of at least three major fires (1851, 1939 and 2009) that have burned hundreds of thousands to millions of hectares of forests in the past 160 years.

Given these multiple, overlapping records of widespread fire throughout the Central Highlands over millennia, it is highly unlikely that areas bearing cool temperate rainforest would not have been directly impacted by fire. However, it is possible that they may have experienced less intense fires than the sclerophyll forest. Even within a landscape subjected to a catastrophic bushfire there is substantial variability in fire intensity at local scales (Figure 3). This variability derives from interactions among topography, microclimatic conditions, variability in weather conditions at the time of the fire (e.g. timing of wind changes, occurrence of rain), and the composition, structure and spatial arrangement of the vegetation. Cool temperate rainforest is often found in cool, moist gullies on southeastern-facing slopes where the lower temperatures and higher relative humidity provide a degree of microclimatic buffering from the fires. However, anecdotal observations from the 2009 bushfires indicate that the fireline intensities within the worst affected areas of the Central Highlands were in excess of 100,000-150,000 kW m^{-1}. It seems unlikely that small differences in microclimatic conditions would have had much potential to stop the fires along the rainforest boundaries under such extreme conditions. However, they may have lowered the amount of radiant heat directed at the rainforest. Another factor that may reduce the intensity of fires burning into cool temperate rainforest is the difference in height of the dominant canopy trees in eucalypt forest and rainforest. In the wet sclerophyll forest, dominant *E. regnans* may be 70-80 m tall, whereas few *N. cunninghamii* reach more than 30 m in height. The preponderance of cool temperate rainforest trees in low-lying areas within the landscape (e.g. gullies and creek lines) further exacerbates differences in canopy height of the rainforest and wet sclerophyll forests. To burn the rainforest canopy, crown fires in the eucalypt forest matrix must drop a considerable distance. However, because heat energy is naturally displaced upwards, burning downwards occurs more slowly and with less energy. Thus, the rainforests may be exposed to less of the energy of the crown fires in the neighbouring tall eucalypt forest as they move towards the rainforest and confine the worst impacts of the fire to the margins of the rainforest. After the 2009 fires, Pappas (2010) found that of the 32 patches of burnt rainforest that she surveyed, the mean distance to the rainforest/eucalypt forest ecotone was only 2.5 m and was heavily skewed towards the edge of the rainforest. Only four patches were more than 15 m into the rainforest, and the one patch farthest into the interior of the rainforest (35 m from the edge) was found in a rainforest patch nearly 200 m wide.

Rainforest species are resilient to fire

The proxy data and direct observations suggest that as catastrophic crown fires race through a wet sclerophyll forest, the interdigitated rainforest patches are subjected to the radiant heat from the fire and may burn. If rainforest tree species have persisted within these landscapes under similar fire regimes over long periods, then they must be relatively resilient to fires. This may be manifest as either the ability to endure fires or the ability to regenerate in the wake of fires. In this section, we consider the potential of rainforest species to endure fire; in the next section, we consider the issue of fire-initiated regeneration in rainforest species.

The tree-ring records demonstrate that rainforest tree species can survive catastrophic fire events. At Bellel Creek in the Central Highlands, the 1939 fire killed eucalypts on both slopes above the creek, as well as rainforest trees within the riparian zone. Yet, a large number of rainforest trees (e.g. *N. cunninghamii*, *A. moschatum*) that were alive at the time of sampling in 2006 were established well before the 1939 fire (Simkin and Baker 2008). In some cases, individual trees were >200 years old, indicating that they had survived both the 1851 and 1939 fires. Direct observations after the 2009 fire showed that the 2009 fires had generally burned up to the margins of the rainforest, but that along several hundred metres of the creek, small patches of rainforest (most <200 m²) had been burned as well. The macro-charcoal record from Bellel Creek shows that fires have occurred within the boundaries of the extant cool temperate rainforest repeatedly over the past 2000 years. Obviously, then, some proportion of the populations of most rainforest taxa must be able to endure fires of moderate to high intensity.

The ability of any plant species to endure fires is highly dependent on the fuel characteristics or flammability of its foliage. Eucalypts are well known for their highly flammable foliage and the role of the volatile oils in their leaves in accelerating foliar combustion. Rainforest species have thicker leaves of higher moisture content and are thus likely to be less flammable (although there are few studies that have examined the flammability of rainforest species). Dickinson and Kirkpatrick (1985) showed that leaves of *Atherosperma moschatum* had lower energy content, had greater moisture and ash contents, and were much slower in propagating fire than eucalypt leaves. Importantly, they found that *A. moschatum* leaves would not ignite unless they had lost >60% of their moisture content. In contrast, eucalypt and other 'dry' forest species would ignite when leaves had lost <40-50% moisture content. We were unable to find equivalent data for the dominant rainforest canopy tree species *N. cunninghamii*, but would expect that the fuel properties of its leaves would be more similar to those of *A. moschatum* than those of eucalypts.

In crown fire ecosystems such as the mixed landscape of the Central Highlands, these differences in the foliar fuel properties of rainforest tree species may play a role in limiting the amount of fire damage to patches of rainforest. With fire spread rates of 5-10 m² sec⁻¹ (20-40 km hr⁻¹), the fire front may pass over the rainforest too quickly to both dry and ignite the foliar biomass in the canopy. In addition, because rainforest patches are often associated with riparian areas, the foliar moisture content of rainforest taxa may remain higher than in upland sites even during the extreme climatic conditions that precede major fire events, further buffering against initiation of crown fires within the rainforest.

The benefits of reduced foliar flammability do not confer immunity from fire, though. Where rainforest tree species are found as scattered individuals in upland sites, fire-induced mortality is much higher. In the 2009 fires, all individuals of *N. cunninghamii* occurring at low densities in the mid-storey of *E. regnans* stands or in narrow rainforest strips were killed. Rainforest taxa only survived where the width of the rainforest strip was >50 m (Pappas 2010), suggesting that there is a threshold level of aggregation above which the benefits of higher foliar moisture content and reduced susceptibility to ignition of rainforest taxa would be able to modify fire behaviour. In large patches of cool temperate rainforest, the greatest heat loads from the adjacent eucalypt forest would only be experienced on the margins of the rainforest, meaning that the necessary pre-drying of the rainforest canopy foliage required for ignition would be less likely to occur in the interior of the rainforest patch. This was precisely what was observed in the 2009 fires. Even in areas of the landscape that experienced the most intense fires in 2009, the larger patches of rainforest survived, although they did not emerge unscathed from the fires. In most of these areas, the fires killed individual trees or small groups of trees, creating small to medium-sized gaps (median gap size ca. 1000 m²; range 160-26,000 m²) in the rainforest canopy and a rich ash bed on the forest floor.

Fire facilitates rainforest regeneration

If fires burn into rainforest areas and the rainforest trees are killed, rainforest can still persist if the fire promotes the regeneration of the rainforest tree species. The conventional wisdom is that rainforest tree species, which typically have heavy, poorly dispersed seeds, do not regenerate well after fires. In mixed forest with a eucalypt overstorey and rainforest understorey, rainforest taxa are often overwhelmed by competition from the faster-growing eucalypt species. However, the tree-ring data demonstrate the ability of at least one rainforest canopy species to regenerate rapidly and profusely after a fire. At Bellel Creek, a pulse of *N. cunninghamii* regeneration established immediately after the 1939 fire, in both the riparian area dominated by cool temperate rainforest and the upland sites dominated by the tall eucalypts. Howard (1973) noted that the age structure of a large stand of *N. cunninghamii* on Mt Donna Buang in the Central Highlands was even-aged, suggesting mass recruitment in the wake of a large, stand-replacing disturbance (presumably fire). Silvicultural experiments in northwestern Tasmania that have manipulated overstorey density and forest floor conditions to evaluate the regeneration response of rainforest taxa support these observations. Hickey and Wilkinson (1999) showed that *N. cunninghamii* regeneration is most abundant and vigorous on sites in which most of the overstorey is removed and the ground is burned and/or mechanically disturbed. Indeed, seedlings in their clearfell (with standards) and burn treatment maintained height growth rates of ca. 45 cm yr^{-1} over nearly two decades (Hickey and Wilkinson 1999). Ellis (1985) showed that after a fire in the 1850s in northeastern Tasmania, diameter growth rates of *N. cunninghamii* were similar to those of sympatric eucalypt species (ca. 3-4 mm yr^{-1}). In a selectively logged area of rainforest in western Tasmania, Jennings et al. (2005) found high levels of *N. cunninghamii* regeneration (>5000 seedlings ha^{-1}), but height growth rates were much slower (ca. 2.5 cm yr^{-1}), although this has been attributed in part to the thin, peaty soils at the site. There was no evidence of *A. moschatum* recruitment despite the presence of adults in the pre-logging rainforest.

In general, the seeds of the dominant tree species in the cool temperate rainforest are relatively large and are dispersed by gravity, water or animals (Howard 1973). In addition, *N. cunninghamii*, like many Fagaceae, is a mast-fruiting tree species, producing large fruit crops at irregular, supra-annual intervals (Howard 1973; Hickey and Wilkinson 1999). In landscapes subjected to rare, but catastrophic, bushfires, mast fruiting would appear to present a serious risk. If mast fruiting occurs in the year or two before a fire, the trees may not have the reproductive capacity to take advantage of the regeneration opportunity presented by a disturbance. However, despite this, *N. cunninghamii* is one of the two dominant canopy tree species in cool temperate rainforest in the Central Highlands and has been for nearly 30,000 years.

Pappas (2010) also noted that almost all of the *N. cunninghamii* that were partially damaged in the 2009 fires showed evidence of vegetation regeneration. Individuals that suffered crown scorch were producing sprouts at the base of the scorched branches, while individuals that suffered heat damage to the main stem were sprouting from the base of the tree. This rapid re-establishment of photosynthetic cover by the rainforest species, independent of sexual reproduction, may be an important mechanism in allowing rainforest taxa to persist in these landscapes where fires are rare, but intense and often damaging (Bond and Midgley 2001).

Conclusion

Cool temperate rainforest and fires have coexisted in the Central Highlands for much of the past 40,000 years. In the past 2500 years, fire activity has been as high or higher than at any other time in the past 40,000 years, yet rainforest is still relatively common across the landscape. Evidence from multiple, overlapping palaeo-proxies, as well as direct observations after the 2009 fires, suggests that the high-intensity crown fires that burn across the Central Highlands once or twice each century do reach into patches of rainforest, but that many of the rainforest tree species are capable of withstanding the impacts of these fires and, at least in the case of *N. cunninghamii*, can quickly respond with increased recruitment soon after the fire has passed. Direct observations demonstrate that patches of rainforest do survive extreme fires; tree rings covering several centuries show that rainforest trees have survived previous fires and that they are capable of recruiting immediately after large fires. The charcoal and pollen evidence, which cover at least 40,000 years, indicate that the cool temperate rainforests of the Central Highlands have survived major fires over that period. Although the dominant tree species in the cool temperate rainforests are not as highly adapted to fire as the tall eucalypts in the adjacent wet sclerophyll forest, they appear to be more resilient to fire than widely believed. This resilience derives from interactions among the nature of the environment in which the taxa typically occur, their reproductive behaviour in relation to fire, the flammability of their foliage and, finally, stand- and landscape-scale heterogeneity in fire intensity.

References

Allen, K.J., Cook, E.R., Francey, R.J. and Michael, K. 2001. The climatic response of *Phyllocladus aspleniifolius* (Labill.) Hook. f in Tasmania. *Journal of Biogeography* 28:305-316.

Allen, K.J., Ogden, J., Buckley, B.M., Cook, E.R. and Baker, P.J. In press. The potential to reconstruct broadscale climate indices associated with southeast Australian droughts from *Athrotaxis* species, Tasmania. *Climate Dynamics*.

Ash, J. 1988. The location and stability of rainforest boundaries in north-eastern Queensland, Australia. *Journal of Biogeography* 15:619-630.

Ashton, D. 2000. The Big Ash forest, Wallaby Creek, Victoria – changes during one lifetime. *Australian Journal of Botany* 48:1-26.

Attiwill, P.M. 1994. Ecological disturbance and the conservative management of eucalypt forests in Australia. *Forest Ecology and Management* 63:301-346.

Bond, W.J. and Midgley, J.J. 2001. Ecology of sprouting in woody plants: The persistence niche. *Trends in Ecology and Evolution* 16:45-51.

Bowman, D. 2000. *Australian Rainforests: Islands of Green in a Land of Fire*. Cambridge: Cambridge University Press.

Brown, P.M. and Wu, R. 2005. Climate and disturbance forcing of episodic tree recruitment in a southwestern ponderosa pine landscape. *Ecology* 86:3030-3038.

Busby, J.R. and Brown, M.J. 1994. Southern rainforests. In: Groves R.H. (ed), *Australian Vegetation* (second ed.), pp. 131-155. Cambridge: Cambridge University Press.

Clark, J.S. 1988. Particle motion and the theory of charcoal analysis: source area, transport, deposition, and sampling. *Quaternary Research* 30:81-91.

Cook, E.R., Bird, T., Peterson, M., Barbetti, M., Buckley, B., D'Arrigo, R., Francey, R. and Tans, P. 1991. Climatic change in Tasmania inferred from a 1089-year tree-ring chronology of Huon Pine. *Science* 253:1266-1268.

Cook, E.R., Buckley, B., D'Arrigo, R. and Peterson, M. 2000. Warm-season temperatures since 1600 BC reconstructed from Tasmanian tree rings and their relationship to large-scale sea surface temperature anomalies. *Climate Dynamics* 16:79-91.

Dickinson, K.J.M. and Kirkpatrick, J.B. 1985. The flammability and energy content of some important plant species and fuel components in the forests of southeastern Tasmania. *Journal of Biogeography* 12:121-134.

Donato, D.C., Fontaine, J.B., Campbell, J.L., Robinson, W.D., Kauffman, J.B. and Law, B.E. 2009. Conifer regeneration in stand-replacement portions of a large mixed-severity wildfire in the Klamath-Siskiyou Mountains. *Canadian Journal of Forest Research* 39:823-838.

Ellis, R. 1985. The relationships among eucalypt forest, grassland and rainforest in a highland area in north-eastern Tasmania. *Australian Journal of Ecology* 10(3):297-314.

Gill, A.M. 1981. Adaptive responses of Australian vascular plant species to fires. In: Gill, A.M., Groves, R.H., and Noble, I.R. (eds), Fire and the Australian Biota, pp. 243-272. Australian Academy of Science, Canberrra.

Gill, A.M. and Ashton, D.H. 1968. The role of bark type in relative tolerance to fire of three central Victorian eucalypts. *Australian Journal of Botany* 16:491-498.

Groves, R.H. 1981. *Australian Vegetation*. Cambridge University Press: Cambridge, UK.

Hickey, J.E. and Wilkinson, G.R. 1999. Long-term regeneration trends from a silivicultural systems trial in lowland cool temperate rainforest in Tasmania. *TasForests* 11:1-22.

Hill, R. and Read, J. 1984. Post-fire regeneration of rainforest and mixed forest in western Tasmania. *Australian Journal of Botany* 32:481-493.

Hill, R.S. 2000. Attempting to define the impossible: a commentary on 'Australian Rain-forests: Islands of Green in a Land of Fire'. *Australian Geographical Studies* 38:320-326.

Howard, T. 1973. Studies in the ecology of *Nothofagus cunninghamii* Oerst. I. Natural regeneration on the Mt. Donna Buang massif, Victoria. *Australian Journal of Botany* 21:67-78.

Jackson, W.D. 1968. Fire, air, water, and earth – an elemental ecology of Tasmania. *Proceedings of Ecological Society of Australia* 3:9-16.

Jennings, S., Edwards, L.G. and Hickey, J.E. 2005. Natural and planted regeneration of huon pine (*Lagarostrobus franklinii*) at Travellers Creek, western Tasmania. *TasForests* 16:61-70.

Kashian, D.M., Tinker, D.B., Turner, M.G. and Scarpace, F.L. 2004. Spatial heterogeneity of lodgepole pine sapling densities following the 1988 fires in Yellowstone National Park, Wyoming, USA. *Canadian Journal of Forest Research* 34:2263-2276.

Kershaw, A.P., Clark, J.S., Gill, A.M. and D'Costa, D.M. 2002. A history of fire in Australia. In: Bradstock, R.A., Williams, J.E. and Gill, A.M. (eds), *Flammable Australia: The Fire Regimes and Biodiversity of a Continent*, pp. 3-25. Cambridge: Cambridge University Press.

Lindenmayer, D.B., Mackey, B., Cunningham, R., Donnelly, C., Mullen, I., McCarthy, M.A. and Gill, A.M. 2000. Factors affecting the presence of the cool temperate rain forest tree myrtle beech (*Nothofagus cunninghamii*) in southern Australia: integrating climatic, terrain and disturbance predictors of distribution patterns. *Journal of Biogeography* 27:1001-1009.

McKenzie, G.M. 1997. The late quaternary vegetation history of the south-central highlands of Victoria, Australia. I. Sites above 900 m. *Austral Ecology* 22:19-36.

McKenzie, G.M. 2002. The late Quaternary vegetation history of the south-central highlands of Victoria, Australia. II. Sites below 900 m. *Austral Ecology* 27:32-54.

McLeod, A.J. 2007. Palaeoenvironmental change in the Central Highlands of Victoria, Australia, interpreted from the analysis of macroscopic soil charcoal. Unpublished PhD thesis, Monash University.

Noble, I.R. and Slatyer, R.O. 1980. The use of vital attributes to predict successional changes in plant communities subject to recurrent disturbances. *Plant Ecology* 43:5-21.

Oliver, C.D., Larson, B.C. 1996. *Forest Stand Dynamics*. New York: John Wiley and Sons.

Pappas, N. 2010. The impacts of the 2009 bushfires on cool temperate rainforest in the Central Highlands of Victoria. Unpublished BSc (Honours) thesis, Monash University.

Read, J. and Hill, R. 1985. Dynamics of Nothofagus-dominated rainforest on mainland Australia and lowland Tasmania. *Plant Ecology* 63:67-78.

Schoennagel, T., Smithwick, E.H. and Turner, M.G. 2008. Landscape heterogeneity following large fires: insights from Yellowstone National Park, USA. *International Journal of Wildland Fire* 17:742-753.

Simkin, R. and Baker, P.J. 2008. Disturbance history and stand dynamics in tall open forest and riparian rainforest in the Central Highlands of Victoria. *Austral Ecology* 33:747-760.

Sniderman, J.M.K., Porch, N. and Kershaw, A.P. 2009. Quantitative reconstructions of Early Pleistocene climate in southeastern Australia and implications for atmospheric circulation. *Quaternary Science Reviews* 28:3185-3196.

Swetnam, T.W. 1993. Fire history and climate change in giant sequoia groves. *Science* 262:885-889.

Swetnam, T.W. and Betancourt, J.L. 1990. Fire-southern oscillation relations in southwestern United States. *Science* 249:1017-1020.

Turney, C.S.M., Kershaw, A.P., Lowe, J.J., van der Kaars, S., Johnston, R., Rule, S., Moss, P., Radke, L., Tibby, J., McGlone, M.S., Wilmshurst, J.M., Vandergoes, M.J., Fitzsimons, S.J., Bryant, C., James, S., Branch, N.P., Comely, J., Kalin, R.M., Ogle, N., Jacobson, G. and Fifield, L.K. 2006. Climatic variability in the southwest Pacific during the Last Termination (20-10 kyr BP). *Quaternary Science Reviews* 25:886-903.

Unwin, G.L. 1989. Structure and composition of the abrupt rainforest boundary in the Herberton Highland, North Queensland. *Australian Journal of Botany* 37:413-428.

Whitlock, C., Bartlein, P., Marlon, J., Brunelle, A. and Long, C. 2003. Holocene fire reconstructions from the northwestern US: an examination at multiple time scales. Fifth Symposium on Fire and Forest Meteorology American Meteorological Society.

Whitlock, C., Skinner, C.N., Bartlein, P.J., Minckley, T. and Mohr, J.A. 2004. Comparison of charcoal and tree-ring records of recent fires in the eastern Klamath Mountains, California, USA. *Canadian Journal of Forest Research* 34:2110-2121.

19

Multi-disciplinary investigation of 19th century European settlement of the Willunga Plains, South Australia

Tim Denham
School of Geography and Environmental Science, Monash University, Clayton, Victoria
Tim.Denham@arts.monash.edu.au

Carol Lentfer
University of Queensland, St. Lucia, Queensland

Ellen Stuart
Flinders University, Adelaide, South Australia

Sophia Bickford
Monash University, Clayton, Victoria

Cameron Barr
University of Adelaide, Adelaide, South Australia

Introduction

The arrival of Europeans in Australia has been described as an 'apocalyptic event for Australian ecosystems' (Adamson and Fox 1982:110). It is generally *assumed* that subsequent transformations of Australian biota and landscapes have been more dramatic than those made by Aborigines over tens of millennia (Young 1996:72). However, there is limited scientific data with meaningful temporal resolution (i.e. decadal or subdecadal) that *shows* the nature, extent and rate of transformation concomitant with European colonisation of landscapes in Australia (e.g. Dodson et al. 1994a, 1994b; Gale et al. 1995; Mooney 1997; Haberle et al. 2006; see review in Dodson and Mooney 2002).

There are several comprehensive reviews of the impacts of European colonisation and agriculture on the Australian landscape (Adamson and Fox 1982; Hobbs and Hopkins 1990; Young 1996). Until recently, there was a dearth of high-resolution records that tracked the environmental effects of European colonisation and changing land uses through time (Lunt 2002). Most palaeoenvironmental studies of landscape change in Australia have tended to rely heavily on palynological and microcharcoal data, which have proven to be problematic (Kershaw et al. 1994) and remain to be more fully explored (Dodson and Mooney 2002:455). Some studies, especially more recent ones, have been broadened to include a wider range of palaeoecological proxies and multi-proxy research (e.g. Lentfer et al. 1997; Haberle et al. 2006; Bowdery 2007).

In an attempt to address these issues, a multi-disciplinary study was undertaken to reconstruct past environments at the California Road Wetland in the Ingleburn Creek catchment on the Willunga Plains. The research was devised to address three inter-related themes: first, to assemble a high-resolution palaeoecological record for the 19th century European colonisation of the Willunga Plains, thereby complementing similar investigations on the Fleurieu Peninsula (Bickford 2001; Bickford and Gell 2005; Bickford et al. 2008); second, to use historical background research to reconstruct the transformation of the Willunga landscape during this period (Stuart 2005, 2006), and to synthesise this historical reconstruction with the palaeoecological record; and, third, to assess the value of multi-proxy palaeoecological investigations for understanding environmental change during the recent past (i.e. past 250 years), with a particular focus on the complementarity of pollen and phytolith analyses for vegetation reconstruction and the identification of recently introduced exotics. In other words, the project was originally intended to address two sets of issues: substantive, i.e. to construct a high-resolution, multi-proxy record of landscape transformation during European colonisation; and, methodological, i.e. to integrate historical-palaeoecological reconstructions and assess the complementarity of pollen-phytolith analyses.

The Willunga Plains

> ... level country stretching for miles; it is of the richest character, and is covered with so long and thick an herbage that it is quite laborious to walk through it. There are numerous woods, of a very open description and some spots where the scenery resembles an Englishman's park... Here was a most luxuriant soil, in some places level and commanding an extensive view; in others having vistas through rows of elegant trees; at others the view is bounded by boldly shaped hills intersected by deep ravines ... (John Morphett, letter home, 1836, describing the plains behind Aldinga while searching South Australia for new places for colonial settlement; cited in Vaudrey and Vaudrey 1991:10).

The Willunga Basin comprises 100,000 ha of relatively low-lying land defined by an escarpment of the Mount Lofty Ranges to the south and east, and includes the Willunga Plains extending west to the coast (Figure 1 upper). On April 13, 1844, the *Adelaide Observer* described surface water on the plains to be scarce, but in 'winter almost every glen and ravine has water in it; but the little rivulets soon run to waste, and after a few warm days they dry up' (unattributed, in Stuart 2005:18). As well as seasonal variability, there was great spatial variability in the availability of water on the plains; some settlers' wells were sunk to 180 feet, while others to only 10 feet (Dunstan 1977:20; see Newman and Lawrence 1999).

Surface soils across the Willunga Basin were described by 19th century surveyors as varying greatly in texture, from sands, sandy loams, black and chocolate loams to yellow clays. The floor of the basin contains 'gilgai' soils, also referred to as 'Bay-of-Biscay' soil (Overton 1993:14). Soils in the region tend to be alkaline and deficient in a wide range of mineral elements, including phosphorus (P), nitrogen (N), zinc (Zn), molybdenum (Mo), sulphur (S) and manganese (Mn) (Northcote 1976:65). Despite these deficiencies, the plains have long-been perceived as fertile and well-suited to horticultural and pastoral uses (e.g. Hawker 1901: 52).

The Willunga Plains have an average rainfall of ca. 650 mm, with highest average rainfall from May to August (ca. 88 mm per month) and lowest average rainfall from December to February (less than 25 mm per month) (Bureau of Meteorology 1985). Seasonally, the coolest and wettest months occur during winter, while summer is generally hotter and drier, with high evapotranspiration, low relative humidity and soil water deficit.

Before European settlement, the Willunga Plains comprised open woodlands dominated by *Eucalyptus leucoxolyn*, *E. odorata* and *Allocasuarina verticillata*, with a herbaceous understorey of grasses, including perennial grass (*Themeda australis*) and scattered low shrubs of *Acacia* spp., *Callitris* spp. and *Melaleuca* spp. (Specht 1972:34; see Newman 1994). The uplands of the watershed supported sclerophyll forests of *E. obliqua* and *E. baxteri* on the steep westward slopes, with associated shrub understorey, including species in the Myrtaceae, Proteaceae and Xanthorrhoeaceae families (Specht 1972:34; Bickford and Gell 2005:201). Following survey in 1839, the plains underwent settlement and transformation into an agricultural landscape. Small pockets of native vegetation communities only survive in some wetlands, gullies and along the top of Sellicks Hill Range.

A brief history of European settlement

The Aborigines – the Kaurna – who lived, foraged and hunted in the region, survived by exploiting food and water resources, of which newly arrived European colonists were largely ignorant. Unfortunately, there is no detailed archaeological information of Kaurna subsistence on the Willunga Plains before European settlement.

In 1839, James Hawker, Colony Assistant Surveyor, was assigned to survey the road from 'Horseshoe' (Noarlunga) to Willunga in a 'special survey' of District C (Hawker 1901). This signalled the first official movement of settlers to the southern region, although unsanctioned and undocumented movements may have occurred earlier (see Bickford 2001; also see Gale and Haworth 2002). The Hundred of Willunga was established as a farming region that came to include the townships of Willunga, Aldinga and McLaren Vale. By 1841, 137 Europeans had already settled in the district and the population consistently rose over the following decades (Lewis 1936).

Clearing the land was a normal activity of farm work and, because demand for timber was high for construction and other uses, tree-felling proved a lucrative occupation for some (Williams 1992:27). Licensed timber-cutters denuded the hills and plains of large eucalypts and shrubs, thereby assisting property owners to clear the land (Dunstan 1977:10). No type of timber was left untouched; even the drooping branches and leaves of *Allocasuarina verticillata* were cut to provide fodder for cattle and sheep (Bickford 2001:33). Land was cleared of less useful timber by ring-barking and burning.

The first settlers on the plains established mixed farms adjacent to water springs and permanent water sources. Cattle and sheep were raised and crops of wheat, barley, oats, maize and potatoes were grown with varying success. Santich (1998:94) and British Parliamentary Papers (BPP 1970 [1843]:101) provide an account of one settler's early years. After arriving in McLaren Vale in January 1840, Charles Hewett and his family took up their sections, named 'Oxenbury Farm', and improved them with post and rail fencing, sheep pens, a dairy and

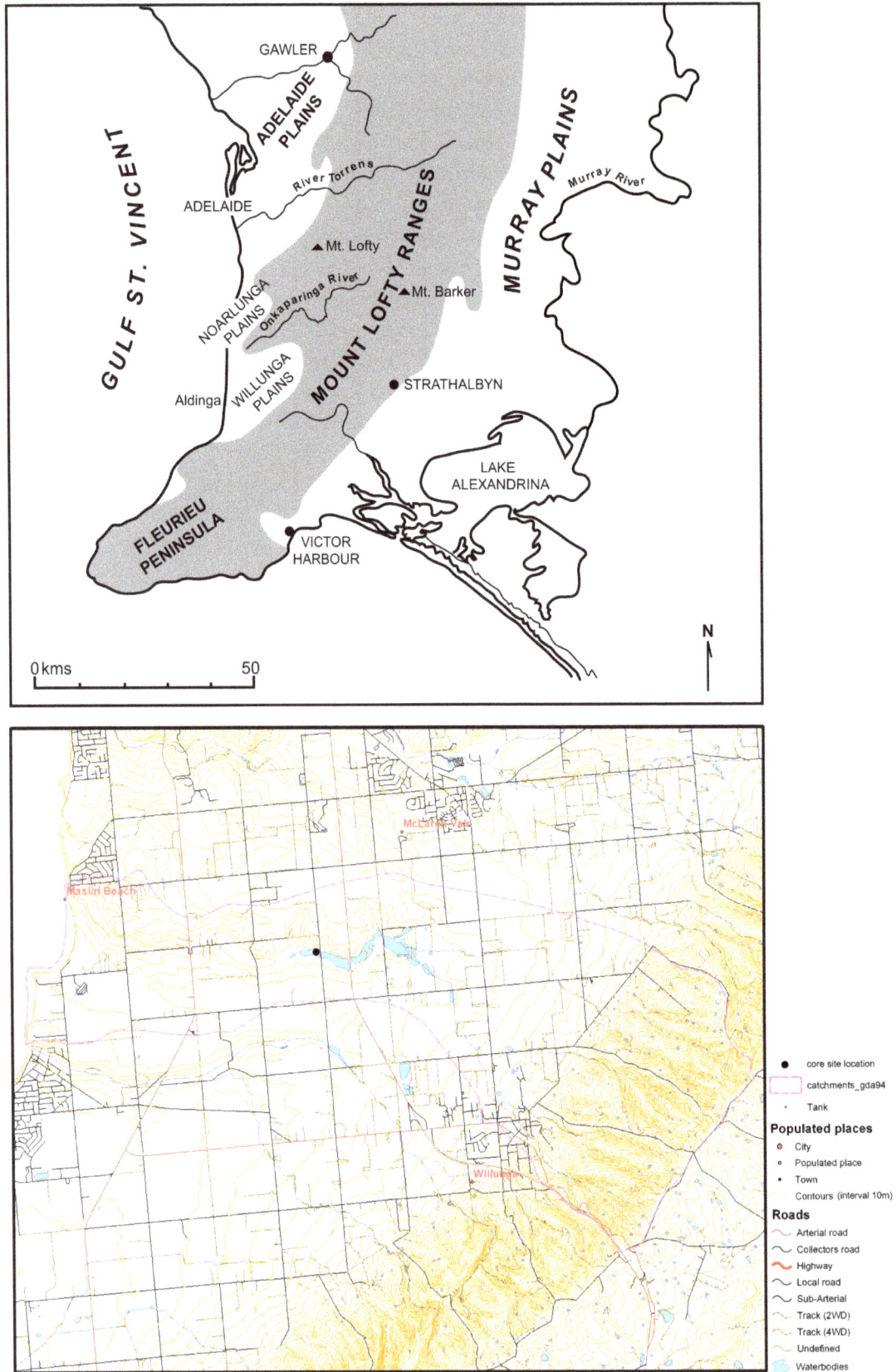

Figure 1. Upper: Map of the Willunga Plains in the Adelaide region of South Australia. Lower: Location of the California Road Wetland coring site on a map of Maslin's Creek catchment (based on information provided by Planning SA and produced by Adelaide Hills Face Zone project, Flinders University, South Australia).

stockyards (BPP 1970 [1843]:101). In the interim, the family made its home in the hollow of a large gum tree (Santich 1998:94). Yields varied, as the crop of wheat sown in June was large, reportedly 40 bushels per acre, while that sown in July was poor, and potatoes planted in July and August failed (BPP 1970 [1843]:101).

Early agricultural production in the Willunga region focused on cereals, primarily wheat (*Triticum aestivum*), barley (*Hordeum distichon*), rye (*Secale cereale*) and oats (*Avena sativa*). Yields were initially good and returns high (Hallack 1892:1). Over the next 20 years, cereal production steadily increased such that surpluses were produced for export from the region. However, continuous wheat cropping rapidly exhausted soil nutrients, which already had low concentrations of phosphorus and nitrogen (Northcote 1976:65). Harvest yields reduced dramatically and sparked an exodus of farmers to seek new pastures (Hallack 1892:1; Whitelock 1985:260; Linn 1991:85). Wheat cultivation decreased rapidly from 1865 to 1889, leading to the closure of flour mills, while coastal jetties became 'tombstones to a departed industry' (Hallack 1892:1). By 1938-39, there were no wheat sales for the region (Charlick 1939), although other fodder and cereal crops remained important: initially hay and then barley from the 1940s being predominant.

Those farmers who remained in Willunga focused on the production of clover (*Trifolium* spp.) and lucerne (*Medicago sativa*) for hay, and oats for sheep and cattle feed, and diversified into fruit and market gardening for local and Adelaide markets (Santich 1998:53). Horticulture was well-established in other colonies by this time, and settlers used their experiences and acquired skills to yield good crops of fruits and vegetables in often unfamiliar environments.

The diversity of fruit production that followed the demise of cereals was on display at the Willunga show in 1883: 'apples, pears, quinces, peaches, damsons, grapes and mulberries, together with dried apricots, peaches, currants and raisins, preserved figs and softshell almonds' (Santich 1998:53). Although olives (*Olea europaea*) were brought to South Australia with the first arrivals on the *Buffalo* in 1836, they were not extensively planted until the mid 20th century (Santich 1998:59), with a marked increase since 2003.

The most significant arboricultural crop associated with the Willunga region was almond (*Prunus amygdalus*). Initially, trees were planted in mixed orchards and small lots, but the size of plantings increased following the development of new varieties. From the early 19th century production increased to meet increasing demand, such that by the 1970s more than half of Australia's almonds were grown there (Santich 1998:58). During the 1990s, the profit potential of wine grapes lured many growers to remove almond trees and other orchards and replace them with vines; the focus of the almond industry has now shifted to the Riverland region of South Australia.

John Reynell established the first South Australian vineyard in 1838 at Reynella on the Noarlunga Plains north of Willunga (Richardson 1936:63). Wine production did not emerge as a significant economic activity until the second half of the 20th century. Since 1980, vines have become the dominant economic crop plant on the Willunga Plains, even though some vineyards in the south have recently given way to olive groves.

In summary, 19th century colonists in South Australia deliberately sought to transpose European land management systems on to the landscape (Hobbs and Hopkins 1990:93; see Williams 1974). The inappropriateness of European crops and farming technology to the Willunga Plains has led, in part, to successive transformations in agricultural practices. As for other parts of the continent, the adjustment of European land-use practices to the Australian environment, or process of landscape learning (see Rockman 2003), has been slow, has been partially imposed through necessity, and is still ongoing.

California Road Wetland

The California Road Wetland is relatively large for the area, permanent and located in the Ingleburn Creek catchment, also known as Maslin's Creek (Figure 1 lower). The site is an open wetland adjacent to a permanently wet and periodically flowing creek. The creek and wetland have been dissected by roadways which, despite the installation of under-road drains, have affected and impeded drainage. Topography is gentle, with low gradient slopes along the edge of the creek and wetland. Sediments and microfossils accumulating in the wetland comprise in situ biogenic, as well as detrital colluvial and fluvial, components.

During McLaren's survey of 1840, the surveyor noted a wetland in this locale, with stands of woolly ti-tree (*Leptospermum lanigerum*), which are still common today with surrounding bulrush (*Typha domingensis*) and common reed (*Phragmites australis*). This shrubland assemblage was characteristic of wetlands on the Willunga Plains at that time, with the California Road Wetland being one of the few places where it has survived today. In and around the wetland are other indigenous plants, such as native water parsley (*Berula erecta*), exotic weeds, such as salvation jane (*Echium plantagineum*), and cultivar escapees, including bearded oat (*Avena barbata*) and olive (*Olea europaea*). Slopes adjacent to the wetland are currently cultivated with vines (*Vitis* sp.).

Field methods

Two 1 m sediment long cores were extracted in the field using a D-section corer. These cores were subsequently stored under refrigeration at the Department of Geographical and Environmental Studies, University of Adelaide. Only Core 1 was subsampled for multiple types of dating, palaeoenvironmental and sedimentological analyses. Subsamples of ca. 1 g (dry weight) were collected at ca. 5 cm intervals for matched, or paired, microfossil analyses. Analyses were undertaken following standard procedures for diatoms (Battarbee 1986), phytoliths (Powers et al. 1989; Parr 2002), pollen (Faegri and Iversen 1989) and microcharcoal (Clark 1982). Moisture content and loss on ignition (LOI) measurements were undertaken on 1 cm thick subsamples collected every 2 cm down the core except at 48-49 cm (Rowell 1994); however, the cores were inadvertently thrown away before detailed stratigraphic descriptions could be completed. Three bulk sediment subsamples were submitted to the Australian Nuclear Science and Technology Organisation (ANSTO) to assess the sediments for ^{210}Pb (lead) and ^{137}Cs (caesium) dating. Additionally, samples were processed to obtain a fine fraction for AMS dating from basal sediments, but insufficient carbon was present (Beta Analytic pers comm. 2007).

Multi-proxy results: General trends and diagnostics

Sedimentary description

Distinct differences in water content and loss on ignition (LOI) occurred down the sedimentary sequence (Figure 2). Organic rich, fibrous, peaty matrices characterised all but one of the upper few samples. Most samples in the field appeared to be organic rich, fibrous, peaty matrices; however, samples had highly variable water contents and LOI suggested generally high, although variable in the upper 10 cm, mineral contents. Mineral inputs were predominantly terrestrial, silt and sand-sized particles. Only the light-coloured sands at the base of the core, below approximately 80 cmbs (cm below surface), appeared of significantly different character in the field.

Distribution and abundance of microfossils

The multi-proxy summary diagram (Figure 3) shows marked differences between the distributions and abundance of pollen, phytoliths and diatoms in the sedimentary sequence. Phytoliths were abundant in all residue samples, with the exception of the four lowermost samples and sample 7. In contrast, most samples had poor pollen preservation, with pollen abundance decreasing down the sequence. Preservation was extremely poor in the lower half of the sequence and pollen was absent in sample 13 and the three lowermost samples 19, 20 and 21. Similarly, diatoms were absent in much of the lower sequence; they occurred intermittently in samples 16 to 5, above which there was a marked increase in abundance and good preservation of unbroken diatoms.

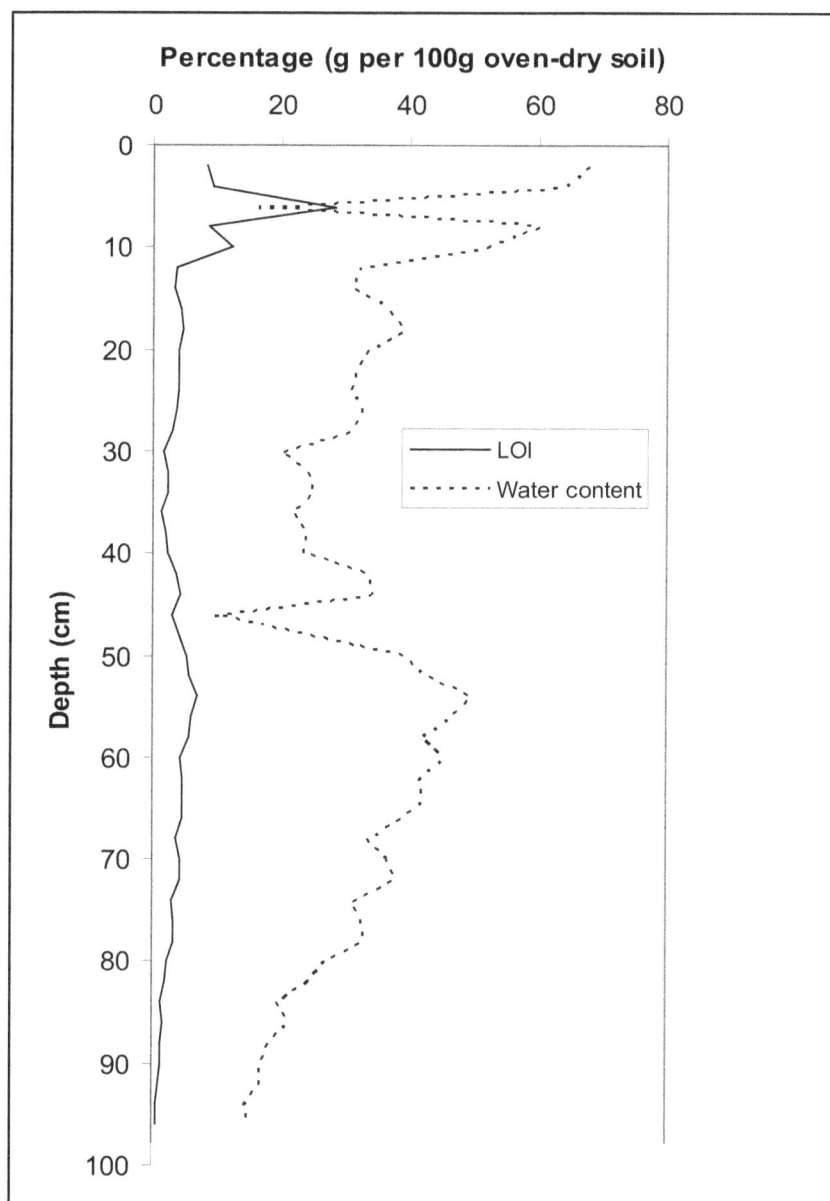

Figure 2. Water content and loss on ignition (LOI) data for subsamples from core 1, California Road Wetland (following methods in Rowell 1994:48).

Pollen

Most samples collected had extremely poor pollen preservation, with pollen abundance decreasing down the core (Table 1; Figure 3). The majority contained too few grains to make analysis meaningful. Wetland species – i.e. derived from on-site vegetation – predominated, with

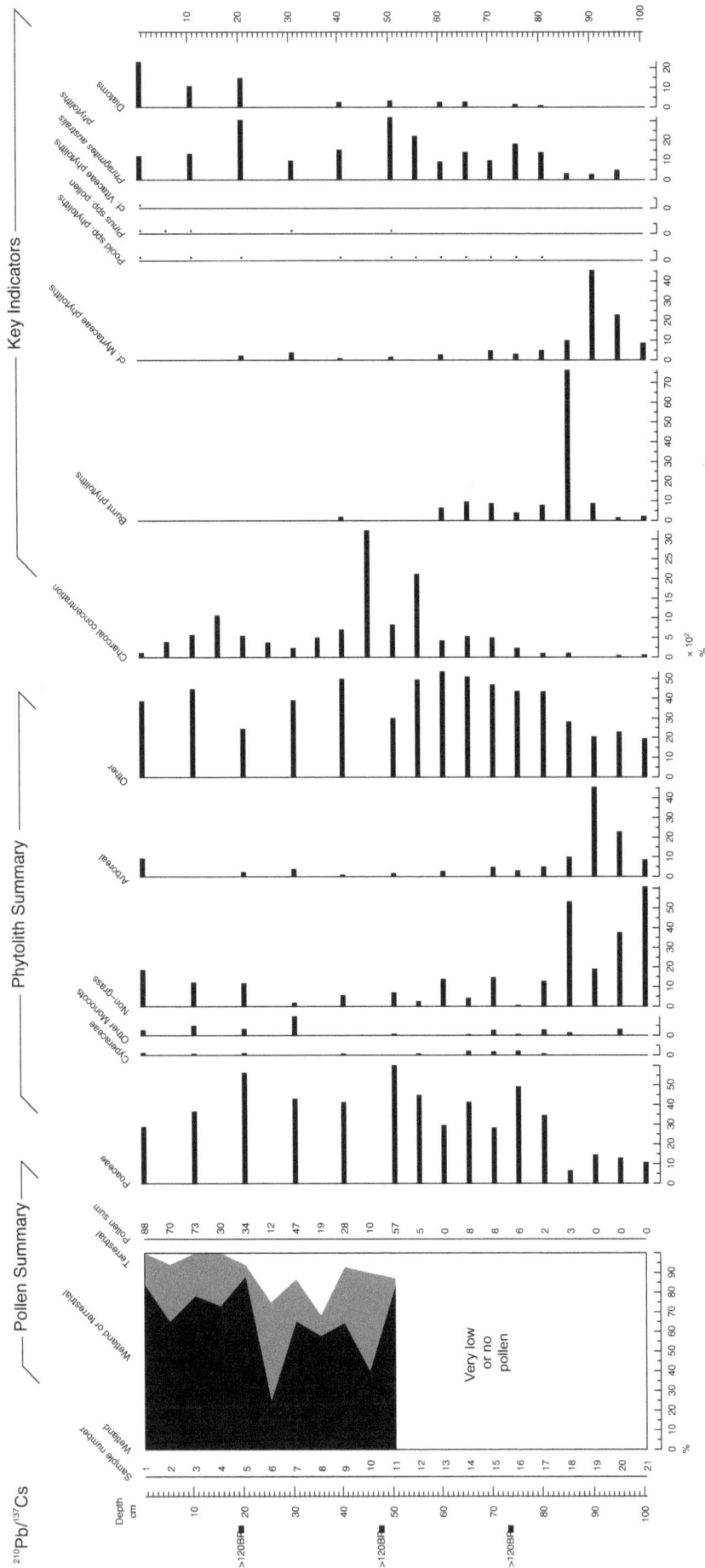

Figure 3. Summary of multi-proxy palaeoecological data (pollen, phytoliths, microcharcoal and diatoms) for core 1, California Road Wetland. Pollen percentages are provided for wetland, wetland or terrestrial, and terrestrial types only when pollen sum ×10. All phytolith and diatom data are given as percentages of total phytolith sum. Dots represent low values <1%. Samples 2, 4, 6, 8 and 10 were not examined for phytoliths.

few off-site terrestrial species present. Although low pollen frequencies occurred throughout the core and hinder interpretation, some highly tentative observations are possible largely based on presence/absence.

The assemblage is dominated by Cyperaceae and Halogoraceae from sample 17 upwards, and indicates on-site vegetation growing in the wetland. *Leptospermum* pollen only occurs in samples 1, 2 and 5 (Table 1). By contrast, the terrestrial component has much lower representation throughout and especially in the upper five samples, where frequencies were recorded as being absent or very low. Moderate terrestrial pollen frequencies occur in samples 6 to 11. Asteraceae pollen is present from sample 10 to the top, and absent in samples below. Poaceae pollen is present in the upper part of the sequence (from sample 6). Most notably, arboreal species are low throughout. Single *Eucalyptus*-type grains representing Myrtaceae occur in only two samples (8 and 11); *Banksia marginata* is very rare and was recorded in the middle of the sequence (samples 7, 8, 11 and 15); and *Allocasuarina verticellata* occurs sporadically throughout

Table 1. Summary of pollen grains and fern spore counts for core 1, California Road Wetland.

Pollen/Spore type	Sample																				
	1	2	3	4	5	6	7	8	9	10	11	12	13	14	15	16	17	18	19	20	21
Wetland																					
Cyperaceae - *Baumea* type	56	25	22	9	16	3	17	8	3	1	19	1		2	2	1	2				
Cyperaceae - *Scirpus* type	9	14	18	11	6		11		13		19	2		3							
Restionaceae		1																			
Halogoraceae *(Myriophyllum/Gonocarpus)*	1		13	2	5		1		1		6			1							
Monolete fern spore			1					2							1						
Trilete fern spore					1			1	2	1	2	1			1		1				
Goodenia		1																			
Geraniaceae							1		1												
Leptospermum	4	4		2																	
Wetland or Terrestrial																					
Asteraceae - Tubuliflorae	2	5	4	2	1	1	2	2	4	2											
Asteraceae - Liguliflorae	6	8	2	6		3	8		4	3											
Poaceae > 50 um	2	1	6						2												
Poaceae < 50 um	3	6	3		1	2															
Terrestrial																					
Myrtacaeae - *Eucalyptus* type							2		1		1										
Banksia marginata							1	2			2				1						
Allocasuarina (prob *verticellata*)		2			2	1	3	3	1		1	1		2	5	3		2			
3C prolate 30µm								1			3										
Chenopodiaceae								1		1											
Dodonaea		2							1												
Pinus	5	1	4				1				2										
Total grains	**88**	**70**	**73**	**30**	**34**	**12**	**47**	**19**	**28**	**10**	**57**	**5**	**0**	**8**	**8**	**6**	**2**	**3**	**0**	**0**	**0**

the middle to lower parts of the sequence (samples 5 to 18) and is absent from all but one of the five uppermost samples.

The European-introduced *Pinus* spp. pollen appears for the first time in sample 11, concomitant with the first major increase in pollen abundance, and it is present sporadically throughout the upper sequence. *Pinus* pollen can be used as a chronological marker, albeit regionally specific, of colonial settlement across Australia (Behre 1986). Pine plantations were not established in the region until the 1930s, which may be reflected in the continuous presence of *Pinus* pollen from sample 3 upwards, with earlier records reflecting planting for ornamentation and landscaping. Given the nature of pollen preservation in samples throughout the core, the absence of *Pinus* pollen in lower samples might not reliably indicate an absence of this introduced genus.

Phytoliths

In contrast to diatoms and pollen, phytoliths were abundant and well preserved in all samples, with the exception of sample 7 and the basal samples 18 to 21, where they were less abundant but still in sufficient frequencies to enable meaningful interpretations of vegetation change for the wetland and its vicinity (Figure 4). Phytolith reference collections for Australian vegetation are currently limited, hindering taxonomic identifications (Clifford and Watson 1977; Hart 1992; Lentfer et al. 1997; Bowdery 1998; Wallis 2000, 2001). Additional reference material was obtained from a number of common plant species currently growing at the California Road Wetland (Table 2); the dry-ashing and self-draining crucible procedure (Lentfer 2006) was used for phytolith extraction from these plant samples. Seventy-six phytolith morphotypes were distinguished from the core samples; diagnostic types were assigned to plant species or groups, and non-diagnostic types were assigned to three general categories: elongates, polyhedrals and stomates.

The phytolith record shows that the lowermost samples, 18 to 21, have a predominance of non-grass and arboreal morphotypes mainly characterised by psilate globular morphotypes, found commonly in Myrtaceae, including *Leptospermum* and *Eucalyptus* spp. (compare Figures 5E and 5F) and other globular morphotypes with nodulose and verrucate surface textures. Grasses, including the arundinoid species *Phragmites australis* (compare Figures 5A and 5B) and possibly *Danthonia*, are present, but compared with the overall sequence, have relatively low frequencies in the basal samples.

There is a distinct and sudden decline in arboreal morphotypes and a marked increase in grass morphotypes directly above sample 18. Morphotypes from *Phragmites australis* become the most prevalent from sample 17 and upwards, persisting throughout the entire sequence. The absence of *Phragmites australis* pollen in these same samples reflects its under-representation, as

Table 2. Phytolith production in some of the most common plants at the California Road Wetland site.

Species	Phytolith production
Leptospermum lanigerum	+
Typha domingensis	-
Carex divisa	+
Gahnia trifida	+
Phragmites australis	+
Avena barbata	+
Bromus catharticus	+
Samolus repens	-
Schoenoplectus pungens	+
Cynara cardunculus	+
Rosa canina	+
Vitis sp.	+
Berula erecta	+

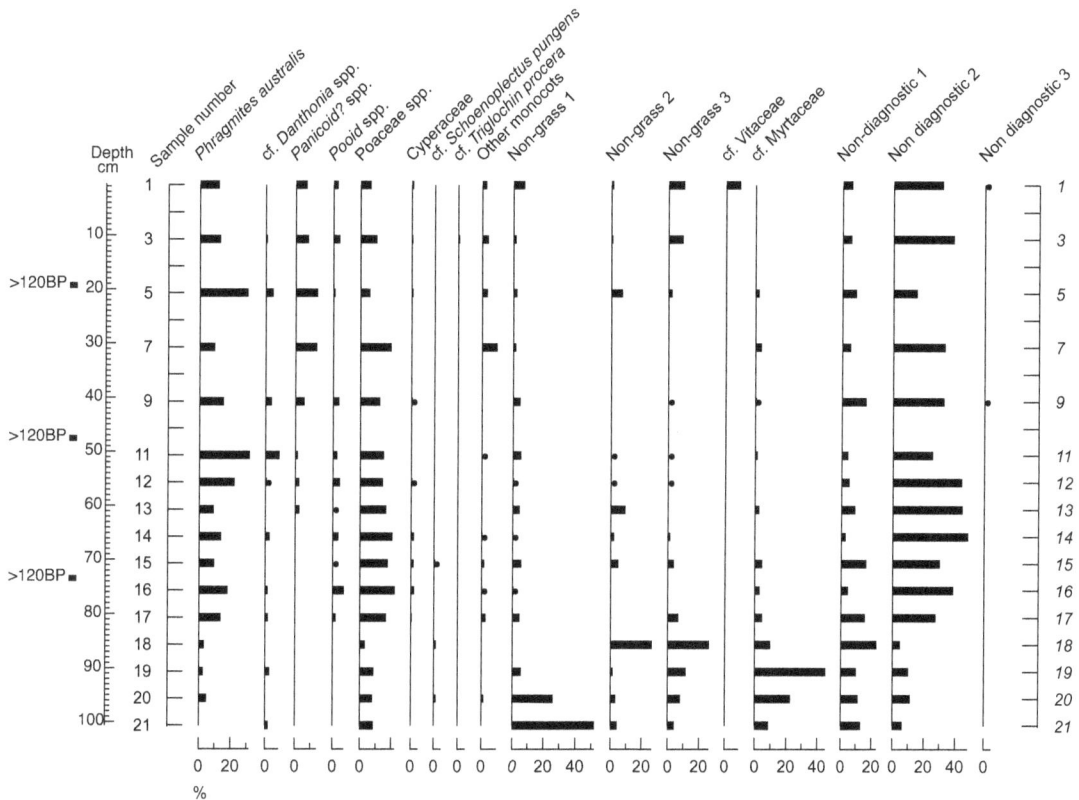

Figure 4. Distribution of phytolith morphotypes in the sedimentary sequence for core 1, California Road Wetland. Percentage values <1 are represented by •.

well as that of *Typha* spp., in pollen spectra generally (Finkelstein and Davis 2005).

Morphotypes typical of pooid grasses including the cereal crops (*Triticum*, *Hordeum* and *Avena* species), related weeds (e.g., *Avena barbata*) and the introduced grass *Bromus catharticus* (Lentfer et al. 1997) appear in the assemblage for the first time in sample 17 and persist throughout the rest of the sequence (Figures 5G and 5H). Larger Poaceae pollen grain types are present from sample 5 upwards, potentially indicative of exotic grasses, and absent from lower contexts due to poor pollen preservation generally. Bilobate morphotypes typical of panicoid grasses, but also present in arundinoid and chloridoid grasses, appear later in the sequence at sample 14, and Cyperaceae morphotypes characteristic of *Ghania* and *Carex* first appear in sample 16 at the same level as the first appearance of diatoms. Finally, morphotypes typical of the grape family Vitaceae (compare Figures 5C and 5D) occur only in the uppermost sample.

Distribution and abundance of microcharcoal and burnt phytoliths

There is a distinct disparity between the distributions and abundance of microcharcoal and burnt phytoliths. Burnt phytoliths, identified from a black, opaque appearance under transmitted light (Lentfer and Torrence 2007; see Figure 6), were recorded in the lower part of the sequence. Most notably, a major peak occurred in sample 18, concomitant with a major decline in Myrtaceae phytoliths. In contrast, microcharcoal was more common in the upper part of the sequence, with lesser and major peaks occurring in samples 12 and 10, respectively.

Dating

The assessment of samples for ^{210}Pb (lead) and ^{137}Cs (caesium) dating suggests that either

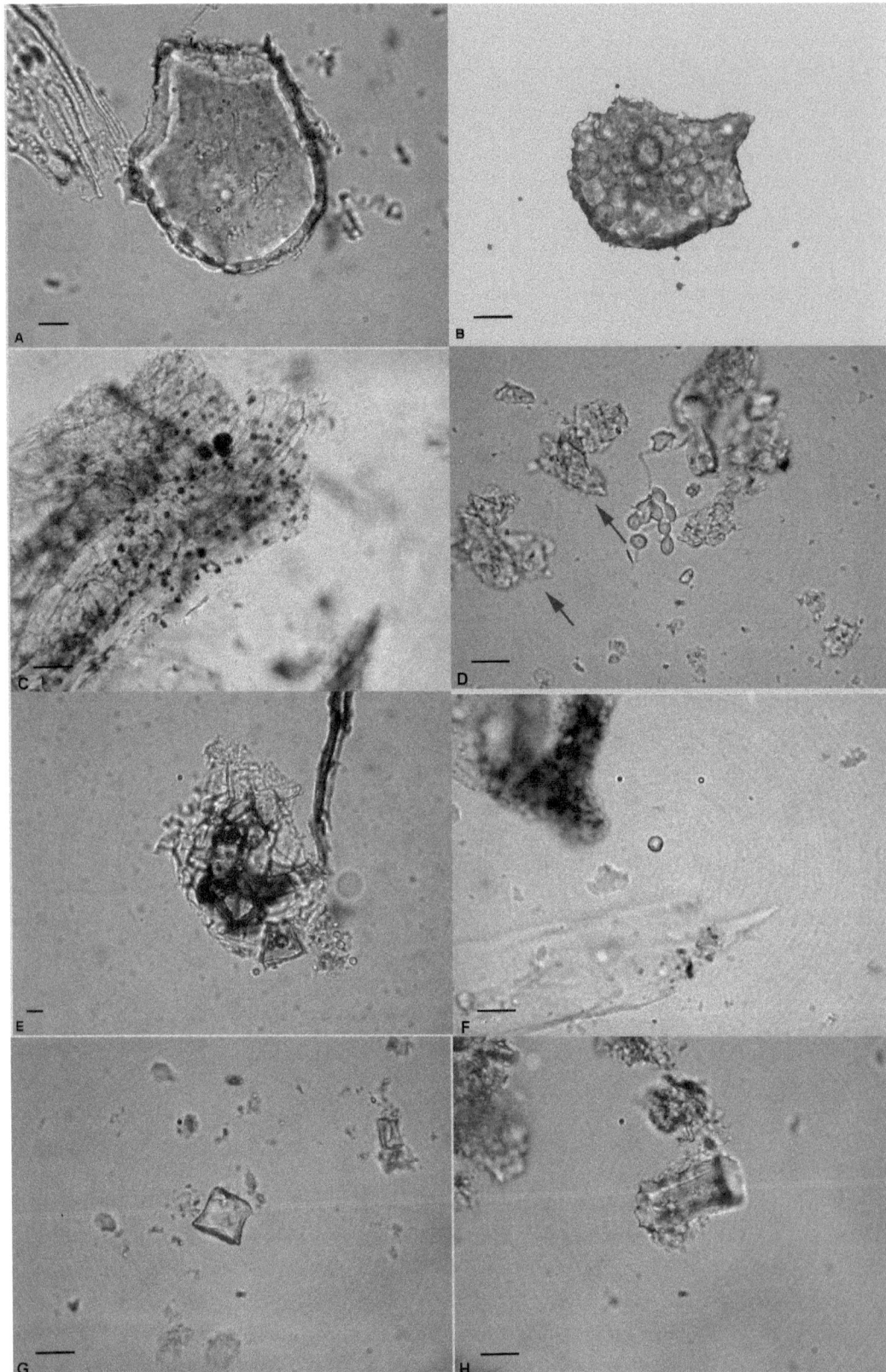

Figure 5. Photomicrographs comparing diagnostic phytoliths from plant reference material and California Road Wetland sediments. Example of large cuneiform (bulliform) phytoliths from *Phragmites australis* **(A)** and weathered cuneiform phytolith extracted from sediment sample 20 **(B)**. Silicified epidermis from cultivated grape vine leaves **(C)** and microfossils found in sediment sample 1 **(D)**. Psilate globular phytoliths from *Eucalyptus* sp. **(E)** and sediment sample 19 **(F)**. Examples of silica bodies (epidermal short cells) typical of pooid grasses, including the cereal crops. Both **(G)** (sample 9) and **(H)** (sample 11) were recovered from California Road Wetland sediments above sample 18 and occur in wheat *(Triticum* spp.) (see Lentfer et al. 1997; Ball et al. 1993, 1999). The scale bar in all panels represents 10 μm.

all three samples chosen were older than the timeframes ordinarily dated by these techniques, or the samples had been diluted with older sediments eroded from the catchment (Jennifer Harrison pers comm. 2005). If taken at face value, this would suggest that all three samples, which were taken above samples 5, 11 and 16 (see Figure 3), were older than at least 120 years. However, the deposition of 'old' sediments in the wetland may have contributed to the greater-than-120-year ages (see Gale et al. 1995 for a detailed examination of this problem). Given uncertainties in the interpretation of the lead and caesium assessments, as well as the

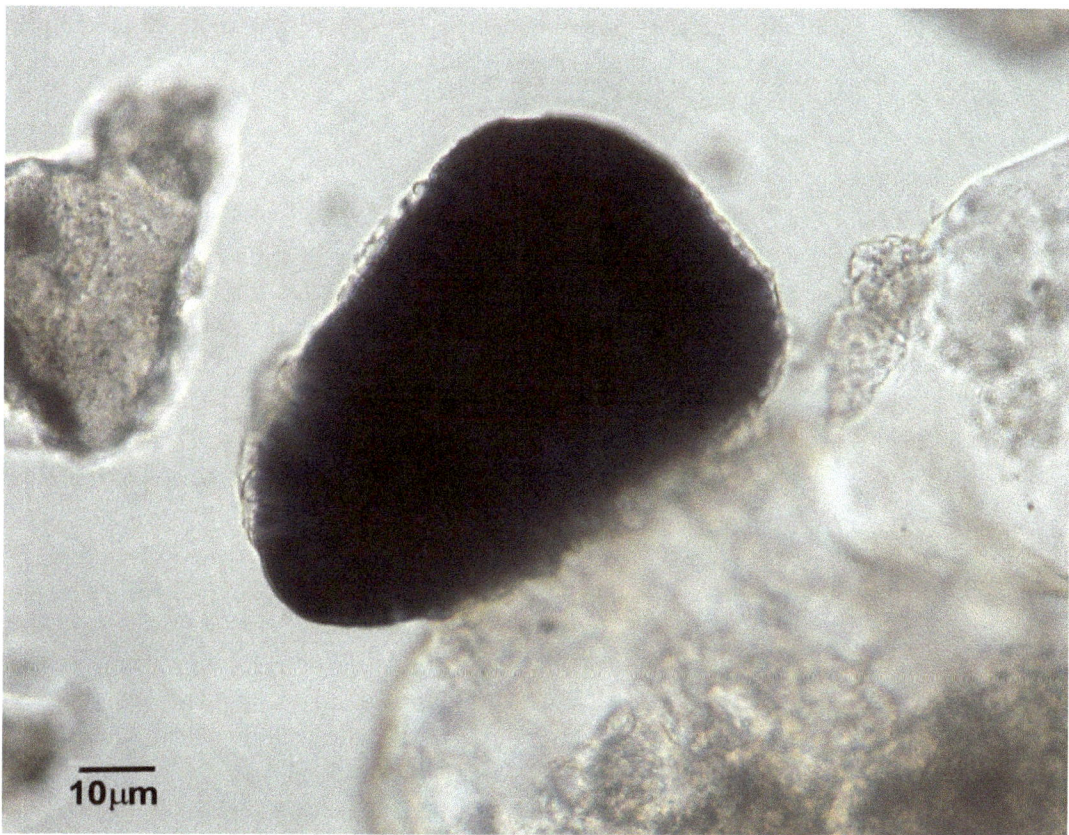

Figure 6. Photomicrograph of a large burnt cuneiform phytolith extracted from sample 14.

inability to obtain a sufficient sample for AMS dating from the basal sediment, relative dating of the stratigraphy has been undertaken using diagnostic pollen and phytolith types. The use of biostratigraphic markers will support an interpretation that all the sediments from directly above sample 5 to the base of the core are at least 120 years old, i.e. pre-date ca. 1880.

Dating, diagnostics and resolution

In the absence of a suitable chronometric method, diagnostic pollen and phytoliths have been relied on to provide approximate and relative indicators of age. *Pinus* pollen, cereal phytoliths and grape phytoliths can be interpreted with respect to a land-use history of the vicinity to provide relative and reasonably reliable chronological markers; they appear reliable because they first occur in the correct chronostratigraphic order as inferred from the land-use history (see Figure 3). Land-use histories for the California Road Wetland vicinity suggest woodland clearance and mixed farming, including cereals, occurred in the 1840s, with cereals persisting into the 20th century. Viticulture did not occur in this area of the Willunga Plains until the late 20th century.

Based on the distribution of pollen, phytoliths and microcharcoal, the majority of the sediment accumulated rapidly in the 19th century. Cereal phytoliths first occur in sample 17, following a major burnt phytolith peak and shift in vegetation, interpreted to represent land clearance in the early 1840s (see below). Sample 11 contains the earliest occurrence of *Pinus* pollen, and grape phytoliths occur in only the uppermost sample (sample 1), which is anticipated given that it is primarily a late-20th century crop in this area. To compare, the ^{210}Pb dating assessment suggests sample 5 dates to at least the 1880s, although it may be more recent if diluted with old sediment. Taking all the chronological markers together, samples 17 to 5 (or above) may represent fewer than 40 years; rapid deposition may account for the rarity of diatoms below sample 5. Thus, the upper four samples (or fewer) represent at least the past 120 years. Consequently, the majority of the stratigraphy probably represents rapid deposition following initial clearance in the mid 19th century.

Other sites in Australia witness similar depositional trends, as well as problematic ^{210}Pb dating and pollen preservation, with rapid sedimentation following initial European colonisation (Gale et al. 1995; Haworth et al. 1999). For example, Gale et al. (1995) show extremely high sedimentation rates during the first decade or two of European settlement for a lake in the New England Tablelands of eastern New South Wales; these are inferred to result from land clearance for pastoralism. At California Road Wetland, rapid sedimentation followed land clearance for mixed farming and very low sedimentation has occurred for the past 120 years. The effects of such rapid erosion on often fragile and ancient soils must have been severe, and might have contributed to the nutrient deficiencies of soils in the Willunga Basin.

Thus, the majority of the core appears to fall into a problematic period; sediments below 20 cmbs seem to be older than the minimum reliability of ^{210}Pb dating in alluvial and colluvial settings, namely ca. 120 years (after Gale et al. 1995), and sediments above at least 80 cmbs contain pollen or phytoliths of exotic species, post-date 1840 and are too young to date with any precision using radiocarbon dating (whether AMS or conventional). Analysis of the California Road Wetland core using diagnostic pollen and phytoliths as chronological markers provides a reasonable, but relatively low resolution, solution to resolving dating problems at this site to clearly show the dramatic transformation of the landscape during the first 40 years or so of European colonisation.

Comparing historical and palaeoecological records

The results of the multi-proxy palaeoenvironmental analysis are indicative of marked changes in vegetation at the California Road Wetland site as a result of European settlement. The distribution of phytoliths shows that the vegetation represented in the lowermost levels was characterised by myrtaceous woody shrubs and trees with little grass ground cover. This type of vegetation cover seems to indicate a more closed woodland locally, perhaps in the proximity of creek-line vegetation, as opposed to the open woodland that is ordinarily considered to have existed on the Willunga Plains before European settlement.

Directly following a major burning event, represented by burnt phytoliths in sample 18, there was a rapid decline in arboreal vegetation and an increased predominance of grasses, dominated by the locally growing common reed *Phragmites australis*, perhaps indicating eutrophic conditions locally. The pollen analysis corroborates the phytolith evidence by showing a predominance of wetland sedge and Halogoraceae species in the middle-to-upper sequence above sample 18. Successive European influences are signalled by the presence of pooid grass phytoliths, *Pinus* spp. pollen, and grape phytoliths.

Pooid grass phytoliths appear for the first time in sample 17, immediately after the first major burning event, and most likely represent introduced cereals – wheat, barley and oats, closely related introduced weeds such as the bearded oat *Avena barbata*, and introduced forage grasses. Unfortunately, inflorescence phytoliths that would have enabled more precise identifications of specific cereal crops and grasses (Lentfer et al. 1997; Ball et al. 1999) were not observed in the phytolith assemblages. In contrast with the poorly preserved palynological record, pollen grains identified as potentially cereal occur first in sample 11. The presence of these pooid grasses accords with the introduction of mixed farming by the first European settlers in the early 1840s.

Pinus pollen occurs in the middle of the sequence (from sample 11), concomitant with the first good preservation of pollen. The possibility that *Pinus* was present earlier but its pollen was not preserved in the sequence is not discounted. Like the introduced pooid grass phytoliths, *Pinus* pollen in this part of South Australia is indicative of settlement during the 19th century. The presence of grape phytoliths in the uppermost sample represents the recent transformation of this area to viticulture. The absence of almond pollen is not necessarily surprising, given poor pollen preservation in most samples, and because, according to historical records, almonds have not been significant on adjacent slopes.

The appearance of introduced species, the peak in burnt phytoliths and the transformation of arboreal vegetation to an open landscape are signatures of intentional clearance of bushland vegetation for farming. Historic records indicate initial clearances for farming in this part of the Willunga Plains occurred in the 1840s. Sample 18 marks the 'apocalyptic' transformation of an open woodland landscape to a farmed landscape.

Fire histories

Flammability has been a feature of the Australian landscape since before human colonisation of the continent (e.g. Bradstock et al. 2001), and burning was characteristic of Aboriginal land management practices, whether to increase foraging opportunities (Jones 1969; Gott 2005) or to increase game and facilitate hunting (Bowman et al. 2001). Given the associations of the burnt phytolith peak with the transformation of an arboreal landscape to an open landscape and with European-introduced exotics, it is taken to be more representative of land clearance associated with initial colonisation of the area in the 1840s than the microcharcoal record. European settlement and clearances in the region entailed clear-felling and the burning of unused vegetation material, with subsequent ploughing.

The burnt phytolith and microcharcoal records reflect varying proxy levels of burning down the sequence, with clear, but asynchronous peaks. The disparity between the two microfossil records is intriguing and two sets of processes can be proffered to account for the observed phenomena. Firstly, the source areas for both phytoliths and microcharcoal are predominantly the same, namely the swamp (in situ), adjacent slopes (colluvial) and the catchment (fluvial). Asynchronous microfossil peaks could plausibly represent differential depositional pathways from local and extralocal sources, perhaps with the phytolith assemblage being anticipated to exhibit a greater local component than microcharcoal.

Secondly, the disparity could indicate distinct fire regimes at different times, most likely associated with changing vegetation and land uses. The burnt phytolith peak may derive from the more intense fires associated with the clearance and burning of myrtaceous woodland, whereas burnt phytoliths are largely absent from the less intense fires occurring in the more open, post-1840 farmed landscape. As to why microcharcoal is absent from the earlier burning event, but more frequent in later, less intense fires, this is uncertain, although it has been noted that 'carbonised particles are particularly well preserved when cool fires burn in grass and sedgelands; fires in forests tend to be more complete in reducing fuel to ash which can leach and wash away' (Gillieson et al. 1989:111).

At present, there is insufficient data on the comparable taphonomies of microcharcoal and burnt phytoliths to interpret what the asynchronies between the two records represent. However, the asynchronous distribution of burnt phytoliths and microcharcoal raises a significant and unresolved methodological issue for the interpretation of fire histories in the Australian landscape. As with previous debates about the relative merits of macrocharcoal (not analysed during this study) and microcharcoal as an indicator of fire regimes (see Whitlock and Larson 2001), the relative worth of burnt phytoliths for the reconstruction of fire histories needs to be more fully explored.

Conclusions

A multi-proxy record indicates the degree of ecological transformation associated with European colonisation of the Willunga Plains during the second half of the 19th century. European colonisation during the 1840s was 'apocalyptic' in terms of vegetation changes and, presumably, for associated soils (although not investigated in detail here); open woodland was rapidly transformed into an open, farmed landscape. Unfortunately, dating problems have enabled only a relatively coarse-grained interpretation of palaeoecological transformation to be reconstructed, a relatively common problem at sites in Australia for the period of early European settlement.

Although of only limited chronological resolution, the study has methodological value. The study reaffirms the value of integrating historical research with multi-proxy palaeoecology in order to understand landscape change in Australia. The use of both approaches provides complementary and mutually corroborating data sets that, on the one hand, enable some scientific measure of historically documented events and, on the other hand, refine and enliven interpretations of palaeoecological signals. Too often, historical background studies are a cursory appendage to palaeoecological research. In this case, the history enables calibration of the relative chronology derived from microfossils. However, history alone does not usually provide a reliable measure of the scale of environmental impacts, which were effectively catastrophic for the ecology of this area in the mid 19th century. This type of study serves as an experimental analogue for places and time periods for which historical records are limited or absent.

Most significantly, the study demonstrates the value of multi-proxy investigations using paired phytolith and pollen analyses in the Australian context. As demonstrated in studies across the world (e.g. Piperno et al. 1991; Denham et al. 2003), but an approach previously not undertaken in Australia, paired phytolith and pollen analyses provide a more robust reconstruction of vegetation and land-use changes in the past than using one method alone. At the California Road Wetland site, and despite limited phytolith reference collections for South Australian flora, phytoliths have proven especially significant because of poor pollen preservation. This study demonstrates the enormous potential of phytolith research to shed light on past human-environment interactions in Australia over the short term (timescales of the past 500 years), as well as its well-documented capacity to shed light on the long term (timescales of millennia and tens of millennia; Bowdery 1998; Wallis 2000).

Multi-proxy (pollen and phytolith) diagnostics are significant for overcoming a 'chronological gap' between the older limits of 210Pb dating and the younger limits and precision of radiocarbon dating. The identification of multiple diagnostic pollen and phytoliths derived from exotics has provided chronological guides – taken together with general palaeoecological trends, fire histories and the historical records – for initial land clearance and mixed farming (pooid cereal-type phytoliths), the early period of European settlement (*Pinus* pollen), and recent viticulture (Vitaceae phytoliths).

Acknowledgements

Denham directed the palaeoecological project and obtained funding from a Flinders University research grant (with Pam Smith) and from the Australian Research Council-funded Adelaide Hills Face Zone project (courtesy of Pam Smith and Donald Pate). Fieldwork was undertaken in 2004 by Barr, Bickford, Denham and Stuart. Microfossil analyses were conducted by Barr (diatoms), Bickford (pollen and microcharcoal) and Lentfer (phytoliths). Stuart undertook a historical archaeology, focused on agricultural history, of the Willunga Basin, as well as the sedimentological analyses. The authors thank Jennifer Harrison (radioanalyst, ANSTO) for undertaking the ^{210}Pb (lead) and ^{137}Cs (caesium) dating; Keryn Walshe (South Australian Museum) for information on the archaeology of the Willunga Plains; Kale Sniderman for preparation of pollen slides, and, with Phil Scamp and Kara Valle (SGES, Monash), for preparation of the figures; John Dodson, Simon Haberle and Scott Mooney for constructive criticism of an earlier draft of this paper; and anonymous reviewers for comments on the final draft.

References

Adamson, D.A. and Fox, M.D. 1982. Change in Australasian vegetation since European settlement. In: Smith, J.M.B. (ed), *A History of Australasian Vegetation*, pp. 109-146. Sydney: McGraw-Hill.

Ball, T.B., Brotherson, J.D. and Gardner, J.S. 1993. A typologic and morphometric study of variation in phytoliths from einkorn wheat (*Triticum monococcum*). *Canadian Journal of Botany* 71:1182-1192.

Ball, T.B., Gardner, J.S. and Anderson, N. 1999. Identifying inflorescence phytoliths from selected species of wheat (*Triticum monococcum*, *T. dicoccon*, *T. dococcoides*, and *T. aestivum*) and barley (*Hordeum vulgare and H. spontaneum*). *American Journal of Botany* 86:1615-1623.

Battarbee, R.W. 1986. Diatom analysis. In: Berglund, B.E. (ed), *Handbook of Holocene Palaeoecology and Palaeohydrology*. Chichester: John Wiley and Sons.

Behre, K.E. 1986. *Anthropogenic Indicators in Pollen Diagrams*. Rotterdam: A.A. Balkema.

Bickford, S.A. 2001. A historical perspective on recent landscape transformation: integrating palaeoecological, documentary and contemporary evidence for former vegetation patterns and dynamics in the Fleurieu Peninsula, South Australia. Unpublished PhD thesis, University of Adelaide.

Bickford, S. and Gell, P. 2005. Holocene vegetation change, Aboriginal wetland use and the impact of European settlement on the Fleurieu Peninsula, South Australia. *The Holocene* 15:200-215.

Bickford, S., Gell, P. and Hancock, G.J. 2008. Wetland and terrestrial vegetation change since European settlement on the Fleurieu Peninsula, South Australia. *The Holocene* 18:425-36.

Bowdery, D. 1998. *Phytolith Analysis Applied to Pleistocene-Holocene Archaeological Sites in the Australian Arid Zone*. BAR International Monograph Series 695. Oxford: Hadrian Books.

Bowdery, D. 2007. Phytolith analysis, sheep, diet and fecal material at Ambathala pastoral station (Queensland, Australia). In: Madella, M. and Zurro, D. (eds), *Plants, Peoples and Places: Recent studies in phytolith analysis*, pp. 134-150. Oxford: Oxbow Books.

Bowman, D.M.J.S., Garde, M. and Saulwick, A. 2001. *Kunj-ken makka man-wurrk* (fire is for kangaroos) landscape burning in central Arnhem land seen through an ethnographic lens. In: Anderson, A., Lilley, I. and O'Connor, S. (eds), *Histories of Old Ages: Essays in honour of Rhys Jones*, pp. 61-78. Canberra: Pandanus Books.

Bradstock, R.A., Williams, J.E. and Gill, M.A. 2001. *Flammable Australia: The fire regimes and biodiversity of a continent.* Cambridge: Cambridge University Press.

British Parliamentary Papers 1970 [1843]. *Colonies Australia.* Vol. 7. Shannon: Irish University Press.

Bureau of Meteorology 1985. *Climatic Statistics for Willunga, Station 023753.* Canberra: Commonwealth of Australia.

Charlick, W. 1939. Wheat buying prior to the Australian Wheat Board. Unpublished office records, State library archival collection, Adelaide.

Clark, R. 1982. Point count estimation of charcoal in pollen preparations and thin sections of sediments. *Pollen and Spores* 24:523-535.

Clifford, H.T. and Watson, L. 1977. *Identifying Grasses: Data, methods and illustrations.* St. Lucia, Queensland: University of Queensland Press.

Denham, T.P., Haberle, S.G., Lentfer, C.J., Fullagar, R., Field, J., Therin, M., Porch, N. and Winsborough, B. 2003. Origins of agriculture at Kuk Swamp in the Highlands of New Guinea. *Science* 301:189-193.

Dodson, J.R., de Salis, T., Myers, C.A. and Sharp, A.J. 1994a. A thousand years of environmental change and human impact in the alpine zone at Mt Kosciusko, New South Wales. *Australian Geographer* 25:77-87.

Dodson, J.R., Roberts, F.K. and de Salis, T. 1994b. Palaeoenvironments and human impact at Burraga Swamp in Montane rainforest, Barrington tops National Park, New South Wales, Australia. *Australian Geographer* 25:161-169.

Dodson, J.R. and Mooney, S.D. 2002. An assessment of historic impact on southeastern Australian environmental systems, using late Holocene rates of environmental change. *Australian Journal of Botany* 50:455-464.

Dunstan, M. 1977. *Willunga Town and District 1837-1900.* Blackwood: Lynton Publications.

Faegri, K. and Iversen, J. 1989. *Textbook of Pollen Analysis.* New York: Hafner Press.

Finkelstein, S.A. and Davis, A.M. 2005. Modern pollen rain and diatom assemblages in a Lake Erie coastal marsh. *Wetlands* 25:551-563.

Gale, S.J. and Haworth, R.J. 2002. Beyond the limits of location: human environmental disturbance prior to official European contact in early colonial Australia. *Archaeology in Oceania* 37:123-136.

Gale, S.J., Haworth, R.J. and Pisanu, P.C. 1995. The ^{210}Pb chronology of Late Holocene deposition in an Eastern Australian Lake Basin. *Quaternary Geochronology* 14:395-408.

Gillieson, D, Hope, G. and Luly, J. 1989. Environmental change in the Jimi valley. In: Gorecki, P.P. and Gillieson, D.S. (eds), *A Crack in the Spine: Prehistory and ecology of the Jimi-Yuat valley, Papua New Guinea*, pp. 105-122. Townsville: Division of Anthropology and Archaeology, James Cook University.

Gott, B. 2005. Aboriginal fire management in south-eastern Australia: aims and frequency. *Journal of Biogeography* 32:1203-1208.

Haberle, S.G., Tibby, J., Dimitriadis, S. and Heijnis, H. 2006. The impact of European occupation on terrestrial and aquatic ecosystem dynamics in an Australian tropical rain forest. *Journal of Ecology* 94:987-1002.

Hallack, E.H. 1892. *Our Townships and Homesteads: Southern district of South Australia.* Adelaide: W.K. Thomas and Co.

Hart, D.M. 1992. A field appraisal of the role of plant opal in the Australian environment. Unpublished Ph.D. thesis, School of Earth Sciences, Macquarie University, Sydney.

Hawker, J.C. 1901. *Early Experiences in South Australia*, second series. Adelaide: E.S. Wigg and Son.

Haworth, R.J., Gale, S.J., Short, S.A. and Heijnis, H. 1999. Land use and lake sedimentation on the New England tablelands of New South Wales. *Australian Geographer* 30:51-73.

Hobbs, R.J. and Hopkins, A.J.M. 1990. From frontier to fragments: European impact on Australia's vegetation. *Proceedings of the Ecological Society of Australia* 16:93-114.

Jones, R. 1969. Fire-stick farming. *Australian Natural History* 16:224-228.

Kershaw, A.P., Bulman, D. and Busby, J.R. 1994. An examination of modern and pre-European settlement pollen samples from southeastern Australia – assessment of their application to quantitative reconstruction of past vegetation and climate. *Review of Palaeobotany and Palynology* 82:83-96.

Lentfer, C.J. 2006. A simple technology for building large phytolith reference collections from modern plant material. *The Phytolitharien* 18:1-6.

Lentfer, C.J., Gojak, D. and Boyd, W.E. 1997. Hope Farm windmill: Phytolith analysis of cereals in early colonial Australia. *Journal of Archaeological Science* 24:841-856.

Lentfer, C.J. and Torrence, R. 2007. Holocene volcanic activity, vegetation succession, and ancient human land use: unravelling the interaction on Garua Island, Papua New Guinea. *Review of Palaeobotany and Palynology* 143:83-105.

Lewis, A.J.P. 1936. *The Official Civic Record of South Australia: Centenary Year, 1936*. Adelaide: The Universal Publicity Company.

Linn, R. 1991. *Cradle of Adversity: A history of the Willunga district*. Adelaide: Historical Consultants.

Lunt, I.D. 2002. Grazed, burnt and cleared: how ecologists have studied century-scale vegetation changes in Australia. *Australian Journal of Botany* 50:391-407.

Mooney, S. 1997. A fine-resolution palaeoclimatic reconstruction of the last 2000 years, from Lake Keilambete, south-eastern Australia. *The Holocene* 7:139-149.

Newman, L.A. 1994. Environmental history of the Willunga Basin 1830s to 1990s. Unpublished BA (Hons) thesis, Department of Geography, University of Adelaide.

Newman, L.A. and Lawrence, R.E. 1999. Hydrological history of the Willunga basin. In: Walker, D. and van der Wel, B. (eds), *Living with Water: Scarcity, security, supply, surplus and sustainability*, pp. 110-117. Adelaide: Hydrological Society of South Australia.

Northcote, K.H. 1976. Soils. In: Twidale, C.V.R., Tyler, M.J. and Webb, B.P. (eds), *Natural History of the Adelaide Region*, pp. 61-74. Adelaide: Royal Society of South Australia.

Overton, I.C. 1993. Willunga Basin geographical information systems report. Unpublished report to Willunga Hills Face Land Care Group, Department of Geography, University of Adelaide.

Piperno, D.R., Bush, M.B. and Colinvaux, P.A. 1991. Palaeoecological perspectives on human adaptation in Central Panama. II. The Holocene. *Geoarchaeology* 6:227-250.

Parr, J.F. 2002. A comparison of heavy liquid flotation and microwave digestion techniques for the extraction of phytoliths from sediments. *Review of Palaeobotany and Palynology* 120:315-336.

Powers, A.H., Padmore, J. and Gilbertson, D.D. 1989. Studies of late prehistoric and modern opal phytoliths from coastal sand dunes and machair in northwest Britain. *Journal of Archaeological Science* 16:27-45.

Richardson, A.E.V. 1936. Vines, fruit and forestry. In: Royal Geographical Society of Australasia, South Australian Branch (ed), *The Centenary History of South Australia*, pp. 163-178. Adelaide: Royal Geographical Society of Australasia, South Australian Branch.

Rockman, M. 2003. Knowledge and learning in the archaeology of colonisation. In: Rockman, M. and Steele, J. (eds), *Colonisation of Unfamiliar Landscapes: The archaeology of adaptation*, pp. 3-24. London: Routledge.

Rowell, D.L. 1994. *Soil Science: Methods and applications*. London: Prentice Hall.

Santich, B. 1998. *McLaren Vale: Sea and vines*. Adelaide: Wakefield Press.

Specht, R.L. 1972. *The Vegetation of South Australia*. Adelaide: A.B. James, Government Printer.

Stuart, E. 2005. Cultural landscape change on the Willunga Plains from 1840. Unpublished Honours thesis, Department of Archaeology, Flinders University.

Stuart, E. 2006. Cultural landscape change in the Willunga Basin from European settlement to present. In: Smith, P., Pate, F.D. and Martin, R. (eds), *Valleys of Stone: The archaeology and history of the Adelaide's Hills Face*, pp. 113-130. Belair (South Australia): Kōpi Books.

Vaudrey, D.P. and Vaudrey, G.C. (eds) 1991. Willunga District: Facts, facets and phases. Two parts. Unpublished manuscript held by authors, Adelaide.

Wallis, L. 2000. Phytoliths, late Quaternary environment and archaeology in tropical semi-arid northwest Australia. Unpublished PhD thesis, Australian National University.

Wallis, L. 2001. Environmental history of northwest Australia based on phytolith analysis at Carpenter's Gap1. *Quaternary International* 83-85:103-117.

Whitelock, D. 1985. *Adelaide from Colony to Jubilee: A sense of difference*. Adelaide: Savvas Publishing.

Whitlock, C. and Larson, C. 2001. Charcoal as a fire proxy. In: Smol, J.P., Birks, H.J.B. and Last, W.M. (eds), *Tracking Environmental Change Using Lake Sediments. Volume 3: Terrestrial, algal and siliceous Indicators*, pp. 75-97. Dordrecht: Kluwer Academic.

Williams, M. 1974. *The Making of the South Australian Landscape: A study in the historical geography of Australia*. London: Academic Press.

Williams, M. 1992. *The Changing Rural Landscape of South Australia*. Adelaide: State Publishing.

Young, A.R.M. 1996. *Environmental Change in Australia since 1788*. Melbourne: Oxford University Press.

20

Modern surface pollen from the Torres Strait islands: Exploring north Australian vegetation heterogeneity

Cassandra Rowe
School of Geography and Environmental Science, Monash University, Clayton, Victoria
cassandra.rowe@monash.edu

Introduction

The Cape York Peninsula region of northeastern Australia supports considerable biological diversity. It is an environment that illustrates stages in evolutionary through to historical biogeography (Mackey et al. 2001; Turner et al. 2001) and a landscape that reflects culturally based value systems, both indigenous and European (Mackey et al. 2001; McNiven et al. 2007).

Of the vegetation, Cape York Peninsula comprises a complex mosaic of plant associations and structural formations (see Neldner and Clarkson 1995). These vegetation communities reflect underlying physical environmental processes and controls (e.g. edaphic conditions), as well as the degree of climatic seasonality and consequential hydroperiod across the region (Walker and Hopkins 1990; Brock 2001). In turn, the vegetation reflects long-standing biological connections with north Australian monsoonal (wet/dry) megatherm flora, and also has links with the tropical lowland megatherm and temperate upland mesotherm flora of New Guinea (Mackey et al. 2001; Turner et al. 2001). In the continental context, Cape York Peninsula is distinctive in the high level of natural integrity exhibited; Cape York has relatively small, isolated human populations, minimal infrastructure development and the land-use activity is either highly localised or extensive rather than intensive (Mackey et al. 2001). The in-situ conservation of biodiversity is high, which has benefited the humid tropics – or wet tropical biome – in particular. A World Heritage listing of the wet tropics was established in the 1980s, securing 90% of Australia's surviving tropical rainforest (Abrahams et al. 1995).

Rainforest habitats occupying the humid-tropics zone have been a major focus of Quaternary research, and profiled extensively through their pollen spectra (e.g. Haberle 2005; Walker 2007). Now-refugial patches of rainforest have been shown to have great antiquity

(Kershaw et al. 2007) and changes in the pollen representation of rainforest through time have provided a guideline chronostratigraphy of other palaeoecological changes in the Cape York Peninsula (and wider) region (e.g. Kershaw 1985; see also Kershaw 1995).

The vegetative heterogeneity – the spatial and temporal connections – at Cape York Peninsula has not been well represented within palynological studies. Many of the monsoonal megatherm vegetation communities have not been captured in the same detail as rainforest, nor palynologically differentiated in terms of structure and composition. This paper seeks to capture and explore some of this missing detail. To quote Kershaw (2007:2), 'pollen analysis on the sclerophyll side of the rainforest boundary has so far proved elusive'. This 'elusiveness' hampers progress in our understanding of Quaternary vegetation-type dynamics, counteracts the up-to-date design and testing of climatic-vegetation models (Harrison et al. 2007; Lynch et al. 2007) of proxy database development/comprehensiveness (Pickett et al. 2004), limits conceptualisations of human-environment interactions and relations (Denham and Mooney 2008), and provides less long-term ecological information relevant to conservationists and land managers (Froyd and Willis 2008).

This paper presents a descriptive study of the modern pollen sampled from numerous vegetation communities within the Torres Strait, the northernmost extension of Cape York Peninsula. Sampling of modern surface (or modern analogue, Overpeck et al. 1985) pollen data is a long-established technique in the interpretative refinement of fossil pollen spectra, and this study accompanies an extensive reconstruction of Quaternary Torres Strait landscapes from fossil pollen assemblages (Crouch et al. 2007; Rowe 2007a, b). The aims are (1) to characterise and differentiate monsoonal vegetation communities of the Torres Strait by their pollen, (2) to identify which pollen taxa (and in what quantities) are indicative of, or most responsible for, vegetation discrimination outside of the north Australian wet-tropical zone, and (3) to gauge the preservation of pollen taxa in the dry sedimentary context of Torres Strait.

Study area

The waters of Torres Strait extend some 45,000 km², separating Cape York Peninsula and northern Australia from New Guinea (Torres Strait Regional Authority 2007). More than 200 islands, coral and sand cays are regionally incorporated. Within this archipelago, the study-site islands of Mua and Badu form part of the southwestern geographical group, centred 70 km north of the Australian mainland (10°08'S, 142°12'E, Figure 1). Climate is strongly seasonal (tropical-isothermal classification, Stern et al. 2000). Spanning 275 km², Mua is separated from Badu by a narrow 2 km passage. Badu spans a smaller, 180 km² area. Each island features a Proterozoic igneous geology, predominantly granitoids and minor interbedded volcanics, with lithosols and texture-contrast low-fertility soils. These hills rise up to 372 m above sea level over low-relief dune systems, plains, swamps and channels of Quaternary alluvium. Mua and Badu are fringed by degraded Quaternary coastal dunes and beaches, extending into off-shore tidal flats (Bain and Draper 1997).

Mua and Badu support 'the broadest array of plant communities and species diversity in Torres Strait' (Wannam 2008:605). Landscape disturbance through economic expansion has been minimal and isolated to the localised loss of lowland woodland and mangrove (although the islander practice of landscape firing is common). At a scale of 1:100,000, Neldner and Clarkson (1995) surveyed, mapped and described the vegetation of Cape York Peninsula, including Torres Strait (20 vegetation communities are recorded for Mua and Badu; see Table 1 and text below). Strong affinities are apparent between Torres Strait's vegetation and that of the western subregion of Cape York Peninsula, particularly in *Eucalyptus* and *Melaleuca* dominated woodlands.

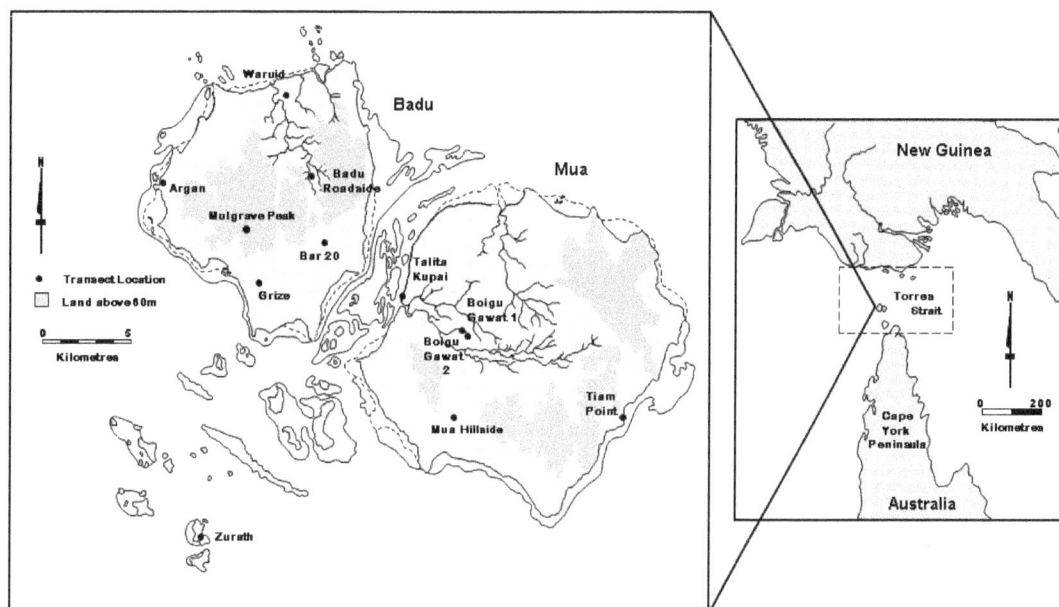

Figure 1. Location map. Position of Mua and Badu with respect to mainland Cape York Peninsula (Australia) and position of island study-site transects.

Throughout the islands' vegetation, the canopy genera *Eucalytpus*, *Corymbia* and *Melaleuca* are common, with the latter more widespread on the low-relief plains. Lowland woodlands incorporate a more prominent graminoid-herbaceous matrix than hill slopes, and more commonly display mixed (sub)canopy components (non-Myrtaceae, non-sclerophyll and/or malacophyll semi-deciduous taxa). Tree height and density increase with moisture availability. Monsoonal closed or vine forest occurs around watercourses or drainage lines, between bases of hills, rocky knolls and established sand ridge/swales. Mixed structured communities, incorporating *Casuarina* and *Cocos*, grow above the coastal high-water mark. Rhizophoraceae-dominated mangroves occupy sheltered, shallow embayments and estuarine floodplains. Communities adjacent to mangrove stands differ according to local variation in elevation, drainage and salinity (swamp conditions, sedgeland or salt flats are observed).

Methodology

Field technique

A total of 126 modern pollen samples were collected. Of these, 50 have been used in this study. The reduced number is primarily a reflection of poor organic preservation in the thin soils or coarse sand-dominated sediments typically encountered throughout each island. Sample numbers were further reduced by subsequently merging subsamples collected within sclerophyll woodland environments. Surface samples were collected by hand and sealed in phials. Approximately 100 g from the uppermost 1.0 cm of sediment was gathered, and, where available, moss polsters were incorporated into these samples. In such cases, the whole moss polster down to, but not including, the soil was collected. At collection, the dominant canopy and understorey taxa were recorded.

Surface sediment sample collection was conducted along 12 transects designed to capture each of the major vegetation communities and plant associations, as described above (Figure 1). In turn, and for ease of presentation, the 12 transects have been grouped into four major environmental zones identified across Mua and Badu: mangrove, coastland lowland, interior swamp and sclerophyll, and rainforest (Table 2).

Table 1. Amalgamated summary of vegetation communities located on Mua and Badu: structural and compositional characteristics (adapted from Neldner and Clarkson 1995; Fox et al. 2001; Beasley 2009).

Vegetation community	Structure	Principal taxa
Corymbia and/or *Eucalyptus* dominated woodlands, open woodlands and open forests	Tall canopy up to 25-30 m. Sparse subcanopy tree layer (15 m) and variable shrub layer (4-5 m). Ground layer sparse to mid-dense.	*Corymbia tessellaris, C. novoguinensis, C. mesophila* and *Eucalytpus* spp. dominate. Secondary tree taxa include *Acacia* (e.g. *A. flavescens*), Fabaceae spp., *Melaleuca viridiflora, Banksia dentate, Pandanus* spp., *Lophostemon* spp. and *Grevillea glauca* occur as secondary tree/subcanopy taxa. Palms include *Livistonia muelleri. Utricularia, Drosera, Tacca* and *Schoenus* are common ground covers.
Melaleuca low woodlands and low-open woodlands	Sparse upper canopy (5-15 m tall) with scattered shrub layer and/or very sparse sub-canopy tree layer (<10 m). Ground cover is short and sparse to moderately dense.	Dominant *Melaleuca viridiflora* and *M. acacioides* with occasional emergent *Corymbia* and *Eucalytpus* spp. Secondary tree taxa include *Banksia dentata, Asteromyrtus symphocarpa, Livistonia muelleri, Leucopogon* and *Acacia* spp. *Heteropogen* and *Aristida* occur as ground cover; *Cyperaceae* (e.g *Fimbristylis* spp.) present where moist.
Grasslands and grassy open woodlands	Grassland to 2 m. Emergent tree scattered and/or rare (2-5 m).	*Heteropogen, Eriachne, Tacca, Evolvules, Sorghum, Euphorbiaceae, Fabcaeae, Convolvulaceae, Liliaceae* and unidentified *Poaceae* spp. occur. *Pandanus* observed as single emergents. *Eucalytpus* common on sites approaching savanna.
Rainforest	Monsoonal closed forest to 30 m. Even, evergreen canopy but sparse sub-canopy and ground-layer. Vine forest and/or thicket with dense, uneven and semi-deciduous canopy.	Diverse. Canopy trees *Welchiodendron longivalve, Acacia polystachya, Terminalia muelleri, Bombax ceriba, Dillenia alata, Ficus* and *Syzgium* spp. are represented. Vines *Flagellaria indica, Similax australis, Dioscorea bulbifera, Abrus precaloris, Pandorea* and *Alyxia* spp. are characteristic.
Wetlands	Semi-permanent swamps and areas subject to inundation, high water tables and/or poor drainage. Structurally variable. Canopy density decreases with tree height and moisture availability. Shallower landscape depressions and/or mudflats favour herbaceous cover.	Dominated by *Melaleuca*, including *M. quinquenervia* and *M. viridiflora*. Subdominant *Pandanus* spp. may be present. Sparse shrub layer typically composed of *Melaleuca* juveniles and Fabaceae. Cyperaceae and Restionaceae spp. well represented. Aquatic groups represented by *Nymphaea* and *Nymphoides*.
Foredune and/or strand	Discontinuous canopy; vegetation often clumped and interspersed with bare areas. Groundcover short, dense, but also patchy. Disturbance may be common.	*Casuarina equisetifolia, Scaevola sericea, Cordia subcordata, Guettanda* and *Eucalytpus* spp. occur as canopy components (palms restricted to areas nearing human habitation). *Thespesia, Sporobolus* spp. and *Ipomoea pes-caprae* dominate groundcover.

Table 2. Characteristics and environmental zone allocation of the modern surface sample transects on Mua and Badu.

Transect name	Central coordinates	Samples	Sample interval (metres)	Transect length (metres)	Vegetation	Environmental zone
Waruid	10°04'N 142°09'E	W1, W2, W3, W4, W5	30	160	Lower-tidal mangrove; upper-tidal mangrove; mudflat; *Corymbia* and *Eucalyptus* woodland	Mangrove
Talita Kupai	10°10'S 142°12'E	TK1, TK2, TK3	30	80	Upper-tidal mangrove; *Corymbia* and *Eucalyptus* woodland	Mangrove
Tiam Point 1	10°12'S 142°18'E	TP1, TP2, TP3, TP4, TP5	20-40	900	Coastal swamp; *Melaleuca* and *Eucalyptus* open woodland/savanna; grassland	Coastal lowland
Tiam Point 2	10°12'S 142°18'E	TP1, TP6, TP7,TP8	50-100	180	Lower-tidal mangrove; upper-tidal mangrove; coastal swamp	Coastal lowland
Argan	10°05'S 142°06'E	AG1, AG2, AG3, AG4, AG5	50-100	350	Upper-tidal mangrove; coastal swamp, dune/strand, *Corymbia* and *Eucalyptus* woodland	Coastal lowland
Zurath	10°16'S 142°06'E	ZU1, ZU2, ZU3, ZU4	40	200	Coastal *Pandanus* swamp, dune/strand	Coastal lowland
Boigu Gawat 1	10°10'S 142°14'E	BG1-1, BG1-2	15	30	Wetland; *Corymbia* and *Eucalyptus* woodland	Inland swamp and sclerophyll
Boigu Gawat 2	10°10'S 142°13'E	BG2-A, BG2-1, BG2-2, BG2-3, BG2-4	25-50	200	Wetland; *Corymbia* and *Eucalyptus* woodland	Inland swamp and sclerophyll
Bar 20	10°08'S 142°09'E	B20-1, B20-2, B20-3, B20-4	50	150	Wetland; *Corymbia* and *Eucalyptus* woodland	Inland swamp and sclerophyll
Grize	10°09'S 142°07'E	GR1	Single	Single	Sedgeland; *Corymbia* and *Eucalyptus* woodland	Inland swamp and sclerophyll
Badu Roadside	10°06'S 142°09'E	BR/s	Single	Single	*Corymbia* and *Eucalyptus* open woodland	Inland swamp and sclerophyll
Mua Hillside	10°12.29'S 142°14.08'E	MH1, MH2, MH3, MH4, MH5 MH6, MH7CHECK	50-100	550	Semi-deciduous vine-thicket, vine forest; *Eucalyptus* and *Melaleuca* open woodland	Rainforest
Mulgrave Peak	10°08.15'S 142°07.87'E	MP1, MP2, MP3, MP4, MP5	50-100	850	Broadleaf, evergreen closed forest; *Eucalyptus* forest and (open)woodland	Rainforest

Pollen preparation and identification

Laboratory preparations followed standard methods of Faegri and Iversen (1989). Chemical preparations successively removed sulphur compounds, carbonates, humic acids, macrofossils, silicates and cellulose, using HCL, NaOH, HF, and an acetolysis mixture consisting of H_2SO_4

and $C_4H_6O_3$. *Lycopodium* markers were added to each sample to facilitate pollen concentration calculations. Pollen residues were suspended in glycerol and counted at x400 magnification under a Zeiss Axiostar compound microscope. Counting continued to a minimum of 200 grains (total pollen sum, including wetland taxa, excluding spore types).

Pollen identification was assisted by field specimen collections, authored reference material (Thanikaimoni 1987; Fuhsiung et al. 1997) and regional reference collections held at the School of Geography and Environmental Science, Monash University. A distinction in pollen of the Myrtaceae and Cyperaceae families was particularly sought after (here, limited island genera characterise different ecologies and/or where inconsistent pollen preservation may enforce a reliance on such major groups) . Based on field reference collections and the discussions of Chalson (1989) and Churchill (1957, quoted in Chalson 1989), myrtaceous pollen was examined according to equatorial diameter, polar axis, polar islands, colpi depth, concavity of sides, comparative thickness of the pollen grain wall and pore characteristics. No distinction was made when fewer than half of the character-states were visible. Observations for *Melaleuca* include smaller size, concave sides, island presence, obviously angled colpi and thin pores. *Eucalyptus* is described as a larger grain, with a heavier, more robust appearance, thickened pores, convex to straight sides, and lack of polar islands. Unspecified grains fall within the category of 'Myrtaceae (undiff.)'. For Cyperaceae, only three types were considered sufficiently distinctive: *Cyperus*, *Eleocharis* and *Schoenus* types. In general, these pollen grains are 1-4 aperturate, elongate or more or less spheroidal, with one broad end. *Eleocharis* presents a more tapered shape, possibly with pores set near the base, a grain size of 30-35 μm and a surface pattern tending towards striate. *Schoenus* is similar, except that the exine is smoother and the grain size 32-38 μm. In comparison, *Cyperus* is a smaller grain (to 25 μm), spheroidal and with a thicker, granular exine surface. Kershaw (1971) recognises a similar *Cyperus* type.

Numerical analyses

In the diagrammatic presentation of the pollen data, the Tilia suite of programs was used, specifically the spreadsheet application Tilia (2.0) and graphing counterpart TGView (v. 2.0.2) (Grimm 1988). Pollen counts are expressed as a percentage of the total pollen sum. Trace pollen is considered as <2% of the sum.

A multidimensional (or multivariate) ordination (MDS plot) was constructed for the 50-sample data set. Count values were square-root transformed and Bray-Curtis coefficients calculated (see Clarke, 1993). The Bray-Curtis index is widely used in ecology (see Faith et al. 1987), and the square-root transformation has the effect of down-weighting the importance of high-abundance taxa, so that sample similarities depend not only on these values, but on those of less common taxa. The robust nature of the ordination is measured through the numerical stress value. A stress less than two corresponds to a good ordination and useful two-dimensional picture of sample similarity (Clarke 1993). An additional analysis-of-similarities (ANOSIM) function test was performed on the 50-sample data set to determine those taxa responsible for sample grouping as observed in the ordination plot, and to reveal those taxa typical of specific vegetation groups (in the sense of making large percentage contributions to the average similarity between group samples) (Clarke 1993; Clarke and Warwick 2001). In seeking out and describing vegetation pattern through pollen, these statistical techniques permit an objective test of the intuitive observations made directly from the pollen diagrams. All calculations and presentations were carried out using the PRIMER v.5 package (Plymouth Routines In Multivariate Analysis, Clarke and Gorley 2001).

Results and discussion

Sixty-three pollen types were identified in the 50 surface sediment samples collected across Mua and Badu. Most pollen types were identified to the genus level, and unidentified taxa represent on average 10.2% of the total pollen sum. Across all samples, the average pollen concentration was 185,610 grains/cm².

Mangrove pollen samples (Figure 2)

Pollen assemblages from the Waruid surface samples reflect a vegetation gradient from low-tide mangrove forest, through back-mangrove mudflat to upland sclerophyll woodland, with a progressive decrease in mangrove representation and increase in non-mangrove components. At Talita Kupai, mangrove representation declines in both landward and seaward directions from the central forest sample of TK2. Eight mangrove taxa are identified in the study samples, five reaching more than trace levels.

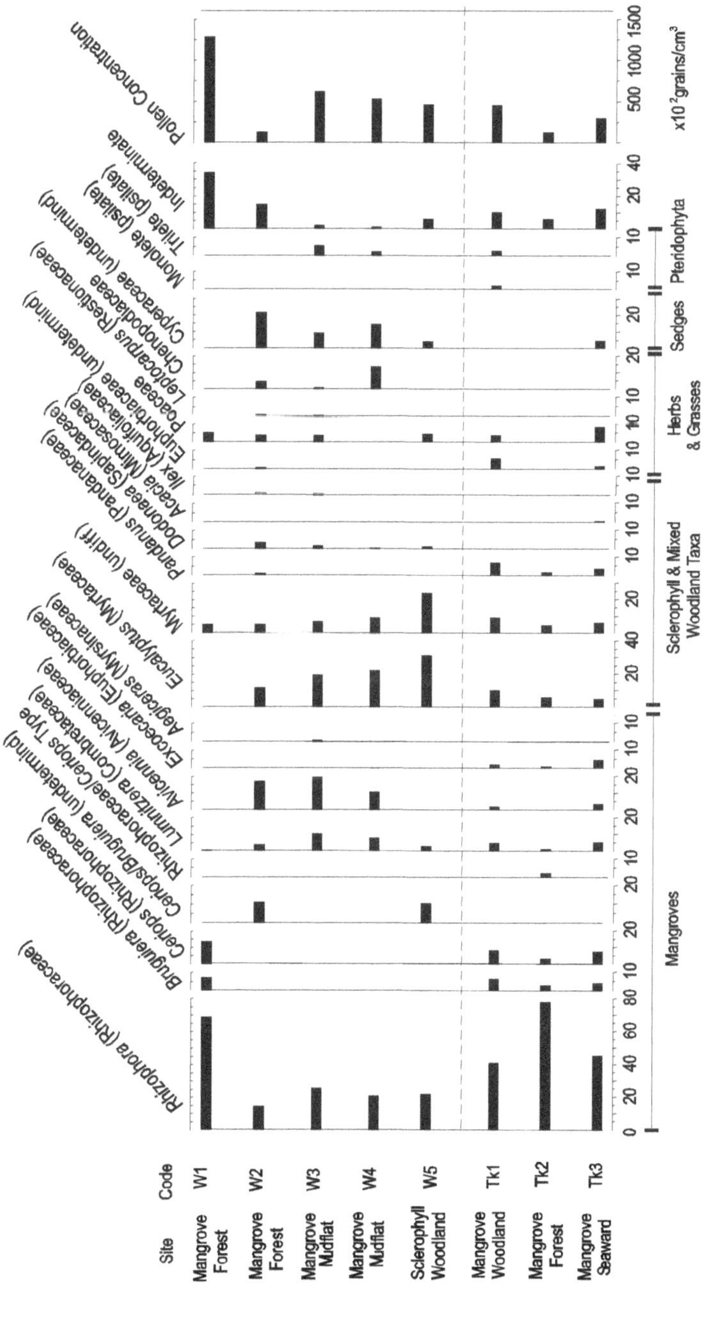

Figure 2. Mangrove pollen diagram. Modern pollen assemblages from Waruid (upper diagram) and Talita Kupai (lower diagram) transects. Pollen are expressed as percentage values of the total pollen sum.

The family Rhizophoraceae comprises the majority of recorded mangrove taxa. Of these, *Rhizophora* is the most abundant. Pollen values exceed 50% in all samples from sites in which *Rhizophora* is found growing, declining from 70% in the seaward mangrove forest sample of W1, to 20-25% of the pollen sum in adjacent samples W2-W5. No other mangrove taxa are characteristic of this seaward forest. A more diverse mangrove pollen flora is represented in sites landward of the *Rhizophora*-dominated system and in close proximity to the terrestrial woodland/mangrove ecotone. Samples W2-W4 and TK1 contain a pollen assemblage in which *Ceriops/Bruguiera*, *Lumnitzera*, *Excoecaria*, and in particular *Avicennia*, are recorded alongside *Rhizophora*. These upper intertidal taxa comprise up to 45% of the total sum in these four samples, with an additional 15-22% represented by Cyperaceae and Chenopodiaceae (Waruid sequence). Poaceae, *Leptocarpus* and Euphorbiaceae are of secondary importance across the intertidal mudflat samples of W4 and W5 (<10%), as a factor of saline tolerances. Clear trends in mangrove to non-mangrove pollen distributions are less evident along the Talita Kupai transect. This may be partly due to the fewer samples in the sequence, but may also relate to less obvious plant species segregation on a drier site more distant from the open coast (and a strong tide).

Myrtaceae and *Eucalyptus* display minor pollen representation in most samples but increase gradually with distance from the central mangrove zone. Highest representation is achieved close to the parent plant source in samples W5 and TK1 (30% and 10% respectively). Remaining pollen types, notably *Dodonaea*, *Ilex*, *Pandanus* and Euphorbiaceae are recorded at low frequencies. The major terrestrial pollen types represented in samples W1 and TK3 – *Eucalyptus*/Myrtaceae, *Pandanus* and Poaceae – are known to be effectively dispersed by wind (Grindrod and Rhodes 1984) and may therefore be expected to show at least minor representation in such seaward situations and reflect regional dispersal. A discontinuous canopy at the seaward edge may also account for terrestrial pollen presence.

Coastal lowland pollen samples (Figure 3)

Pollen spectra from coastal lowland study environments fall into three groups: samples from sites within the Rhizophoraceae-dominated mangrove system, samples approaching the limit of saltwater influences, and samples extending into the surrounding upland catchment. The difference here is mangrove forest habitats defined through local mangrove representation exceeding 80% up to 100%, and mangrove/terrestrial sites with longer-distance mangrove taxa contributing only 30-40% of the total pollen sum. Samples Ag3 to Ag5, Zu2 and Zu4 are noted for their absence of mangrove pollen taxa.

With the exception of *Rhizophora*, mangrove pollen types are poorly represented. This overriding percentage dominance of *Rhizophora* is directly comparable to seaward sites in the above mangrove study. *Rhizophora* shows strongest representation in samples fringing the seaward edge of collection (93%) and decreases sharply landward (25%). In general, samples TP6-TP8 and Ag1 display relatively low pollen diversity when compared with the number of pollen taxa identified in transect samples across the remainder of this study. *Ceriops/Brugueria*, *Avicennia*, *Excoecaria* and *Xylocarpus* are most common (up to 12%) to the rear of the mangrove forest. *Scyphiphora* is locally abundant (16%) adjacent to the *Rhizophora*-dominated forest at Argan. A diverse mangrove pollen flora associated with an increasing terrestrial presence appears an important indicator of upper-intertidal conditions and of a coastal site which has a more open canopy habit.

High pollen percentage values for Poaceae (32%) and Cyperaceae types (10%) occur in both coastal swamp and coastal woodland samples. Local distributions of *Freycinetia* and Asteraceae are reflected in surprisingly low percentage pollen values (<5%) across swamp samples TP1-TP3. *Leptocarpus* has a wider, although similarly minor (<5%), pollen distribution through swamp-

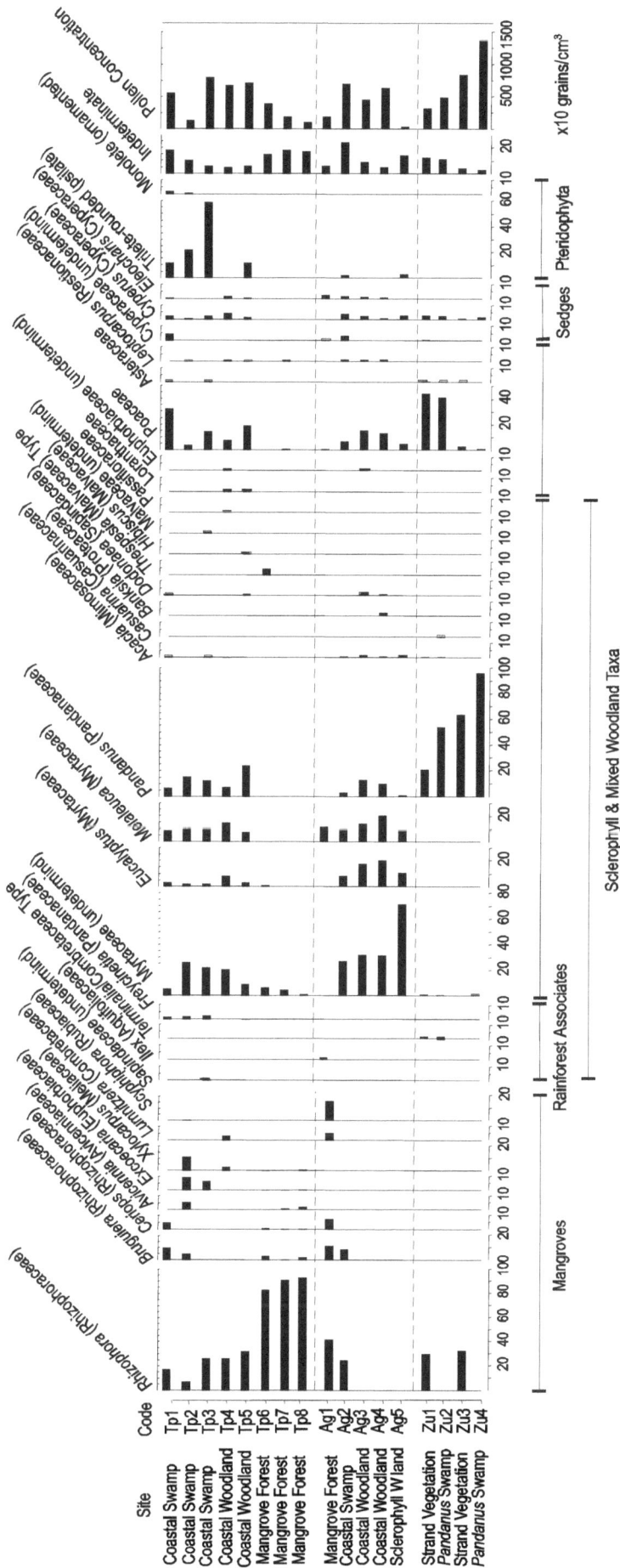

Figure 3. Coastal lowland pollen diagram. Modern pollen assemblages from Tiam Point (upper diagram), Argan (mid diagram) and Zurath (lower diagram) transects. Pollen are expressed as percentage values of the total pollen sum.

edge and woodland locations. Samples from coastal swamp habitats are also noted in particular for their abundant pteridophyta and/or spore morphologies, dominated by the rounded-trilete type (59%) and representative of local sources. Myrtaceae dominates the coastal woodland canopy component, the highest relative pollen values for these taxa recorded with increasing distance from the seaward mangrove limit (71% at Ag5). In general, *Melaleuca* pollen accounts for a greater representation than *Eucalyptus*, particularly surrounding the coastal swamp sites. *Pandanus* pollen comprises a further 7-24% of the total assemblage, with remaining woodland types being minimally represented.

A comparison between samples Zu1-Zu4 and observations made on the parent vegetation of Zurath Islet indicates the absence of numerous taxa. The islet is fringed by strand vegetation dominated by dune flora. Seaward, mangroves are patchy and as such *Rhizophora* (ca. 30%) is a minor component of pollen assemblages in comparison with Mua and Badu. Inland on Zurath, a narrow vine thicket becomes associated with granite outcrops but this community

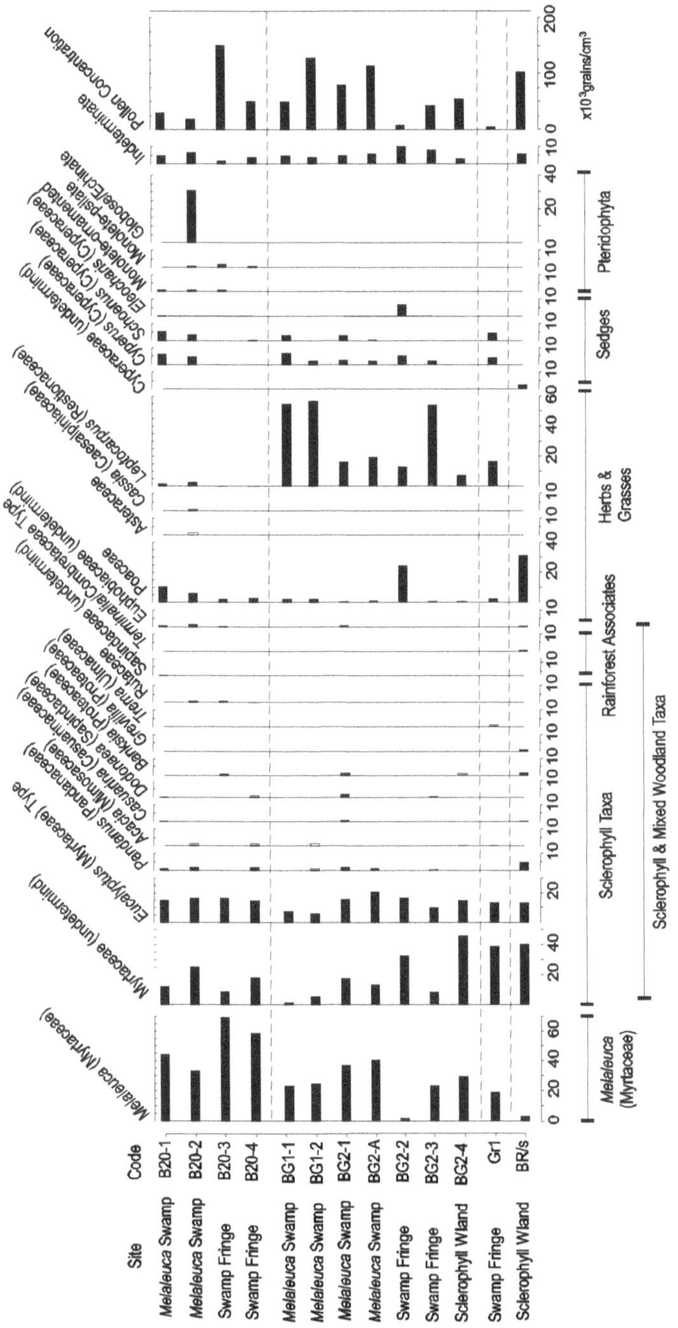

Figure 4. Inland swamp and sclerophyll pollen diagram. Modern pollen assemblages from Boigu Gawat 1 and 2 (upper diagram) Grize and Badu Roadside (lower diagram) transects. Pollen are expressed as percentage values of the total pollen sum.

does not appear to be at all represented through pollen (see below for local deposition trends in rainforest associated pollen). High values for Poaceae pollen (40%), recorded with herbaceous components such as Asteraceae and Cyperaceae, reflect extra-local dispersal from grassland across central islet sandy deposits. Where samples were collected from dune environments fringing Zurath Swamp, *Pandanus* pollen dominates the terrestrial pollen taxa at up to 96%. The Zurath pollen samples also present an interesting near-absence of Myrtaceae.

Inland swamp and sclerophyll pollen samples (Figure 4)

Modern pollen content of inland swamps and sclerophyll woodlands is broadly governed by the family Myrtaceae. The dominance of *Melaleuca* in forming a canopy around each swamp is reflected in the high counts of both open-swamp and swamp-fringe samples; *Melaleuca*-type pollen generally maintains a representation of 23% to 70% and becomes increasingly important with distance from the open water. *Melaleuca* values exceed 60% of the pollen sum in samples B20-3 and B20-4, 25% in BG1-2, and more than 40% in BG2-A. The transition to open woodland can be seen in increased representation of sclerophyll canopy dominants. Myrtaceae (undiff.) achieves up to 40% of the pollen sum in samples BG2-4, GR1, and BR/s. Positioned away from swampy habitats, BR/s is distinctive, with a combined dominance of *Eucalyptus/Myrtaceae* (40%) and Poaceae (32%), with higher percentage values for subcanopy sclerophyll taxa such as *Pandanus*, *Banksia* and *Grevillia*.

Reflecting the regional importance of the surrounding woodland, *Eucalyptus* maintains a consistent percentage contribution of 15-21% of the total pollen sum in most samples. Myrtaceae (undiff.) shows the same trend, but with less consistency. Other taxa represented are *Acacia*, *Casuarina*, *Dodonaea*, Sapindaceae and *Terminalia*.

A reduced surface area of open water appears to have affected the proportions of locally versus regionally derived pollen components. The major difference between swamp Bar-20 samples and those from the smaller Boigu Gawat Swamp sites is the increased representation of *Leptocarpus* pollen to ca. 70% in the latter; *Leptocarpus* pollen effectively reflecting high local abundance. A similar trend is a consistent appearance of Cyperaceae pollen, incorporating *Cyperus*, *Schoenus* and *Eleocharis*. Pollen percentages of herbaceous taxa therefore appear more closely related to local presence or absence of parent sources at all sites, and sampling across the smaller swamp site has resulted in high percentages of the local swamp-forest understorey pollen in association with *Melaleuca*. Low levels of sclerophyll components in samples BG1-1 and BG1-2 are explained as dilution from local pollen. The background component of regional sclerophyll pollen is therefore generally better represented in the open-water samples where it is not filtered by dense swamp fringe vegetation (cf. Tauber 1967).

Rainforest pollen samples (Figure 5)

Rainforest taxa do not comprise the bulk of the pollen recorded in rainforest surface samples. Rainforest and allied taxa represent, as a maximum, 53% of the total pollen sum in sample MP1. At the remaining rainforest sites, this component varies from 14% to 41%. In contrast, sclerophyll woodland pollen, particularly Myrtaceae and *Acacia*, make up at least 45% of the pollen sum, increasing to ca. 90% outside the rainforest boundary (although it is noted that the woodland component may be inflated by pollen from local rainforest variants of *Acacia* and Myrtaceae, whose presence could not be differentiated from the sclerophyll category).

Along the two transects, no single rainforest pollen component predominates consistently within any sample. A large number of plant taxa are therefore under-represented within the assemblage. Rainforest samples are marked by numerous poorly represented taxa demonstrating low and often sporadic occurrence (<5%). As best represented, values for *Freycinetia* range from 2-16%, 1-11% for *Croton*, and 1-10% for *Aidia*. *Mangifera*, *Alstonia*, Tilliaceae and Fabaceae are represented by values up to 7%, and *Ficus* is locally significant at MP4 (18%).

Open woodland taxa reflect many of the trends already described. They are almost entirely composed of Myrtaceae (up to 70%) and the canopy genera *Eucalyptus* and *Melaleuca*. Myrtaceae has background values in sites at some distance from a source, increasing in representation towards and across the forest-open woodland boundary (samples MP5 and MH3). Sample MH4 presents the only variation from this pattern, dominated by *Pandanus* (35%) and significant in its representation of *Banksia* (10%) and *Leptocarpus* (5%). *Dodonaea* and other Sapindaceae are minor dryland components. Although Poaceae does travel into forest samples,

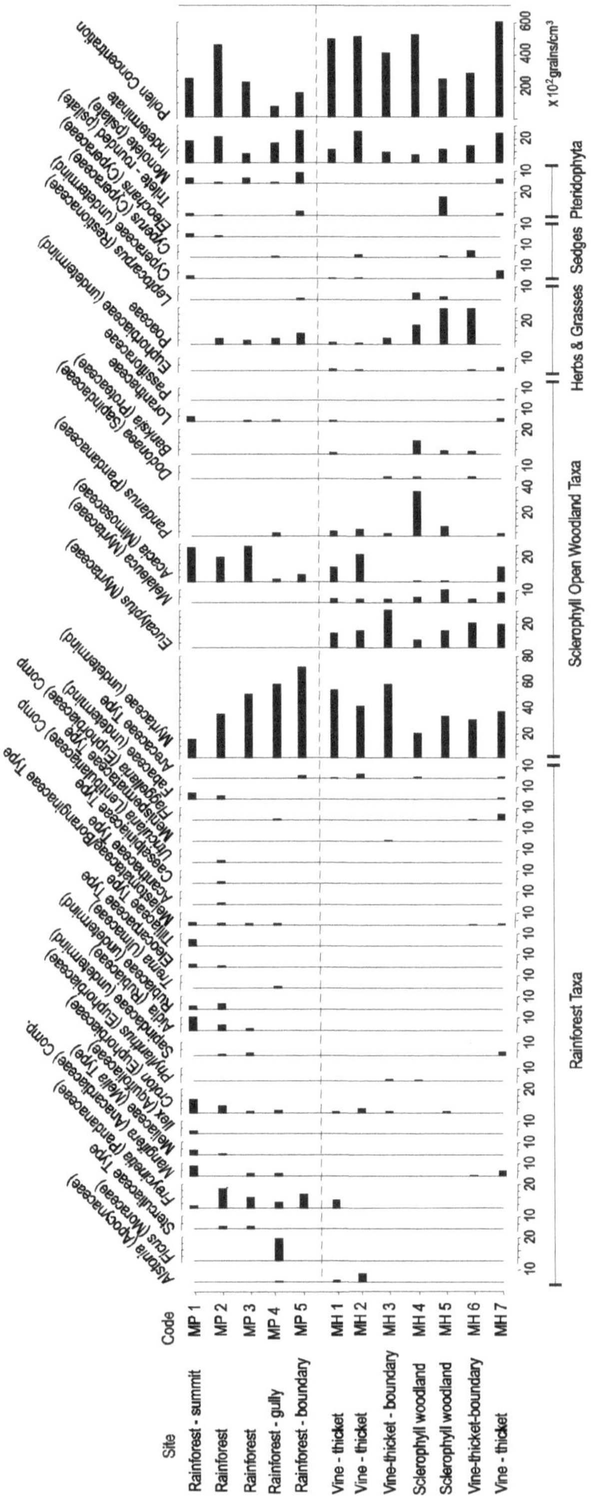

Figure 5. Rainforest pollen diagram. Modern pollen assemblages from Mulgrave Peak (upper diagram) and Mua Hillside (lower diagram) transects. Pollen are expressed as percentage values of the total pollen sum.

it does not penetrate in great quantities (<5%). Rather, Poaceae increases at the forest boundary (MH5) and makes a marked contribution to the open woodland samples of MH4-MH6 (up to 28%).

Taxon representation

The relative abundances of mangrove pollen types reflect the present zonation of plant species and provide a useful index of depositional environments and distance to the open coast. Identification of mangrove habitat by the dominance of Rhizophoraceae is a consistent observation across both Mua and Badu, and directly comparable to other regional mangrove palynological studies (e.g. Grindrod 1985; Crowley et al. 1994; Ellison 2005). High values of mangrove taxa (up to 95% of the total pollen sum) are recorded for lower-tidal mangrove sites, compared with the lower values of 30-60% for upper-tidal and lowland coastal regions. Lowest mangrove pollen values are found at the landward extremity of intertidal mudflats. No mangrove pollen taxa are represented in terrestrial pollen samples associated with sclerophyll, inland *Melaleuca*-swamp, rainforest or vine-thicket communities, thus demonstrating a clear relationship between pollen source and mangrove distribution. Strong representation of *Rhizophora* pollen indicates frequently inundated saline environments. The representation of *Rhizophora* across the Waruid, Talita Kupai and Tiam Point transects reflects the near-monospecific, local nature of the lower-tidal community. A lack of mangrove pollen diversity, with very low representation of grasses, sedges and herbaceous taxa, confirms the closed-canopy nature of this forest type.

Upper-tidal, landward mangrove vegetation is pollen distinguished from lower-tidal mangrove-forest on the basis of greater individual representation of taxa such as *Ceriops* and *Bruguiera*, *Avicennia* and *Lumnitzera* relative to *Rhizophora* (each of these taxa have restricted pollen dispersal, Grindrod 1985). In turn, the transition to saline mudflat or coastal back-swamp surfaces incorporates increased terrestrial pollen components, reflecting reduced marine inundation and lower salinity of site samples. In agreement, Grindrod (1985, 1988) describes the strong representation of *Avicennia* pollen as particularly indicative of mainland Australian back-mangrove, brackish environments. On Mua and Badu, *Ceriops* and *Bruguiera*, *Avicennia* and *Lumnitzera* combine with Cyperaceae, Poaceae and herbaceous pollen such as Asteraceae and *Freycinetia*. The modern pollen data suggests that high values of Cyperaceae, as well as Chenopodiaceae, in fossil samples are locally derived, i.e. from a swamp surface or nearby. Crowley et al. (1994) confirm that values of Chenopodiaceae exceeding trace pollen levels indicate the presence of chenopods in local or extra-local vegetation, but warn also that absence cannot be assumed from lower values. High fern values, represented by monolete and trilete spores, are also indicative of coastal swamp surfaces, owing to the predominantly seasonal freshwater accumulation experienced at such sites. *Acrostichum* is the likely species recorded at Tiam Point, where the spores are trilete. Canopy vegetation in the mangrove/terrestrial transition and coastal swamp samples tends to be dominated by *Melaleuca*/Myrtaceae pollen derived from swamp *Melaleuca* species. A proportion of the *Pandanus* pollen rain also appears to be widely dispersed, but *Pandanus* dominates pollen assemblages where present in the vegetation (Crowley et al. 1994 also discuss *Pandanus* pollen deposition, both wetland and dryland).

The sclerophyll woodland sites are characterised by the pollen types of their dominant and sub-dominant tree species. *Eucalyptus*/Myrtaceae and *Pandanus* pollen indicate the principal canopy composition at the site. Other components such as *Casuarina*, *Dodonaea*, *Acacia* and *Terminalia*/Combretaceae and various herbs and lianes have poor, sporadic representation. Each is only likely to appear in a pollen sample when it is abundant in the vegetation of the area. Given the mosaic, open nature of the woodland/savanna vegetation, as well as its regional pollen distribution characteristics, it is probable that only large-scale canopy changes would be detected in the palaeoecological record. Grasses are consistently represented but, considering that they dominate the understorey of woodland and savanna areas, their pollen contribution is less than expected. Bush (2002) suggests the representation of tropical Poaceae pollen is influenced by numerous factors, including the pollination strategies of other members of their community and the scale of disturbance accompanying human activities. Bush (2002) cautions, in particular, against the interpretation of Poaceae pollen abundance without due regard to local hydrology. Using Poaceae as a simple indicator of regional vegetation change may overstate trends or boundary transitions towards aridity, and disregard seasonal variation in swamp communities. For environments such as Torres Strait, the persistence and timing of fire across woodland samples may also affect the Poaceae pollen record.

Inland swamp forest shows similar Poaceae pollen values to the sclerophyll surrounds, though they combine with high values and diversity within Cyperaceae and the presence of semi-aquatics such as *Leptocarpus* to indicate wetter conditions. Sites in which *Leptocarpus* and Cyperaceae are prominent in the vegetation are clearly identified by the domination by these taxa of the pollen spectra. Deduction of water depth and permanency on the basis of the pollen assemblage is, however, more difficult. The majority of samples comes from sites of fluctuating water levels. It could be concluded that Cyperaceae and *Leptocarpus* pollen are better indicators of shallow (or seasonal) surface water than *Melaleuca*. Higher pollen values of dryland taxa with low pollen production or poor dispersal may also signify shallow water or a high water table. Here, the associated moisture and organic sediment type would allow for greater preservation and entrapment of such pollen. *Melaleuca* pollen associated with a greater regional component is therefore more likely to reflect prominent areas of open water. The improved identification of *Melaleuca* will be important for the refined recognition of free water, particularly so in the observed absence of pollen from *Nymphoides* and the like. Overall, the inland surface study gives confidence that local swamp pollen can be separated from regional pollen sources.

There is a comparatively high level of diversity among the monsoonal-rainforest and vine-thicket pollen samples. No one pollen type dominates; numerous taxa were found to contribute to the pollen signature and, although widely separated, the two rainforest transect studies show similar pollen spectra. On Mua and Badu, rainforest pollen behaviour resembles analyses from mainland humid tropical locations (Kershaw and Strickland 1990; Kershaw and Bulman 1994; Walker and Sun 2000). Linking trends include pollen diversity, but also a penetration of non-rainforest tree pollen to a point exceeding rainforest pollen itself. Similarly, many taxa will not be represented in pollen diagrams given that, within closed forest zones, long-distance pollen dispersal is of little importance; taxa are typically recorded only if growing in proximity to the pollen collection site (i.e. dominant local deposition). However, the two locations vary in their pollen rain composition. Major canopy or emergent tree taxa (e.g. *Stenocarpus*, *Acmena*) are high pollen contributors to the mainland humid tropical records, as are palms and ferns (tree, epiphytic and/or ground ferns). A lack of distinction within the Myrtaceae and *Acacia* pollen at both Mulgrave Peak and Mua Hillside has hampered the isolation of rainforest vegetation, and the monsoonal communities are differentiated by the presence of *Ficus*, *Aidia*, *Croton* and *Freycinetia* pollen. Although each taxon appeared in extremely low quantities, these low pollen values do not necessarily mean insignificant results. The simple presence/absence of pollen taxa is considered important in differentiating monsoonal rainforest vegetation, and in representing forest patchiness, and may provide a data set more likely to be comparable with a fossil record than percentile data. Burn et al. (2010) make a similar observation in the pollen differentiation of Amazonian rainforest communities. Plants may be well represented in floristic inventories, but under-represented in the pollen because their pollen is less well dispersed. Where such pollen is found within the pollen rain, its presence can provide significant ecological information with

A

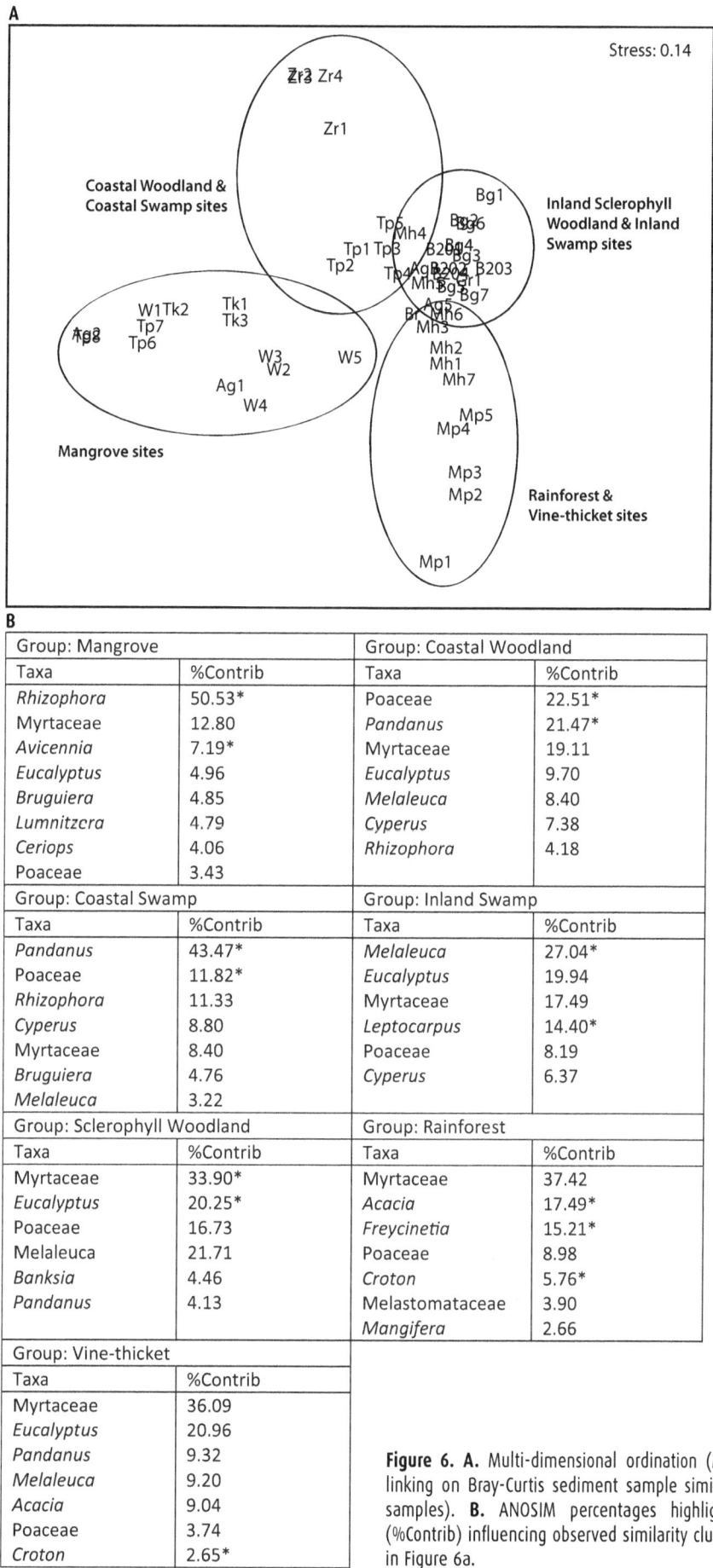

B

Group: Mangrove		Group: Coastal Woodland	
Taxa	%Contrib	Taxa	%Contrib
Rhizophora	50.53*	Poaceae	22.51*
Myrtaceae	12.80	*Pandanus*	21.47*
Avicennia	7.19*	Myrtaceae	19.11
Eucalyptus	4.96	*Eucalyptus*	9.70
Bruguiera	4.85	*Melaleuca*	8.40
Lumnitzcra	4.79	*Cyperus*	7.38
Ceriops	4.06	*Rhizophora*	4.18
Poaceae	3.43		

Group: Coastal Swamp		Group: Inland Swamp	
Taxa	%Contrib	Taxa	%Contrib
Pandanus	43.47*	*Melaleuca*	27.04*
Poaceae	11.82*	*Eucalyptus*	19.94
Rhizophora	11.33	Myrtaceae	17.49
Cyperus	8.80	*Leptocarpus*	14.40*
Myrtaceae	8.40	Poaceae	8.19
Bruguiera	4.76	*Cyperus*	6.37
Melaleuca	3.22		

Group: Sclerophyll Woodland		Group: Rainforest	
Taxa	%Contrib	Taxa	%Contrib
Myrtaceae	33.90*	Myrtaceae	37.42
Eucalyptus	20.25*	*Acacia*	17.49*
Poaceae	16.73	*Freycinetia*	15.21*
Melaleuca	21.71	Poaceae	8.98
Banksia	4.46	*Croton*	5.76*
Pandanus	4.13	Melastomataceae	3.90
		Mangifera	2.66

Group: Vine-thicket	
Taxa	%Contrib
Myrtaceae	36.09
Eucalyptus	20.96
Pandanus	9.32
Melaleuca	9.20
Acacia	9.04
Poaceae	3.74
Croton	2.65*

Figure 6. A. Multi-dimensional ordination (MDS) using group-average linking on Bray-Curtis sediment sample similarities (50 modern pollen samples). **B.** ANOSIM percentages highlighting taxon contribution (%Contrib) influencing observed similarity cluster through the ordination in Figure 6a.

regards to local vegetation and aid palynological characterisation (Burn et al. 2010). Drier rainforest formations, therefore, are best separated on the basis of the pollen assemblage, rather than any one taxon.

Multivariate analysis of modern pollen

The pollen samples have been highlighted according to their respective environmental zone (mangrove, coastal lowland, inland swamp and sclerophyll, and rainforest). The ordination (MDS plot Figure 6a) highlights the relationships between vegetation groups while still acknowledging a degree of within-group variability. It is encouraging that no scores are scattered throughout the ordination. Rather, the tight grouping observed within and around inland swamp and sclerophyll samples is identified as the predominant feature. In essence, these open-canopy vegetation sites show a greater similarity within group than is evident in the mangrove and rainforest groups. A similar, although lesser, trend through the coastal lowland samples is also apparent. Here, the increased importance of regional pollen rain (and its uniformity) in open vegetation sites is a significant factor in reducing variability. The mangrove forest system and rainforest group are clearly distinguishable, and the Zurath samples form a cluster separate from all other samples. This pattern is driven largely by the presence of *Pandanus* and absence of Myrtaceae at Zurath. Coastal swampland is appreciably different from both mangrove and coastal woodland habitats, though closer in approximation to the latter. Vine thicket separates from monsoonal rainforest, approaching sclerophyll woodland in its similarity. The positioning of vine-thicket samples is indicative of the more open nature of its canopy, allowing for a greater regional signal and shared flora, particularly in the presence of sclerophyll emergents.

When considering Figure 6a, strength of the ordination lies in displaying a gradation in vegetation formation across the set of samples. A changing pollen assemblage can be traced from the mangrove forest through the wider coastal environment (firstly incorporating coastal swampland, then coastal woodland), to the more coherent sclerophyll interior. Likewise, a separation of the rainforest samples extends stepwise from the inland swamp to sclerophyll woodlands, blending with vine thicket and stretching through the rainforest at high elevations.

Determining discriminating taxa

The contribution each pollen taxon makes to the observed ordination and/or average sample similarity within the vegetation groups of Figure 6a is shown in Figure 6b. Figure 6b reveals species which are statistically typical of a group; pollen taxa are ordered by their average contribution (%Contrib) and those considered good discriminators in this study are indicated by an asterisk.

In many cases, close to half the average similarity value is accounted for by only two pollen taxa, with 90% of a group's distinction represented by no more than seven taxa. Undoubtedly, the mangrove pollen group is distinguished through *Rhizophora*, independently accounting for 51% of average within-group similarity. *Avicennia* is also significant, at 7%. Poaceae and *Pandanus* contribute equally to the coastal woodland group, at 22%; *Pandanus* is a better representative of the coastal swamp: 44%. The Myrtaceae family maintains a significant presence within the sclerophyll woodland, inland swamp, vine-thicket and rainforest groups. While *Melaleuca* is particular to the inland swamp, contributing 34%, the remaining groups are perhaps best described through other non-Myrtaceae taxa specific to their list. In this respect, sclerophyll woodland is defined via Poaceae as well as *Eucalyptus* pollen, and rainforest through *Acacia*, *Freycinetia* and *Croton*. A lack of distinction within the Myrtaceae family remains a problem when identifying vine-thicket communities. *Croton* is therefore the only pollen taxon highlighted.

Table 3. Guiding principles for the interpretation of fossil pollen assemblages, as specified from modern pollen samples in Torres Strait.

Pollen sum observation	Pollen dispersal	Interpretation
Rhizophora >50%	Regional	Lower-tidal, inundated mangrove
Rhizophora <50%	Regional	Upper-tidal mangrove & adjacent mangrove-terrestrial transition
Avicennia >10%	Local	Upper-tidal mangrove
Present – *Ceriops, Bruguiera, Lumnitzera, Avicennia, Xylocarpus, Excoecaria*	Local/Extra-local	Upper-tidal mangrove and adjacent mangrove-terrestrial transition
Chenopodiaceae > trace level	Local	Mudflat, saltmarsh
Spore types (cumulative) >20%	Local	Swamp surface
Spore types (cumulative) <10%	Local	Myrtaceous open forest to woodland
Pandanus >20%	Local	Localised stand; swamp nearby
Present – *Pandanus, Melaleuca*	Extra-local/Regional	Swamp surface and/or swamp nearby
High *Cyperaceae*; Present – *Cyperus, Eleocharis, Schoenus, Poaceae*	Local/Extra-local	Swamp surface; inundated sediments/depression
Present – *Cyperaceae, Leptocarpus*	Local/Extra-local	Moist sediment surface, localised poor drainage
Leptocarpus >20%	Local/Extra-local	Wet swamp surface
Poaceae >10%	Extra-local/Regional	Grasses abundant in local vegetation
Melaleuca >20%	Extra-local/Regional	Wet swamp surface nearby; open water
Eucalyptus >20%	Regional	Open forest; woodland
Myrtaceae >25%	Regional	Open forest; woodland
Ficus >trace level	Local	Monsoonal rainforest (established closed forest)
Present – *Freycinetia, Aidia, Mangnifera*	Local	Monsoonal rainforest
Acacia >20%	Extra-local/Regional	Monsoonal rainforest
Freycinetia >10%	Local	Monsoonal rainforest – Sclerophyll transition
Present – *Myrtaceae, Terminalia, Ilex, Hibiscus, Acacia, Alstonia*	Local/Extra-local	Mixed canopy forest in region

A total of 11 pollen taxa have been identified as principally responsible for the observed clustering in Figures 6a and 6b. Between them, *Rhizophora*, *Avicennia*, *Pandanus*, Poaceae, Myrtaceae, *Eucalyptus*, *Melaleuca*, *Leptocarpus*, *Acacia*, *Freycinetia* and *Croton* capture the full multivariate pattern. However, identifying taxa as typical of a vegetation group does not necessarily correspond to an equal importance in discriminating from one group to another. Taxa may be typical of a number of groups, and the example of Myrtaceae has affected the degree to which environments across island interiors can be statistically separated.

Concluding discussion

As a first study of its kind for Torres Strait, the modern pollen spectra have allowed for the establishment of guidelines for the interpretation of fossil pollen diagrams (Table 3). This research has shown the importance of independently describing the modern pollen rain from monsoonal Cape York Peninsula, and testing pollen's diagnostic qualities in capturing vegetation heterogeneity. The approach incorporated a visual interpretation of broad trends in the pollen data with statistical analyses of these trends and their significance. The analyses of similarity support each percentage pollen diagram in describing the pollen assemblages and in isolating useful indicators of vegetation and ecological settings. Together, the results show that the modern pollen rain does vary and vegetation groups (structurally and in composition) can be differentiated.

The pollen from a wide range of vegetation associations contained within the mangrove, coastal, inland swamp and sclerophyll, and rainforest environmental zones can be explained in terms of local, extra-local and regional components. The majority of taxa recorded were local or extra-local. Locally produced and dispersed pollen types are highly variable between samples and sites, while regional pollen types have relatively even fallout. Key examples of local taxa include *Avicennia*, *Leptocarpus*, *Freycinetia* and *Croton*. Regional dispersal, while contributing considerably to the pollen sum, particularly in open environments, is restricted to a few canopy taxa such as *Eucalyptus* and Myrtaceae, and notably *Rhizophora* across coastal landscapes. Based on the pollen diagrams, *Pandanus* behaves in a similar regional manner, but one important result that follows on from the ordination is that palynological dominance of *Pandanus* is indicative of local swamp conditions, especially within coastal lowlands. From the pollen diagrams, Cyperaceae is a minor component of terrestrial communities, but is also diagnostic of swamp communities in high percentages and diverse genus types. When palaeoecological discussions combine with archaeological interpretations, as has occurred in the Torres Strait (Crouch et al. 2007), such swamp (i.e. freshwater availability) indicators are noteworthy. This modern pollen study also suggests that the overall assemblage, as opposed to exclusive reliance on individual pollen types and irrespective of grain percentages, is a useful indicator – valuable in the documentation of monsoonal rainforest as well as upper-tidal mangrove environments, for example. Diverse pollen assemblages have been comprehensively used in the documentation of Torres Strait Holocene coastal changes (Rowe 2007b), but require stronger use inland to be considered outside of Myrtaceae pollen trends. An important lesson for the interpretation of Torres Strait palaeovegetation (e.g. Rowe 2008) further rests within the described gradation of samples across Figure 6a. Boundaries between vegetation communities will not always be sharp, introducing the concept of extensive ecotones as important components of the landscape and indicators of spatial vegetation shift. Once again, the smaller understorey and/or secondary plant growth indicators – the so-called micro-trends – exist as valuable explanatory components.

While the guidelines in Table 3 can assist in the interpretation of fossil pollen assemblages, the study has also highlighted certain difficulties in obtaining precise signatures. Upon inspection of the pollen diagrams, there is considerable similarity between the overall pollen assemblage of coastal woodland, inland sclerophyll woodland and vine-thicket communities, exemplified by the tight clustering observed in the ordination. High levels of Myrtaceae pollen are one reason it is difficult to separate these three different vegetation types. It should also be considered that regular burning activities may serve to create more uniform vegetation and corresponding pollen assemblages across dryland environments, fire blurring the distinction between sediment samples collected within different vegetation communities. In future analyses, modern pollen and charcoal counts should be used together to determine the level of fire disturbance evident in any given sample. Further research also needs to examine modern pollen representation across environmental gradients. Altitude, soil moisture, nutrient status and salinity could be incorporated, in addition to charcoal. Such measurement would improve an understanding of sample ordination, for example, and assist in documenting wider spatial and temporal connections between vegetation groups.

The fact that fewer than half the surface samples collected were used in this study highlights some limitations in litter collections across seasonal environments (sandy and thin, dry sediments). The sedimentary surface is an active, oxidative environment and pollen was found in varied degradative states in all sediment types and locations. This contrasts with wet sedimentary environments, where sediments become anoxic and abiotic as they are buried and where pollen recovery tends towards relatively fresh preservation. At the surface, differential preservation and degradation, and differential recognition of poorly preserved grains might have resulted in some bias, robust grains assuming a more important interpretative role than might otherwise occur in damp sedimentary fossil collections (Hall 1981; Orvis 1998). As such, there is also a need to adopt additional sampling strategies, such as pollen traps, if this work is to be extended.

In conclusion, the usefulness of modern pollen data from Torres Strait is not limited to this region alone. Demonstrated characterisation of island monsoonal, megatherm vegetation pollen signatures can be usefully applied to a refined understanding of seasonal tropical mainland Australia, a region in which few investigations of modern and fossil pollen rain have been made.

Acknowledgements

This research was supported by Monash University's School of Geography and Environmental Science, with special thanks to the Western Torres Strait Cultural Heritage Project team for their assistance and encouragement. Warm thanks to the people of Mua and Badu for permission to undertake island palaeoecological research, and for their hospitality during fieldwork. The author also acknowledges the Mura Badulgal Corporation which kindly provided ongoing permission to publish. Thanks are also extended to the editors of this volume and to the anonymous referees for feedback and comment.

References

Abrahams, H., Mulvaney, M., Glasco, D. and Bugg, A. 1995. An assessment of the Conservation and Natural Heritage Significance of Cape York Peninsula. *Cape York Peninsula Land Use Study*. Queensland Environment Protection Agency – Brisbane, Australian Government – Department of Environment and Heritage, Canberra, ACT.

Bain, J.H.C. and Draper, J.J. 1997. *North Queensland Geology*. AGSO Bulletin 240/Queensland Geology 9, Queensland Department of Mines and Energy.

Beasley, J. 2009. *Plants of Cape York*. Everbest Publishing, Queensland.

Brock, J. 2001. *Native plants of northern Australia*. Reed New Holland, Sydney.

Burn, M.J. Mayle, F. and Kileen, T.J. 2010. Pollen-based differentiation of Amazonian rainforest communities and implications for lowland palaeoecology in tropical South America. *Palaeogeography, Palaeoclimatology, Palaeoecology* 295:1-18.

Bush, M.B. 2002. On the interpretation of fossil Poaceae pollen in the lowland humid neotropics. *Palaeogeography, Palaeoclimatology, Palaeoecology* 177:5-17.

Chalson, J.M. 1989. *The late Quaternary vegetation and climatic history of the Blue Mountains, New South Wales, Australia*. Unpublished PhD thesis, University of NSW.

Clarke, K.R. 1993. Non-parametric multivariate analysis of changes in community structure. *Australian Journal of Ecology* 18:117-143.

Clarke, K.R. and Gorley, R.N. 2001. *PRIMER v5: User Manual/Tutorial*. PRIMER-E Ltd, Plymouth Marine Laboratory, Plymouth.

Clarke, K.R. and Warwick, R.M. 2001. *Change in marine communities: an approach to statistical analysis and interpretation. 2nd edition*. PRIMER-E Ltd. Plymouth Marine Laboratory, Plymouth.

Crouch, J., McNiven, I., David, B., Rowe, C. and Weisler, M. 2007. Berberass: marine resource specialisation and environmental change in Torres Strait during the last 4000 years. *Archaeology in Oceania* 42:49-62.

Crowley, G.M., Grindrod, J. and Kershaw, A.P.K. 1994. Modern pollen deposition in the tropical lowlands of northeaster Queensland, Australia. *Review of Palaeobotany and Palynology* 83:299-327.

Denham, T. and Mooney, S. 2008. Human-environment interactions in Australia and New Guinea during the Holocene. *The Holocene* 18:365-371.

Department of Environment and Heritage 2007. *Australia's Native Vegetation: Australian Major Vegetation Groups, 2007.* Australian Government, Canberra, ACT.

Ellison, J. 2005. Holocene palynology and sea-level change in two estuaries in Southern Irian Jaya. *Palaeogeography, Palaeoclimatology, Palaeoecology* 3-4:291-309.

Faegri, K. and Iversen, J. 1989. *Textbook of Pollen Analysis (4th edition).* Wiley, Chichester.

Faith D.P., Minchin P.R. and Belbin, L. 1987. Compositional dissimilarity as a robust measure of ecological distance. *Vegetatio* 69:57–68.

Fox, I.D., Neldner, V.J., Wilson, G.W. and Barrick, P.J. 2001. *The Vegetation of Australian Tropical Savannas.* Environmental Protection Agency, Brisbane, Queensland.

Froyd, C.A. and Willis, K.J. 2008. Emerging issues in biodiversity and conservation management: the need for a palaeoecological perspective. *Quaternary Science Reviews* 27:1723-1732.

Fuhsiung, W., Nanfen, C., Yolong, Z. and Huiqiu, Y. 1997. *Pollen flora of China (2nd edition).* Institute of Botany, Academia Sinica.

Grindrod, J. 1985. The palynology of mangroves on a prograded shore, Princess Charlotte Bay, north Queensland, Australia. *Journal of Biogeography* 12:323-348.

Grindrod, J. 1988. *Holocene mangrove history of the South Alligator River estuary, Northern Territory, Australia.* Unpublished PhD Thesis, Australian National University, Canberra.

Grindrod, J. and Rhodes, E.G. 1984. Holocene sea level history of a tropical estuary: Missionary Bay, north Queensland. In: Thom, B.G. (ed), *Coastal Geomorphology in Australia.* Academic Press, London, pp. 151-178.

Grimm, E. 1988. Data Analysis and Display. In: Huntley, B. and Webb, T. (eds), *Handbook of Vegetation Science, Volume 7.* Kluwer Academic Publisher, Dordrecht, pp. 43-76.

Haberle, S.G. 2005. A 23,000-yr pollen record from Lake Euramoo, Wet Tropics of NE Queensland, Australia. *Quaternary Research* 64:343-356.

Hall, S.A. 1981. Deteriorated pollen grains and the interpretation of Quaternary pollen diagrams. *Review of Palaeobotany and Palynology* 32:193-206.

Harrison, S.P., Zhao, Y., Rowe, C. and Marshall, A. 2007. Climates, fire regimes and vegetation patterns of Australia during the past 70,000 years: observations and model results. *Quaternary International* 167:156-157.

Kershaw, A.P. 1971. A pollen diagram from Quincan Crater, north-east Queensland, Australia. *New Phytologist* 70:669-681.

Kershaw, A.P. 1985. An extended late Quaternary vegetation record from north-eastern Queensland and its implications for the seasonal tropics of Australia. *Proceedings of the Ecological Society of Australia* 13:179-189.

Kershaw, A.P. 1992. The development of rainforest-savanna boundaries in tropical Australia. In: Furley, P.A., Proctor, J. and Ratter, J.A. (eds), *Nature and dynamics of forest-savanna boundaries.* Chapman and Hall, London, pp. 255-271.

Kershaw, A.P. 1995. Environmental change in Greater Australia. *Antiquity* 69:656-675.

Kershaw, A.P. and Strickland, K.M. 1990. A 10 year pollen trapping record from rainforest in northeastern Queensland, Australia. *Review of Palaeobotany and Palynology* 64:281-288.

Kershaw, A.P. and Bulman, D. 1994. The relationship between modern pollen samples and environment in the humid tropical region of northeastern Australia. *Review of Palaeobotany and Palynology* 83:83-96.

Kershaw, A.P., Moss, P. and van der Kaars, S. 1997. Environmental change and the human occupation of Australia. *Anthropologie* 35:35-43.

Kershaw, A.P. 2007. Preface: Environmental History of the Humid Tropics region of north-east Australia. *Palaeogeography, Palaeoclimatology, Palaeoecology* 251:1-3.

Kershaw, A.P., Bretherton, S. and van der Kaars, S. 2007. A complete pollen record of the last 230 ka from Lynch's Crater, north-eastern Australia. *Palaeogeography, Palaeoclimatology, Palaeoecology* 251:23-45.

Lynch, A.H., Beringer, J., Kershaw, A.P., Marshall, A., Mooney, S., Tapper, N., Turney, C. and van der Kaars, S. 2007. Using the Palaeorecord to Evaluate Climate and Fire Interactions in Australia. Annual *Review of Earth and Planetary Sciences* 35:215-239.

Mackey, B.G., Nix, H.A. and Hitchcock, P. 2001. *The natural heritage significance of Cape York Peninsula.* ANUTECH Pty. Ltd. and Queensland Environmental Protection Agency, Brisbane.

McNiven, I., David, B. and Barker, B. 2007. The social archaeology of Indigenous Australia. In: David, B., Barker, B. and McNiven, I. (eds), *The Social Archaeology of Australian Indigenous Societies.* Aboriginal Studies Press, Canberra, Australia, pp. 2-20.

Neldner, V.J. and Clarkson, J.R. 1995. *Vegetation Survey and Mapping of Cape York Peninsula.* Cape York Peninsula Land Use Study, Queensland Environment Protection Agency – Brisbane, Australian Government – Department of Environment and Heritage, Canberra, ACT.

Orvis, K.H. 1998. Modern surface pollen from three transects across the southern Sonoran Desert margin, Northwestern Mexico. *Palynology* 22:197-211.

Pickett, E.J., Harrison, S.P., Hope, G., Harle, K., Dodson, J., Kershaw, A.P., Prentice, I.C., Haberle, S., Hassell, C., Kenyon, C., Macphail, M., Martin, H., Martin, A.H., McKenzie, M., Newsome, J.C., Penny, D., Powell, J., Raine, J.I., Southern, W., Stevenson, J., Sutra, J.P., Thomas, I., van der Kaars, S. and Ward, J. 2004. Pollen based reconstructions of biome distributions for Australia, Southeast Asia and the Pacific (SEAPAC) region at 0, 6000 and 18,000 14C yr BP. *Journal of Biogeography* 31:1381-1444.

Overpeck, J.T., Webb, T. and Prentice, I.C. 1985. Quantitative interpretation of fossil pollen spectra: dissimilarity coefficients and the method of modern analogues. *Quaternary Research* 23:87-108.

Rowe, C. 2007a. A palynological investigation of Holocene vegetation change in Torres Strait, seasonal tropics of northern Australia. *Palaeogeography, Palaeoclimatology and Palaeoecology.* 25:83-103.

Rowe, C. 2007b. Vegetation change following mid-Holocene marine transgression of the Torres Strait shelf: a record from Mua. *The Holocene* 17:917-927.

Rowe, C. 2008. Holocene vegetation change on Mua. In: David, B., Tomsana, D. and Quinnell, M. (eds), *Gelam's Homeland: cultural history on the island of Mua, Torres Strait.* Memoirs of the Queensland Museum: Cultural Heritage Series 4:593-604.

Stern, H., de Hoedt, G. and Ernst, J. 2000. Objective classification of Australian climates. *Australian Meteorological Magazine* 49:87-96.

Tauber, H. 1967. Investigations of the mode of pollen transfer in forested areas. *Review of Palaeobotany and Palynology* 3:277-286.

Thanikaimoni, G. 1987. *Mangrove Palynology* UNDP/UNESCO Regional Project on training and research on mangrove ecosystems. RAS/79/002, French Institute.

Turner, H., Hovenkamp, P. and van Welzen, P.C. 2001. Biogeography and Southeast Asia and the West Pacific. *Journal of Biogeography* 28:217-230.

Torres Strait Regional Authority 2007. *The Torres Strait: Torres Strait Annual Report 2006-2007.* http://www.tsra.gov.au

Walker, J. and Hopkins, M.S. 1990. Vegetation. In: McDonald, R.C., Isbell, R.F., Speight, J.G., Walker, J. and Hopkins, M.S. (eds), *Australian Soil and Land Survey. Field Handbook (2nd edition).* Australian Collaborative Land Evaluation Program, CSIRO, Canberra. pp. 58-86.

Walker, D. and Sun, X. 2000. Pollen fall-out from a tropical vegetation mosaic. *Review of Palaeobotany and Palynology* 110:229-246.

Wannam, B. 2008. Terrestrial Vegetation of Gelam's homeland, Mua. *Memoirs of the Queensland Museum (Cultural Heritage Series)* 4:605-615.

21

Surface $\partial^{13}C$ in Australia: A quantified measure of annual precipitation?

Chris S.M. Turney

Climate Change Research Centre and School of Biological, Earth and Environmental Sciences,
University of New South Wales, Sydney
c.turney@unsw.edu.au

Introduction

Since the 1960s, scientific understanding of our global environment and its climate has undergone a remarkable transformation. We are now increasingly aware that the world around us is dynamic, and quasi-stable only in the short term. Recognising the challenge of human-induced climate change, the Intergovernmental Panel on Climate Change (IPCC) was established in 1988 and released its most recent Fourth Assessment Report (AR4) in 2007. The AR4 conclusions are startling: By 2100, global temperatures are estimated to increase between 1°C and 6.5°C compared with 1990, accompanied by a sea level rise of between 0.18 m and 0.58 m. This relatively large range in projections is partly due to chaotic climate variability and to uncertainties in emissions, but another significant factor is the paucity of instrumental data with which to test the estimates. A major source of concern is the extent to which the historical record captures processes representative of future change.

The above issues are of particular concern for Australia, a country distinguished by lack of water and high interannual climate variability, but with historical records extending back to only 1880 (Nicholls et al. 2006). Future expectations for increasing aridity, variability and population concentration in urban and coastal areas represent a complex, uncertain and potentially dangerous challenge to Australian society, for which historical records are insufficient to capture the full range of the climate system. For instance, climate models predict that subtropical regions will expand with an increase in global temperatures (Held and Soden 2006), bringing more arid conditions to heavily populated areas (Bengtsson et al. 2006). Recent data, however, indicate expansion over the past few decades is of the same order of magnitude (5-8° of latitude) as that predicted for the end of this century (Seidel et al. 2008). This shift is associated with a reduction of ca. 20% in winter rainfall over the southwest of Western Australia, and the development of new water sources for Perth estimated to have already cost more than $500 million (IOCI

2002). Within the AR4, perhaps most critical of all for Australia, the future absolute amount and seasonality of rainfall across the region are highly uncertain, but seem likely to decline by the end of the current century (Christensen et al. 2007).

Past climate change provides a critical baseline against which to compare present and future warming by encompassing a broader range of extremes. Most climate reconstructions obtained from geological, chemical and biological proxies have published relationships with temperature (Mann et al. 1998, 2008; Esper et al. 2002; Moberg et al. 2005). Critically, a few measures of precipitation have been reported (e.g. Kershaw et al. 1994; Bowler 1998; Cook and van der Kaars 2006; Lough 2007; Cullen and Grierson 2009), but most are from individual sites and largely of a qualitative nature, limiting our ability to generate a long-term spatially robust reconstruction of past rainfall within Australia. One possibility for resolving this apparent impasse is the exploitation of stable isotopes in terrestrial plant material, particularly species- and tissue-specific $\partial^{13}C$, an approach that has been demonstrated to provide a measure of the moisture-related conditions under which the tissues formed (e.g. Ehleringer and Cooper 1988; Farquhar et al. 1989; Turney et al. 1999, 2002). Unfortunately, few plant macrofossils (including wood) are found within terrestrial and marine sedimentary sequences across and adjacent to the mainland of Australia (D'Costa et al. 1989; Bohte and Kershaw 1999; Moss and Kershaw 2007), precluding continuous $\partial^{13}C$ measurements of material through profiles. One alternative is charcoal (e.g. Ferrio et al. 2005; Turney et al. 2006).

Charcoal has considerable potential for developing long-term climate reconstructions. Firstly, charcoal is a common product from biomass burning and largely recalcitrant in lake (Kershaw 1971, 1974, 1975, 1976, 1995; Kershaw et al. 2004; Turney et al. 2004), marine (Kershaw et al. 1993; Wang et al. 1999; van der Kaars et al. 2000; Moss and Kershaw 2007) and soil (Hopkins et al. 1993; Bird et al. 1999; Lehman et al. 2008) environments, allowing preservation on geological timescales (Lynch et al. 2007; Power et al. 2008). Secondly, if charcoal is finely disseminated with sediments, its $\partial^{13}C$ composition should reflect the proportions of C3 and C4 plants within the local vegetation (primarily controlled by the most effective season of rainfall; Hattersley 1983; Polley et al. 1993; Ehleringer et al. 1997) and/or the degree of physiological stress on C3 plants as a result of changing moisture availability (Ehleringer and Cooper 1988; Turney et al. 1999; Turney et al. 2002). To date, however, although $\partial^{13}C$ of charcoal has been measured through selected sedimentary sequences within the Australian region (Wang et al. 1999; Turney et al. 2001), demonstrating a quantitative relationship with any moisture-related variable has proved elusive.

Methods

To test the relationship between charcoal isotopic content and moisture, surface soil samples were collected from a network of 17 sites spanning a large precipitation gradient across Australia (Figure 1 and Table 1), ranging from Buderim and Darwin at >1500 mm/ year, to Marla in South Australia at <200 mm/year. Samples were taken down to a depth of 2 cm below the surface, in an attempt to provide a long-term average isotopic composition of charcoal from each site. In the laboratory, the samples were sieved through a series of meshes to isolate the fraction 2 mm and 125 µm. Using a biological microscope, individual fragments of charcoal were hand picked.

Importantly, because particulates produced during combustion are a complex mix of variably carbonised material (some of which can undergo further oxidation during diagenesis), the direct measurement of charcoal particles for $\partial^{13}C$ composition is not appropriate, as incompletely carbonised material may distort the values obtained during analysis. Here, we have applied the method outlined by Bird and Gröcke (1997) for isolating oxidation resistant elemental carbon

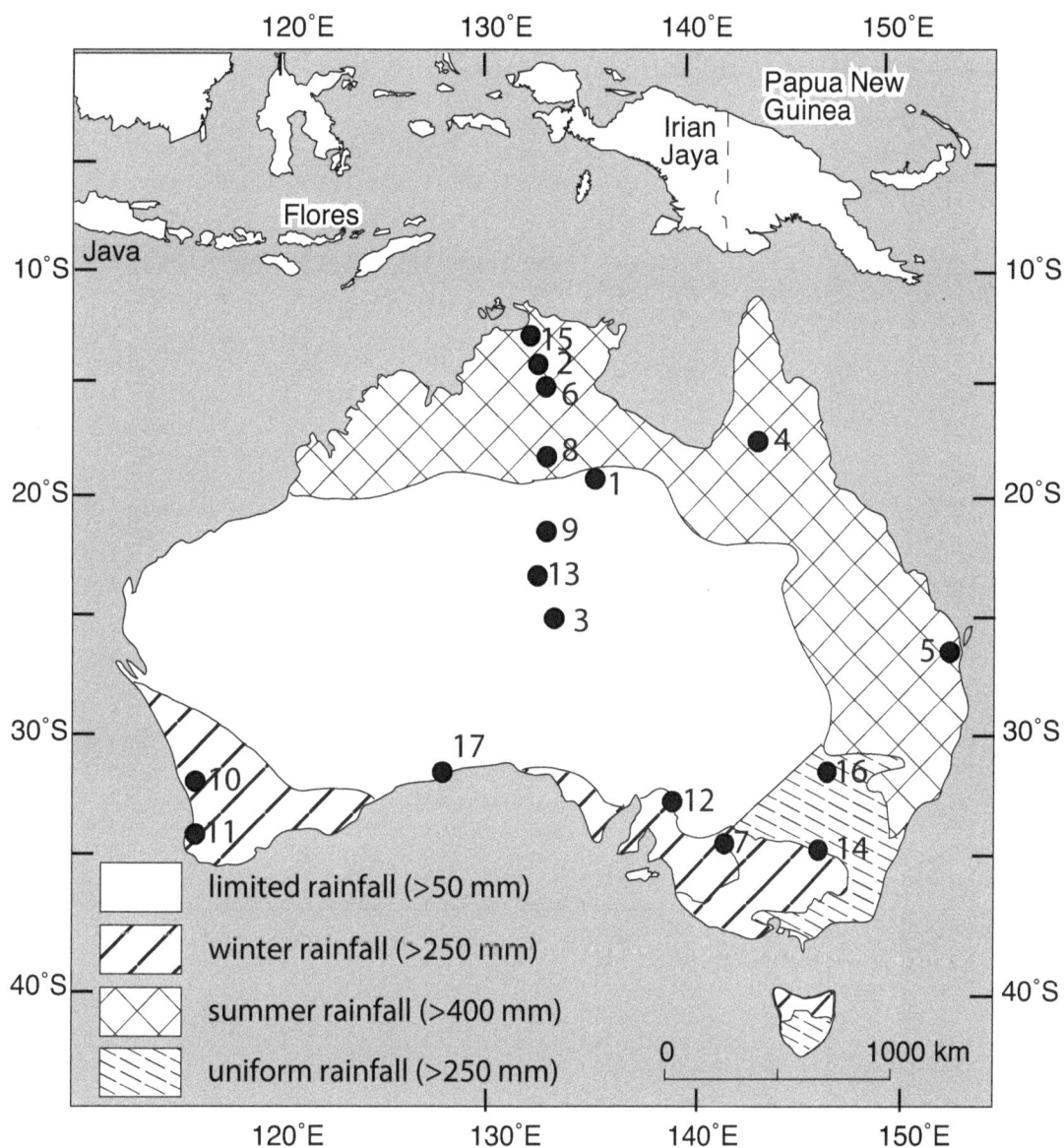

Figure 1. Surface charcoal sampling locations across Australia. Details of numbered sites are given in Table 1. Modified from Williams et al. 2009.

Table 1: Site locations, surface charcoal $\partial^{13}C$ and annual mean precipitation values, with summary statistics.

Site number	Site name	Latitude, °S	Longitude, °E	$\partial^{13}C$, ‰ (VPDB)	Annual precipitation, mm
1	Barkly Station	19°42″	135°49″	−23.1	332
2	Pine Creek	13°49″	131°49″	−27.5	1178
3	Marla	27°08″	133°30″	−24.9	190
4	Mount Garnet	17°47″	144°57″	−25.4	759
5	Buderim	26°42″	153°04″	−26.6	1712
6	Mataranka	14°56″	133°04″	−25.2	800
7	Renmark	34°14″	140°37″	−23.1	257
8	Renner	18°19″	133°47″	−24.7	407
9	Ti Tree	22°07″	133°25″	−22.6	299
10	Freemantle	32°02″	115°45″	−25.2	838
11	Margaret River	34°09″	115°02″	−25.8	1163
12	Pt Augusta	32°28″	137°44″	−24.7	241

Table 1: *Continued*

Site number	Site name	Latitude, °S	Longitude, °E	∂¹³C, ‰ (VPDB)	Annual precipitation, mm
13	Erlunda	25°11″	133°12″	−23.6	203
14	Dubbo	32°26″	148°21″	−25.6	590
15	Darwin	12°40″	131°04″	−26.5	1521
16	Goondawindi	28°44″	150°16″	−24.0	577
17	Allan's Cave	31°36″	129°06″	−25.5	248

Statistics	
R^2 (R^2_{adj})	0.57 (0.54)
F value	20.01
P value	0.0004469

(OREC). Elemental carbon is defined here as carbon that survives the chemical isolation procedure outlined below.

Charcoal samples extracted from between 2 mm and 125 µm were decarbonated overnight using 1N HCl, washed with MilliQ™ water, centrifuged and then placed in concentrated HF overnight at 60°C to remove silicate material. The remaining material was then washed again in MilliQ™ water, centrifuged and placed in 0.1N NaOH for three hours at room temperature to remove humic acids. Samples were then washed repeatedly in MilliQ™ water, until the solution became clear, and placed in a $K_2Cr_2O_7/H_2SO_4$ solution at 60°C for 14 hours (Bird and Gröcke 1997). The OREC samples were again washed with MilliQ™ water, then freeze-dried.

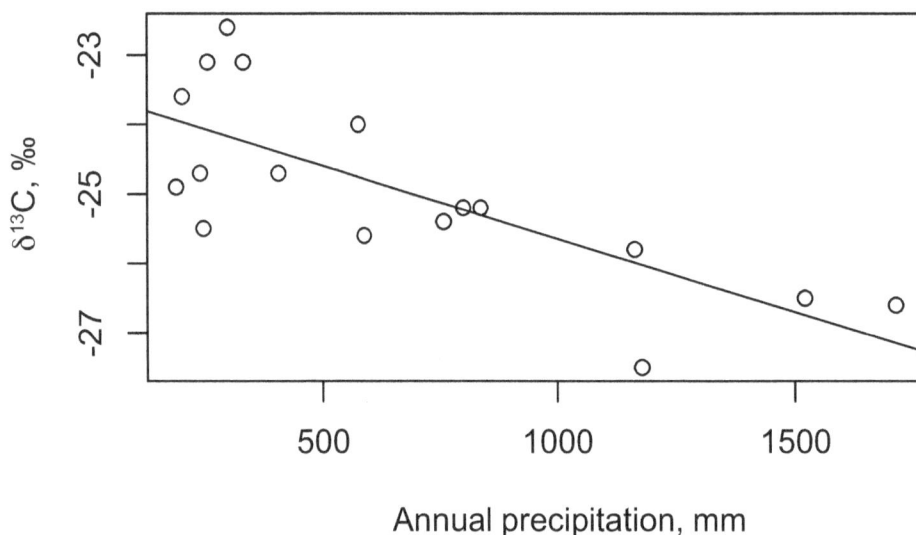

Figure 2. Relationship between surface charcoal ∂¹³C and annual precipitation in Australia.

The stable carbon isotope (∂¹³C) composition of OREC samples was determined using an elemental analyser coupled to a Micromass Prism III mass spectrometer operated in continuous flow mode. ∂¹³C values are expressed as per mille (‰) relative to the international V-PDB standard, with a precision of 0.15‰ at 1σ. Duplicate measurements were made and an average value taken.

Results and discussion

The isotopic values obtained during this study range from −23.1‰ at Renmark in the arid interior of Victoria, to −27.5‰ at Pine Creek in the tropical Northern Territory (Table 1). Overall, there appears to be a general trend to heavier values in the interior of Australia (Figure 1). Assuming the carbonised plant material from the surface soil samples reflects local vegetation, the OREC $\partial^{13}C$ values can be used for comparison against long-term climate data.

Typical C3 and C4 $\partial^{13}C$ values range from −22 to −33‰, and −9 to −16‰ respectively (Deines 1980). The results obtained fall almost entirely within the range of values expected for C3 vegetation and suggest little (if any) carbon derived from C4 photosynthesis is present in the OREC samples collected across Australia. If correct, the enriched ^{13}C values are typical of C3 plants in moisture-limited environmental conditions (Ehleringer and Cooper 1988; Turney et al. 1999), as a result of reduced stomatal conductance and/or altered net assimilation.

To test the relationship between OREC $\partial^{13}C$ values and climate, data were compared with bioclimate estimates obtained from each site generated from the prediction system BIOCLIM (Busby 1991). BIOCLIM produces up to 35 bioclimatic parameters based on long-term climate measurements of maximum and minimum temperature, rainfall, solar radiation and pan evaporation. Comparisons were made with all parameters. Although several of the climate variables proved highly correlated with surface charcoal OREC $\partial^{13}C$ values, the most robust and significant relationship was that obtained against annual precipitation across a range of 260 mm to 1200 mm (Figure 2 and Table 1).

We observe a linear and negative correlation between annual precipitation and OREC $\partial^{13}C$ (Figure 2) (F = 20.01, p < 0.0004469), explaining more than half of the variance (R^2 = 0.57, R^2_{adj} = 0.54). The correlation is highly significant, suggesting that the $\partial^{13}C$ in vegetation of the immediate area (as represented by the surface soil charcoal) is strongly influenced by the amount of rainfall over the year. This result is consistent with previous studies, which have identified the importance of moisture availability in controlling stomatal conductance (Ehleringer and Cooper 1988) and the composition of individual Australian species (Miller et al. 2001) in the Queensland plant community $\partial^{13}C$ (Stewart et al. 1995). For instance, over a 1100 mm annual rainfall range, the Queensland study demonstrated a mean 4‰ shift (Stewart et al. 1995), comparable to the mean 3‰ difference observed in the charcoal samples collected across Australia (Figure 2).

The mechanism for the changes in $\partial^{13}C$ may be best explained by stomatal conductance responses to moisture availability. During growth, under low moisture availability, plant stomatal conductance will decrease to minimise water loss, reducing the exchange of carbon dioxide between the substomatal cavity and the surrounding atmosphere, thereby decreasing the discrimination against ^{13}C relative to ^{12}C (Farquhar et al. 1989). The preliminary results reported here, therefore, provide strong support that charcoal $\partial^{13}C$ may offer considerable potential for quantifying past changes in precipitation, and suggest the observation made within community-averaged $\partial^{13}C$ observed across a rainfall gradient in Queensland (Stewart et al. 1995) may be extended to the fossil record.

The above relationship should only be considered a first-order estimate, however. During heating, the cellulose and hemicellulose content of plant material form mainly volatile products due to the thermal cleavage of sugar units, while lignin dominates the production of charcoal since it is not so easily cleaved to lower-molecular-weight fragments. As a result, during carbonisation of woody material, increasing temperature progressively depletes the ^{13}C content of bulk charcoal by up to 1.3‰ (Turney et al. 2006), consistent with the greater susceptibility of cellulose to thermal degradation relative to lignin (Czimczik et al. 2002). Although the OREC is most likely dominated by lignin (Bird and Gröcke 1997), it is unclear whether the isotopic fractionation observed in bulk charcoal reflects an increasing proportion of this component in the final char and/or there is a genuine fractionation within lignin with changing temperature. Another potentially significant limitation of this study is the uncertain age range of the charcoal obtained from the surface soil samples. Although the sampling strategy adopted here had the advantage of providing an average estimate of surface vegetation $\partial^{13}C$ values, the duration represented is unknown and may be of the order of centuries. Remarkably, in spite of these issues, there still remains a statistically significant correlation between isotopic content and climate, suggesting that if samples were obtained over the same period as meteorological data, a more robust relationship may be quantified.

Conclusions

There is a statistically significant relationship between elemental carbon $\partial^{13}C$ obtained from 'modern' surface charcoal and annual precipitation in Australia. Such a relationship is expected because of the important role moisture availability plays in the distribution and response of flora. In spite of the uncertainties associated with comparing climate parameters derived from historic meteorological data and surface charcoal of unknown age, the relationship suggests this approach might be used to quantify past changes in rainfall across Australia. Future studies focusing on comparing charcoal samples of known age with meteorological data over a common period should improve the robustness of future reconstructions. This finding is of particular importance in Australia, a country distinguished by lack of water and where few quantified methods of precipitation are available to extend historical records beyond 1880.

Acknowledgements

I would like to thank Peter for all his help and encouragement over the years. I've been extremely privileged to work with Peter and have hugely benefited from our collaboration. It's been an absolute pleasure and I feel sure I've gained far more from working with him than he has from me! This work was made possible thanks to the support of the Australian Research Council through a QEII and Laureate Fellowship. Many thanks to Joan Cowley at the Research School of Earth Sciences who kindly helped with the isotopic analysis of the samples and Nick Porch who extracted the BIOCLIM data for me. Thanks also to Mike Smith and Charlie Dortch who kindly provided samples from the Nullarbor Plain and southwest Australia respectively. Michael Bird and one anonymous reviewer kindly helped improve the text.

References

Bengtsson, L., Hodges, K.I. and Roeckner, E. 2006. Storm tracks and climate change. *Journal of Climate* 19:3518-3543.

Bird, M.I. and Gröcke, D.R. 1997. Determination of the abundance and carbon-isotope composition of elemental carbon in sediments. *Geochimica et Cosmochimica Acta* 61:3413-3423.

Bird, M.I., Moyo, C., Veenendaal, E.M., Lloyd, J. and Frost, P. 1999. Stability of elemental carbon in a savanna soil. *Global Biogeochemical Cycles* 13:923-932.

Bohte, A. and Kershaw, A.P. 1999. Taphonomic influences on the interpretation of the palaeoecological record from Lynch's Crater, northeastern Australia. *Quaternary International* 57/58:49-59.

Bowler, J.M. 1998. Willandra Lakes revisited: environmental framework for human occupation. *Archaeology in Oceania* 33:120-155.

Busby, J.R. 1991. BIOCLIM – a bioclimatic analysis prediction system. In: Margules, C.R. and Austin, M.P. (eds), *Nature Conservation: Cost Effective Biological Surveys and Data Analysis*. CSIRO, Melbourne, pp. 64-68.

Christensen, J.H., Hewitson, B., Busuioc, A., Chen, A., Gao, X., Held, I., Jones, R., Kolli, R.K., Kwon, W.-T., Laprise, R., Magaña Rueda, R., Mearns, L., Menéndez, C.G., Räisänen, J., Rinke, A., Sarr, A. and Whetton, P. 2007. Regional Climate Projections. In: Solomon, S., Qin, D., Manning, M., Chen, Z., Marquis, M., Averyt, K.B., Tignor, M. and Miller, H.L. (eds) 2007. *Climate Change 2007: The Physical Science Basis. Contribution of Working Group I to the Fourth Assessment Report of the Intergovernmental Panel on Climate Change*. Cambridge University Press, Cambridge, UK.

Cook, E.J. and van der Kaars, S. 2006. Development and testing of transfer functions for generating quantitative climatic estimates from Australian pollen data. *Journal of Quaternary Science* 21:723-733.

Cullen, L.E. and Grierson, P.F. 2009. Multi-decadal scale variability in autumn-winter rainfall in south-western Australia since 1655 AD as reconstructed from tree rings of Callitris columellaris. *Climate Dynamics* 33:433-444.

Czimczik, C.I., Preston, C.M., Schmidt, M.W.I., Werner, R.A. and Schulze, E.-D. 2002. Effects of charring on mass, organic carbon, and stable carbon isotope composition of wood. *Organic Geochemistry* 33:1207-1223.

D'Costa, D.M., Edney, P., Kershaw, A.P., de Deckker, P. 1989. Late Quaternary palaeoecology of Tower Hill, Victoria, Australia. *Journal of Biogeography* 16:461-482.

Deines, P. 1980. The isotopic composition of reduced organic carbon. In: Fritz, P. and Fontes, J.Ch. (eds), *Handbook of Environmental Isotope Geochemistry 1A*. Elsevier, Amsterdam, 329-406.

Ehleringer, J.R., Cerling, T.E. and Helliker, B.R. 1997. C4 photosynthesis, atmospheric CO_2 and climate. *Oecologia* 112:285-299.

Ehleringer, J.R. and Cooper, T.A. 1988. Correlations between carbon isotope ratio and microhabitat in desert plants. *Oecologia* 76:562-566.

Esper, J., Cook, E.R. and Schweingruber, F.H. 2002. Low-frequency signals in long tree-ring chronologies for reconstructing past temperature variability. *Science* 295:2250-2253.

Farquhar, G.D., Ehleringer, J.R. and Hubick, K.T. 1989. Carbon isotope discrimination and photosynthesis. *Annual Review of Plant Physiology and Plant Molecular Biology* 40:503-537.

Ferrio, J.P., Araus, J.L., Buxó, R., Voltas, J. and Bort, J. 2005. Water management practices and climate in ancient agriculture: inferences from the stable isotope composition of archaeobotanical remains. *Vegetation History and Archaeobotany* 14:510-517.

Hattersley, P.W. 1983. The distribution of C3 and C4 grasses in Australia in relation to climate. *Oecologia* 57:113-128.

Held, I.M. and Soden, B.J. 2006. Robust responses of the hydrological cycle to global warming. *Journal of Climate* 19:5686-5699.

Hopkins, M.S., Ash, J., Graham, A.W., Head, J. and Hewett, R.K. 1993. Charcoal evidence of the spatial extent of the Eucalyptus woodland expansions and rainforest contractions in North Queensland during the late Pleistocene. *Journal of Biogeography* 20:357-372.

IOCI. 2002. Climate variability and change in south west Western Australia. Indian Ocean Climate Initiative Panel, c/o Department of the Environment, Water and Catchment Protection, Hyatt Place, 3 Plain St, East Perth, WA, 6004.

Kershaw, A.P. 1971. A pollen diagram from Quincan Crater, north-east Queensland, Australia. *New Phytologist* 70:669-681.

Kershaw, A.P. 1974. A long continuous pollen sequence from north-eastern Australia. *Nature* 251:222-223.

Kershaw, A.P. 1975. Stratigraphy and pollen analysis of Bromfield Swamp, North Eastern Queensland, Australia. *New Phytologist* 75:173-191.

Kershaw, A.P. 1976. A late Pleistocene and Holocene pollen diagram from Lynch's Crater, north-eastern Queensland, Australia. *New Phytologist* 77:469-498.

Kershaw, A.P. 1995. Environmental change in Greater Australia. *Antiquity* 69:656-675.

Kershaw, A.P., McKenzie, G.M. and McMinn, A. 1993. A Quaternary vegetation history of northeastern Queensland from pollen analysis of ODP site 820. *Proceedings of the Ocean Drilling Program, Scientific Results* 133:107-114.

Kershaw, A.P., Bulman, D. and Busby, J.R. 1994. An examination of modern and pre-European settlement pollen samples from southeastern Australia – assessment of their application to quantitative reconstruction of past vegetation and climate. *Review of Palaeobotany and Palynology* 82:83-96.

Kershaw, A.P., D'Costa, D.M., Tibby, J., Wagstaff, B.E. and Heijnis, H. 2004. The last million years around Lake Keilambete, western Victoria. *Proceedings, Royal Society of Victoria* 116:95-106.

Lehman, J., Skjemstad, J., Sohi, S., Carter, J., Barson, M., Falloon, P., Coleman, K., Woodbury, P. and Krull, E. 2008. Australian climate-carbon cycle feedback reduced by soil black carbon. *Nature Geoscience* 1:832-835.

Lough, J.M. 2007. Tropical river flow and rainfall reconstructions from coral luminescence: Great Barrier Reef. *Paleoceanography* 22:doi:10.1029/2006PA001377.

Lynch, A.H., Beringer, J., Kershaw, P., Marshall, A., Mooney, S., Tapper, N., Turney, C. and van der Kaars, S. 2007. The scope for the palaeorecords to evaluate climate and fire interactions in Australia. *Annual Review of Earth and Planetary Sciences* 35:215-239.

Mann, M.E., Bradley, R.S. and Hughes, M.K. 1998. Global-scale temperature patterns and climate forcing over the past six centuries. *Nature* 392:779-787.

Mann, M.E., Zhang, Z., Hughes, M.K., Bradley, R.S., Miller, S.K., Rutherford, S. and Ni, F. 2008. Proxy-based reconstructions of hemispheric and global surface temperature variations over the past two millennia. *Proceedings of the National Academy of Sciences* 105:13252-13257.

Miller, J.M., Williams, R.J. and Farquhar, G.D. 2001. Carbon isotope discrimination by a sequence of Eucalyptus species along a subcontinental rainfall gradient in Australia. *Functional Ecology* 15:222-232.

Moberg, A., Sonechkin, D.M., Holmgren, K., Datsenko, N.M. and Karlén, W. 2005. Highly variable Northern Hemisphere temperatures reconstructed from low- and high-resolution proxy data. *Nature* 433:613-617.

Moss, P.T. and Kershaw, A.P. 2007. A late Quaternary marine palynological record (oxygen isotope stages 1 to 7) for the humid tropics of northeastern Australia based on ODP Site 820. *Palaeogeography, Palaeoclimatology, Palaeoecology* 251:4-22.

Nicholls, N., Collins, D., Trewin, B. and Hope, P. 2006. Historical instrumental climate data for Australia – Quality and utility for palaeoclimatic studies. *Journal of Quaternary Science* 21:681-688.

Polley, H.W., Johnson, H.B., Marino, B.D. and Mayeux, H.S. 1993. Increase in C3 plant water-use efficiency and biomass over glacial to present CO2 concentrations. *Nature* 361:61-64.

Power, M.J., Marlon, J., Ortiz, N., Bartlein, P.J., Harrison, S.P., Mayle, F.E., Ballouche, A., Bradshaw, R.H.W., Carcaillet, C., Cordova, C., Mooney, S., Moreno, P.L., Prentice, I.C., Thonicke, K., Tinner, W., Whitlock, C., Zhang, Y., Zhao, Y., Ali, A.A., Anderson, R.S., Beer, R., Behling, H., Briles, C., Brown, K.J., Brunelle, A., Bush, M., Camill, P.,

Chu, G.Q., Clark, J., Colombaroli, D., Connor, S., Daniau, A.-L., Daniels, M., Dodson, J., Doughty, E., Edwards, M.E., Finsinger, W., Foster, D., Frechette, J., Gaillard, M.-J., Gavin, D.G., Gobet, E., Haberle, S., Hallett, D.J., Higuera, P., Hope, G., Horn, S., Inoue, J., Kaltenrieder, P., Kennedy, L., Kong, Z.C., Larsen, C., Long, C.J., Lynch, J., Lynch, E.A., McGlone, M., Meeks, S., Mensing, S., Meyer, G., Minckley, T., Mohr, J., Nelson, D.M., New, J., Newnham, R., Noti, R., Oswald, W., Pierce, J., Richard, P.J.H., Rowe, C., Sanchez Goñi, M.F., Shuman, B.N., Takahara, H., Toney, J., Turney, C., Urrego-Sanchez, D.H., Umbanhowar, C., Vandergoes, M., Vanniere, B., Vescovi, E., Walsh, M., Wang, X., Williams, N., Wilmshurst, J. and Zhang, J.H. 2008. Changes in fire regimes since the Last Glacial Maximum: an assessment based on a global synthesis and analysis of charcoal data. In: *Climate Dynamics*. pp. 887-907.

Seidel, D.J., Fu, Q., Randel, W.J. and Reichler, T.J. 2008. Widening of the tropical belt in a changing climate nature. *Nature Geoscience* 1:21-24.

Stewart, G.R., Turnbull, M.H., Schmidt, S. and Erskine, P.D. 1995. 13C natural abundance in plant communities along a rainfall gradient: a biological integrator of water availability. *Australian Journal of Plant Physiology* 22:51-55.

Turney, C.S.M., Barringer, J., Hunt, J.E. and McGlone, M.S. 1999. Estimating past leaf-to-air vapour pressure deficit from terrestrial plant ∂13C. *Journal of Quaternary Science* 14:437-442.

Turney, C.S.M., Bird, M.I. and Roberts, R.G. 2001. Elemental ∂13C at Allen's Cave, Nullarbor Plain, Australia: Assessing post-depositional disturbance and reconstructing past environments. *Journal of Quaternary Science* 16:779-784.

Turney, C.S.M., Hunt, J.E. and Burrows, C. 2002. Deriving a consistent ∂13C signature from tree canopy leaf material: implications for palaeoclimatic reconstruction. *New Phytologist* 155:301-311.

Turney, C.S.M., Kershaw, A.P., Clemens, S.C., Branch, N., Moss, P.T. and Fifield, L.K. 2004. Millennial and orbital variations of El Niño/Southern Oscillation and high-latitude climate in the last glacial period. *Nature* 428:306-310.

Turney, C.S.M., Wheeler, D. and Chivas, A.R. 2006. Carbon isotope fractionation in wood during carbonization. *Geochimica et Cosmochimica Acta* 70:960-964.

van der Kaars, S., Wang, X., Kershaw, P., Guichard, F. and Setiabudi, A. 2000. Late Quaternary palaeoecological record from the Banda Sea, Indonesia: Patterns of vegetation, climate and biomass burning in Indonesia and northern Australia. *Palaeogeography, Palaeoclimatology, Palaeoecology* 155:135-153.

Wang, X., van der Kaars, S., Kershaw, P., Bird, M. and Jansen, F. 1999. A record of fire, vegetation and climate through the last three glacial cycles from Lombok Ridge core G6-4, eastern Indian Ocean, Indonesia. *Palaeogeography, Palaeoclimatology, Palaeoecology* 147:241-256.

Williams, M., Cook, E., van der Kaars, S., Barrows, T., Shulmeister, J. and Kershaw, P. 2009. Glacial and deglacial climatic patterns in Australia and surrounding regions from 35 000 to 10 000 years ago reconstructed from terrestrial and near-shore proxy data. *Quaternary Science Reviews* 28:2398-2419.

22

Palaeoecology as a means of auditing wetland condition

Peter Gell

Centre for Environmental Management, University of Ballarat, Ballarat, Victoria
p.gell@ballarat.edu.au

The line it is drawn, the curse it is cast
The slow one now, will later be fast
As the present now, will later be past
The order is rapidly fadin'
And the first one now will later be last
For the times they are a changin'

Bob Dylan

Introduction

One could line up a suite of palaeoecological research papers published about Australian sites and, while they would not extend from Lake Wangoom to Lynch's Crater, they would fill much of the pollen microscope laboratory at Monash University. In one way, that, in fact, would be the best place to start to assemble the bibliography, as many of the papers have emanated from Peter Kershaw and the long list of honours and postgraduate students he has supervised, his post-doctoral fellows and the palaeoecological diaspora that is the legacy of this legend from Littleborough. Of course, all of these students would suggest it be assembled elsewhere, as they know too well that it would take many years to find all of the papers in Peter's office.

If this list was separated into those with a long-term focus and those with a direct management focus, there would be a clear bias to the former. While Peter's supervisor patiently examined detailed records of change in fine temporal (and spatial) resolution (Walker et al. 2000), and first coined the term 'fine-resolution pollen analysis', his student's focus was clearly on the ecological response of vegetation communities to Milankovich-scale climatic fluctuations. The pollen record from Lynch's Crater is progressively developed in an ever increasing number of publications (Kershaw 1978, 1986, 1993; Turney et al. 2006; Kershaw et al. 2007) and this is complemented by long offshore records (Harle 1997; Moss and Kershaw 2007) and those from the western plains (Kershaw et al. 1991; Kershaw, 1998; Harle et al. 2004), and uplands (McKenzie and Kershaw 1997; Kershaw et al. 2007) of Victoria. While Peter Kershaw's website observes that his research focus is on "Environmental Change … as a basis for understanding present landscapes and contributing to their future management", Peter has only occasionally ventured into the dark side of environmental management. Perhaps he shied away after the Queensland Forestry Commission used his Lynch's Crater pollen diagram in a brochure justifying rainforest logging (Figure 1) and, in doing so, changed the chronology from thousands to millions of years, reversed the time frame and postulated that humans (*Homo erectus*!) arrived in Australia at 1.5 million years ago (Kershaw and Gell 1990). Despite this, Peter rightly holds a firm view that these long-term records are relevant to management, but his passion for the large, deep-in-time changes kept him largely temporally isolated from those doing the managing. The charge of examining more recent time frames, in fine detail, was left to his students (Gell et al. 1993), and his students' students (Bickford et al. 2008), and these have, more or less, helped or hindered managers, depending on your point of view.

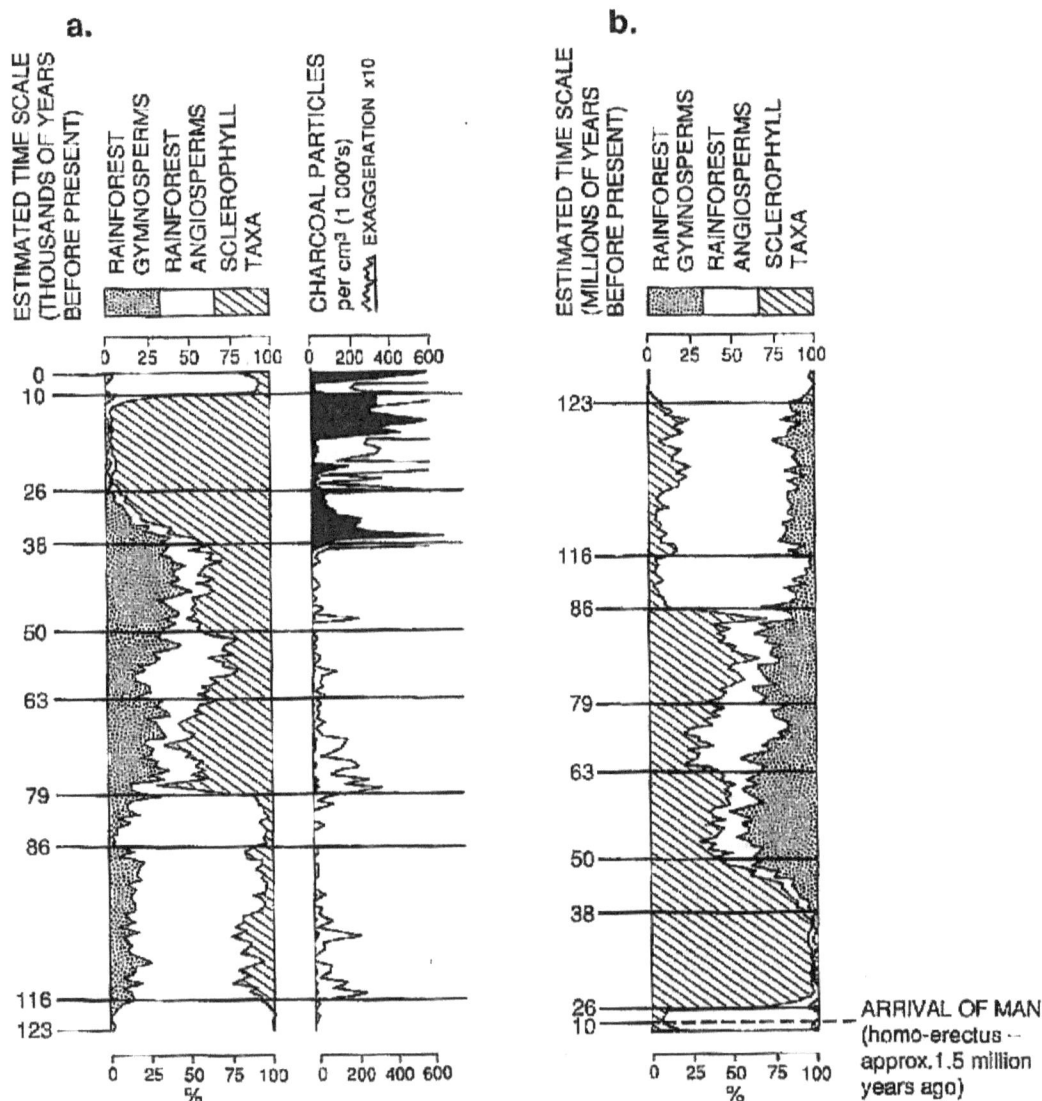

Figure 1. The pollen record from Lynch's Crater as (re)interpreted by the Queensland Forestry Commission (from Kershaw and Gell 1990).

Management questions

Particularly over the past decade, there has been an explosion of research directed towards applied ecology, undertaken with a view to generating evidence to assist environmental managers make effective decisions. Increasingly, these have been fully replicated research designs dedicated to establishing sufficient power of analysis to demonstrate both significance and repeatability. Also, they have been analysed or modelled using increasingly sophisticated statistical techniques that reveal, at least to the statistician, what the data is showing in terms of identifiable changes or causal relations. However, these are contemporary studies, and so, by their very nature, do not address the forces that lead to the management issues at hand. As Weatherhead (1986) observed, managers tend to use the right techniques to address the wrong questions. In Australia, the evidence of very early, post-European settlement landscape destabilisation (Gale and Haworth 2002), salinisation (Gell et al. 2005a), eutrophication (Gell and Little 2006) and sediment flux (Reid et al. 2002, 2007) is testament to the fact that contemporary ecologists are researching already disturbed, and sometimes degraded, ecosystems.

There are several management questions that require a historical, and palaeoecological, approach. For wetland managers, palaeolimnological approaches can provide critical evidence relating to:

- The heritage status of a wetland by assessing its present condition against its long-term range of variability.
- The sensitivity or resilience of a wetland by retrospectively assessing its ecological response to past perturbations.
- The prognosis of a wetland by identifying its trajectory of change or infilling.
- The drivers of wetland change by associating past ecological shifts with documented shifts in climate or management regime.
- The health of the wetland by identifying the degree to which it has departed from its historical range of variability (Bennion and Battarbee 2007).

In all cases, in Australia, where the shift in ecological condition predated the 1970s, these questions cannot be answered by contemporary ecological research.

For wetlands and their managers, the magnitude of the recent 'Big Dry' very much sharpened, perhaps for the first time, their focus on the history of drought and the trajectories of wetland condition subjected to reduced effective rainfall within substantially disturbed catchments. The types of questions asked included:

- What are the effects of clearfell harvesting on aquatic systems?
- What is the natural ecological character of this wetland (e.g. for reviews of status under the Ramsar protocol)?
- What are the main drivers of river and wetland change and what should be the management target for our wetland?
- Is this drought unusual and if so what is the cyclicity of events of this magnitude?

Impact of clearfell harvesting

The lag in vegetation, and so pollen, response to catchment changes often means that palynological evidence is not a decisive measure of the impact of direct human impacts, such as clearfell harvesting. The more responsive nature of the aquatic biota, however, has proved to be incisive, as demonstrated in the case of Tea Tree Swamp on the Delegate River (Gell and

Stuart 1989). The Delegate River drains the northern slopes of the Errinundra Plateau, East Gippsland, which is a large area of Victoria's forest estate that was subject to considerable environmental debate in the 1980s (Mercer 1995). Short, fine-resolution pollen and charcoal records were generated with a view to assessing the changes to the catchment through the European period, in association with an ethnohistoric study of catchment settlement and use. While some forest understorey taxa (e.g. *Tasmannia*) were shown to be sensitive to the post-settlement fire regimes, the most dramatic changes in the pollen flora were identified to be in the uppermost sediments. In three different cores, taxa with reproductive life histories sensitive to swamp drying (*Sphagnum* and *Myriophyllum pedunculatum*) increased in representation by orders of magnitude in the upper sediments deposited after 1970 AD (Gell et al. 1993). The timing of this change coincided with the commencement of forest harvesting in the catchment. By the time of the study, almost half the forest had been clearfelled. While the 12,000-year record of Ladd (1979) was of coarser resolution, this very recent change was unprecedented, and attested to the significance of the hydrological impact of ash forest regeneration on this scale (Figure 2). This phenomenon was documented in the classic Correnderrk catchment study using space-for-time substitution (Kuczera 1985), and the combination of this modelling approach and the palaeoecological record (Wilby and Gell 1994) was able to document conclusively the opportunity cost in terms of water yield of intensively harvesting tall, ash forests. The outcome also identified clearfell harvesting as the driver of change that led a state-listed wetland outside its range of historic variability.

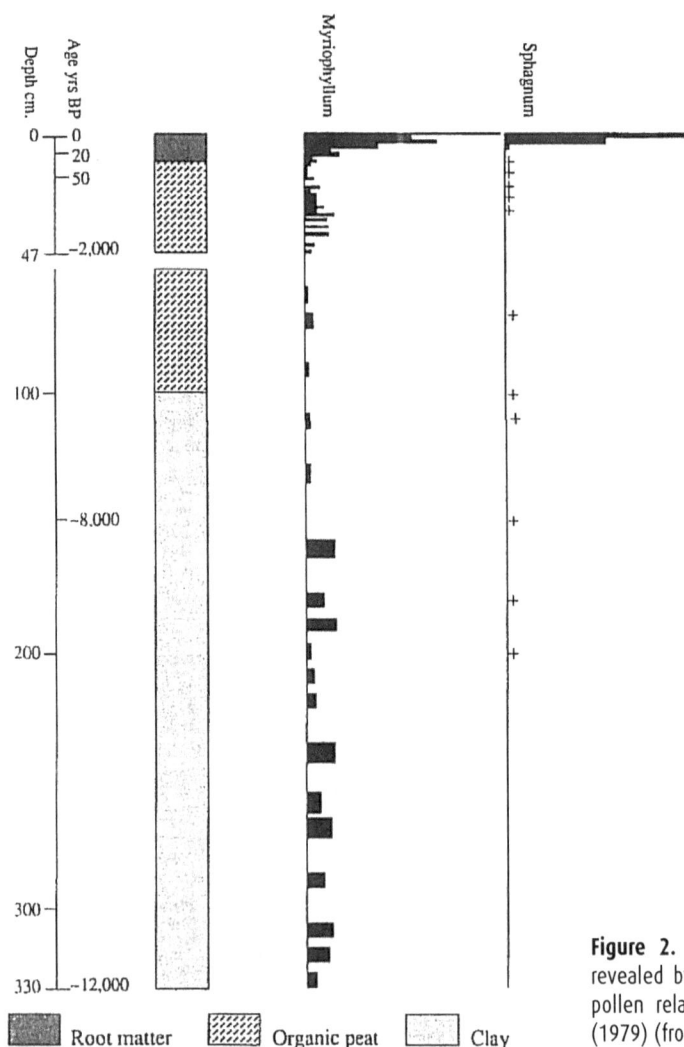

Figure 2. The impact of clearfell harvesting as revealed by Sphagnum spores and *Myriophyllum* pollen relative to the 12,000-year record of Ladd (1979) (from Wilby and Gell 1994).

Assessment of natural ecological character

The International Treaty on Wetlands of International Significance was signed in Ramsar, Iran, in 1971. Australia, as a signatory, listed many wetlands in the early years. As part of this process, nominations were required to state the natural ecological character of the wetland, as well as the attributes for which it qualified under a range of criteria. One such wetland was the Coorong, a 110 km long back-dune lagoon adjacent to the mouth of the River Murray. It was listed in 1985 and its ecological character was described as (DEH 2000):

The Coorong, Lake Alexandrina & Lake Albert. *01/11/85;* South Australia; 140,500 ha; 35°56'S 139°18'E. National Park, Game Reserves. Shorebird Network Site. A saline to hypersaline lagoon separated from the ocean by a dune peninsula and connected to two lakes forming a wetland system at the river's mouth. The lakes contain fresh to brackish water. The site is of international importance for migratory waterbirds, providing habitat for more than 30% of the waders summering in Australia. The site includes important nesting colonies of waterbirds. The globally endangered Orange-bellied Parrot over-winters on the reserve. The area is noted for its extensive aboriginal, historic and geological sites.

To preserve this saline to hypersaline state, an embargo was placed on releases of freshwater from the upper southeast of South Australia, via Salt Creek. Effectively, no more than 45 ML was to be released lest the condition of the lagoon become too fresh. As the rate of water abstraction across the Murray-Darling Basin increased, and the regional climate shifted from a flood to a drought-dominated regime, the mouth of the Murray River closed and the salinity of the Coorong rapidly increased to as much as 220 g/L in its southern waters. As the salinity increased, the populations of fish and waterbirds, which underpinned the initial nomination, declined. This lead to calls for the dedication of 700 GL of river water into the system to save the Coorong and adjacent lakes (Gell 2010). Palaeolimnological evidence from cores taken the length of the Coorong revealed it to have a subsaline history with little direct contribution from the river (Fluin et al. 2007), as revealed by very low levels of the river plankter *Aulacoseira granulata* (Figure 3). In fact, it was highly reliant on freshwater contributions from Salt Creek. The embargo on these, based on the misidentification of its natural ecological character during its Ramsar listing, as well as the commissioning of weirs to limit the tidal prism into the lakes, was a driver of its hypersalinity. This lead to a depletion in its decomposer flora, which drove a net accumulation of carbon (McKirdy et al. 2010). The elevated carbon oxygen demand led to sediment anoxia which extinguished much of its invertebrate biota and its functional *Ruppia* autotroph community (Krull et al. 2009; Dick et al. In press). Clearly, the lack of historical context was fatal to the state of this internationally significant wetland and brings into question the merits of the expenditure of several million dollars of research on the contemporary ecology of this degraded system.

Managing floodplain wetlands

The records of wetland change along the floodplain of the River Murray and its tributaries provide evidence for management to understand the drivers of ecological change and the natural ecological condition of wetlands, which may form 'aspirational' targets of restoration efforts. While they clearly contradict the views of the irrigation lobby that the case for degradation has been exaggerated (Benson et al. 2003), they also provide clear warnings for managers that the mere provision of environmental flows is not sufficient to alleviate the problems associated with

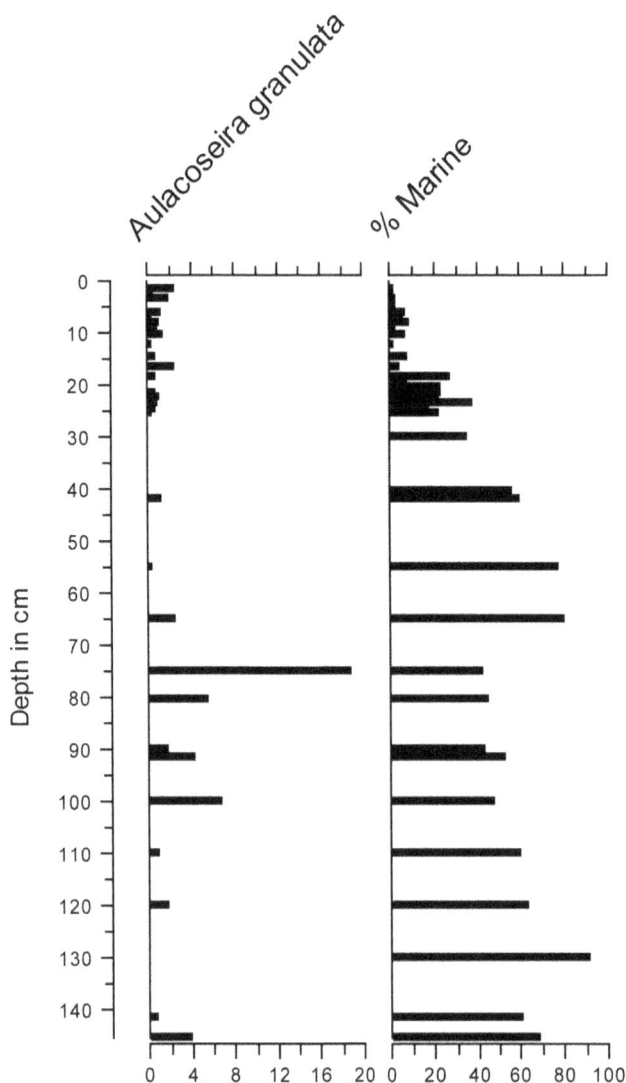

Figure 3. The record of *Aulacoseira granulata* from the north lagoon of the Coorong, revealing the lack of direct influence of the River Murray. The proportion of marine taxa reveals the influence of the tidal prism on the ecological character of the lagoon. Data from Fluin et al. 2007.

excessive sedimentation, salinisation and eutrophication.

Several of the Monash diaspora have undertaken research on the limnology (Gell et al. 2002; Tibby et al. 2003; Tibby 2004; Tibby and Reid 2004; Philibert et al. 2006) and palaeolimnology of floodplain wetlands from the upper (Reid et al. 2002; Tibby et al. 2003) to the lowermost reaches above the terminal lakes (Gell et al. 2005a, b, 2007a; Gell and Little 2006; Fluin et al. 2010; Gell 2010). The palaeodiatom records were assembled in a larger database as part of the Environmental Futures Network working group OZPACS (Figure 4), which endeavoured to provide temporal context to the management of Australian ecosystems (Fitzsimmons et al. 2007). A metadata analysis of the Murray Darling Basin sites reveals that considerable change has taken place, and that the diatom assemblages of all MDB sites differ markedly from their 'natural' historical range of variability. At a superficial level, this contradicts the claims of Benson et al. (2003), and suggests that the degraded state of the wetlands of the MDB has been underestimated.

Ogden and Reid (Thoms et al. 1999; Ogden 2000; Reid et al. 2002) were the first to demonstrate the demise of the basin's productive wetlands by illustrating the shift in diatom and cladoceran assemblages from predominantly epiphytic forms to communities dominated by plankton. This shift is a consequence of a change in the light regime, itself impacted by the increased flux in fine sediments, leading to turbid waters. This effect has similarly been revealed in wetlands down-catchment, mainly revealed through the widespread colonisation of sites by

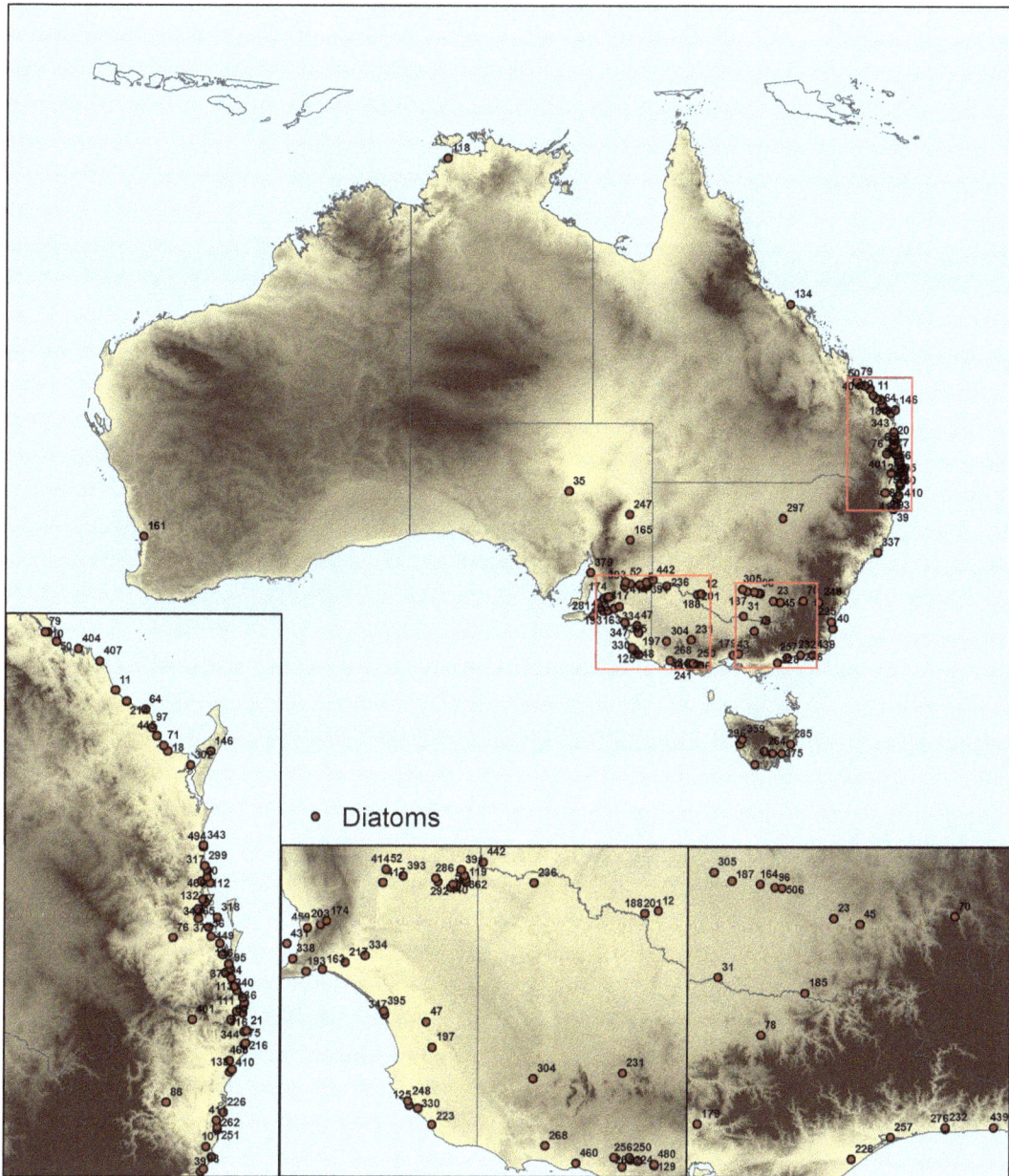

Figure 4. The metadatabase of diatom records across eastern Australia from the OZPACS website.

low-light-tolerant, tychoplanktonic diatoms within the Fragilariaceae (Gell et al. 2009a). This has been supported by consistent evidence for increased sediment accumulation rates. These have increased five to 80 fold, attaining up to 40mm/year, raising the prospect that these shallow wetlands are at risk of complete infilling. This process is further accelerated through sediment accretion by aquatic plants that have increased their abundance since regulation (Figure 5) and as wetlands infill (Gell et al. 2006). Additionally, there is evidence for early salinisation (Gell et al. 2005b) and eutrophication (Gell and Little 2006), and, more recently, acidification (Gell 2010). Several of these stressors appear to have commenced concurrently, prompting Gell et al. (2007b) to suggest co-variation between drivers of change and stressors on condition. In effect, however, calls for water allocations to reverse the degradation of the wetlands of the Murray Darling Basin are somewhat facile when faced with the body of evidence now assembled more fully revealing their plight.

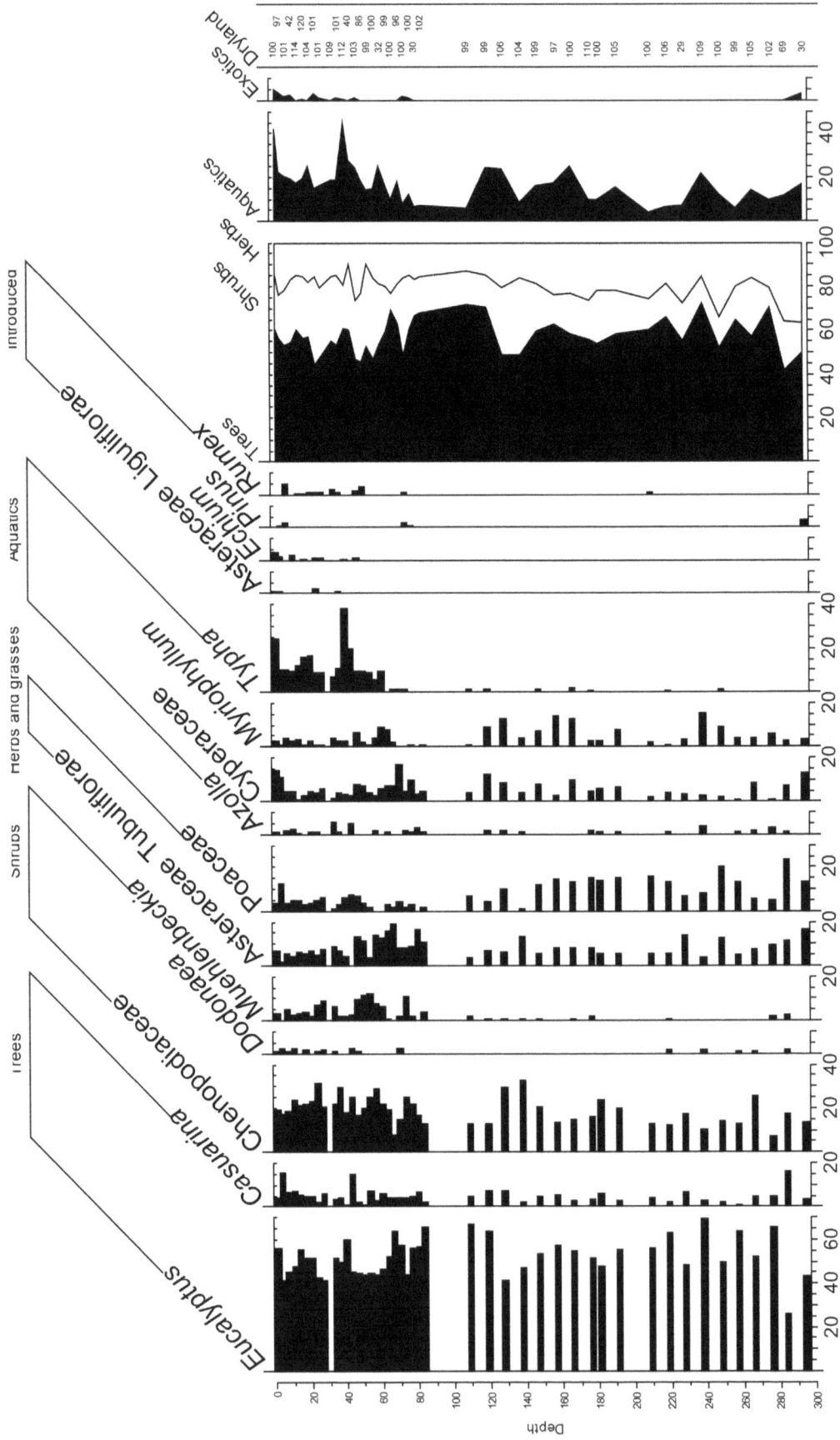

Figure 5. The rise of *Typha* after the regulation of the Murray River relative to the past 2700 years.

Drought

It is pleasing to muse that the goals of those who drove the wealth of long-term, palaeoclimate research in Australia tended to shy from the environmental debate, and that now, past climates are central to the greatest environmental challenge of our time. It is amusing also to recognise that this has drawn the focus to examining the past 2000 years in detail to provide context, in the manner of the 'Hockey Stick' of Mann et al. (1998), to the present climatic circumstances. The assemblage of the record of data of Australia's climate over this time frame is being advanced by several means, including through the IGBP PAGES-supported AUS2k workshop (Gergis and Turney 2010) and the *Palaeoclimates relevant to NRM in the MDB* workshop and report (Gell et al. 2009b). While the goal of deriving climate metrics from sediment records remains a challenge, it is clear from the emerging records that droughts of considerable duration have occurred in the recent past (Barr 2010), that wetlands have changed significantly in the absence of industrialised people across this time frame (Mills et al. 2010), and that the recent 'big dry' is unusual at the millennial timescale.

Conclusion

While not necessarily the primary goal of Peter Kershaw, a clear legacy of his contribution to palaeoecological science is the production of a diverse array of research, undertaken in fine resolution, that has been directed at addressing recent environmental change with the deliberate and direct intention of informing natural resource management. From this research, we understand that most of our aquatic ecosystems are degraded and are under considerable ongoing stress through increasing fluxes of salts, nutrients and sediments under hydrological dry conditions wrought by a drought of unusual duration and depth. We can say with confidence that many wetlands are now in unprecedented condition and that the demands of managers have never been so acute.

Some of the lessons for managers include the reality that some resource-extraction activities have a clear opportunity cost in terms of water yield. In the context of the recent drought, it is even clearer now that catchment activities should be focussed firstly on prioritising water yield over other consumptive activities. There is a lesson from the Coorong that, where goals for restoration or rehabilitation are to be set, the identification of those goals should include a palaeoecologically derived understanding of the natural condition and audit of the present status before management measures are implemented. A further and perhaps final lesson is that palaeoecology can provide the only means of understanding the drivers of wetland change as contemporary ecological approaches can only experiment on derived systems. The evidence from the MDB wetlands reveals that these sites changed very early after settlement, have been stressed by multiple drivers of change and are unlikely to be remediated in a sustainable way by the mere provision of river water. In this instance, we can advocate that the allocation of scarce water resources should be limited to those sites that have already implemented measures to control the influx of salts, nutrients and sediments.

As a postscript, it can be noted that much of this research emerged in the 20th century but has rarely been incorporated into restoration plans or management measures. It remains true that 'Despite the obvious importance of the historical approach, there is a reluctance on the part of planners and managers to take full account of its implications' (Kershaw and Gell 1990:19). It is clear, therefore, that there remains considerable scope for an improved program of the extension of this knowledge to the broader natural resource management community. As with all science, there is both a reluctance to, and considerable incentives not to, engage with research users through the production of outputs and presentation at fora that do not provide good return to the academic's institution. While the national research priorities continue to draw research into applied priorities, and 'Water – a critical resource' is a clear, relevant example, there remains an obligation for outputs to be palatable for end users so that the management measures implemented are targeted, timely and effective.

References

Barr, C. 2010. *Droughts and Flooding Rains: A fine-resolution reconstruction of climatic variability in southeastern Australia over the last 1500 years*. Unpublished PhD thesis, The University of Adelaide.

Bennion, H. and Battarbee, R.W. 2007. The European Union Water Framework Directive: opportunities for paleolimnology. *Journal of Paleolimnology* 38:285-295.

Benson, L., Markham, A. and Smith, R. 2003. *The Science Behind the Living Murray Initiative*. Deniliquin: Murray Irrigation Limited.

Bickford, S., Gell, P. and Hancock, G.J. 2008. Wetland and terrestrial vegetation change since European settlement on the Fleurieu Peninsula, South Australia. *The Holocene* 18:425-436.

Department of Environment and Heritage 2000. *Coorong, and Lakes Alexandrina and Albert Ramsar Management Plan*. Adelaide: Government of South Australia.

Dick, J., Haynes, D., Tibby, J., Garcia, A. and Gell, P. 2011. A history of aquatic plants in the Ramsar-listed Coorong wetland, South Australia. *Journal of Paleolimnology*. Doi: 10.1007/s10933-011-9510-4.

Fitzsimmons, K.E., Gell, P.A., Bickford, S., Barrows, T.T., Mooney, S.D. and Denham, T.P. 2007. The OZPACS database: A resource for understanding recent impacts on Australian ecosystems. *Quaternary Australasia* 24:2-6.

Fluin, J., Gell, P., Haynes, D. and Tibby, J. 2007. Paleolimnological evidence for the independent evolution of neighbouring terminal lakes, the Murray Darling Basin, Australia. *Hydrobiologia* 591:117-134.

Fluin, J., Tibby, J. and Gell, P. 2010. Testing the efficacy of electrical conductivity (EC) reconstructions from the lower Murray River (SE Australia): a comparison between measured and inferred EC. *Journal of Paleolimnology* 43:309-322.

Gale, S.J. and Haworth, R.J. 2002. Beyond the limits of location: human environmental disturbance prior to official European contact in early colonial Australia. *Archaeology in Oceania* 37:123-136.

Gell, P. 2010. With the benefit of hindsight: The utility of palaeoecology in wetland condition assessment and identification of restoration targets. In: Batty, L. and Hallberg, K. (eds), *Ecology of Industrial Pollution*, pp. 162-188. Cambridge: CUP and the British Ecological Society.

Gell, P.A. and Stuart, I.-M. 1989. Human settlement history and environmental impact: the Delegate River catchment, East Gippsland. *Monash Publications in Geography* no. 36.

Gell, P.A., Stuart, I.-M. and Smith, J.D. 1993. The response of vegetation to changing fire regimes and human activity in the Delegate River catchment, East Gippsland, Victoria. *The Holocene* 3(2):150-160.

Gell, P.A., Sluiter, I.R. and Fluin, J. 2002. Seasonal and inter-annual variations in diatom assemblages in Murray River-connected wetlands in northwest Victoria, Australia. *Marine and Freshwater Research* 53:981-992.

Gell, P., Bulpin, S., Wallbrink, P., Bickford, S. and Hancock, G. 2005a. Tareena Billabong – A palaeolimnological history of an everchanging wetland, Chowilla Floodplain, lower Murray-Darling Basin. *Marine and Freshwater Research* 56:441-456.

Gell, P., Tibby, J., Fluin, J., Leahy, P., Reid, M., Adamson, K., Bulpin, S., MacGregor, A., Wallbrink, P., Hancock, G. and Walsh, B. 2005b. Accessing limnological change and variability using fossil diatom assemblages, south-east Australia. *River Research and Applications* 21:257-269.

Gell, P., Fluin, J., Tibby, J., Haynes, D., Khanum, S., Walsh, B., Hancock, G., Harrison, J., Zawadzki, A. and Little, F. 2006. Changing Fluxes of Sediments and Salts as Recorded in lower River Murray wetlands, Australia. *International Association of Hydrological Sciences* 306:416-424.

Gell, P. and Little, F. 2006. Water Quality History of Murrumbidgee River Floodplain Wetlands. Wagga Wagga: Murrumbidgee Catchment Management Authority.

Gell, P., Baldwin, D., Little, F., Tibby, J. and Hancock, G. 2007a. The impact of regulation and salinisation on floodplain lakes: the lower River Murray, Australia. *Hydrobiologia* 591:135-146.

Gell, P., Jones, R. and MacGregor, A. 2007b. The sensitivity of wetlands and water resources to climate and catchment change, south-eastern Australia. *PAGES News* 15 (1):13-15.

Gell, P., Fluin, J., Tibby, J., Hancock, G., Harrison, J., Zawadzki, A., Haynes, D., Khanum, S., Little, F. and Walsh, B. 2009a. Anthropogenic Acceleration of Sediment Accretion in Lowland Floodplain Wetlands, Murray-Darling Basin, Australia. *Geomorphology* 108:122-126.

Gell, P., Gergis, J., Mills, K., Baker, P., De Deckker, P., Finlayson, M., Hesse, P., Jones, R., Kershaw, P., Pearson, S., Treble, P., Barr, C., Brookhouse, M., Drysdale, R., Haberle, S., Karoly, D., McDonald, J., Thoms, M. and Tibby, J. 2009b. *Palaeoclimates relevant to NRM in the MDB*. Canberra: Murray-Darling Basin Authority.

Gergis, J. and Turney, C. (eds) 2010. The First Australian 2k (AUS2K) PAGES Regional Workshop: Towards Data Synthesis. Unpublished workshop programme and abstract booklet.

Harle, K. 1997. Late Quaternary vegetation and climate change in southeastern Australia: palynological evidence from marine core E55-6. *Palaeogeography, Palaeoclimatology, Palaeoecology* 161:465-483.

Harle, K.J., Kershaw, A.P. and Clayton, E. 2004. Patterns of vegetation change in southwest Victoria (Australia) over the last two glacial/interglacial cycles. *Proceedings, Royal Society of Victoria* 116:107-139.

Kershaw, A.P. 1978. Record of last interglacial-glacial cycle from north-eastern Queensland. *Nature* 272:159-161.

Kershaw, A.P. 1986. Climatic change and Aboriginal burning in north-east Australia during the last two glacial/interglacial cycles. *Nature* 322:47-49.

Kershaw, A.P., D'Costa, D.M., McEwan-Mason, J.R.C. and Wagstaff, B.E. 1991. Palynological evidence for Quaternary vegetation and environments of mainland southeastern Australia. *Quaternary Science Reviews* 10:391-404.

Kershaw, A.P., McKenzie, G.M., Porch, N., Roberts, R.G., Brown, J., Heijnis, H., Orr, L.M., Jacobsen, G. and Newall, P.R. 2007. A high resolution record of vegetation and climate through the last glacial cycle from Caledonia Fen, south-eastern highlands of Australia. *Journal of Quaternary Science* 22:481-500.

Kershaw, A.P. 1993. Palynology, biostratigraphy and human impact. *The Artefact* 16:12-18.

Kershaw, A.P. 1998. Estimates of regional climatic variation within south-eastern mainland Australia since the Last Glacial Maximum from pollen data. *Paleoclimates: Data and Modelling* 3:107-134.

Kershaw, A.P. and Gell, P.A. 1990. *Quaternary vegetation and the future of the forests*. In: Bishop, P. (ed), Lessons for human survival: nature's record from the Quaternary. *Geological Society of Australia Symposium Proceedings* 1:11-20.

Krull, E., Haynes, D., Lamontagne, S., Gell, P., McKirdy, D., Hancock, G., McGowan, J. and Smernik, R. 2009. Changes in the chemistry of sedimentary organic matter within the Coorong over space and time. *Biogeochemistry* 92:9-25.

Kuczera, G. 1985. *Predictions of water yield reductions following bushfire in ash-mixed species eucalypt forest.* Melbourne: Melbourne and Metropolitan Board of Works report No. MMBW-W-0014.

Ladd, P.G. 1979. Past and present vegetation on the Delegate River in the highlands of eastern Victoria, II. Vegetation and climate history from 12 000 BP to present. *Australian Journal of Botany* 27:185-202.

Mann, M.E., Bradley, R.S. and Hughes, M.K. 1998. Global-scale temperature patterns and climate forcing over the past six centuries. *Nature* 392:779-787.

McKenzie, G.M. and Kershaw, A.P. 1997. A vegetation history and quantitative estimate of Holocene climate from Chapple Vale, in the Otway Region of Victoria, Australia. *Australian Journal of Botany* 45:565-581.

McKirdy, D.M., Thorpe, C.S., Haynes, D.E., Grice, K., Krull, E.S., Halverson, G.P., Webster, L.J. 2010. The biogeochemical evolution of the Coorong during the mid- to late Holocene: An elemental, isotopic and biomarker perspective. *Organic Geochemistry* 41:96-110.

Mercer, D. 1995. *A Question of Balance.* Sydney: Federation Press.

Mills, K., Gell, P., Kershaw, A.P., MacKenzie, M. and Lewis, T. 2010. Evidence for a pre-European drought in the western Victorian lakes: a historical context for the recent Victorian drought? In: Gergis, J. and Turney, C. (eds), The First Australian 2k (AUS2K) PAGES Regional Workshop: Towards Data Synthesis. 27: Melbourne University: Unpublished workshop programme and abstract booklet.

Moss, P.T. and Kershaw, A.P. 2007. A late Quaternary marine palynological record (oxygen isotope stages 1 to 7) for the humid tropics of northeastern Australia based on ODP site 820'. *Palaeogeography, Palaeoclimatology, Palaeoecology* 251:4-22.

Ogden, R.W. 2000. Modern and historical variation in aquatic macrophyte cover of billabongs associated with catchment development. *Regulated Rivers: Research and Management* 16: 487-512.

Philibert, A., Gell, P., Newall, P., Chessman, B. and Bate, N. 2006. Development of diatom-based tools for assessing stream water quality in south eastern Australia: Assessment of environmental transfer functions. *Hydrobiologia* 572:103-114.

Reid, M., Fluin, J., Ogden, R., Tibby, J. and Kershaw, P. 2002. Long-term perspectives on human impacts on floodplain-river ecosystems, Murray-Darling Basin, Australia. *Verhandlungen der Internationalen Vereinigung für Theoretische und Angewandte Limnologie* 28(2):710-716.

Reid, M.A. and Ogden, R.W. 2009. Factors affecting diatom distribution in floodplain lakes of the southeast Murray Basin, Australia and implications for palaeolimnological studies. *Journal of Paleolimnology* 41(3):453-470.

Reid, M.A., Sayer, C.D., Kershaw, A.P. and Heijnis, H. 2007. Palaeolimnological evidence for submerged plant loss in a floodplain lake associated with accelerated catchment erosion (Murray River, Australia). *Journal of Paleolimnology* 38:191-208.

Thoms, M.C., Ogden, R.W. and Reid, M.A. 1999. Establishing the condition of lowland floodplain rivers: a palaeo-ecological approach. *Freshwater Biology* 41:407-423.

Tibby, J. and Reid, M. 2004. A model for inferring past conductivity in low salinity waters derived from Murray River (Australia) diatom plankton. *Marine and Freshwater Research* 55:597-607.

Tibby, J., Reid, M.A., Fluin, J., Hart, B.T. and Kershaw, A.P. 2003. Assessing long-term pH change in an Australian river catchment using monitoring and palaeolimnological data.

Environmental Science and Technology 37(15):3250-3255.

Tibby, J. 2004. Development of a diatom-based model for inferring total phosphorus in south-eastern Australian water storages. *Journal of Paleolimnology* 31:23-36.

Turney, C.S.M., Kershaw, A.P., Lowe, J.J., van der Kaars, S., Johnston, R., Rule, S., Moss, P., Radke, L., Tibby, J., McGlone, M.S., Wilmshurst, J.M., Vandergoes, M.J., Fitzsimons, S.J., Bryant, C., James, S., Branch, N.P., Cowley, J., Kalin, R.M., Ogle, N., Jacobsen, G. and Fifield, L.K. 2006. Climatic variability in the southwest Pacific during the Last Termination (20-10 kyr BP). *Quaternary Science Reviews* 25:886-903.

Walker, D., Head, M.J., Hancock, G.J. and Murray, A.S. 2000. Establishing a chronology for the last 1000 years of laminated mud accumulation at Lake Barrine, a tropical upland maar lake, northeastern Australia. *The Holocene* 10:415-427.

Weatherhead, P.J. 1986. How unusual are unusual events? *American Naturalist* 128:150-154.

Wilby, R.L. and Gell, P.A. 1994. The impact of forest harvesting on water yield: modelling hydrological changes detected by pollen analysis. *Journal of Hydrological Sciences* 39 (5):471-486.

23

Regional genetic differentiation in the spectacled flying fox (*Pteropus conspicillatus* Gould)

Samantha Fox
School of Marine and Tropical Biology and School of Earth and Environmental Sciences, James Cook University, Townsville, Queensland

Michelle Waycott
James Cook University, Townsville, Queensland

David Blair
James Cook University, Townsville, Queensland

Jon Luly
James Cook University, Townsville, Queensland

Introduction

Climatic excursions in the late Pleistocene dramatically reduced habitat available to organisms dependent on forested landscapes (Hopkins et al. 1993; Kershaw 1994; Kershaw et al. 2007; VanDerWal et al. 2009). Pollen analysis and bioclimatic modelling of rainforest in northeastern Queensland indicate the region was subject to massive change during Quaternary glaciations. The consequences for rainforest-dependent species were severe, especially for organisms with limited mobility or adaptability (Schneider et al. 1998). We report here on present-day regional-scale genetic structure in the spectacled flying fox (*Pteropus conspicillatus*), generally assumed to be a rainforest specialist, and on the insights modern-day processes may provide for understanding responses of an extremely mobile animal to Pleistocene habitat contraction and fragmentation.

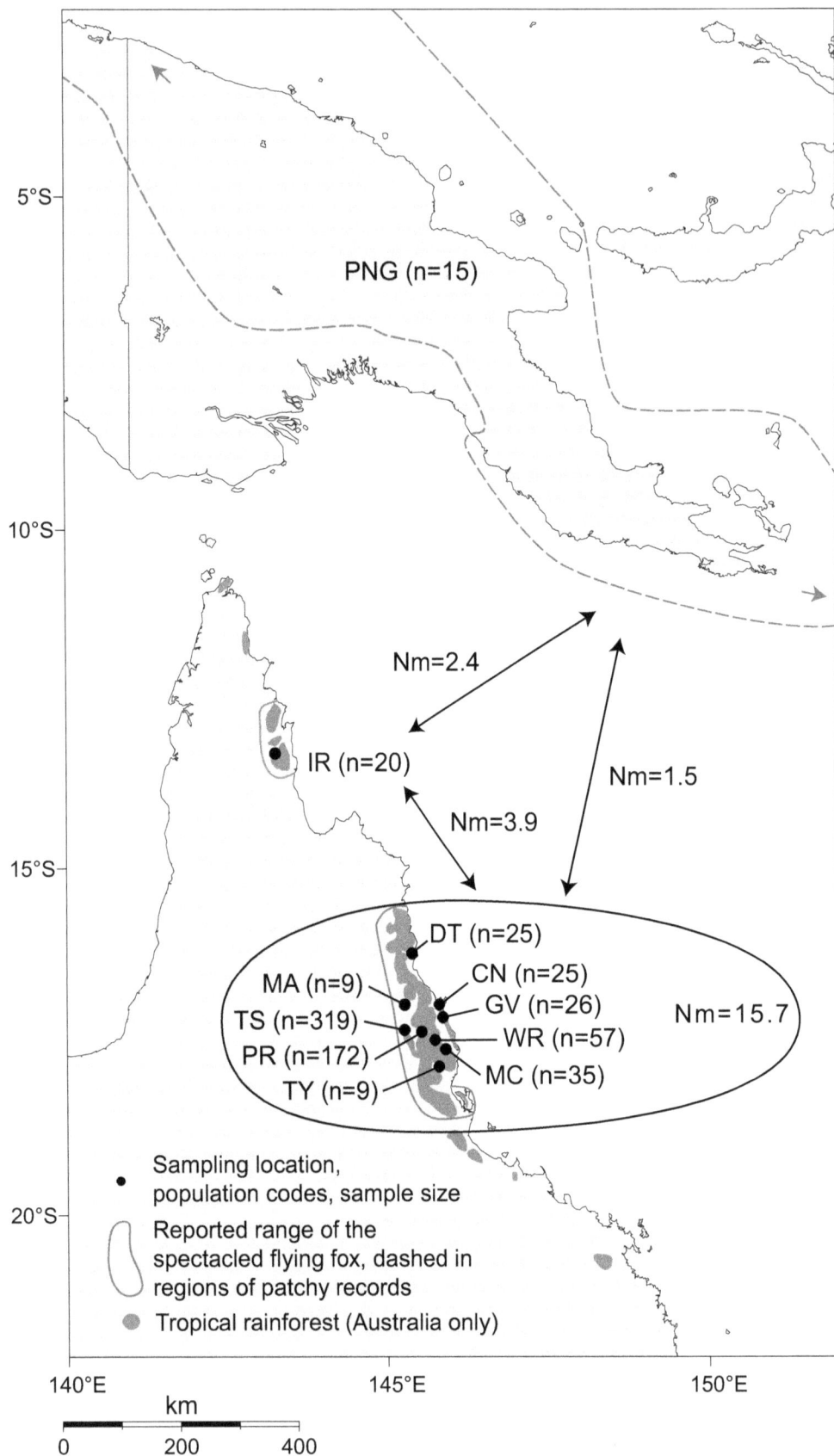

Figure 1. Sampling locations of Pteropus conspicillatus, indicating sample sizes and gene flow estimates (Nm) among regions and within the Wet Tropics region. Locality codes: PNG = Papua New Guinea, IR = Iron Range, DT = Daintree, CN = Cairns, GV = Gordonvale, WR = Whiteing Road, MC = Mena Creek, TY = Tully, PR = Powley Road, TS = Tolga Scrub, MA = Mareeba.

All flying foxes are potentially extremely mobile and some species make seasonal migrations across hundreds of kilometres, following cycles of fruiting and flowering of favoured food trees (Eby 1991; Tidemann and Nelson 2004). Species such as the little red flying fox (*Pteropus scapulatus*) and grey-headed flying fox (*P. poliocephalus*) traverse many hundreds of kilometres in the course of a year and are apparently panmictic (Sinclair et al. 1996; Webb and Tidemann 1996: Luly et al. 2010). In contrast, the spectacled flying fox is considered to be closely associated with rainforest (Richards 1990a, b) and currently has a discontinuous distribution in northeastern Queensland, New Guinea and adjacent islands. By far the largest known populations are associated with coastal and upland rainforest in the Wet Tropics World Heritage Area (hereafter the Wet Tropics) between about Townsville and Cooktown in northeast Queensland (Westcott et al. 2001). A small colony – a few hundred strong at most (Fox 2006) – is found approximately 400 km north of the Wet Tropics in the Iron Range National Park on Cape York Peninsula, where isolated pockets of wet tropical rainforest provide habitat. The species occurs widely and patchily outside Australia (Figure 1), but little is known about population sizes. It is found in the Molucca Islands in Indonesia, in lowland New Guinea from approximately Agats in Irian Jaya, around the north coast to the Port Moresby district in the east and offshore in the D'Entrecasteaux Islands and the Louisiade Archipelago. There is a break in the range on the drier southern coast of New Guinea closest to the tip of Cape York Peninsula (Bonaccorso 1998). The New Guinea distribution is, at its closest, approximately 600 km from the northernmost permanent Australian colony at Iron Range. Radio telemetry and satellite tracking have shown that individual spectacled flying foxes in the Wet Tropics region undertake regular movements between colonies (Shilton pers. comm.). It is not known whether any movement occurs between the Wet Tropics, Iron Range (IR) and Papua New Guinea (PNG), or what movements occur within New Guinea and Indonesia.

Given the assumed habitat specialisation of the spectacled flying fox, the species might be expected to exhibit stronger genetic differentiation across its disjunct range than apparently less specialised congeners. Examining the extent of differentiation, if any, in the modern range of the spectacled flying fox may help inform thinking on the capacity of flying fox species to maintain genetic integrity in the even more severely fragmented habitats that prevailed during glacial episodes. To investigate this, we set out to determine the extent of gene flow between colonies in the Wet Tropics, and between the Wet Tropics as a whole and other regions (IR, PNG) where spectacled flying fox populations occur.

Materials and methods

Samples

Samples were collected from 718 individual spectacled flying foxes distributed widely across the range of the species (see Figure 1). Sampling effort across the species' range was very uneven because of the difficulties of gaining access to elusive animals living in remote areas. Many of our samples were obtained from tick-paralysed bats rescued from camps on the Atherton Tableland (location abbreviations as in Figure 1). Other samples came from bats taken for veterinary attention or to be raised by wildlife care groups, the location assigned to such animals being the camp nearest to where they were found. Samples from IR were obtained from bats captured in mist nets. Wing tissue was obtained from living bats following Worthington-Wilmer and Barratt (1996). A small piece of wing membrane was removed from dead bats with scissors. Tissue samples were stored in 5 M NaCl-saturated 20% DMSO (Dimethylsulphoxide) and refrigerated at 4°C until processed.

Samples from PNG were small pieces of wing membrane obtained from skins held in the Australian Museum. Most (seven) were from Hull Island, Milne Bay Province (museum accession numbers EBU23156, 23157, 23159, 23162, 23164, 23171, 25578). Two came from elsewhere in Milne Bay Province (EBU23179, 26345) and one from West Sepik Province (EBU25020). The remaining five PNG samples were collected in Madang Province courtesy of the Wildlife Conservation Society. The unevenness of sampling from across the species' range constrained the level of analyses attempted.

DNA extraction and amplification

DNA extraction was carried out using a QIAGEN DNeasy™ tissue extraction kit according to the manufacturer's instructions. DNA was used at a final concentration of 3 ng/μL for amplification by polymerase chain reaction (PCR).

Six microsatellite loci were amplified and scored: one dinucleotide (Ph9) and one trinucleotide (C6) locus characterised from *Pteropus rodricensis* (O'Brien et al. 2007); three dinucleotide loci (PC25b6, PC26a7, PC31h4) and one with a compound repeat (PC36c2) characterised from *P. conspicillatus* by Fox et al. 2007. Primers were labelled with 5' fluorophores HEX, FAM or TET (Geneworks Ltd). Concentrations of reagents in the PCR reaction mix varied by locus (see Table 1). PCR products were purified by centrifugation through 300 μL of Sephadex (G-50), before being analysed on a MegaBace 1000 Genetic Analyser (Amersham BioSciences™) at the Advanced Analytical Centre, James Cook University. Allele sizes were estimated using the ET 400-Rox (Amersham BioSciences ™) internal size standard and the program Fragment Profiler v 1.2 (Amersham BioSciences™).

Table 1. Final concentration of reagents used in amplification of each microsatellite locus for 15 μL PCR reactions (including 1.5 mM MgCl2 and 0.4 units Taq DNA polymerase).
1. 94°C 3 mins; 4 cycles (94°C 30 secs, 58°C 40 secs, 72°C 1 min), repeated four times with a reduction in temperature of 2°C each step down; 25 cycles (94°C 30 secs, 50°C 40 secs, 72°C 1 min); 72°C 10 mins.
2. 95°C 5 mins; 35 cycles (95°C 45 secs, 55°C 30 secs, 72°C 45 secs); 72°C 10 mins.
3. 94°C 3 mins; 4 cycles (94°C 30 secs, 50°C 40 secs, 72°C 1 min), repeated three times with a reduction in temperature of 2°C each step down; 25 cycles (94°C 30 secs, 44°C 40 secs, 72°C 1 min); 72°C 10 mins.

Locus	MgCl$_2$ (mM)	dNTPs (mM)	Primer (μM)	DNA (ng)	Thermocycler profile
PC25b6	1.5	0.1	0.3	7.5	1
PC26a7	1.5	0.1	0.6	7.5	2
PC36c2	-	0.2	0.4	7.5	1
PC31h4	1.5	0.2	0.2	4.5	2
c6	1.5	0.2	0.2	7.5	3
ph9	-	0.1	0.2	3.0	2

Data analyses

All PNG samples were treated as a single 'colony'. Hardy Weinberg equilibrium (HWE) expectations were tested for each locality by locus using GenAlEx v6.1 (Peakall and Smouse 2005). Micro-Checker (van Oosterhout et al. 2006) was used to detect the presence of null alleles, large allele dropout and stuttering. GenAlEx v 6.1 was used to calculate mean and effective numbers of alleles and expected and observed heterozygosities. Allelic richness, calculated in FSTAT (Goudet 1995), was averaged across loci to give mean allelic richness by locality. Numbers of private alleles, and the number of migrants (N_m) between regions per generation using the private alleles method (Slatkin 1985), were calculated using GenePop (Raymond and Rousset 1995).

Two approaches were used to measure genetic differentiation among localities. Firstly, an analysis of molecular variance (AMOVA), which accounts for gene frequencies and number of mutations, was calculated for F_{ST} as well as R_{ST} using GenAlEx. Pairwise F_{ST} (following Peakall et al. 1995), and R_{ST} (following Michalakis and Excoffier 1996) values were also calculated, and a principal coordinates analysis (PCA) was performed using a distance matrix. All significance tests were based on 999 random permutations. Secondly, an exact test based on an unbiased estimate of the p-value of a log-likelihood (G) was performed using a Markov Chain method in GenePop. This is a more powerful way of testing for panmixia using multiple loci and unbalanced sample sizes (Ryman and Jorde 2001; Waples and Gaggiotti 2006) Genotypic differentiation across population pairs was calculated using a Markov Chain of 1000 dememorisations, 1000 batches and 10,000 iterations per batch. F_{IS} (*f*) by locality and locus was calculated following Weir and Cockerham (1984) in Fstat.

An analysis of isolation-by-distance (IBD) was performed using the program isolde in GenePop. This uses a regression of $F_{ST}/(1- F_{ST})$ against the shortest distance (in kilometres) between any two localities. A Bayesian population assignment protocol implemented in structure v2.1 (Pritchard et al. 2000) was used to infer the number of populations represented in the data and to assign individuals to those populations. Models were run for 1-15 putative populations. Conditions for running structure included a model run burn-in procedure of 100,000 replicates, followed by 100,000 Markov Chain Monte Carlo (MCMC) simulations, using the admixture model with allele frequencies correlated between populations. Longer burn-in and MCMC trials were performed but likelihood values were not improved up to 250,000 replicates for each. Ten iterations were performed for each putative number of populations (K).

Evidence of population expansion/contraction was tested using the program Bottleneck v 1.2.02 (Piry et al. 1999). This program evaluates deviation from a theoretical mutation-drift equilibrium and is expressed as the difference between the measured heterozygosity (H_e, defined and discussed further in Piry et al. 1999), and the heterozygosity expected at mutation-drift equilibrium (H_{eq}). Luikart and Cornuet (1998) suggested that the Wilcoxon's test is appropriate for data sets with fewer than 20 loci. Although only six loci were used in this analysis, the highly polymorphic nature of all loci will increase the power of the test. The one-tailed Wilcoxon's test for heterozygosity excess and the sign test were used for both mutation models (infinite alleles model IAM, and step-wise mutation model SMM). Estimations were made over 1000 replicates. Each locality was tested individually and then all Wet Tropics localities were combined and treated as a single population.

Results

Genetic diversity, Hardy Weinberg equilibrium and linkage disequilibrium

All microsatellite loci exhibited high levels of polymorphism, with the number of alleles recorded ranging from 14 (locus C6) to 22 (locus Ph9). Twenty-three private alleles were detected across all localities and loci. Significant departures from Hardy Weinberg expectations were recorded in the PR, TS and WR localities for one, four and two loci respectively. There was no pattern to the loci that were out of Hardy Weinberg equilibrium. Loci were known not to be linked (Fox et al. 2007).

Evidence of locus stutter, which was attributed to a possible single base mutation but not further verified, was detected by Micro-Checker. No loci showed evidence of large allele drop-out. Several loci showed a homozygous excess. However it was not possible to discriminate between normal population processes and the presence of null alleles. We have assumed the former in our analyses.

Table 2 presents allele frequency statistics by locality. The mean number of alleles per population (across all loci) ranged from 10.0 to 16.2, except for three localities (PNG, MA and TY), two of which had small sample sizes, where the mean number of alleles per population ranged from 6.7 to 7.5 (Table 2). Mean allelic richness by locality over all loci ranged from

5.7 to 7.2 (Table 2). There was no significant difference across localities, but mean allelic richness was significantly lower, at loci PC36c2, C6 and PC31h4 (data not shown). Overall, mean expected and observed heterozygosity was high (H_e 0.79; H_o 0.78). Across localities, H_e ranged from 0.83 in DT to 0.75 in PNG, while H_o ranged from 0.82 in TY to 0.74 in PNG (Table 2). Weir and Cockerham's f (F_{IS}) by locality and locus indicated that some populations exhibit excess heterozygosity or a deficit at some loci, although no single population or locus consistently deviated from zero. Averaged over all localities and loci (0.051) f was low and not significant.

Table 2. Summary of mean standard allele frequency statistics by locality. Na = mean number of alleles across loci; Ne = effective number of alleles; Ho = observed heterozygosity; He = expected heterozygosity; A. Rich = allelic richness across loci. Refer to Figure 1 for locality codes.

	n	Na	Ne	Ho	He	A. Rich.
PNG	15	7.5	4.6	0.74	0.75	5.74
IR	20	10.0	6.3	0.79	0.79	6.91
CN	25	10.3	6.0	0.81	0.81	6.31
DT	25	11.5	7.4	0.78	0.83	7.21
GV	26	11.0	6.6	0.78	0.82	6.81
MA	9	6.7	4.5	0.80	0.75	6.02
MC	35	10.8	6.7	0.78	0.82	6.65
PR	172	14.0	6.6	0.78	0.82	6.59
TS	319	16.2	6.8	0.78	0.82	6.61
TY	9	7.5	5.1	0.82	0.77	6.68
WR	57	12.3	6.2	0.76	0.81	6.48

Genetic differentiation

When the three regions (PNG, IR, Wet Tropics) were compared, AMOVA generated a low but significant F_{ST} value of 0.041 (p=0.001) (Table 3), with most variability found among individuals within populations (96%). Four percent of the total variation occurred among regions (p=0.001). Using R_{ST} values, a larger proportion (17%) and a significant amount of the variability was found among regions (p=0.01). Neither F_{ST} nor R_{ST} pair-wise comparisons of populations were significant after Bonferroni correction, but many were significant before this correction was applied (Table 4). It is worth noting that all between-locality R_{ST} comparisons, including IR, are high, supporting and possibly driving the greater regional R_{ST} result. In the PCA, 90% of the variation could be explained by the first axis and a further 6% was described by the second axis (Figure 2). The PCA highlights the close relationship between localities within the Wet Tropics region, and a distinct difference between Iron Range and the Wet Tropics. The PNG samples are separated from all other localities, mainly on the second axis.

calculated using Fisher's method of combining probabilities across independent tests, revealed 20 locality pairs (out of 55 possible comparisons) as significantly different (before correction for multiple tests), with the majority of those 20 pairs including either PNG or IR. Once the level of significance was adjusted, seven pairs remained significant, six containing PNG and one of the Wet Tropics populations, the last pair containing IR and the Cairns (Wet Tropics) population.

The number of migrants per generation (Nm), estimated using the conservative method of private alleles, was low between the Wet Tropics and PNG (*Nm* = 1.5), greater between IR and PNG (*Nm* = 2.4), and greater again between IR and the Wet Tropics (*Nm* = 3.9). These figures suggest low levels of movement between regions. A relatively high gene flow was inferred

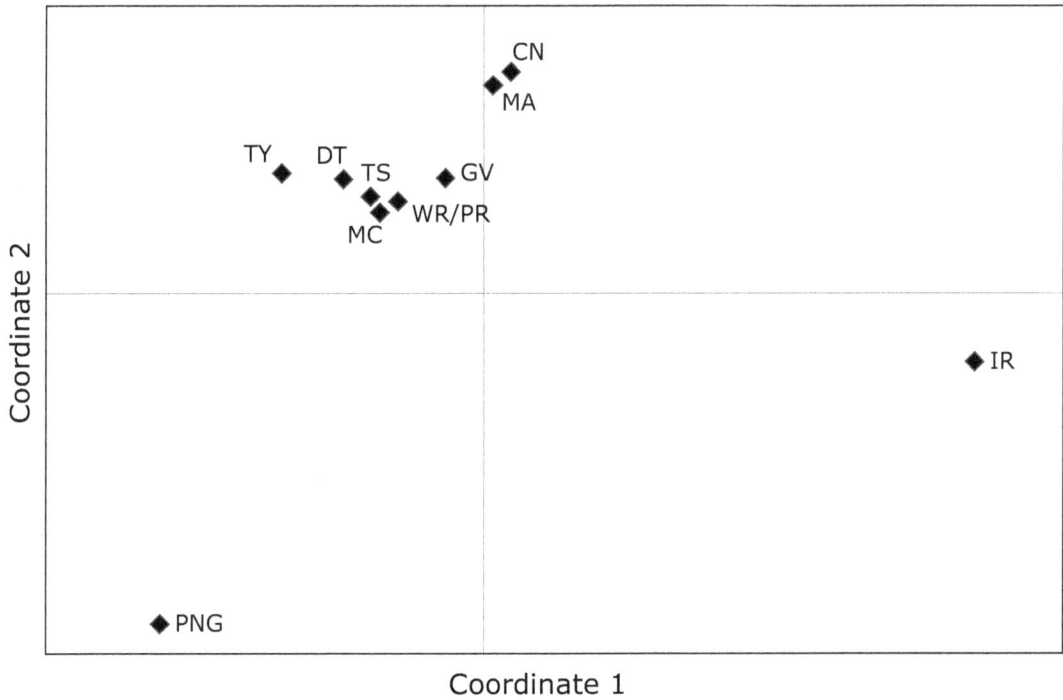

Figure 2. Two-dimensional plot of principal co-ordinates analysis (PCA) based on population pairwise Fst values for sampled locations of *Pteropus conspicillatus*, using a standardised distance analysis. The first two axes explain 96% of the variation of the data set (axis 1 = 90% and axis 2 = 6%). Refer to Figure 1 for locality codes. Note that points for PR and WR are superimposed.

Table 3. Amova estimates of genetic variation among three regions (WT, IR and PNG), among localities and among individuals in the spectacled flying fox.

Source of variation	d.f.	Sum of squares	Variance	Fixation indices	P value	Percentage of variation
F-statistics						
Among regions	2	19.57	0.106	0.041	0.001	4.0
Among populations within regions[1]	8	27.66	0.008	0.003	0.001	0.0
Among individuals within pops	1423	3558.94	2.501	0.044	0.001	96.0
R-statistics						
Among regions	2	3986.94	27.92	0.170	0.01	17.0
Among populations within regions[1]	8	1755.59	0.73	0.006	0.04	0.0
Among individuals within pops	1423	187176.63	131.54	0.18	0.01	82.0

[1] Note that regions PNG and IR are each assumed to consist of one population for the purposes of this analysis.

Table 4. Pairwise F$_{ST}$ values (below diagonal) and R$_{ST}$ (above diagonal). None was significant after Bonferroni correction for multiple comparisons (Dunn Sidak method – [Sokal and Rohlf 1995]). * significant before Bonferroni correction. Pairwise R$_{ST}$ values with IR as one of the localities are shown in bold. Refer to Figure 1 for locality codes.

Pop	PNG	IR	CN	DT	GV	MA	MC	PR	TS	TY	WR
PNG		0.366*	0.086*	0.010	0.050*	0.156*	0.085*	0.036*	0.006	0.113*	0.044*
IR	0.090*		0.152*	0.340*	0.222*	0.112*	0.332*	0.290*	0.219*	0.287*	0.256*
CN	0.056*	0.042*		0.069*	0.006	0.000	0.065*	0.020*	0.017	0.062*	0.014
DT	0.033*	0.062*	0.012*		0.047*	0.107*	0.041*	0.020*	0.005	0.079*	0.027*
GV	0.036*	0.041*	0.005	0.009		0.016	0.029*	0.004	0.002	0.008	0.000
MA	0.054*	0.043*	0.021*	0.015	0.011		0.075*	0.015	0.004	0.074	0.019
MC	0.027*	0.053*	0.012*	0.007	0.000	0.010		0.013*	0.007	0.000	0.033
PR	0.029*	0.049*	0.012*	0.008*	0.003	0.005	0.001		0.000	0.014	0.000
TS	0.028*	0.054*	0.012*	0.004	0.005	0.006	0.001	0.001		0.000	0.001
TY	0.031*	0.071*	0.018*	0.012	0.000	0.013	0.000	0.000	0.003		0.034
WR	0.029*	0.047*	0.010*	0.005	0.001	0.006	0.002	0.000	0.001	0.004	

among localities within the Wet Tropics region (Nm=15.7).

A significant IBD effect was found throughout the range of *P. conspicillatus* (R^2=0.47, p=0.003) (Figure 3a). To ensure that this pattern was not solely created by differences between Australia and PNG, the data from PNG samples were removed and the test repeated using only Australian samples. A non-significant result was obtained R^2=0.17, p=0.10) (Figure 3b), indicating that there is weak differentiation between the Wet Tropics localities and Iron Range. Structure results (not shown) did not support the occurrence of multiple populations, as all individuals shared proportional assignment to multiple populations when K>1.

Statistical analysis of allele frequencies using the program Bottleneck indicated that the majority of localities conformed to the Infinite Alleles Model (IAM). A significant excess of heterozygotes relative to that expected under mutation-drift equilibrium was exhibited by TS, PR and WR. A significant result under the IAM for the sign test and the one-tailed Wilcoxon test was detected in PR and WR (Table 5). One locality, PR, was only significant for the one-tailed Wilcoxon test under the IAM model. When combined, all Wet Tropics populations showed a significant result for a bottleneck under the Wilcoxon test.

Table 5. Significance values (*p*-values) for the sign and Wilcoxon's tests for a heterozygosity excess relative to that expected under mutation-drift equilibrium in BOTTLENECK. * Marginally significant (p = 0.05). **Significant (p <0.05) *** Highly significant (p <0.01). Refer to Figure 1 for locality codes. Shaded locality codes are from the Wet Tropics region.

Locality	Sign test	Wilcoxon test (One-tailed for H excess)
PNG	0.204	0.945
IR	0.502	0.922
CN	0.203	0.922
DT	0.196	0.922
GV	0.209	0.922
MA	0.514	0.578
MC	0.481	0.656
PR	0.047*	0.008***
TS	0.249	0.023**
TY	0.412	0.945
WR	0.244	0.039**
All Wet Tropics	0.229	0.016**

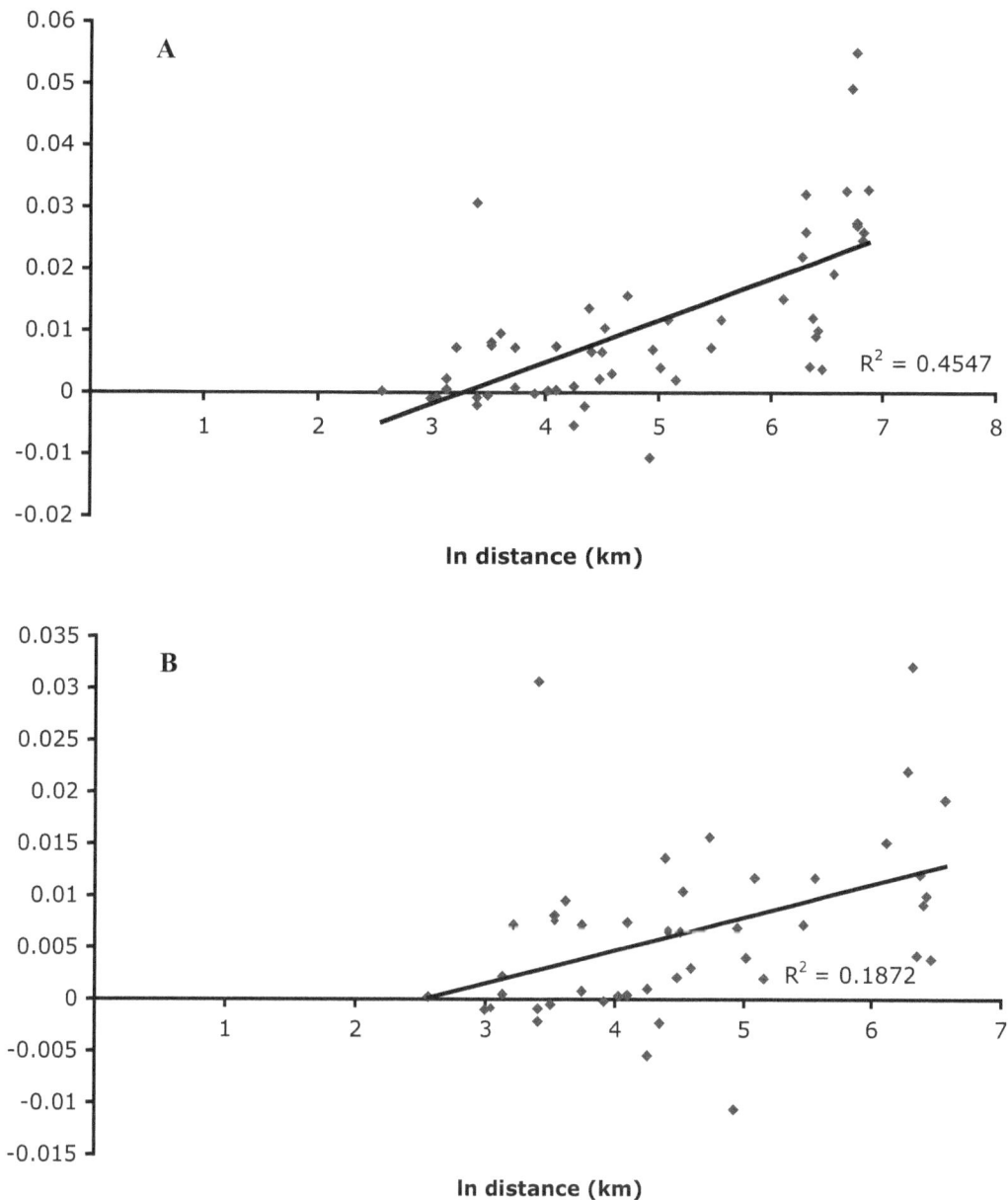

Figure 3. Relationship between genetic distance, calculated using $F_{ST}/(1-F_{ST})$, and the natural logarithm of geographical distance (km). **A.** Test of isolation-by-distance between localities throughout the entire distribution of the spectacled flying fox (*Pteropus conspicillatus*) (R^2 = 0.47, p = 0.003) (including two separate PNG sampling locations). **B.** Test of isolation-by-distance for colonies of spectacled flying foxes within Australia (R^2 = 0.17, p = 0.104).

Discussion

In the Wet Tropics region, the spectacled flying fox is panmictic, with no impediments to gene flow detected among colonies throughout the region. Little genetic differentiation was observed between Iron Range and the Wet Tropics region. At the very broadest scale, there is weak genetic structuring across the range of the species. Bats in PNG differ from those in Australia, reflecting the effects of IBD across spatial and topographic barriers, but differences are slight, suggestive either of continuing but rare exchange of individuals between the regions or of relatively recent fragmentation of a previously continuous range.

The Wet Tropics World Heritage Area is repository for the greatest genetic diversity in the spectacled flying fox, suggesting that colonies in this area are part of a large, long-established metapopulation. This finding contrasts with suggestions that the species is a relatively recent immigrant from PNG or southeast Asia (McKean 1970; Schodde and Calaby 1972). Our conclusion of panmixia is consistent with radio and satellite tracking of individual bats moving throughout the Wet Tropics region (Shilton pers. comm.) and suggests that these movements may underlie the high rate of gene flow within this region. No bats have yet been detected moving between the Wet Tropics and Iron Range, but the low genetic differentiation between these regions suggests either that occasional reproductively effective movements occur, or that there have been few generations since colonisation and isolation of IR.

Application of BOTTLENECK analysis to samples from the Wet Tropics region suggests a marked reduction in spectacled flying fox numbers at some time in the past. The effect is discernable when BOTTLENECK is applied individually to the largest sample groups (PR, TS and WR) and when all Wet Tropics samples are pooled. Sample sizes from PNG and IR were inadequate for a meaningful test. The timing of the bottleneck in the Wet Tropics cannot be determined as we are unable to estimate effective population size. However, palaeoenvironmental reconstructions (Hopkins et al. 1993; Graham et al. 2006; Hilbert et al. 2007; Kershaw et al. 2007; VanDerWal et al. 2009) show that rainforest in the Wet Tropics Region was severely reduced through the last glacial cycle, and recovered to approach its pre-European extent in the early Holocene. Habitat contraction provides a potential cause for the decreased populations suggested by BOTTLENECK. Taken at face value, the genetic data appear to indicate that habitat reductions, and the climate changes that caused them, must have been modest rather than severe in amplitude and that a substantial amount of suitable habitat persisted through the Last Glacial Maximum. This interpretation is at variance with palaeoenvironmental reconstructions of LGM climate and vegetation and illustrates the need for multi-disciplinary perspectives on interpretation of the past, no matter what proxies are employed.

The effect of the unavoidably uneven sampling effort among localities limits the extent to which some relationships can be analysed and conclusions drawn. In particular, interpretations of weak differences in genetic structure at the broadest sampling scale should be made cautiously. The geographical gap between PNG and Australian bats is reflected in a measurable IBD effect. However, analyses in STRUCTURE did not differentiate between regions but instead suggested that all *P. conspicillatus* samples, including those from PNG and IR, came from a single genetic population. Small sample sizes from some localities (Waples and Gaggiotti 2006) and the frequency of common alleles (data not shown) among the PNG samples would have reduced the ability of STRUCTURE to resolve what may be subtle differences in population genetic structure. Thus, population similarity between PNG and Australia might reflect a type II error rather than a meaningful biological finding, and further sampling in PNG and elsewhere is needed to confirm these results.

The weak genetic distinction between bats in Australia and PNG suggests that there is at least some interchange between regions. Estimated numbers of migrants between Australia (especially the Wet Tropics) and PNG are low. Although one migrant per generation is theoretically sufficient to offset genetic drift between populations (Mills and Allendorf 1996; Whitlock and McCauley 1999), in 'real world' scenarios, especially where population sizes fluctuate (Vucetich and Waite 2000), 10-20 migrants per generation may be needed to slow drift-induced divergence. Movement of spectacled flying foxes between Australia and PNG has not yet been recorded but, based on the apparent interchange between the Wet Tropics and Iron Range, the distance is not insurmountable. Return flights do occur to PNG by the black flying fox (*Pteropus alecto*), an apparent habitat generalist, via Torres Strait (Breed et al. 2010). The limited genetic differentiation between Australia and PNG would be consistent with occasional cross-Torres Strait flights of spectacled flying foxes. It would also be consistent with a geologically recent disjunction between Australian and PNG bats. A much more intensive sampling effort in PNG will be required to draw further conclusions with confidence.

In Australia, weak inter-regional differentiation is readily understandable in the light of the capacity of the spectacled flying fox for strong and sustained flight and its apparent plasticity of habitat use at times of environmental stress. Although the spectacled flying fox rarely roosts more than 6 km from wet tropical rainforest, and has long been assumed to feed primarily on rainforest species (Richards 1990a, b), in reality individuals regularly feed on a wide variety of non-rainforest species, including eucalypts (*Eucalyptus* spp., *Corymbia* spp.) in tall open forests adjoining rainforest communities and in tropical woodland and savanna ecosystems (Parsons 2005; Parsons et al. 2006). When circumstances require it, the species is able to cross, or survive within, substantial tracts of sclerophyllous vegetation, as illustrated by the dispersal response elicited by severe tropical cyclone Larry (Shilton et al. 2008). This event occurred in March 2006, caused massive damage to rainforest across a broad swathe of the Wet Tropics, and was followed by near total evacuation of known haunts of the spectacled flying fox in the region (Shilton et al. 2008). A year after the cyclone, the majority of animals had returned to the Wet Tropics from wherever it was that they had found refuge.

Such catastrophic events might be the trigger for colonisation of distant patches of suitable habitat, such as IR. There is no evidence that the Wet Tropics animals moved *en masse* to IR after Cyclone Larry, as no field surveys were conducted. However, any such dispersal could lead to levels of gene flow between Australian regions, and conceivably beyond, sufficient to offset divergence through genetic drift. Iron Range and PNG are themselves prone to natural disasters (cyclones, perhaps also volcanic eruptions in PNG) which could disperse populations. Given that severe tropical cyclones are frequent on a micro-evolutionary time-scale, these could be potent drivers of dispersal and population mixing in a volant species such as the spectacled flying fox. Fleeing a disaster is not an option open to an overwhelming majority of rainforest inhabitants of the Wet Tropics region: localised extinction is a more likely event (Schneider et al. 1998).

We have established that there is substantial gene flow between colonies in the Wet Tropics, but some, albeit rather weak, genetic structure when the three regions are considered (Wet Tropics, IR and PNG). Does current gene flow between widely separated habitats provide a model for survival of the species during glacial periods? It appears that under present-day conditions, the spectacled flying fox is able to maintain functional gene flow across distances of several hundred kilometres of apparently unsuitable habitat. When considered against the modelled extent of potential Pleistocene habitat in Australia (VanDerWal et al. 2009), we conclude that the mobility of this species, in combination with its habitat plasticity, evident under both stressed and normal circumstances, would allow gene flow to occur reasonably freely throughout the region and between isolated rainforest patches at that time.

Acknowledgements

We wish to thank the Australian Museum and Andrew Mack for samples of spectacled flying fox tissue from PNG, Jenny Maclean, the Tolga Bat Hospital and the many volunteers who helped collect samples from tick-affected bats on the Atherton Tableland. SF was supported by an ARC Linkage grant in partnership with the Tolga Bat Hospital and the Queensland Fruit and Vegetable Growers Association. The manuscript was greatly improved by comments from Andrew Lowe and Dominique Thiriet and two anonymous reviewers.

References

Bonaccorso, F.J. 1998. *Bats of Papua New Guinea*. Washington DC: Conservation International.

Breed, A., Field, H.E., Smith, C., Edmanston, J. and Meers, J. 2010. Bats without borders: Long distance movements and implications for disease risk management. *EcoHealth* 7:204-212.

Eby, P. 1991. Seasonal movements of grey-headed flying foxes, *Pteropus poliocephalus* (Chiroptera: Pteropodidae), from two maternity camps in northern New South Wales. *Wildlife Research* 18:547-559.

Fox, S. 2006. Population structure in the spectacled flying fox, *Pteropus conspicillatus*: a study of genetic and demographic factors. Unpublished PhD thesis, James Cook University, Townsville.

Fox, S., Waycott, M. and Dunshea, G. 2007. Isolation and characterisation of polymorphic microsatellite loci in the vulnerable spectacled flying fox, *Pteropus conspicillatus*. *Conservation Genetics* 8:1013-1016.

Goudet, J. 1995. FSTAT (version 1.2): a computer program to calculate F-statistics. *Journal of Heredity* 86:485-486.

Graham, C., Moritz, C. and Williams, S.E. 2006. Habitat history improves prediction of biodiversity in rainforest fauna. *Proceedings of the National Academy of Sciences USA* 103:632-636.

Hilbert, D.W., Graham, A. and Hopkins, M.S. 2007. Glacial and interglacial refugia within a long-term rainforest refugium: The Wet Tropics Bioregion of NE Queensland, Australia. *Palaeogeography, Palaeoclimatology, Palaeoecology* 251:104-118.

Hopkins, M.S., Ash, J., Graham, A.W., Head, J.W. and Hewett, R.K. 1993. Charcoal evidence of the spatial extent of *Eucalyptus* woodland expansions and rainforest contractions in north Queensland during the late Pleistocene. *Journal of Biogeography* 20:357-372.

Kershaw, A.P. 1994. Pleistocene vegetation of the humid tropics of northeastern Queensland, Australia. *Palaeogeography, Palaeoclimatology, Palaeoecology* 109:399-412.

Kershaw, A.P., Bretherton, S.C. and van der Kaars, S. 2007. A complete pollen record of the last 230 ka from Lynch's Crater, north-eastern Australia. *Palaeogeography, Palaeoclimatology, Palaeoecology* 251:23-45.

Luikart, G. and Cornuet, J.M. 1998. Empirical evaluation of a test for identifying recently bottlenecked populations from allele frequency data. *Conservation Biology* 12:228-237.

Luly, J.G., Blair, D., Parsons, J.G., Fox, S. and VanDerWal, J. 2010. Last glacial maximum habitat change and its effects on the grey-headed flying fox (*Pteropus poliocephalus* Temminck 1825). In: Haberle, S., Stevenson, J. and Prebble, M. (eds), *Altered ecologies: fire, climate and human influence on terrestrial landscapes*, pp. 83-100. Canberra: ANU E Press (Terra Australis vol. 32).

McKean, J.L. 1970. Geographical relationships of New Guinean Bats (Chiroptera). *Search* 1:244-245.

Michalakis, Y. and Excoffier, L. 1996. A generic estimation of population subdivision using distances between alleles with special reference for microsatellite loci. *Genetics* 142:1061-1064.

Mills, L.S. and Allendorf, F.W. 1996. The one-migrant-per-generation rule in conservation and management. *Conservation Biology* 10:1509-1518.

O'Brien, J., McCracken, G.F., Say, L. and Hayden, T.J. 2007. Rodrigues fruit bats (*Pteropus rodricensis*, Megachiroptera: Pteropodidae) retain genetic diversity despite population declines and founder events. *Conservation Genetics* 8:1073-1082.

Parsons, J. 2005. Spatial and temporal patterns of resource use by spectacled flying foxes *(Pteropus conspicillatus)*. Unpublished Honours thesis, James Cook University, Townsville.

Parsons, J.G., Cairns, A., Johnson, C.N., Robson, S.K.A., Shilton, L.A. and Westcott, D.A. 2006. Dietary variation in spectacled flying foxes (*Pteropus conspicillatus*) of the Australian Wet Tropics. *Australian Journal of Zoology* 54:417-428.

Peakall, R. and Smouse, P.E. 2005. GenAlEx v6: Genetic analysis in Excel. Population genetic software for teaching and research. Canberra: The Australian National University.

Peakall, R., Smouse, P.E. and Huff, D.R. 1995. Evolutionary implications of allozyme and RAPD variation in diploid populations of dioecious buffalo grass *Buchloe dactyloides*. *Molecular Ecology* 4:135-147.

Piry, S., Luikart, G. and Cornuet, J.M. 1999. BOTTLENECK: A computer program for detecting recent reductions in the effective population size using allele frequency data. *Journal of Heredity* 90:502-503.

Pritchard, J.K., Stephens, M. and Donnelly, P. 2000. Inference of population structure using multilocus genotype data. *Genetics* 155:945-959.

Raymond, M. and Rousset, F. 1995. GENEPOP: population genetics software for exact tests and ecumenicism. *Journal of Heredity* 86:248-249.

Richards, G. 1990a. The Spectacled flying-fox, *Pteropus conspicillatus*, (Chiroptera: Pteropodidae) in North Queensland. 1. Roost sites and distribution patterns. *Australian Mammalogy* 13:17-24.

Richards, G. 1990b. The Spectacled flying-fox, *Pteropus conspicillatus* (Chiroptera: Pteropodidae), in north Queensland. 2. Diet, seed dispersal and feeding ecology. *Australian Mammalogy* 13:25-31.

Ryman, N. and Jorde, P.E. 2001. Statistical power when testing for genetic differentiation. *Molecular Ecology* 10:2361-2373.

Schneider, C.J., Cunningham, M. and Moritz, C. 1998. Comparative phylogeography and the history of endemic vertebrates in the Wet Tropics rainforests of Australia. *Molecular Ecology* 7:487-498.

Schodde, R. and Calaby, J.H. 1972. The biogeography of the Australo-Papuan bird and mammal faunas in relation to Torres Strait. In: Walker, D. (ed), *Bridge and Barrier: the natural and cultural history of Torres Strait*, pp. 257-300. Canberra: Australian National University.

Shilton, L.A., Latch, P.J., McKeown, A., Pert, P. and Westcott, D.A. 2008. Landscape-scale redistribution of a highly mobile threatened species, *Pteropus conspicillatus* (Chiroptera, Pteropodidae), in response to Tropical Cyclone Larry. *Austral Ecology* 33:549-561.

Sinclair, E.A., Webb, N.J., Marchant, A.D. and Tidemann, C.R. 1996. Genetic variation in the little red flying fox *Pteropus scapulatus* (Chiroptera: Pteropodidae): implications for management. *Biological Conservation* 76:45-50.

Slatkin, M. 1985. Rare alleles as indicators of gene flow. *Evolution* 39:53-65.

Sokal, R.R. and Rohlf, F.J. 1995. *Biometry: the principles and practice of statistics in biological research*. New York: W.H. Freeman and Company.

Tidemann, C.R. and Nelson, J.E. 2004. Long-distance movements of the grey-headed flying fox (*Pteropus poliocephalus*). *Journal of Zoology* 263:141-146.

VanDerWal, J., Shoo, L.P. and Williams, S.E. 2009. New approaches to understanding late Quaternary climate fluctuations and refugial dynamics in Australian wet tropical rain forests. *Journal of Biogegraphy* 36:291-301.

van Oosterhout, C., Weetman, D. and Hutchinson, W.F. 2006. Estimation and adjustment of microsatellite null alleles in nonequilibrium populations. Molecular Ecology Notes 6:255-256.

Vucetich, J.A. and Waite, T.A. 2000. Is one migrant per generation sufficient for the genetic management of fluctuating populations? *Animal Conservation* 3:261-266.

Waples, R.S. and Gaggiotti, O. 2006. What is a population? An empirical evaluation of some genetic methods for identifying the number of gene pools and their degree of connectivity. *Molecular Ecology* 15:1419-1439.

Webb, N.J. and Tidemann, C. 1996. Mobility of Australian flying-foxes, *Pteropus* spp.

(Megachiroptera): evidence from genetic variation. *Proceedings of the Royal Society of London B* 263:497-502.

Weir, B.S. and Cockerham, C.C. 1984. Estimating *F*-Statistics for the analysis of population-structure. *Evolution* 38:1358-1370.

Westcott, D.A., Dennis, A.J., Bradford, M.G. and Margules, C.R. 2001. The Spectacled Flying fox, *Pteropus conspicillatus*, in the context of the world heritage values of the Wet Tropics World Heritage Area. Atherton: Environment Australia.

Whitlock, M.C. and McCauley, D.E. 1999. Indirect measures of gene flow and migration: FST ≠ $1/(4Nm+1)$. *Heredity* 82:117-125.

Worthington-Wilmer, J. and Barratt, E. 1996. A non-lethal method of tissue sampling for genetic studies of chiropterans. *Bat Research News* 37:1-3.(Megachiroptera): evidence from genetic variation. *Proceedings of the Royal Society of London B* 263:497-502.

www.ingramcontent.com/pod-product-compliance
Lightning Source LLC
Chambersburg PA
CBHW051308270326
41929CB00029B/3454